Environmental Meteorology

Environmental Meteorology

Proceedings of an International Symposium
held in Würzburg, F.R.G.
29 September - 1 October 1987

edited by

K. Grefen
and
J. Löbel

Verein Deutscher Ingenieure,
Kommission Reinhaltung der Luft,
Düsseldorf, F.R.G.

Kluwer Academic Publishers
Dordrecht / Boston / London

Library of Congress Cataloging in Publication Data

CIP

Environmental meteorology.

"The VDI Commission on Air Pollution Prevention--
in cooperation with the German Meteorological Society--
presents in this book the proceedings of the first
International Symposium on "Environmental Meteorology,"
held in Würzburg (West Germany) from 29 September to 1
October 1987"--Pref.
1. Air--Pollution--Meteorological aspects--
Congresses. I. Grefen, Klaus. II. Löbel, J.
III. VDI-Kommission Reinhaltung der Luft.
IV. Deutsche Meteorologische Gesellschaft.
V. International Sympsium on "Environmental Meteorology"
(1st : 1987 : Würzburg)
TD881.E58 1988 628.5'3 88-3078

ISBN 90-277-2697-3

Published by Kluwer Academic Publishers,
P.O. Box 17, 3300 AA Dordrecht, The Netherlands.

Kluwer Academic Publishers incorporates
the publishing programmes of
D. Reidel, Martinus Nijhoff, Dr W. Junk and MTP Press.

Sold and distributed in the U.S.A. and Canada
by Kluwer Academic Publishers,
101 Philip Drive, Norwell, MA 02061, U.S.A.

In all other countries, sold and distributed
by Kluwer Academic Publishers Group,
P.O. Box 322, 3300 AH Dordrecht, The Netherlands.

Printed in The Netherlands

PREFACE

The VDI Commission on Air Pollution Prevention - in cooperation with the German Meteorological Society - presents in this book the proceedings of the first International Symposium on "Environmental Meteorology", held in Würzburg (West Germany) from 29 September to 1 October 1987.

The primary goal was to get together scientists, experts of the meteorological services, specialists of environmental boards, and consulting engineers of the European countries. An equally important objective was to provide a bench mark document in the resulting proceedings publication.

The 1987 symposium shall start a series of symposia on all fields of environmental meteorology to be held once in three or four years in one of the European countries. We are full of hope to come to an intense cooperation with the national meteorological and environmental societies in the countries with this concern.

We like to express our sincere appreciation to the authors for their efforts and attention to the quality shown herein. The credit must be extended to the session Chairmen and to the advisory committee for the selection of the papers.

We think the book contributes substantially to a better understanding of meteorology being the link between emission and deposition of atmospheric pollutants.

Düsseldorf, November 1987

K. Grefen J. Löbel

CONTENTS

ADEQUATE METEOROLOGICAL MEASUREMENTS AS A PREREQUISITE FOR THE ASSESSMENT OF AIR POLLUTION

K. Höschele
Institut für Meteorologie und Klimaforschung
Universität Karlsruhe
D - 7500 Karlsruhe

ABSTRACT. The paper deals with meteorological measurements as a part of measuring programs for air pollutants. The importance of atmospheric conditions for the assessment of an air pollution situation is demonstrated by their influence on the various compartments of an air pollution system, comprising all stages from emission to effects. The main purpose of this review is a discussion of how to get representative data in various scales of time and space. A concluding survey contains various recommendations concerning the design of measuring programs.

1. INTRODUCTION

Concentrations of air pollutants, as a measure for ambient air quality, vary in space and time over one ore two orders of magnitude in the range of $\mu g/m^3$, with periods from the time resolution of measuring instruments in the range of minutes or seconds up to years. On the other hand the concentrations of water vapor, a typical meteorological element, in the higher range of g/m^3, have much smaller variations in space and time, typical are 10 % of the daily mean. The primary reason for the irregular behaviour of pollutants is the uneven distribution and changing activity of sources, in combination with the varying meteorological conditions for the distribution, the transformation and the removal of substances, emitted into the atmosphere. The widely adopted conception, restricting the role of meteorology to transport and diffusion of gases and particles, is too simple and incomplete. A more thorough analysis reveals, that atmospheric conditions control all stages and branches of a pollutants way from sources to sinks (Figure 1).

2. AIR QUALITY SYSTEM, THE ROLE OF METEOROLOGY

Starting with sources: how far are emissions dependent on meteorological conditions? If we consider emissions released near the ground, say below 20 or 50 m - these are mainly responsible for local pollution

K. Grefen and J. Löbel (eds.), Environmental Meteorology, 1–8.

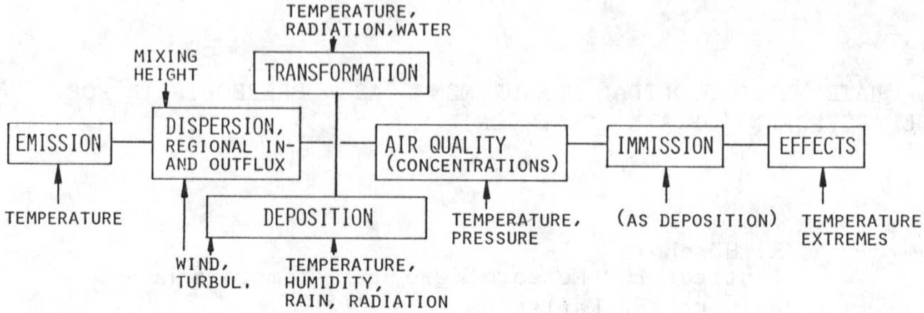

Figure 1. Air Quality System, Compartments with some Relevant Meteorological Conditions.

in cities - the prevailing sources are 'road traffic' and 'space heating'. As a whole the emissions of these two groups of sources have about the same importance. The bulk of sulfur dioxide from these two source groups (90 - 95 %) is due to domestic heating, but this is equalled by nitrogen oxides, carbon monoxide and hydrocarbons with a larger contribution of road traffic (typical are 60 - 90 % of emissions near the ground). As for particulate emissions the relation is about 50:50. All these relations depend on the kind of heating systems in use and on the specific emissions of motor cars.

It is a well known fact that the consumption of fuel for space heating over a certain period may be assessed very well by a measure like 'heating degree days', based on mean daily ambient temperatures. As a consequence, a large portion of emissions into the lower layers of the urban atmosphere increases linearly with mean temperatures falling beneath for instance 15 ^{o}C. For single days the estimation of fuel consumption may be improved by additional meteorological parameters as wind velocity, rain, humidity and solar radiation. Even emissions from road traffic may be increased in severe cold weather because of lower temperatures in engines and exhaust systems and reduced velocities on roads, covered with ice or snow. Under very heavy winter conditions motorists may switch over to public transportation and so reduce car emissions. An accurate emission inventory for a case study has to allow for these modifications of source activities by changing atmospheric conditions.

After leaving the source pollutants will be distributed in the atmosphere. Transport and diffusion are governed by the wind field and its turbulent components up to a distance from the source, relevant for dispersion. The volume, available for dispersion, may be limited in height by an inversion layer, preventing further vertical turbulent transports. In an extremely simplified box model for area sources concentrations are inversely proportional to the product of the mean wind speed and the height of a layer with turbulent mixing. Small values of wind speed result in very large values of concentration. This is one simple reason for the extreme skewness of frequency distributions for observed concentration values and the resulting difficulties with arithmetic means of concentrations. For the linkage between source activities on the one side and concentration fields on the other side, wether observed or calculated, we need representative data for the turbulent wind field, possibly with complementary data of thermal stratification, interacting with turbulence.

Immediately after release into the atmosphere and parallel to dispersion primary pollutants are exposed to physical and chemical transformations. An example for very fast transformations, resulting in a very short 'life time' of minutes is the oxidation of NO to NO_2. Decisive for reaction or transformation rates are, besides the concentrations of participating substances, meteorological parameters as solar radiation for photochemical reactions, or the existence of liquid water for reactions in cloud or fog droplets. Besides this temperature has a rather large effect on rates of bimolecular reactions or rates of disintegration of single molecules. As for solar radiation, the photolytic reaction rates depend on the irradiation of a volume of gas or a sphere from all directions and not on the radiation striking a horizontal plane surface, we normally measure as 'global radiation'. For a sphere an increasing solar zenith angle will only reduce radiation intensity by an elongated path through the atmosphere, the so-called air mass. In the case of global radiation there is an additional reduction by the changing incident angle. So the irradiance changes less with growing solar zenith angles up to about 50°, it depends on the albedo of the ground and on the height above ground or above sea level.

Transformation or decomposition of a certain pollutant will limit its lifetime in the atmosphere, such it is one process contributing to the removal of pollution, but possibly it is at the same time a source for secondary pollutants. For the atmosphere of a given region or for a limited airshed a transport to neighbouring regions or to higher levels may be regarded as another process of removal, but this output from one region will be an input for another region or atmospheric layer. The genuine processes of removal from the air are dry and wet deposition at the ground. It is obvious, that for the precipitation-related process of scavenging, resulting in wet deposition, parameters as cloud and precipitation droplet spectra and rain intensity as well as duration are of interest. Especially the role of vegetation in dry deposition makes it more difficult, to quantify the depletion of pollutants at the lower limit of the atmosphere. Present or past weather events, responsible for the condition of vegetation and soil, for instance wetness or opening of stomata, play a role. So deposition velocities change under the combined effects of rain, evaporation, radiation, temperature. Another essential meteorological parameter controlling deposition velocity is wind speed and turbulence.

A result of all processes described until now, from source emission to deposition is ambient air quality as another compartment in the air pollution system. Monitoring concentration fields as a measure for air quality is in many investigations the first step into the matter, because air quality and its control is in the focus of general interest. This paper also started from this point with a view on changing concentrations. Meteorological parameters play a role at this stage as well, especially for the comparison of measurements of gaseous pollutants in different heights. Temperature and pressure are needed for calibration of instruments and for the conversion of units, for instance mixing ratios of molecule numbers or volumes (in ppm or ppb) in mass per volume (mg or $\mu g/m^3$). The same is true for comparing measured data with the set points of pollution control, the air quality standards.

The last compartments in this systems approach are the immission and the effects of air quality or the impact on receptors like humans, animals, plants, soil, water, materials or the effects in the atmosphere itself (reduction of visibility, attenuation of radiation, promotion of fog formation). The transfer of pollutants to receptors is on the one side a removal or deposition process, but here the primary interest lies in the effects on receptors. Compared with soil, water and vegetation man will play a poor role in removing pollutants, but concerning impact man is in the centre of interest and effects on other receptors are considered as impairment of man's environment. Obviously the same atmospheric parameters as in the deposition process contribute to the variable impact on receptors for a given ambient air quality situation or concentration field. For soil, vegetation or materials we have already mentioned these parameters, which control the mass transfer to the receptor or the poisoning dosage. In addition, the effect of the same dosage may be changed in living organisms by meteorological conditions as extreme heat or cold. For man the transfer of pollutants to skin or eyes is governed by the rules of deposition too. The transfer to lungs by active breathing will depend mainly on activity or metabolic rate; meteorological conditions will normally play a minor role, except for extreme conditions. An example: the same concentration in units of mass per volume will result in an about 20 % higher dosage in 2000 m height than in sea level, for the same activity.

3. GATHERING METEOROLOGICAL AND CLIMATOLOGICAL DATA

There is no basic difference between meteorological or climatological observations or data but we may distinguish the application of these data in mainly meteorological or climatological aimed studies. As a typical example for the first kind appear single episodes of acute pollution as the accidental release of toxic gases. Forecasting the path and dispersion of released toxicants needs actual and prognostic informations about the state of the atmosphere. In most cases however, studies in air pollution problems will have a statistical character, the aim is a probalitistic prediction. Even examining pollution episodes as a consequence of extreme weather conditions, as smog periods, has not the purpose to give a detailed reconstruction of a certain case. This is only needed for the description and classification of samples of past cases, for an extraction of common properties and relevant preconditions with the aim to take better precautions for the future. The bulk of air pollution monitoring systems is organized to make measurements in a quasi climatological time scale and to render informations about air quality over periods of typically one year.

Reliable meteorological measurements require as much diligence as measurements of pollutants; the expenditure necessary for every single parameter at any point is often underestimated. Therefore one of the first steps in developing a measuring program for air pollutants has to be to prepare an inventory of already existing meteorological observations in the survey area and to make sure, that the data will be available for the provided purpose. This is not only for saving additional surplus measurements: long time series of existing observations are a useful and scarcely replaceable means to normalize new observations over short periods by a regression analysis between parallel measurements.

Observations at single stations are very small samples in space of parameters with a high spatial variability. A random choice of sites for a station would transfer this variability as uncertainty of the observation results concerning conclusions for other sites. A way out would be to install a great number of stations in a regular or irregular network. Our knowledge of reasons for variations of atmospheric conditions in space enables us to reduce this uncertainty to a large extent in a less expensive way: the random sample is replaced by a representative sample, at a station with standardized environmental conditions.

The more our knowledge of the reasons for spatial variations of air pollutants is growing, the more we could dispense with monitoring networks and work with representative stations as in climatology. One danger however is, that our models, explaining the variations, are incomplete or wrong for the given situation.

A network of representative stations, together with topographic data, as height above sea level, exposition to sun radiation or prevailing winds, land use and vegetation cover, enable us to draw or to calculate and plot climate maps with a much higher, perhaps up to one order of magnitude better resolution that we could with respect to the grid size. Nevertheless the distances between stations of the meteorological services are too large, to cover the needs of air pollution climatology. In the southern hilly part of the Federal Republic of Germany climate stations have a distance of 20 - 25 km, a map of 1:1 000 000 with picture elements of 2 km in nature, corresponding with 2 mm on the map, will cover an area of about 400 km x 500 km in the upper mesoscale, Southern Germany for instance. Many air pollutant problems however belong mainly to the local or microscale, below an extent of 20 km down to some meters, for the dispersion from single stacks or from a highway.

Another problem is that reaching this high resolution above an irregular topography is possible with some climatological elements as temperature, but not as easy for the direction of surface wind, the most decisive parameter for dispersion of pollutants in the local and microscale.

The next step, after making an inventory of existing meteorological observing stations is therefore a careful examination of the survey area on topographic maps and on up-to-date aerial phorographs. Slopes, valleys and escarpments indicate the possibility of mesoscale wind systems with a change of direction in a daily course. Inspections of typical sites during a critical time of the day will help to decide, if further investigations or additional measurements are needed. Using the knowledge from existing investigations about similar topographic features we choose a place, representative for the range of the respective mesoscale wind system and manage to get along with only one or few stations.

For microscale pollution studies around buildings we mainly refer to wind measurements above roof tops. The correlation with air motion between obstacles can be derived from existing examples of results calculated with numerical models or from wind tunnel experiments. For verification one or two low level measuring points are helpful.

Parallel with the growing horizontal scale there is a slowly increasing height of the atmospheric layer, relevant for the dispersion of pollutants. At the small scale end the height of sources above ground plays a role too. For meteorological parameters as temperature, humidity, wind velocity and direction, the autocorrelation in space increases too with height, the small scale properties of the surface loose their influence more and more. This justifies large distances as 200 km in central Europe between soundings in the free atmosphere; the gaps can be filled by interpolation methods. An approximation to the wind at the base of the free air is the geostrophic wind, calculated from pressure data of synoptic stations. This geostrophic level, about 1,5 km above ground, is relevant for dispersion in the upper mesoscale, over more than 200 km horizontal distance. For the lower mesoscale (< 20 km) it is possible to calculate wind fields for the boundary layer with a numerical model, using the geostrophic wind and topography data. In Session 4 Dr. Adrian, Karlsruhe, will present results of a wind climatology with a 5 km distance of grid points.

4. SURVEY AND CONCLUSIONS

The guidelines VDI 3786 contain a comprehensive compilation of instructions for standard meteorological measurements, concerning air pollution: wind, temperature, humidity, radiation and precipitation at surface stations, as well as aerological measurements with free and tethered balloons. They deal with measuring methods, instruments, installation and site selection, calibration, maintenance and evaluation in detail; but only in a general way or very little with the necessity and the form of a measuring program. Not yet available is a guideline for more modern methods, competing with standard profile measurements, as SODAR. The possible use of such a system has to be examined, replacing a meteorological tower, especially for short term measurements in connection with dispersion problems in the lower mesoscale. If large scale pollution studies use aircraft measurements for monitoring pollutants in higher levels, these often are combined with meteorological measurements of temperature, humidity, turbulence, thus completing the results of upper air soundings from ground stations.

Some more remarks have to be made, concerning the measuring of meteorological elements as described in the guidelines. Ambient air temperature and humidity is in many cases with enough precision available from existing stations. Temperature profiles at towers demand a rather high effort for the accuracy needed for the assessment of thermal stratification. For the determination of stability classes this can be replaced by measuring vertical wind fluctuations in one height. Solar radiation, in the time scale of hours or days, does not change very much over large horizontal distances. On the other hand poor maintenance will produce considerable deviations. If possible the data from an existing meteorological station should be used. We have already mentioned, that the reference for radiation in photochemistry is a sphere, not a horizontal plane. Radiation balance is a basic quantity in boundary layer meteorology, but not recommendable as a means for the determination of stability classes.

Finally summing up we recommend

- to adjust the frequency and accuracy of measurements to the demands of resolution in time and space and to feasibility

- to consider the application of remote sensing methods

- to use the existing knowledge for the selection of representative sites for measuring points and representative terms for measuring periods

- to combine results from field observations with data from model calculations or wind tunnel experiments

- to couple short term measurements (about for instance 6 month or 1 year) with existing long time series of observations (10 or 30 years) and/or

- to relate results of a short time series with dominating large scale weather patterns and transform for a normal long time frequency distribution of these patterns.

REFERENCES

Finlayson-Pitts, B.J., and Pitts, J.N., 1986: Atmospheric Chemistry, Fundamentals and Experimental Techniques. J. Wiley, New York

Höschele, K., 1974: 'Die Repräsentanz zeitlicher Stichproben einiger für Immissionsbetrachtungen wesentlicher meteorologischer Größen'. Meteorol. Rdsch. **27**, 5-10

Nester, K. and Hübschmann, W., 1985: 'Anwendung klimatologischer Daten bei der Ermittlung der Umweltbelastung durch Großemittenten'. promet 1'85, 6-13

Verein Deutscher Ingenieure (VDI): Meteorological Measurements concerning Questions of Air pollution

Part 2, 1980: Wind
Part 3, 1985: Air Temperature
Part 4, 1985: Air Humidity
Part 5, 1986: Global radiation, direct solar radiation and net total radiation
Part 6, 1983: Turbidity of ground-level Atmosphere, Standard visibility
Part 7, 1985: Precipitation
Part 8, 1987: Aerological Measurements

Weber, E. (ed), 1982: Air Pollution, Assessment Methodology and Modelling. Plenum Press, New York

Yoshino, M.M., 1975: Climate in a Small Area, an Introduction to Local Meteorology. University of Tokio Press

OPTICAL PARAMETERS, SOOT CONTENT, AND ABSORPTION RATES OF URBAN PARTICLES

R. Busen, D. Weidert, G. Hänel
Institute for Meteorology and Geophysics
University of Frankfurt
Feldbergstraße 47, 6000 Frankfurt 1, FR Germany

ABSTRACT. Inside an urban area there are many sources of highly absorbing atmospheric particles. To a considerable amount these particles consist of soot generated by domestic heating, traffic, and industrial combustion. For the last years samples of particles together with radiation measurements have been taken at Frankfurt/Main. From photometric measurements complete and consistent sets of the optical parameters of dry particles are obtained for the first time. The results are discussed with reference to seasonal variations and varying weather situations. Results obtained at other measuring sites are given for comparison. Within the whole atmosphere up to 20 % of the solar energy impinging at the top of the atmosphere can be absorbed by particles. The mean value of the apparent imaginary part of the complex refractive index of dry particles is about 0.03, that of the apparent soot content is about 0.07 per volume. By all means the results point towards the significant climatological relevance of the absorption of solar radiation in particles especially in polluted urban areas.

1. INTRODUCTION

Particles influence the atmospheric radiation budget and thus climate by scattering and absorbing solar electromagnetic radiation. Thereby absorption leads to an increase of the internal energy of the atmosphere resulting in a warming of the air. This effect has been demonstrated to be most important for urban areas (Blanchet and Leighton, 1981; Blanchet et al., 1986; Busen and Hänel, 1987).

In the present paper we will report about in situ measurements of the absorption of solar radiation in particles near the ground. The resulting heating rate of the atmosphere is presented for a large number of days of measurements for different sites and seasons. Besides the absorption of solar radiation by particles within the whole atmosphere is estimated from the measurements. The corresponding quantity referred to the absorption by water vapor is given for comparison.

K. Grefen and J. Löbel (eds.), Environmental Meteorology, 9–21.

Scattering and absorption of solar radiation by particles depends on their size distribution, their shape and structure, as well as on the complex refractive indices of the materials they are composed of. Although these properties dont have to be known for the evaluation of the measured data, the apparent complex refractive index and the apparent volume fraction of soot within the dry particulate material can be derived from the measured value of the single scattering albedo. Like the single scattering albedo the apparent complex refractive index is a spectral mean value for the solar region. Accuracy and use of it are discussed in this paper.

2. MEASURING TECHNIQUES

The optical properties of the particles can be determined by using a photometric method. The aerosol particles are collected on nuclepore filters. During the daylight period several consecutive samples are taken. The filters have to be evaluated twice, before and after the sampling. The radiation transmitted through and scattered by the clean and the particle loaded filter is measured photometrically between 0 and 160 degrees of the scattering angle (Figure 1).

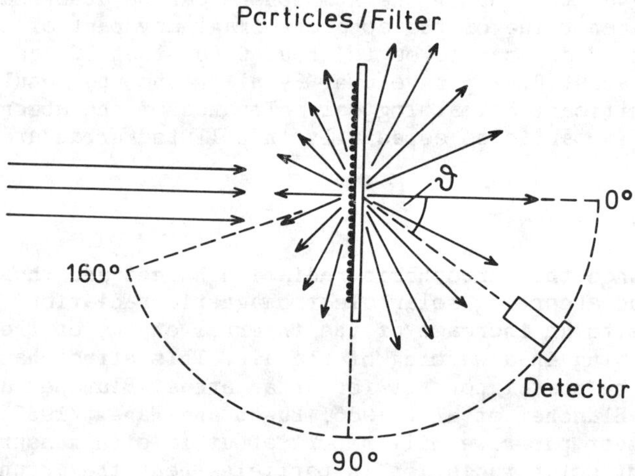

Figure 1. Schematic diagram of the photometric measurement.

The theory (Hänel, 1987) is based upon budget equations for the radiative energy of the clean and the loaded filter. The radiative interaction between the particles and the filter is described by the adding method and the radiative transfer within the layer of particles by a two stream approximation (Coakley and Chylek, 1975). The phase function of the dry particles is assumed to be of Henyey-Greenstein type.

From this theory the following optical properties can be derived, which have to be understood as spectral mean values for the solar spectral range (0.3 to 3.5 μm wavelength) and are valid for particles in dry state:

σ_{EP}	:	mean solar volume extinction coefficient of the particles
σ_{AP}	:	mean solar volume absorption coefficient of the particles
$\omega = 1 - \sigma_{AP}/\sigma_{EP}$	:	single scattering albedo of the particles
g	:	asymmetry parameter of the particles

The errors for the mean solar volume absorption and extinction coefficients of the particles are about ± 10 % and ± 10 to ± 20 % respectively. The mean experimental error of $1 - \omega$, i.e. of the ratio σ_{AP}/σ_{EP}, amounts to ± 16 % . The mean experimental error of the asymmetry parameter g is ± 0.1 .

The photometric method has been tested against an independent calorimetric absorption measurement on airborne particles (compare Hänel et al., 1982) and against visibility observations yielding a very good agreement.

In addition measurements of the atmospheric radiation quantities are performed, especially measurements of the solar radiation supply S. The solar radiation supply is defined to be the integral over the radiances N converging from all spatial directions to the point of measurement (Busen and Hänel, 1987). Therefore the integration runs over the total solid angle of $4\pi\ \omega_1$.

$$S = \int_{4\pi\ \omega_1} N\, d\omega \qquad (1)$$

3. THEORY

3.1. The Data Processing

By combining both measurements, that of the solar radiation supply and that of the mean solar volume absorption coefficient of the particles, the following quantities can be derived, which are significant with reference to climatological considerations. First the mean values of the optical properties for the consecutive samples of particles are converted to smooth temporal lapses (Busen and Hänel, 1987) and corrected to ambient relative humidity (Hänel, 1976).

Next the solar volume absorption rate A_{SP} due to particles can be calculated from minute to minute. This quantity denotes the rate, by which solar electromagnetic energy is absorbed within particles per unit time and unit volume of air. The common unit is W/m^2.

$$A_{SP} = \sigma_{AP} \cdot \int_{4\pi\ \omega_1} N \, d\omega = \sigma_{AP} \cdot S \qquad (2)$$

The solar volume absorption rate due to particles may also be expressed as the appertaining solar heating rate $(dT/dt)_{SP}$. For that purpose the solar volume absorption rate has to be related to the product of the density ρ of the air and the specific heat c_P of air at constant pressure.

$$(dT/dt)_{SP} = A_{SP} \; / \; (\rho \cdot c_P) \qquad (3)$$

By integrating the solar heating rate due to particles over the whole daylight period the daily heating ΔT_{SP} due to particles can be found, which can easily be compared for different seasons and locations.

$$\Delta T_{SP} = \int_{\text{Sunrise}}^{\text{Sunset}} (dT/dt)_{SP} \, dt' \qquad (4)$$

By integrating the volume absorption rate over the total height of the atmosphere from the ground (z_G) to the upper boundary (z_O) and by referring the result to the solar energy entering the atmosphere perpendicularly, the column absorptivity a_P due to particles is derived.

$$a_P = \frac{\int_{z_G}^{z_O} A_{SP}(z) \, dz}{F_{SO} \cos \theta_O} \qquad (5)$$

F_{SO} is the solar constant, and θ_O is the zenith distance of the sun at the top of the atmosphere. The evaluation of the integral in the numerator of (5) is described in detail by Busen und Hänel (1987).

The corresponding column absorptivity a_w due to water vapor can be calculated using a parameterization formula proposed by Lacis and Hansen (1974).

3.2. The Calculation of Apparent Properties of the Particles

In the following section a procedure will be described, which allows for calculating the complex refractive index of the particles and the volume fraction of soot within the dry particulate material. But these quantities are thought to be apparent ones, where apparent has to be understood in the sense of "appearing (but not necessarily) real or true". There are three main reasons for this restriction:

- To calculate the complex refractive index of the particles the theory of Mie (1908) has to be inverted. This homogeneous sphere model for the particle shape and structure does not meet reality. Therefore the result may not be a true property of the particulate material.
- For each optical property there is a specific weighting function referred to different sized particles. Consequently the shape of the size distribution of the particles, which in most cases has to be assumed, may become an important factor.
- One complex number is used for a mixture of many substances. Thus an apparent value of the complex refractive index has to be used with care.

The computations are based on two equations. The particles are assumed to be homogeneous spheres having all the same apparent complex refractive index n - ik. Thus from Mie theory the property $1 - \omega$ can be expressed as a function of the shape of the aerosol size distribution and of the effective mean shortwave values of the real and imaginary part of the refractive index.

$$1 - \omega = \frac{\sigma_{AP}}{\sigma_{EP}} = \text{function (n, k, size distribution)} \tag{6}$$

with n = real part of the apparent complex refractive index
k = imaginary part of the apparent complex refractive index

Secondly a refractive index mixture rule is used, which describes the complex refractive index of a composite material. The particles are assumed to be composed of non-soot material an soot. Then the quantity (Van de Hulst, 1957)

$$X = \frac{(n-ik)^2 - 1}{(n-ik)^2 + 2} \tag{7}$$

for the composite material is equal to the sum of the weighted individual properties. The weighting factors are the volume fraction α_S of soot and $1 - \alpha_S$ of the non-soot material.

$$X = X_O (1-\alpha_S) + X_S \alpha_S \tag{8}$$

The index 'S' stands for soot, the index '0' stands for the non-soot material.

Separation of this complex equation leads to two equations containing the three unknown properties n, k, and α_s. Elimination of α_s leads to an equation for n as a function of k. Elimination of n results in an equation for α_s as a function of k. These equations read:

$$n = \left[\{ 1.5756+k(1.8599-0.4k) \} / \{ 0.7122+0.2k \} \right]^{1/2} \quad (9)$$

$$\alpha_s = k\,(0.7122+0.2k^2) / (0.2627+0.1629k) - A \quad (10)$$

with A = 0.01 for non-maritime particles
and A = 0.005 for maritime particles

k itself is calculated from (6) as a function of 1-ω. n is removed by k using (9). Then (6) can be solved numerically using different size distributions. In each case the result is

$$\ln k = a_o + \sum_{i=1}^{5} a_i \ln^i (1-\omega) \; . \quad (11)$$

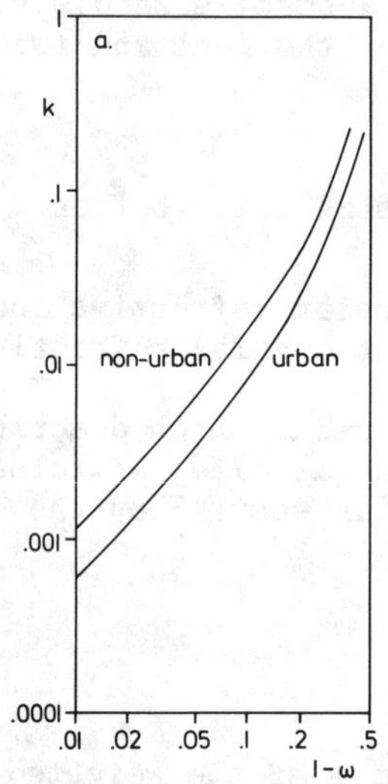

Figure 2. Mean shortwave value of the imaginary part k of the apparent complex refractive index versus the absorption to extinction ratio 1-ω. Mean results for urban and non-urban particles.

The results show that a reduction to two sets of coefficients a_i ($i=0,\dots,5$) for non-urban and urban particles is possible. The relevant curves are plotted in Figure 2. The maximum influence of the aerosol size distribution on k is ± 17 % for the mean non-urban and ± 38 % for the mean urban aerosol. Since 1-ω is independent from the particle number concentration, this scattering comes from the different shapes of the size distributions used. For both types of aerosols k increases with increasing 1-ω. But for a given value of 1-ω the imaginary part k becomes larger for non-urban than for urban particles.

4. RESULTS

4.1. Locations of Measurements

Collectively 53 measurements have been performed each of them lasting a whole day. In total 275 samples of particles have been taken. Most of these measurements have been executed at Frankfurt. Therefore for this station (FFM) a subdivision into summer and winter values has become possible.

For comparison the other sites of measurements are specified: Piumazzo (PIU) is a rural station in the Po valley between Modena and Bologna. There is only light industry, but during the period of measurements (October 1985) the surrounding fields got burned down, which temporary resulted in very high soot concentrations. Monte Ombraro (MTO), 19 km south of and about 650 m higher than Piumazzo, and also Garmisch -Partenkirchen (GAP) and the nearby Wank Mountain (WK) are examples for higher located and less polluted stations. TUC stands for Tucson/Arizona.

In the following tables only mean values or ranges of the daily means of the quantities discussed above are given. This kind of compilation in the best way allows to compare different measuring sites and various seasons. The number of days and the number of individual samples of aerosol particles are indicated for every station to confirm the significance of the mean values.

4.2. Measured and Apparent Properties of the Aerosol Particles

Mean values of the ratio 1-ω between the absorption and the extinction coefficients and of the asymmetry parameter g are compiled in Table 1. The highest values for 1-ω are found at Frankfurt during winter and at Piumazzo caused by domestic heating and fires from the surrounding fields, respectively. Distinctly lower values must be stated at Frankfurt during summer, at Monte Ombraro, and also at Tucson. The lowest values appear at Garmisch and on top of the Wank. During these last measurements there were general cyclonic weather conditions with temporal rain and strong winds.

For Garmisch and Wank and also for Piumazzo and Monte Ombraro, where in each case the measurements took place at the same time, a strong decrease of 1-ω with height is obvious.

The corresponding values of g are available only for Frankfurt during winter and for Piumazzo, and only for a subset of the samples. The technique for obtaining g is very new and cannot be applied to the older samples. The mean value for g is about 0.4 for both stations. This value is much smaller than Mie theory predicts for urban particles. This may be explained by the influence of the soot on the scattering function of the particles. Soot aggregates consist of very small primary particles with very low asymmetry parameters and therefore strongly influence the total asymmetry parameter of the particles. This holds especially for urban particles containing a considerable amount of soot.

TABLE 1 Absorption to extinction ratio 1-ω, asymmetry parameter g, apparent imaginary part k of the complex refractive index, and apparent soot content α_s of dry aerosol particles at different measuring sites.

Station	Number (#) of Samples	1-ω	g (#)	k	α_s
FFM (summer)	59	.13		.018	.036
FFM (winter)	110	.22	.38 (31)	.038	.088
PIU (Oct. 1985)	46	.18	.40 (46)	.027	.062
MTO (Oct. 1985)	14	.12		.014	.028
GAP (Sep. 1984)	14	.075		.011	.020
WK (Sep. 1084)	10	.05		.0064	.007
TUC (Oct. 1982)	22	.11		.010	.017

The mean values of the imaginary part k of the apparent complex refractive index are strongly related to 1-ω, thus the maximum values again appear for Frankfurt during winter and for Piumazzo. From the total of 275 samples only 11 values of k exceed 0.1 .

The results for the apparent volume fraction α_s of soot within the total dry particulate material show the same features like those of k. The mean volume fraction of soot from all our measurements is 0.058 .

In addition to Table 1 two interesting features concerning the measured data of 1-ω and g shall be mentioned. In Figure 3 the mean temporal variation of 1-ω is plotted for FFM(summer), FFM(winter), and for TUC. The temporal lapses can be explained having in mind that the sources of soot are near the ground. Due to convection the air near the ground containing a large amount of soot is mixed with the air from higher levels of the atmosphere. Therefore the soot content and also 1-ω should reach their minimum when the convection is strongest. During

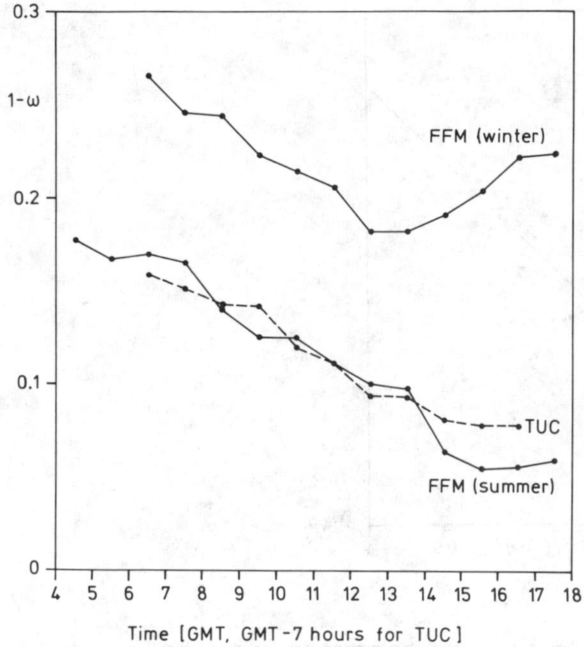

Figure 3. Mean variation of the ratio 1-ω of the absorption and the extinction coefficients of dry atmospheric particles with time.

winter in urban areas this happens to be shortly after local noon, in the summertime this may happen late in the afternoon or even at night. In Figure 4 the decrease of the asymmetry parameter with decreasing single scattering albedo ω is plotted. The regression line $g = \omega - 0.4$ is statistically highly significant. It is important to note that if ω approaches unity g approaches 0.6 with a scattering of ± 0.15 . This agrees with direct measurements by Grams and Colletti (1982), who found $0.5 \leqslant g \leqslant 0.7$ for polar and normal background particles.

4.3. Daily Heating and Column Absorptivity

The mean value of the daily heating due to particles is highest at Frankfurt during winter and at Piumazzo, somewhat lower at Frankfurt during summer and at Tucson. It must be recognized, that during winter the daylight period is shorter and the radiation supply is lower than during summer. But the higher absorption coefficients of the particles during winter compensate these effects, resulting in a slightly higher winterly mean daily heating. At the rural or background stations the ranges and also the mean values of the daily heating due to particles are lower due to the generally lower absorption coefficients.

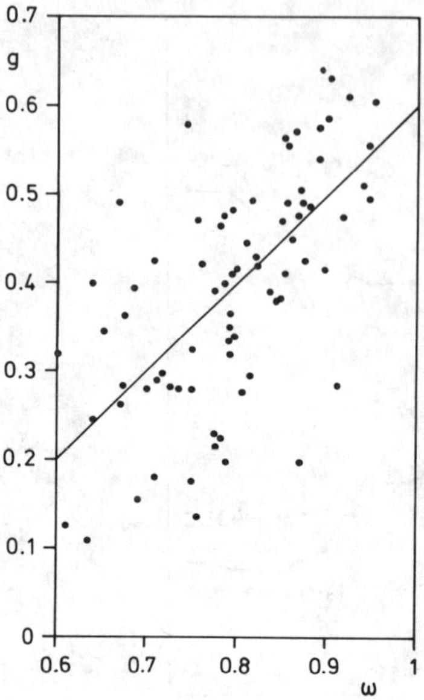

Figure 4. Relation between the asymmetry parameter g and the single scattering albedo ω of dry atmospheric particles. The solid line is $g = \omega - 0.4$.

TABLE 2 Daily heating due to particles ΔT_{SP} and column absorptivities a_P due to particles and a_W due to water vapor at different measuring sites.

Station	Number of days	ΔT_{SP} Range [K]	ΔT_{SP} Mean [K]	a_P [%]	a_W [%]
FFM (summer)	10	.07 - .96	.54	5.9	13.9
FFM (winter)	19	.06 - 1.32	.60	12.1	11.3
PIU (Oct. 1985)	10	.09 - 1.31	.68	10.6	13.7
MTO (Oct. 1985)	5	.08 - .58	.32	3.0	11.8
GAP (Sep. 1984)	2	.22 - .30	.26	2.8	
WK (Sep. 1984)	3	.05 - .17	.11	.3	
TUC (Oct. 1982)	4	.49 - .69	.56	4.2	11.6

The column absorptivities have been calculated for particles and also for water vapor. The results compiled in Table 2 point to the relative importance of these two main atmospheric absorbers of solar radiation under different conditions. The column absorptivity due to water vapor does not vary much for different measuring sites or seasons. But for particles even the mean values show strong variations related to the optical properties and the local concentrations of the particles.

At Frankfurt during winter and at Piumazzo the effects of water vapor and particles are comparable and nearly equivalent, whereas for the other situations the column absorptivity due to water vapor is more or less pronounced. The lowest influence of the particles is found at mountaineous stations.

5. DISCUSSION

At this point one may ask for the meaning of the apparent complex refractive index and of the apparent soot content of the particles. To decide, whether the results are physically relevant, the values of α_s have been compared with different measurements of this quantity reported from some other groups (Countess et al., 1980; Ferman et al., 1981; Shah et al., 1985; Shah et al., 1986; Wolff et al., 1986). These measurements are based on chemical analyses and therefore totally different from our method. Nevertheless the results coincide very well, for urban areas like Denver and New York as well as for continental background stations. Therefore α_s really states the physical volume fraction of soot within the total particulate material. In this sense the imaginary part k of the apparent complex refractive index, which is used to derive α_s, should be physically meaningful, too.

But in spite of these results there are considerable restrictions regarding the use of the apparent complex refractive index. The maximum error of the imaginary part is very large. Fluctuations of the shape of the particles size distribution lead to errors of ±17% for non-urban and of ±38% for urban particles. The experimental error in the absorption to extinction ratio is ±16% . This yields an additional error in the imaginary part. Therefore the typical error in the imaginary part of the apparent complex refractive index is about ±40% for non-urban particles and ranges from ±60 to ±80% for urban ones.

In principle the apparent complex refractive index should not be used to calculate the scattering function of dry atmospheric particles using Mie theory. The sideward scattering from this theory is far too low.

The apparent complex refractive index can be applied only to calculate the attenuation coefficients of particles. If a reliable size distribution is available it may be used to calculate the extinction and the scattering coefficients of dry atmospheric particles with Mie theory. However, the absorption coefficient should not be evaluated because of the large uncertainty in the imaginary part. For this quantity direct measurements are much more accurate. For the same reason the single scattering albedo should not be calculated if it shall be used for climate considerations. In this case 1-ω is the relavant property and the same difficulty arises as with the absorption coefficient.

The results of our measurements indicate that the model for urban aerosols by Shettle and Fenn (1979) should be modified. A mean solar single scattering albedo of about 0.6 as these authors indicate has been found only in 4 cases. According to the present results a value of about 0.8 would be more realistic. Secondly the mean solar asymmetry parameter should be in the order of 0.4 instead of about 0.65 .

As a conclusion it may be stated, that in this paper complete sets of optical properties of aerosol particles are presented. The measurements and the evaluations are based on the theory of radiative transfer and therefore the results can be stated to be physically meaningful and consistent. The column absorptivities point towards the significant climatological relevance of the absortion of solar radiation in particles especially in polluted urban areas.

6. REFERENCES

Blanchet, J.P., and H.G. Leighton, 1981:
'The Influence of the Refractive Index and Size Spectrum on Atmospheric Heating due to Aerosols', *Contr. Atm. Phys.* ***54***, *143.*

Blanchet, J.P., J. Heintzenberg, and P. Winkler, 1986:
'Radiative Heating During an Intense Pollution Episode in Hamburg, FRG', *Contr. Atm. Phys.* ***59***, *359.*

Busen, R., and G. Hänel, 1987:
'Radiation Budget of the Boundary Layer. Part 1: Measurement of Absorption of Solar Radiation by Atmospheric Particles and Water Vapor', *Contr. Atm. Phys.* ***60***, *229.*

Coakley, J.A.Jr., and P. Chylek, 1975:
'The two-stream approximation in radiative transfer: Including the angle of the incident radiation', *J. Atm. Sci.* ***32***, *409.*

Countess, R.J., G.T. Wolff, and S.H. Cadle, 1980:
'The Denver winter aerosol: A comprehensive chemical characterization', *J. Air Poll. Control Ass.* ***30***, *1194.*

Ferman, M.A., G.T. Wolff, and N.A. Kelly, 1981:
'The nature and sources of haze in the Shenandoah valley / Blue Ridge Mountains area', *J. Air Poll. Control Ass.* ***31***, *1074.*

Grams, G.W., and A. Colletti, 1982:
'Analysis of polar nephelometer data obtained at the First International Workshop on Light Absorption by Aerosol Particles', in : H.E. Gerber and E.E. Hindman (editors), 1982: *Light absorption by aerosol particles.* Spectrum Press, Hampton.

Hänel, G., 1976:
'The Properties of Atmospheric Aerosol Particles as Function of the Relative Humidity at Thermodynamic Equilibrium with the Surrounding Moist Air', *Advances in Geophysics* ***19****, 73.*

Hänel, G., 1987:
'Radiation Budget of the Boundary Layer. Part 2: Simultaneous Measurement of the mean Solar Volume Absorption and Extinction Coefficients of Particles', *Contr. Atm. Phys.* ***60****, 241.*

Hänel, G., R. Busen, Ch. Hillenbrand, and R. Schloß, 1982:
'Light absorption measurements: New techniques', *Appl. Opt.* ***21****, 382.*

Lacis, A.A., and J.E. Hansen, 1974:
'A Parameterization for the Absorption of Solar Radiation in the Earth's Atmosphere', *J. Atm. Sci.* ***31****, 118.*

Mie, G., 1908:
'Beiträge zur Optik trüber Medien, speziell kolloidaler Metalllösungen', *Annalen der Physik* ***25****, 377.*

Shah, J.J., T.J. Kneip, and J.M. Daisey, 1985:
'Source apportionment of carbonaceous aerosol in New York city by multiple linear regression', *J. Air Pollut. Control Ass.* ***35****, 541.*

Shah, J.J., R.L. Johnson, E.K. Heyerdahl, and J.J. Huntzicker, 1986:
'Carbonaceous Aerosol at urban and rural sites in the United States', *J. Air Pollut. Control Ass.* ***36****, 254.*

Shettle, E.P., and R.W. Fenn, 1979:
'Models for the aerosols of the lower atmosphere and the effects of humidity variations on their optical properties', *Report AFGL-TR-79-0214, Hanscon AFB.*

Van de Hulst, H.C., 1957:
Light scattering by small particles.
Wiley, New York, and Chapman and Hall, London.

Wolff, G.T., J.-S. Siak, T.L. Chan, and P.E. Kersog, 1986:
'Multivariate statistical analyses of air quality data and bacterial mutagenicity data from ambient aerosols', *Atmosperic Environment* ***20****, 2231.*

OPTICAL AND PHYSICO-CHEMICAL PROPERTIES OF TROPOSPHERIC AEROSOLS IN THE LEE OF THE CITY OF KARLSRUHE

G. Metzig and G. Poß
Kernforschungszentrum Karlsruhe GmbH
Laboratorium für Aerosolphysik und Filtertechnik I
Postfach 3640
7500 Karlsruhe 1
West-Germany

ABSTRACT. Within the framework of the POETA program (a German acronym for physico-chemical and optical properties of tropospheric aerosols) the aerosol near Karlsruhe is chemically described by several analytics. The particles are collected on bulkfilter. A high volume sampler in connection with a five-stage impactor (Andersen, particle size ranging from 0.5 to 15 µm) is also used. The particle composition is analysed as well as the element distribution on the particles' surface, and the element composition of single particles. A set of chemical analyses is applied to determine the portion of free acid, elemental carbon, sulfuric acid on the particles, and some important anions and cations. The spectral extinction coefficient of ambient aerosol particles is measured using a White cell. Reliable data about the extinction coefficient is obtained, even under clean air conditions, i.e. after rainfall. The data base represents seven field experiments of several days' duration at KfK site. In some cases aerosol samples were taken at the top of the meteorological tower at KfK in a height of 200m.

1. INTRODUCTION

The physical classification of particles is carried out with regard to their formation process and/or size (Junge, 1963; Whitby, 1978). The atmospheric aerosol cannot be characterized completely only by mass/number-concentration and particle size but also by the chemical composition of particles. First chemical investigations on atmospheric particles were carried out by Lodge (1962) using the reaction spot technique. With the upcoming of modern chemical analytics, partly based on nuclear technology, analyses could be performed with less material required. By the help of methods such as X-ray fluorescence analysis (Junge, 1963; Gillette and Blifford, 1971) or neutron activation analysis (Duce et al., 1966; Rahn et al., 1971) an early overview of the element composition of strato-spheric and tropospheric aerosol particles was obtained. Hidy et al. (1975) described the particulate matter measured during the California Aerosol Characterization Experiment. In the accumulation mode particles consist mainly of carbon, nitrogen and sulfur compounds. Lead, vanadium,

K. Grefen and J. Löbel (eds.), Environmental Meteorology, 23–37.

and bromine are found in the range of fine particles. The coarse mode consist mainly of sea salt components, iron, aluminium, titanium, and silicon. Background aerosol was first analysed by Junge. He compared the aerosol measured at Frankfurt with those at maritime locations in eastern USA. Later on, the Hawaii measurements followed. The analysis of the water soluble portion of the aerosols showed a deviation between the large and the giant size range. Junge found out that the main part of the large size range consists of ammonium sulfate, in Frankfurt as well as in the eastern USA. In more maritime environments, a strong increase of sea salt particles was detected in the giant size range. More and more chemical information on aerosols has become available dependent on particle size and sampling location. Special analytic methods have been developed, e.g. sulfuric acid measurements of the particles (Penzhorn and Panter, 1982).

Although numerous chemical measurements on aerosols are documented in literature, knowledge of the chemical composition of aerosols is still uncomplete. The main reason for this is the fact that for most of the investigators only one or two different analytical equipments are available. This means, a single analysis of the element concentration (Vogg and Härtel, 1977) does not describe all chemical characteristics of the aerosol particles, but only about two percent of the total aerosol mass. Of course, these results can be helpful when dealing with toxicity etc. They are, however, insufficient for a complete chemical characterization of the aerosol. Furthermore, the results of chemical analyses of aerosols taken at different times and locations cannot simply be mixed because of the great temporal and spatial variations of the aerosol. Nevertheless, basic aerosol types were classified during the Williamsburg meeting (Deepak and Gerber, 1983). Six basic aerosol components were defined (water soluble, dust-like, oceanic, soot, volcanic ash, and 75% H_2SO_4 droplets) and mixed together. From this mixture five basic aerosol models (continental, urban/industrial, maritime, stratospheric, and volcanic) were determined. This may be used as a first approximation. However, no reference to the range of fluctuations is given.

Up to now, there has only been one long-term data series: the 13-year measuring series at Wank near Garmisch-Partenkirchen, FRG (Reiter et al., 1984a,b; Sladkovic et al., 1986). On this mountain station filters are exposed over short intervals of one to a maximum of four days, depending on the weather condition. By means of wet-chemical methods, the following major and trace elements of the samples can be detected: SiO_2, Al_2O_3, Fe_2O_3, CaO, SO_4^{2-}, NO_3^-, Cl^-, NH_4^+, Na^+, Ca^{2+}, K^+, PbO, ZnO, Cu, and Cd. All meteorological parameters are determined and an additional measurement to obtain aerosol-physical data is carried out. In 1985, filters were also used in Garmisch-Partenkirchen. Between these two stations is a difference in altitude of about 1 km.

In the present paper the aerosol samples are analysed using eight different analytical methods. The layout, of the sampling equipment for bulk and size dependent measurements is designed so that the minimum mass required for all chemical analyses is sampled within in sampling time of 3 to up to 6 hours. The only reason for the application of a five-stage Anderson impactor in connection with a high volume sampler was the mere existence of such a sampler. The Watman back-up filter cannot be used for all chemical analyses. A back-up filter consisting of various filter materials which would be necessary for these analyses cannot be used in this case, because such a filter would influence the flow conditions within the sampler in an undefined manner. The samples were taken at the meteorological station at KfK and in some cases at the top of the meteorological tower (200m high) where only bulkfilters are used. The spectral extinction coefficient is measured at ground only.

2. THE SET OF ANALYTICS

2.1. Measuring Chemical Properties

It is important that qualified filter material is used for each analysis. For bulkfilters this does not constitute a problem, whereas for the impactor this is the case. At each stage, stripes of different filter materials have to be mounted without overlapping and changing the flow conditions in the stages. While mounting is the easier part the main problem is the frequent handling of the filter material itself. The cutting of stripes is an enormous possible source of pollution. For our analyses, six different filter materials are used.On the basis of underground value and its variation the following materials were selected: silver, borosilicate, quartzglassfiber, teflon, nuclepore, and cellulose acetate.

The following set of chemical analytics is applied to bulkfilters as well as to impactor samples:

- Measurement of the water soluble and insoluble portion of the aerosol particles.
- Measurement of the pH-value and the specific conductivity of the water-soluble part. Using these two values: calculation of the free acid (after Winkler, 1980).
- Measurement of the sulfuric acid on the particles by means of gaschromatography (after Penzhorn and Panter, 1982).
- Measurement of SO_4^{2-}, NO_3^-, F^-, Cl^-, Na^+, K^+, and NH_4^+ of the water-soluble part by means of ionchromatography.
- Determination of soot by measuring the total amounts of elemental carbon by means of thermal decomposition analysis.
- Bulk analysis of elements of the aerosol particles by total reflection X-ray fluorescence analysis (TRFA) for all elements heavier than P.
- Analysis of compounds on the particles' surface using electron spectroscopy (ESCA).
- Determination of elements of single particles by REM using an energy-dispersive X-rays analysis (EDAX).

2.2. Measuring Optical Properties

One part of the POETA program includes the measurements of volume extinction, scattering, and absorption coefficient. The instruments for the extinction and scattering measurements have been newly developed with the objective of measuring optical properties of airborne aerosol particles at typical tropospheric mass/number-concentrations. Besides, the instruments had to be suitable for field experiments.

The spectral extinction coefficient of ambient aerosol particles

composition	at ground	at 200m height
average aerosol mass [µg/m^3]	24.2	28.6
average analysed mass [%]	71.3	83.3
average soluble part [%]	55.8	67.7
pH-value	5.6	5.4
free acid [%]	2.9	*
H_2SO_4 [% of average aerosol mass]	10.5	10.8
SO_4^{2-}	21.5	20.0
NO_3^-	10.6	6.4
Cl^-	6.0	2.5
Na^+	2.1	3.1
K^+	3.0	2.0
NH_4^+	7.9	13.1
C	17.9	36.2
elements	2.3	*
unidentified	28.7	16.7

Table 1: Average values of the relative composition of aerosol particles sampled on bulkfilter during seven field experiments of several days' duration at KfK site (1985-1987). Comparison of the measured values. The average aerosol mass is related to 1m^3 sampled air. *: not yet measured.

Element:	minimum [ng/m^3]	average [ng/m^3]	maximum [ng/m^3]
Ag	0.12	0.34	0.59
Ba	13	39	68
Ca	50	120	170
Cd	0.01	0.13	0.15
Cr	1	1.5	2
Cu	35	120	265
Fe	46	112	140
K	30	220	740
Mn	2	3	5
Na	200	510	6810
Ni	5	5.5	7
Pb	12	15	20
Sr	0.12	0.25	0.40
Ti	1	5	8
V	1	4	7
Zn	12	35	79

Table 2: Element concentration of particles sampled at ground on bulkfilter during seven field experiments of several days' duration at KfK site (1985-1987). The results are related to 1m^3 sampled air.

is measured using the 'White Cell technique' (Poß and Metzig, 1987) which is usually applied for measuring gas absorption in the IR region. The 'White Cell' has been adapted to particle measurements.

The absorption and scattering measurements are not yet completely available for presentation in this paper.

2.3. Measuring Physical Properties

Physical characterization of the aerosol particles is carried out by electron microscope analysis and size dependent mass concentration measurements using a ten stage QCM cascade impactor. Furthermore, total mass concentration is measured too, by weighing the teflon bulkfilters.

2.4. The Meteorological Data

Except for the meteorological data distributed by the German Weather Service, the following data is measured and recorded by the KfK institute for meteorology: wind velocity and direction, temperature, humidity, pressure, short-wave and global radiation. These data items are determined in ten minutes' intervals and for different heights of the KfK meteorological tower. The measuring positions are: ground, 20m, 30m, 40m, 50m, 60m, 80m, 100m, 130m, 160m, and 200m height.

3. RESULTS

3.1. The Chemical Analyses

In this chapter mainly average values obtained by aerosol analysis during the 7 campaigns are presented, although not all analyses are mentioned. At ground the average values are based on 130 samples and at 200 m height on 40 samples. Up to now, it was not possible to take all weather conditions into consideration, e.g. sampling during a stable high pressure condition is missing. In order to obtain real average values, a weighing by the frequency of meteorological events has to be carried out.

In the obtained data no clear correlation between the chemical composition of the aerosol particles and the classified air mass type was found. There is only one exception. During the fourth campaign (13.3.1987-27.3.1987) fresh maritime air masses have been transported via France to Karlsruhe for 15 hours. Compared with the days before the concentrations of sodium and chlorine were increased by a factor of ten.

The analyses of the compounds on the particles' surface have identical results for each measured sample: oxygen, carbon, nitrogen, sodium, chlorine, and sulfur are found only. This differs from the results from samples taken in a plume just behind the stack (Dlugi et al., 1981).

3.1.1. Analyses of Bulkfilters

The main results of the comparison between the average aerosol characteristics at ground and at 200m height are shown in table 1. At ground, the average mass concentration is about 20% lower than at 200m height. The average analyzed mass and the average soluble part are also smaller at ground. As a consequence of measuring elemental carbon by means of thermal decomposition analysis (some organics), most of the unidentified mass is water. At ground, the nitrate and chlorine concentrations are approximately twice as high whereas the ammonia and elemental carbon concentrations are approximately reduced by half. The lower aerosol mass at ground can be explained by the cleaning effect of the forest area around the sampling place.

Table 2 shows the average element concentration obtained by the bulkfilter analyses. The minimum and maximum values are also indicated. Although more elements were analyzed they are not presented due to minor values. Compared with the Wank data, more sodium and chlorine, but less lead were detected. The latter one is consistent with the lead reduction in gasoline. Calcium and iron can be compared with the Wank data. Al, Br, and Si were not analysed.

3.1.2. Analyses of Impactor Measurements

The average size-dependent aerosol composition at ground is shown in fig. 1. In some plots data measured at the Wank (Reiter et al., 1984a,b) and at Hanover (Georgi, 1986) is added. In future measurements the impactor will have to be changed in order to obtain information on particles smaller than .5 µm in diameter.

3.1.3. Time Series

As an example for time series, sulfate concentrations on bulkfilters at ground and at 200 m height during the third campaign (25.11.85-13.12.85) are represented in fig. 2. The main wind directions are indicated as well as two periods of snowfall. The cleaning effect of the snowfalls can easily be noted both at ground and at heigher up. A strong increase is recorded when the wind turns to south-west (11.12.85), i.e. it comes directly via Karlsruhe. The figures at the bars indicate the measured concentration in ug/m^3 sampled air. The width of the bars represents the sampling time.

3.2. The Extinction Coefficient

The extinction coefficient of ambient aerosol particles is measured using a multiple traverse cell (White Cell) with an effective path length of 100m. Measurements are performed at 7 fixed wavelengths in the visible region using a white light source and two interference filter sets of 2 and 25 nm bandwidth and centerwavelengths of 405.5, 450, 500, 550, 600, 650, and 692.5 nm. The results obtained during two measuring campaigns are shown in fig. 3a-d. During these field experiments typical mass concentrations amounted to 2-25 µg/m^3, i.e. clean air conditions.

When reproducing the measured extinction coefficient by straightforward MIE calculations particle number distribution and complex index of refraction of the aerosol must be known. In order to obtain information on the number distribution, simultaneous measurements of the mass distribution is carried out by using a ten stage QCM cascade impactor. If particle bulk density and sphericity are assumed, with aerodynamic particle size cut offs as well as the broadness of the distribution on each stage being known, the number distribution can be calculated. It is difficult to obtain representative effective complex index of refraction. This value can be approximated when the aerosol composition is known from chemical analysis. On the basis of a net of isolines of extinction and absorption coefficients for a typical number distribution and a fixed wavelength, it can be estimated when absorption starts to become a dominant effect. The imaginary index > 0.1 is rarely observed (Deepack and Gerber, 1983). This index value represents the maximum for urban aerosol with a higher soot content.

Extinction coefficient calculations are performed at the above mentioned 7 wavelengths using a complex index of refraction of (1.54, $-i\ 8.3\cdot10^{-3}$). The number distribution is calculated from measured mass distributions assuming a particle bulk density from 1.4 as the minimum to 2.0 g/cm^3 as the maximum. Consequently a different number per size functions as multiplier for the respective extinction cross sections is obtained. The different cut off sizes of the QCM impactor, which assumes the possible bulk density variations of the sampled particles are also taken into consideration. Thus, the initially measured mass distribution is converted into a number size distribution.

3.2.1. Comparison Between Measurements and Calculations

The experience made during several measuring campaigns is that agreement between calculations and measurements of the extinction coefficient better than a factor of 2 is hardly achievable. On the one hand, calculation and measurement results are identical concerning single measurements(see fig. 3b at 692.5 nm). On the other hand, great differences occur as can be seen in fig. 3b at 550 and 600 nm.

Usually an overlapping of the two bandwidths within a variation of the calculated σ_{ex} (dashed lines in fig. 3a-d) and measured σ_{ex} (solid lines in fig. 3a-d) can be observed. During the extinction measurements and size dependent mass concentration measurements too many conditions have to be taken into account (sample technique, accordance in time, unknown time scale of the aerosol concentration fluctuations). Consequently a more reliable agreement between theory and measurement can hardly be achieved. The bandwidths of the measurements represent the standard deviation of σ_{ex} including measuring errors. The corresponding bandwidth of the theoretical values results from calculations with different number distributions which were obtained by assuming a particle bulk density of 2 g/cm^3 (minimum: B in fig. 3) and 1.4 g/cm^3 (maximum: A in fig. 3).

The left number of the solid lines' minimum indicates the quantity of extinction coefficient measurements at the respective wavelength. The left number of the dashed lines' minimum represents the number of distributions which are used to obtain a theoretical mean extinction coefficient.

Besides the cases of good agreement shown in fig. 3a, c, and d, great deviations appear in fig. 3b at a wavelength of 500 - 650 nm. It is striking that these deviations occur in connection with bi-modal distributions. However, compared with all the other measured mass concentrations during the same period no significant change in the total mass concentration can be noted. Although the second mode occurs at larger particle sizes at the expense of a drastic reduction of the number of fine particles, the measured extinction is unexpectedly high. This leads to the assumption that the second mode contains high-absorber material. The only obvious difference found in these chemical analyses is a significantly smaller soluble portion of the total of sampled particles on this day (about 50% instead of 70-80% during this campaign).

4. CONCLUSIONS

The aerosol particles are described by means of several analytics. Both, bulk and size-dependent samples are analysed. The sampling time is limited to 6 hours. Therefore high volume sampling is required in order to obtain the minimum masses needed for the analsyses. It was not possible to analyze the back-up filter of the Anderson impactor. Thus no information is available on particles smaller then .5 µm in diameter. In future, the use of a Berner impactor is planned.

The presented average values obtained by the chemical analyses can only be used for a first description of the measured aerosol. There is no statistical correlation between the frequency of air mass events, wind direction etc. and the 7 measuring campaigns. Real average values can only be deduced from long term series only. However, the present data base can be used for analyzing diurnal variations in the particles composition which depend on seasons, air mass type, meteorological data etc.

The accordance of the extinction coefficient measurement and calculation within a factor of two can be regarded as the maximum achieveable result for average values. Taking into account the various influencing parameters such as mean size and broadness of the used particle distribution, achieveable accuracies of flow control, uncertainties of the assumed particle bulk density and complex index of refraction, the margin of the variations cannot be further confined. However, improvements if possible at all, can be achieved by more accurate measurements of the above mentined parameters. In many cases this leads to the development of new measuring techniques. Nevertheless, the results obtained using the White Cell technique are sufficient to describe the optical behaviour of the ambient aerosol in reliable limits. Furthermore, the suitability of the cell for routine measurements was demonstrated. In future, a continuous measurement of the extinction coefficient is necessary in order to obtain information on time fluctuations of the aerosol.

5. ACKNOWLEDGEMENT

The White Cell has been modified for aerosol measurements in cooperation with FHG-IMP Freiburg (Dr.Riedel). The element analyses were performed at KfK-LIT (Dr. Schneider) and GKSS-IP (Dr. Knoth). The analyses of compounds at the particles surface were carried out at KfK-IRCh (Dr. Klewe-Nebenius).

6. REFERENCES

Deepak, A. and Gerber H.E., 1983: 'Report of the Experts Meeting on Aerosols and their Climatic Effects'. Williamsburg, Virginia, 28-30 March 1983, WMO Report WCP-55

Dlugi, R., Jordan, S., Lindemann, E., 1981: 'Experimentelle Untersuchungen zur Entstehung von schwefelhaltigen Aerosolen unter atmosphärischen Bedingungen'. KfK-Bericht Nr. 3187, Kernforschungszentrum Karlsruhe

Duce, R.A., Winchester, J.W. and van Nahl, T.W., 1966: 'Iodine, Bromine, and Chlorine in Winter Aerosols and Snow from Barrow, Alaska'. Tellus **18**, p.238-248

Georgi, B., Giesen, K.P., Müller, W.J., 1986: 'Measurements of Airborne Particles in Hannover'. AEROSOLS: Formation and Reactivity. 2nd Int. Aerosol Conf. Berlin, Pergamon Journals Ltd.

Gillette, D.A. and Blifford, I.H., 1971: 'Composition of Tropospheric Aerosols as a Function of Altitude'. J. Atmosph. Sci. **28**, p.1199-1210

Hidy, G.M., Appel, B.R., Charlson, R.J., Clark, W.E., Friedlander, S.K., Hutchison, D.H., Smith, T.B., Suder, J., Weselowski, J.J. and Withby, K.T., 1975: 'Summery of the California Aerosol Characterization Experiment'. J. Air. Poll. Control Ass. **25**, p.1106-1114

Junge, C.E., 1963: Air Chemistry and Radioactivity. Academic Press, New York, London

Lodge, J.P., 1962: 'Identification of Aerosols'. Adv. in Geophys. **9**, p.97-130

Penzhorn, R.D. und Panter, R., 1982: 'Untersuchung zur Entstehung von schwefelhaltigen Aerosolen in der Atmosphäre und deren Bestimmung (Teilprojekt B)'. KfK-Bericht Nr. 3440, Kernforschungszentrum Karlsruhe, FRG

Poß, G. and Metzig, G., 1987: 'Extinction Coefficient Measurements of Ambient Aerosol Particles Using a White Cell'. Paper presented at: Urban and Environmental Air Pollution, 9-11 September 1987, Hannover, FRG, accepted for publication in J. Aerosol Sci.

Rahn, K.A., Dams,R., Robbins, J.A. and Winchester, J.W., 1971: 'Diurnal Variations of Aerosol Trace Elements Concentrations as Determined by Non-Destructive Neutron Activation Analysis'. Atmosph. Environment **5**, p.413-422

Reiter, R., Pötzl, K. and Sladkovic, R., 1984a: 'Determination of the Concentration of Chemical Main and Trace Elements (Chemical Matrix) in the Aerosol from 1972 to 1982 at a North-Alpine Pure Air Mountain Station at 1780 m a.s.l.; Part I: Methods of Measurement, Total Concentrations, Families of Elements, Annual Variations, Trend Analysis'. Arch. Met. Geophys. Bioklim., **Ser. B34**, p.215-241

Reiter, R., Pötzl, K. and Sladkovic, R., 1984b: 'Determination of the Concentration of Chemical Main and Trace Elements (Chemical Matrix) in the Aerosol from 1972 to 1982 at a North-Alpine Pure Air Mountain Station at 1780 m a.s.l.; Part II: Parametric Correlation Analysis of the Chemical Matrix with Consideration of Meteorological Quantities'. Arch. Met. Geophys. Bioklim., **Ser. B35**, p.1-30

Sladkovic, R., Reiter, R. and Pötzl, K., 1986: 'Chemical Components of the Atmospheric Aerosol at a High Mountain Station as a Function of Temporal Trend and Meteorological Parameters, Based on a 13-Year Measuring Series'. J. Aerosol Sci. **17**, p.231-239

Vogg, H., Härtel, R., 1977: 'Experience in the Analysis of Atmospheric Aerosols at the Karlsruhe Nuclear Research Center'. Journal of Radioanalytical Chemistry **37**, p.857-866

Whitby, K.T., 1978: 'The Physical Characteristics of Sulfur Aerosols'. Atmospheric Environment **12**, p.135-159

Winkler, P.,: 'Observations on Acidity in Continental and in Marine Atmospheric Aerosol and in Precipitation'. J. of Geophysical Research **85**, p.4481-4486

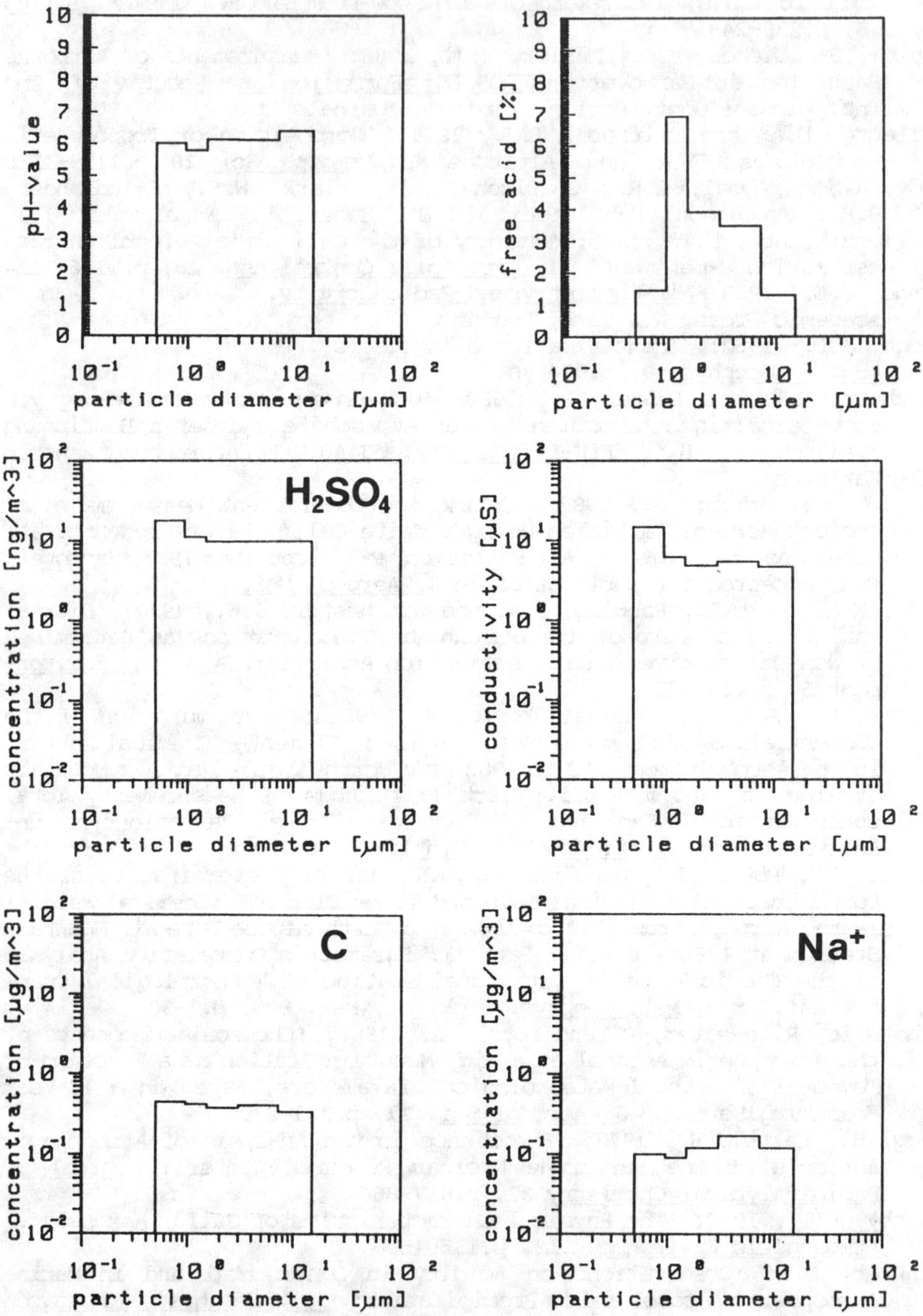

Figure 1: Explanation see next page.

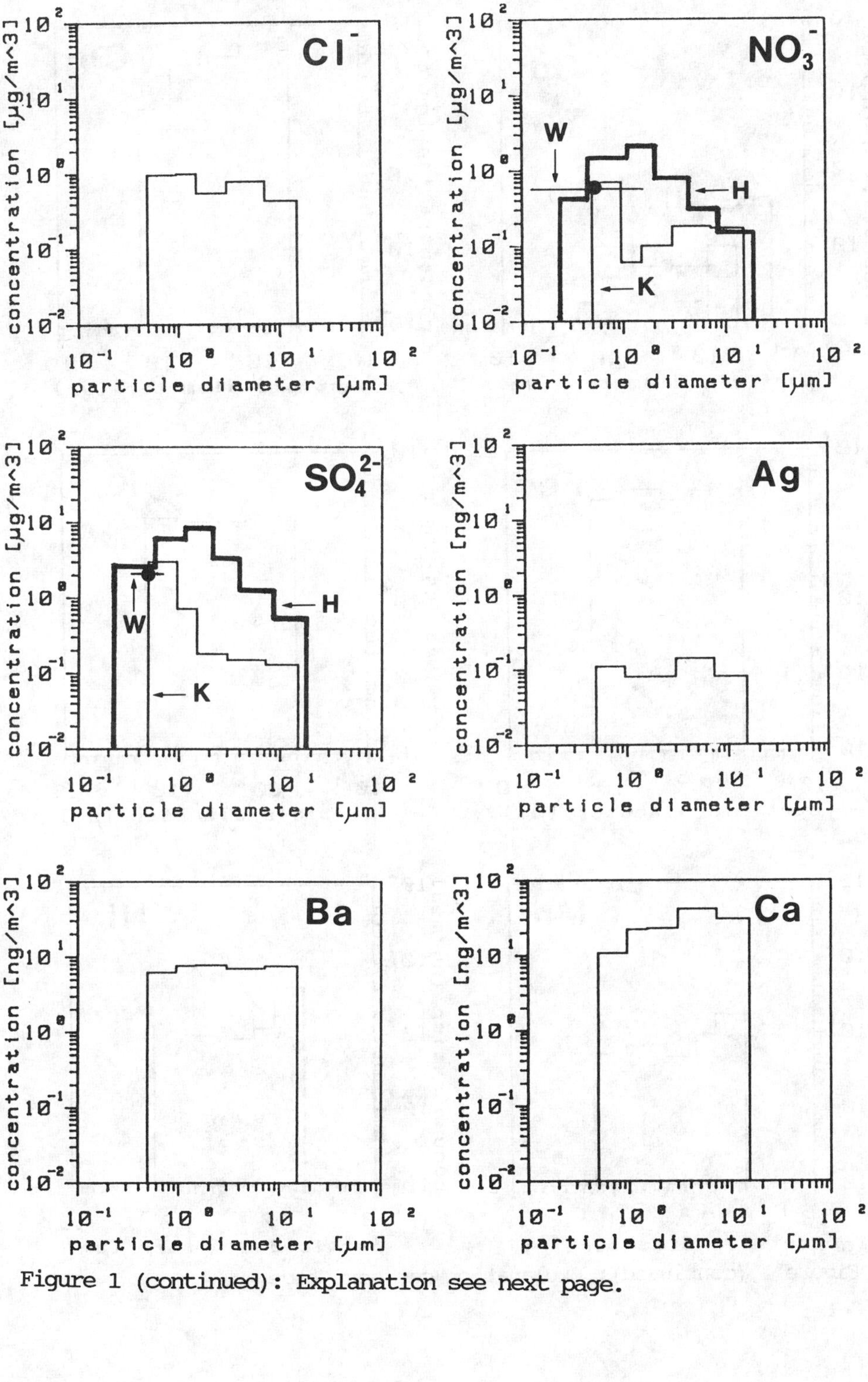

Figure 1 (continued): Explanation see next page.

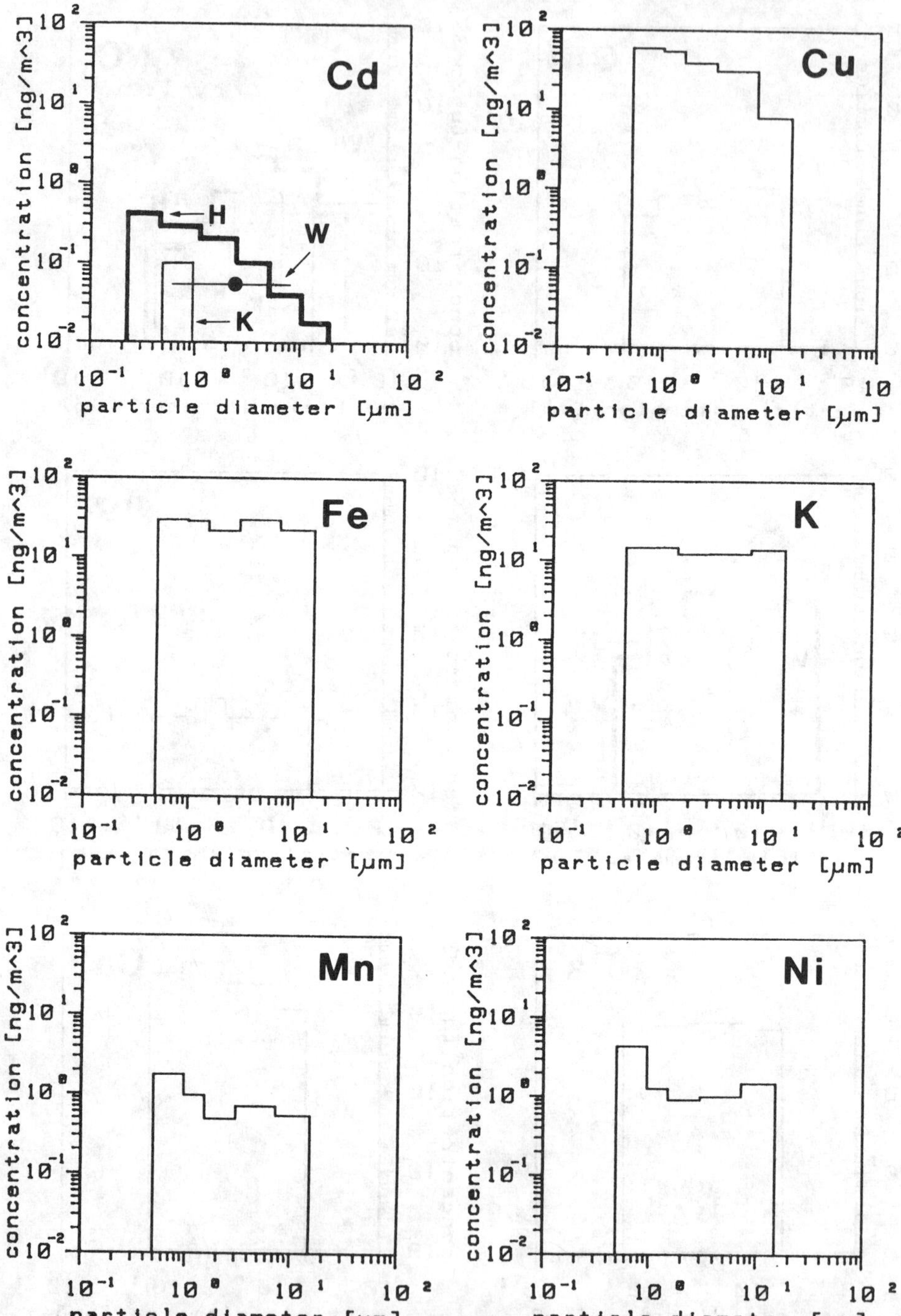

Figure 1 (continued): Explanation see next page.

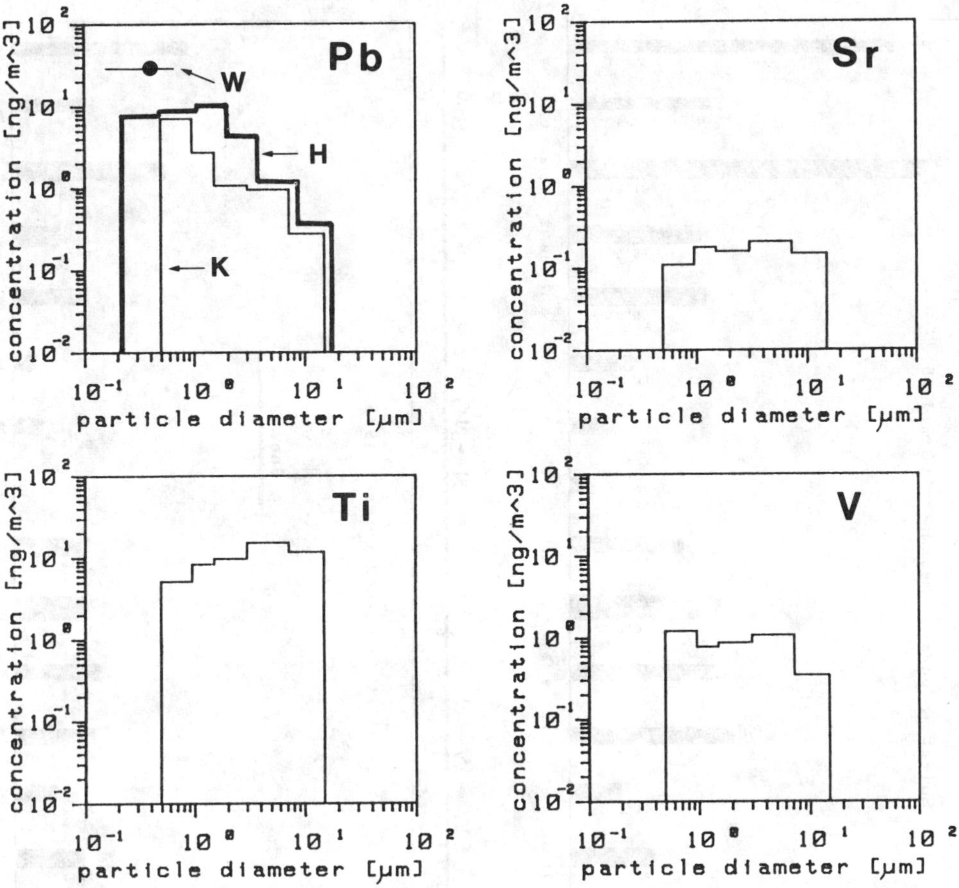

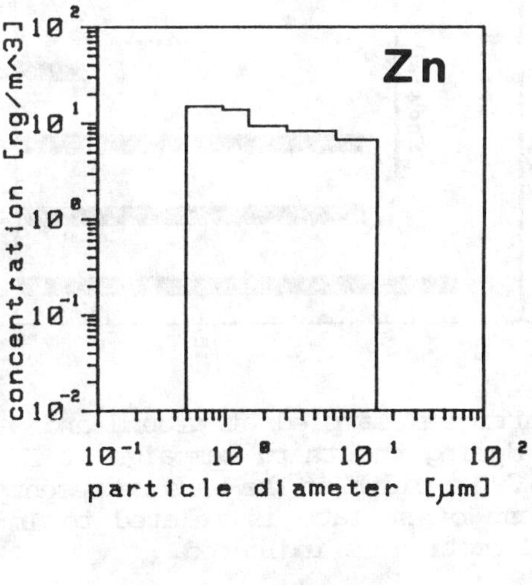

Figure 1 (continued): Average size dependent aerosol composition at ground during seven field experiments of several days' duration at KfK site (1985-1987).
Notation: H - Hanover data
W - Wank data
K - Karlsruhe (KfK)

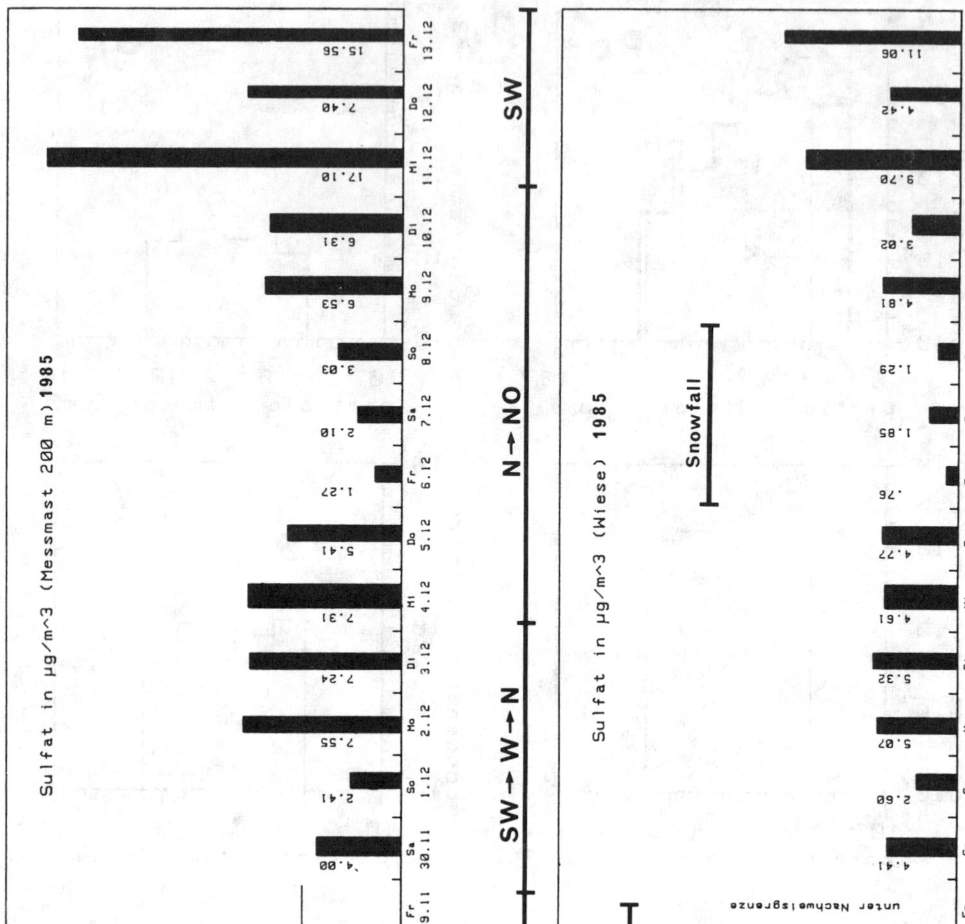

Figure 2: Sulfate concentration of particles sampled at ground and at 200m height on bulkfilters during the third campaign at KfK site (25.11.1985-13.12.1985). The width of the bars represents the sampling time. The amount of sulfate is related to $1m^3$ sampled air. The main wind direction is indicated.

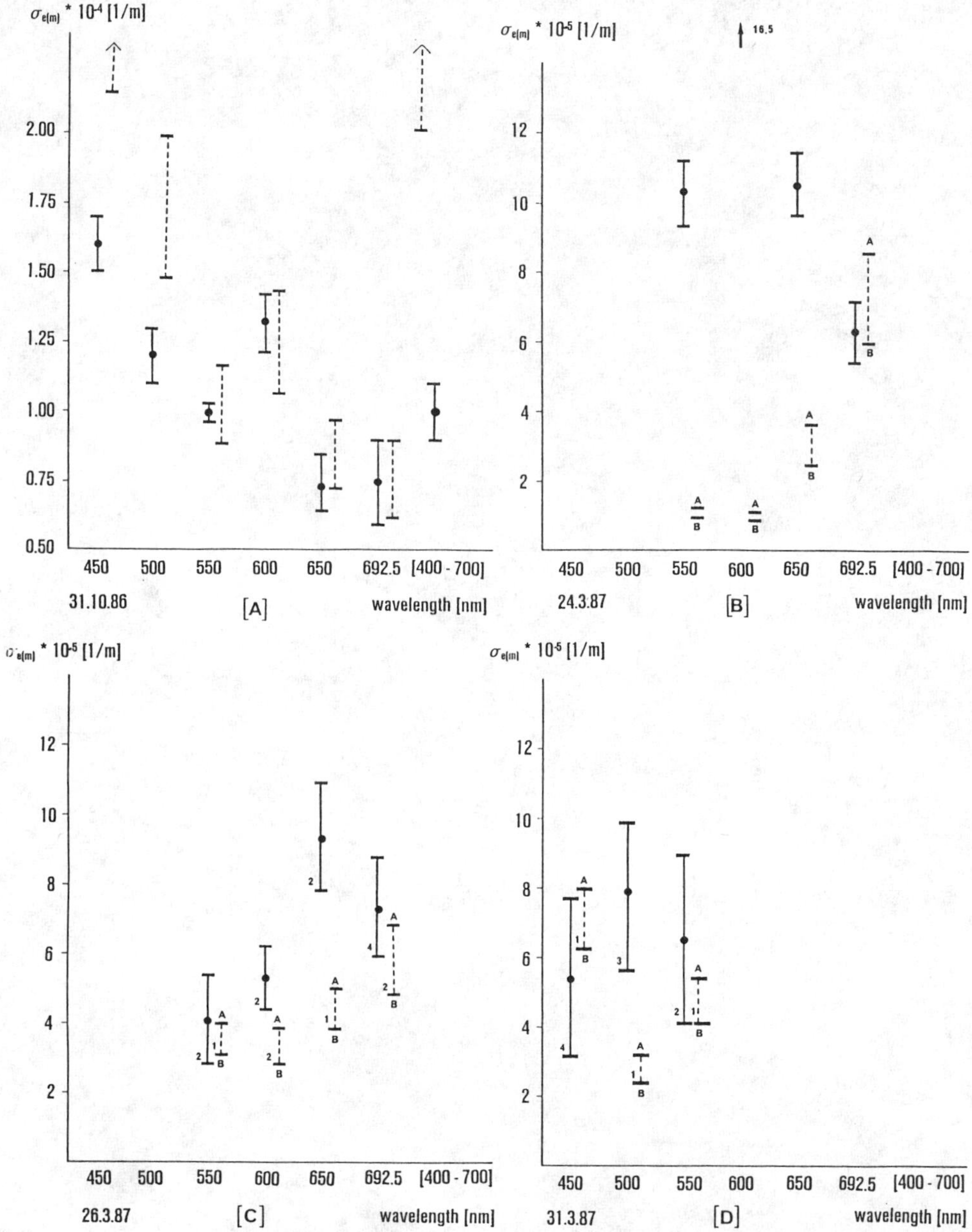

Figure 3 a - d: Comparison between measured (solid lines) and calculated (dashed lines) extinction coefficients. Explanation see text.

DISPERSION OF CAR EMISSIONS IN THE VICINITY OF A HIGHWAY

M. Kuhler, J. Kraft	W. Koch, H. Windt
Volkswagen AG	Fraunhofer Institut für
Forschung Meßtechnik	Toxikologie und Aerosolforschung
Postfach	Nikolai-Fuchs-Straße 1
D-3180 Wolfsburg	D-3000 Hannover 61
Germany	Germany

ABSTRACT. Emissions from highways are said to contribute to forest damage. Measurements of air quality at highways performed so far were mostly confined to distances close to the lanes. Relatively few is known for the range of several hundred metres away from the lanes, where emissions from the highway are expected to be still predominant.

Therefore the state of Lower Saxony and Volkswagen run a joined project to investigate the dispersion of exhaust emissions from a motorway in a rural surrounding. Measurements are performed by the "Fraunhofer-Institut für Toxikologie und Aerosolforschung" in Hannover. NO_x, CO, and O_3 are continuously monitored at different distances up to 600 m togehter with meteorological data. A single SO_2 monitor indicates if pollution from more distant stationary sources occurs.

Diurnal variations of excess concentrations of NO_x und CO were found to correlate well with traffic density. Measurements at different heights frequently showed the existence of vertical gradients indicating deposition of NO_2 and SO_2.
For Ozone, reduced levels were found on the leeward side of the highway.

The applicability of different dispersion models for describing the dispersion of car exhaust emissions and fitting the measured concentrations is discussed.

1. INTRODUCTION

Emissions from cars driving on highways are suspected to contribute to forest damage. Therefore the state of Lower Saxony and Volkswagen started a joined project to investigate the dispersion of exhaust emissions from a motorway in a rural surrounding. The site is situated at motorway A7 near Seesen, about 80 km south of Hannover. The measurements are performed by the "Fraunhofer-Institut für Toxikologie und Aerosolforschung" in Hannover. So far, measurements of air quality at highways were mostly confined to distances close to the lanes /1,2/. Relatively few is known for the range of several hundred metres away from the lanes. So we extended our measurements to

K. Grefen and J. Löbel (eds.), Environmental Meteorology, 39–47.

a distance of 600 m, where the emissions from the highway are expected to be still predominant, but effects of deposition and chemical reactions might be detectable.

2. INSTRUMENT LAYOUT

Three sets of analyzers are installed for nitrogen oxides, carbon monoxide and ozone. Up to three sampling points are multiplexed to each analyzer. As SO_2 emissions from the highway are small, SO_2 is regarded as an indicator for pollution from more distant stationary sources. SO_2 is registered by a single monitor. Figure 1 shows the instrument layout.

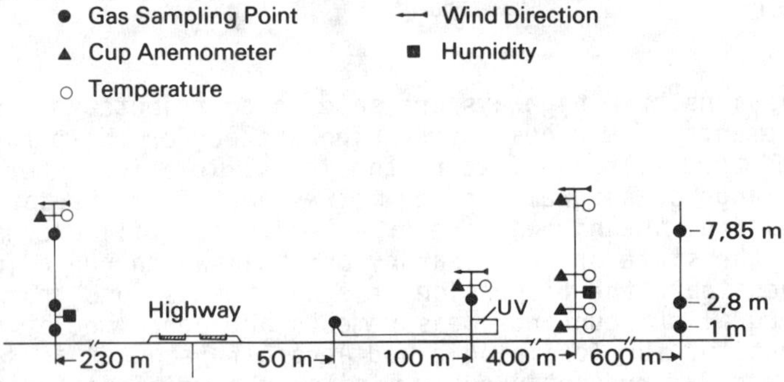

Figure 1: Instrument Layout

At a separate tower temperature and windspeed are measured at 4 different heights in order to get vertical profiles.

Furthermore, windspeed and direction are recorded at the reference station and at 100 m distance. Additionally an Eppley radiometer and a rain indicator are installed. The terrain is not completely flat, the station at 600 m lies about 20 m higher than the highway, but it is not a complex terrain. The traffic is counted by means of induction coils in the lanes, a rough differentiation between passenger cars and trucks is possible.

3. BEHAVIOUR OF EMISSIONS

3.1 Correlation of exhaust gas concentrations and traffic.

During steady crosswind, the diurnal variations of traffic and ΔNO_x apparently are in a good agreement as can be seen from figure 2, thus indicating a close correlation between traffic and ΔNO_x. " Δ " means the difference of concentrations measured at the leeward and the windward side of the highway.

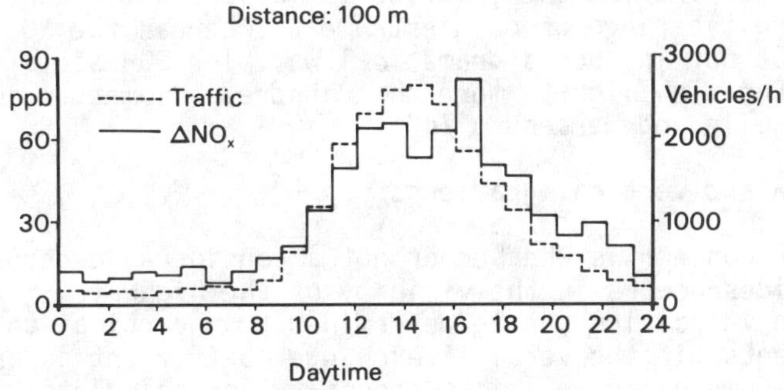

Figure 2: Diurnal variation of ΔNO_x and traffic

To optimize this correlation, it has to be taken into account that the NO_x emissions from trucks are different from those of passenger cars. The best fit was obtained if the NO_x emissions of a truck were assumed to be about 3 times the emissions of a passenger car. More specialized investigations on the contribution of trucks performed by other authors showed comparable results /3/. For carbon monoxide a similar correlation is found.

3.2 Advection of polluted air

The site is normally dominated by emissions from the highway. There are no major other emission sources in the vicinity.

Normally the SO_2 level is low (~ 5 ppb), but there were some pronounced smog episodes during winter, in connection with easterly winds and closed snowcover. Figure 3 shows the registrations of SO_2 and NO_2 during such an episode.

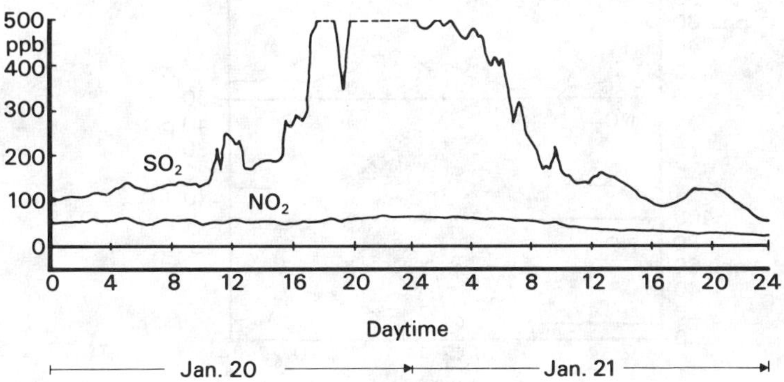

Figure 3: Concentrations of SO_2 and NO_2 in background air during a smog episode

The SO_2 concentration shows strong variations and reaches exessive levels exceeding the range of our instrument, whereas the NO_2 level is elevated, but not in such a dramatical way. The SO_2 stems from the burning of sulfur-rich lignite more than hundred kilometres away in the region of Halle and Merseburg /4/.

3.3 Deposition and vertical gradients

One question of concern is whether or not a considerable deposition of nitrogen oxides occurs in the vicinity of the highway.

Deposition velocities can be determined from vertical concentrations gradients, if the vertical exchange coefficient is known. Figure 4 shows some basic relations necessary for calculating deposition velocities from vertical gradients.

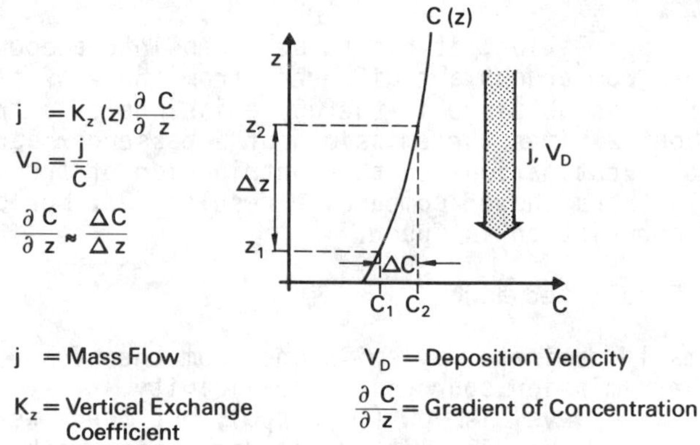

Figure 4: Relations for calculating deposition velocities from vertical concentration gradients

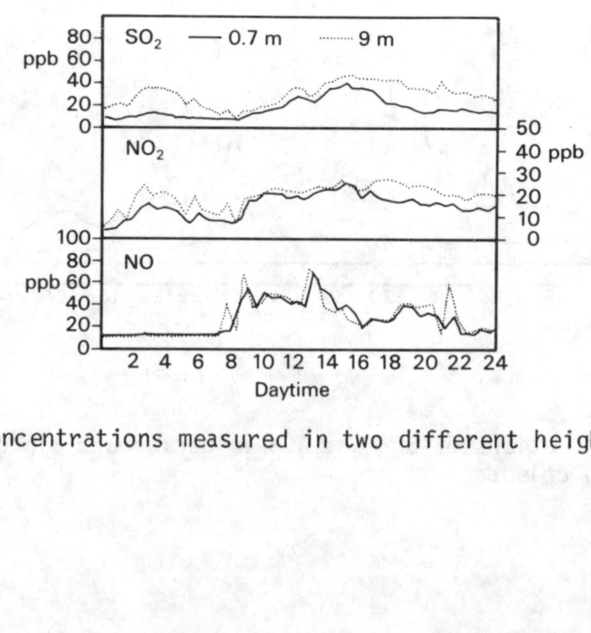

Figure 5: Concentrations measured in two different heights

The exchange coefficient itself can be calculated from the vertical profiles of windspeed and temperature /5,6/. Figure 5 shows concentrations measured in two different heights during stable atmospheric conditions.

For SO_2 and NO_2 the concentrations at the upper sampling point are significantly higher. For NO, only random differences are observed. This behaviour indicates that deposition occurs only for SO_2 and NO_2 but is negligible for NO.

During a year, the gradients show considerable variations. In January, there are nearly no gradients at all.
But in June, there is a pronounced diurnal variation for the gradients of SO_2 and NO_2, whereas for CO there are only some small deviations from zero, with questionable significance.

Figure 6 shows averaged diurnal variations of the vertical gradients in relative units for June 1986.

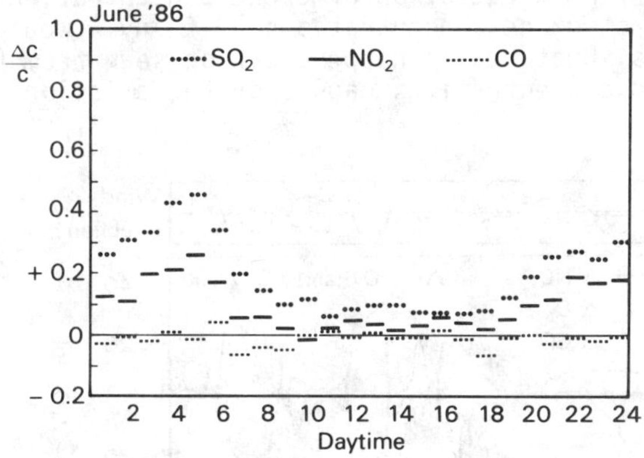

Figure 6: Averaged diurnal variations of vertical gradients in relative units

This picture reflects to some extent the diurnal variation of atmospheric stability: The heating by the sun's radiation causes turbulence, which hinders the formation of vertical gradients. Nevertheless, the deposition does not necessarily behave in the same way: The increase of the vertical exchange coefficient caused by turbulence can over-compensate for the smaller gradients. Summing up, we can expect that depostion of NO_2 and SO_2 takes place during summertime, but is at least strongly reduced in the winter.
On the other hand, the lack of gradients for CO indicates that no deposition of CO occurs.

4. CHEMICAL REACTIONS AND OZONE

Another point of interest is the influence of emissions from the highway on the ozone level. With respect to the 600 m-distance, only effects of fast reactions can be detectable. Figure 7 shows some important reactions.

$$NO + O_3 \rightarrow NO_2 + O_2 \quad \sim 1\ \text{min}$$
$$NO_2 + O_2 + h\nu \rightarrow O_3 + NO \quad > 2\ \text{min}$$
$$NO + R\text{-}O_2 \rightarrow NO_2 + R\text{-}O$$

$R\text{-}O_2$: Peroxy-Radical
R-O : Oxy-Radical

Figure 7: Some reactions of ozone and oxides of nitrogen

The fastest reaction is the oxidation of NO by ozone, followed by the photolytical formation of ozone. Oxidation of NO by radicals, which is essential for producing high ozone levels, is even slower. So we can expect to find a reduction of ozone concentrations whereas a possible increase of ozone concentrations - if any - would occur at greater distances. What really happens can be seen from figure 8; where the changes in concentrations caused by the emissions from the highway are shown.

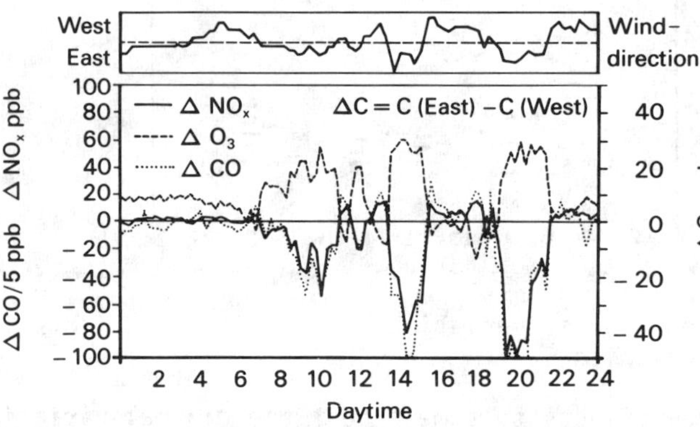

Figure 8: Change in concentrations caused by the emissions from the highway and wind direction

At first glance this figure is a little bit sophisticated as the differences change their signs when the wind direction changes from one side of the highway to the other one. On the respective leeward side we always find an increase in concentrations of CO and NO_x, but a reduction of ozone. This behaviour indicates that in the wake of the highway ozone has been consumed for oxidizing NO. This assumption is confirmed by a strong correlation of the changes in concentrations of NO_2 and ozone. The coefficient of correlation was found to be $r = -0.86$ (calculated from $n > 2000$ data points). Furthermore it justifies the assumption that ozone is the main oxidant for the formation of NO_2 from NO.

If we assume a simplified Gauss-type model for dispersion, plotting concentration versus distance in a representation with both scales in logarithmic units will give a straight line. As figure 9 shows, for NO_x, this straight line is an acceptable first approximation.

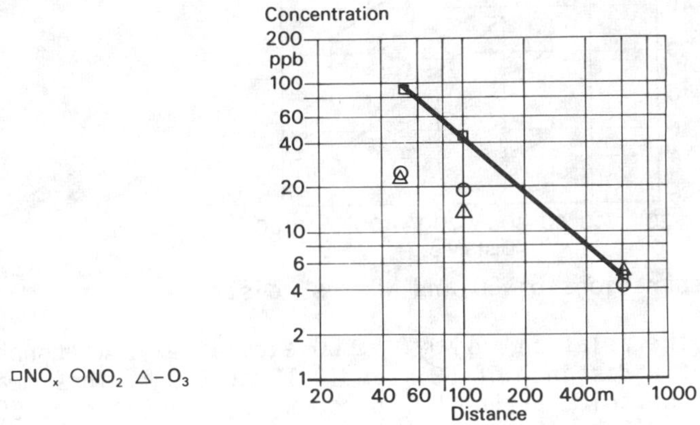

Figure 9: Excess of NO_x and NO_2 and deficit of ozone vs. distance. Both scales logarithmic

In addition we can see that the excess NO_2 equals quite well the ozone deficit, once more backing the assumption that the NO from the highway is oxidized mainly by O_3. At a distance of 600 m, nearly all NO_x is NO_2, and the ozone consumed for oxidizing NO has not yet been replaced.

5. RELATION OF NO_x AND CO

As already mentioned, CO behaves as a nearly inert gas, a consequence of its long atmospheric lifetime of more than a month. If NO_2 is deposited or NO_x is removed in another way, the relation between NO_x and CO should change with distance. If we look at the correlation of ΔCO and ΔNO_x, we will find it significantly worse than for NO_2 and O_3.

But this can also be due to variations in the fraction of trucks on the highway.

Plotting ΔNO_x and ΔCO in logarithmic scales, figure 10, gives 2 straight lines with different slopes, the line for ΔNO_x being steeper. This indicates a change in the ratio of ΔNO_x and ΔCO, most probably caused by a loss of NO_x.

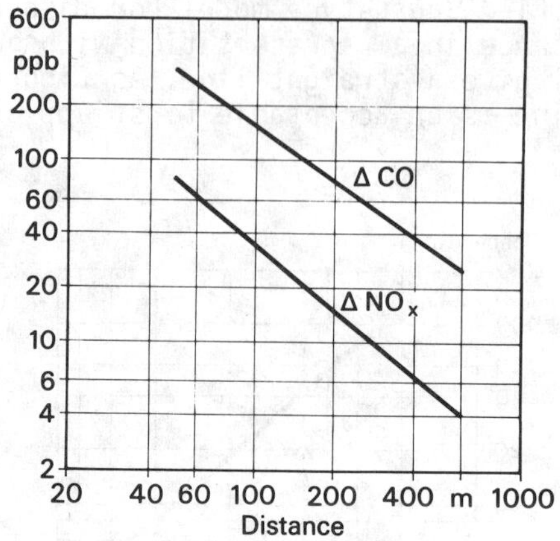

Figure 10: Concentrations of CO and NO_x vs. distance

Looking at the relation $\Delta NO_x / \Delta CO$ explicitely, we found mean values of 0.29 at a distance of 50 m and 0.18 at 600 m. This means a loss of about one third of the NO_x present at a distance of 50 m from the motorway, if we assume that the change in the relation is completely attributable to NO_x.

Anyway, this figure is based on a limited set of data and needs confirmation by further measurements.

6. DISPERSION MODELS

For interpreting the measurements in a more detailed way, it is necessary to use dispersion models. As a first approximation, a Gauss-type line source model with partial reflection can be used to describe dilution and deposition /7,8/.

To include chemical reactions as well, a different approach is necessary.

A Lagrange-type model might be a suitable way. Problems may arise from the unusually small dimensions of the moving air parcel necessary for a sufficient spatial resolution.

The project is still running. We expect to be able to give more quantitative results as well as more detailed interpretations in the near future.

LITERATURE

/1/ Esser, J: 'Schadstoffkonzentration im Nahbereich von Autobahnen in Abhängigkeit von Verkehr und Meteorologie', in: Abgasbelastungen durch den Kraftfahrzeugverkehr, Verlag TÜV Rheinland, Köln 1982.

/2/ Chock, D.P: 'Pollutant Dispersion near Roadways - Experiments and Modeling', in: Abgasbelastungen durch den Kraftfahrzeugverkehr, Verlag TÜV Rheinland, Köln 1982.

/3/ VdTÜV: Abgas-Großversuch, Abschlußbericht, Im Auftrage des Bundesministers für Verkehr, Verlag TÜV Rheinland GmbH, Köln 1986.

/4/ Bruckmann, P, Borchert, H, Külske, S, Lacombe, R, Lenschow, P, Müller, W.J and Vitze, W: 'Die Smog-Periode im Januar 1985', Staub Reinhaltung der Luft, Band 46 (1986), p. 334-342.

/5/ Businger, J.A, Wyngaard, J.C, Izumi, Y, Bradley, E.F: 'Flux-Profile Relationships in the Atmospheric Surface Layer,' Journal of the Atmospheric Sciences, Vol 28 (1971), p. 181-189.

/6/ Koch, W: 'Bestimmung von horizontalen und vertikalen Schadstoffgradienten im Einflußbereich einer Autobahn', in: Aktuelle Aufgaben der Meßtechnik in der Luftreinhaltung, VDI Berichte 608, VDI Verlag, Düsseldorf, 1987.

/7/ Chock, D.P: 'A simple Line-Source Model for Dispersion near Roadways', Atmospheric Environment, Vol. 12. p. 823-829, 1978.

/8/ Rao, K.S: 'Analytical Solutions of a Gradient-Transfer Model for Plume Deposition and Sedimentation', NOAA Technical Memorandum ERL ARL-109, 1981, NTIS Acc. Nr. PB 82-215153.

POLLUTION LEVELS NEAR A MOTORWAY IN A CUTTING

V H Brennan° and I S McCrae+
°Transport and Road Research Laboratory, Crowthorne, Berkshire, RG11 6AU, United Kingdom
+ Middlesex Polytechnic, Queensway, Enfield, Middlesex, EN3 4SF, United Kingdom

ABSTRACT. A computer method of estimating traffic generated air pollution near to roads has been developed at TRRL for use in the environmental appraisal of new road schemes. A series of pollution surveys is being carried out to provide data to enable a modification of the method for applications in topographically complex locations. One survey took place near to a heavily trafficked section of the M25 London Orbital motorway, at a site where the road passes through a 7m deep cutting, during Autumn 1986. A range of pollutants was monitored at a mobile laboratory parked close to the hard shoulder and carbon monoxide concentrations were measured at a number of locations inside and outside the cutting using portable analysers. Traffic flows and meteorological conditions were also recorded. The concentrations at the mobile laboratory were typical of a motorway site. Carbon monoxide levels were generally lower and oxides of nitrogen higher than in urban locations, consistent with their relative emission rates from high and low speed traffic. Spatial variations in the carbon monoxide concentrations measured within the cutting, in comparison with those measured outside and with those forecast for a similarly trafficked site in level ground, demonstrated the effect of the cutting in modifying the local dispersion characteristics. At road level the concentration in the cutting was substantially higher than outside, suggesting a restriction to the dispersion. High concentrations were found on the side of the road opposing the prevailing regional wind, probably because of the formation of turbulent eddies in the cutting.

1. INTRODUCTION

In the United Kingdom, the Department of Transport evaluates proposals for construction and major modification of trunk roads with respect to their likely economic and environmental effects. The procedure for the economic appraisal is described in the Cost-Benefit Analysis Manual (COBA) (Department of Transport 1981) and that for the environmental analysis is described in the Manual of Environmental Appraisal (MEA) (Department of Transport 1983).

K. Grefen and J. Löbel (eds.), Environmental Meteorology, 49–61.

Air pollution from traffic is one of the factors considered as part of the environmental appraisal. The MEA describes a method by which any potential air pollution problem may be identified and states that a more detailed evaluation should be carried out in problem cases. The method is based on a computer model that was originally designed for use in situations where the area around the road system is level. However, the increased number of urban motorways now being built are often topographically or structurally complex, and therefore a need has arisen for a more comprehensive model able to take these factors into account.

This paper describes a survey undertaken between 24 September and 27 November 1986, at a monitoring site on the M25 London Orbital Motorway where it passes through a cutting. The survey is one of a series that will be used to modify the model so that more accurate results may be obtained for sites of complex topography.

2. SITE LOCATION

A mobile laboratory, equipped to measure levels of air pollutants was installed adjacent to the hard-shoulder of the northbound carriageway of M25 near to junction 11 (see Fig 1). This location was approximately halfway along a cutting that is 2 km long, between 5.5 and 7 m deep and 35 m wide at road level. In addition to the main sampling point at the mobile laboratory, carbon monoxide monitors were sited at a number of locations within and near to the cutting.

This stretch of motorway carried a daily traffic flow of between 70,000 and 122,000 vehicles/day. Peak traffic flows occurred between 0700 and 1000 hours on weekdays, when there were typically around 8,700 vehicles/hour. Manual counts showed that diesel vehicles constituted between 13 and 24 per cent of the total flow.

3. POLLUTION MONITORING METHODS

The mobile laboratory was equipped to measure continuously a range of gaseous pollutants in air samples taken via an inlet 1.6 m above the ground (see Table I). Meteorological equipment was installed on the roof of the laboratory to measure wind speed, wind direction, temperature and humidity. Output signals from the equipment were sampled at one second intervals by an on-board computer. Ten minute averages were calculated and stored on flexible disk and hourly averages and daily summaries were written to a printer.

Particulate samples for lead analysis were collected hourly using a modified sequential smoke sampler with an inlet at a height of 2 m above the road surface. Additional particulate samples were collected using high volume samplers with the filter heads 2.5 m above the ground and 0.4 m from the side of the mobile laboratory under a protective canopy. One of these samples was collected over 24 hour periods and was analysed for lead and three others were collected over eight hour periods to be analysed for particulate associated polycyclic aromatic hydrocarbons (PAH).

The lead samples were analysed using atomic absorption spectophotometry and the PAH samples by gas chromatography-mass spectrometry in the selective ion monitoring mode.

TABLE I Pollutant analysers installed in the mobile laboratory

POLLUTANT	DETECTION PRINCIPLE
Carbon monoxide	
portable	Electrochemical
mobile laboratory	Infra red
Oxides of nitrogen	Chemiluminescent
Sulphur dioxide	UV fluorescence
Total and non-methane hydrocarbons	Flame ionization
Ozone	UV absorption

Carbon monoxide was also monitored at nine other positions in and around the cutting to investigate the effects of the cutting on pollutant dispersion. The monitors were placed on both sides of the road (at road level, halfway up the embankment and at the top of the embankment both in front of and behind a noise barrier) and at road level outside the cutting approximately 1.5 km south of the mobile laboratory. They were left in position for 24 hour periods and the data were used to calculate averages comparable with those obtained at the mobile laboratory. The locations of these remote monitors are shown in Figure 1.

4. RESULTS

4.1. At the mobile laboratory

Summaries of the pollutant levels recorded at the mobile laboratory are given in Table II and Figure 2.

In general, the levels of the various pollutants corresponded with those expected at a roadside location of this type, and were well below generally recommended maximum levels for exposure (for example World Health Organization 1977, 1979). In comparison with those at urban sites (see, for example, Hickman and Lunn 1981) the levels of carbon monoxide and hydrocarbons were lower and the levels of oxides of nitrogen higher. It is likely that this was because of the emission characteristics of the traffic on the motorway. The rates of emission of carbon monoxide and hydrocarbons are strongly influenced by the speed of the vehicle, with the highest rates at

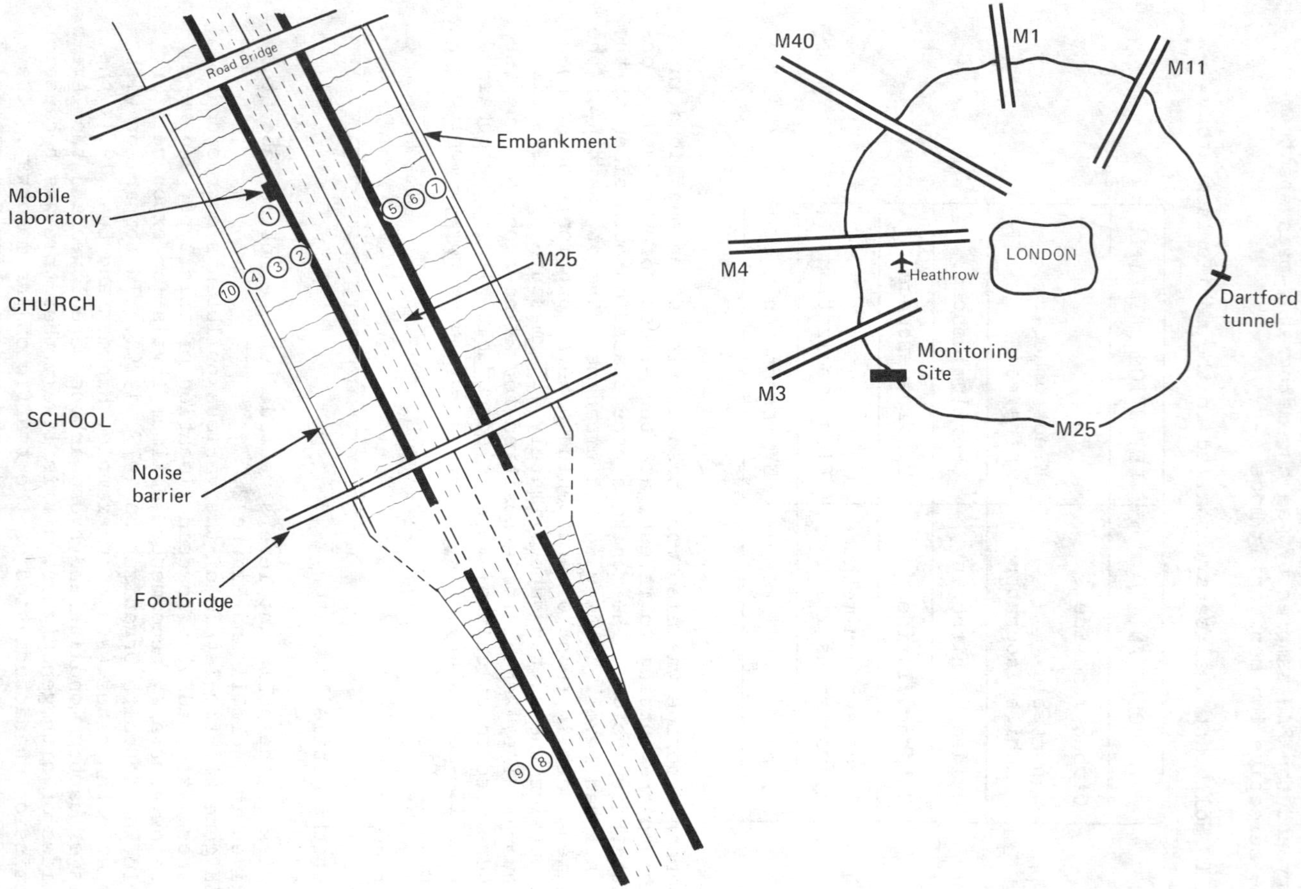

Fig. 1 Position of the site and location of the portable monitors on the M25

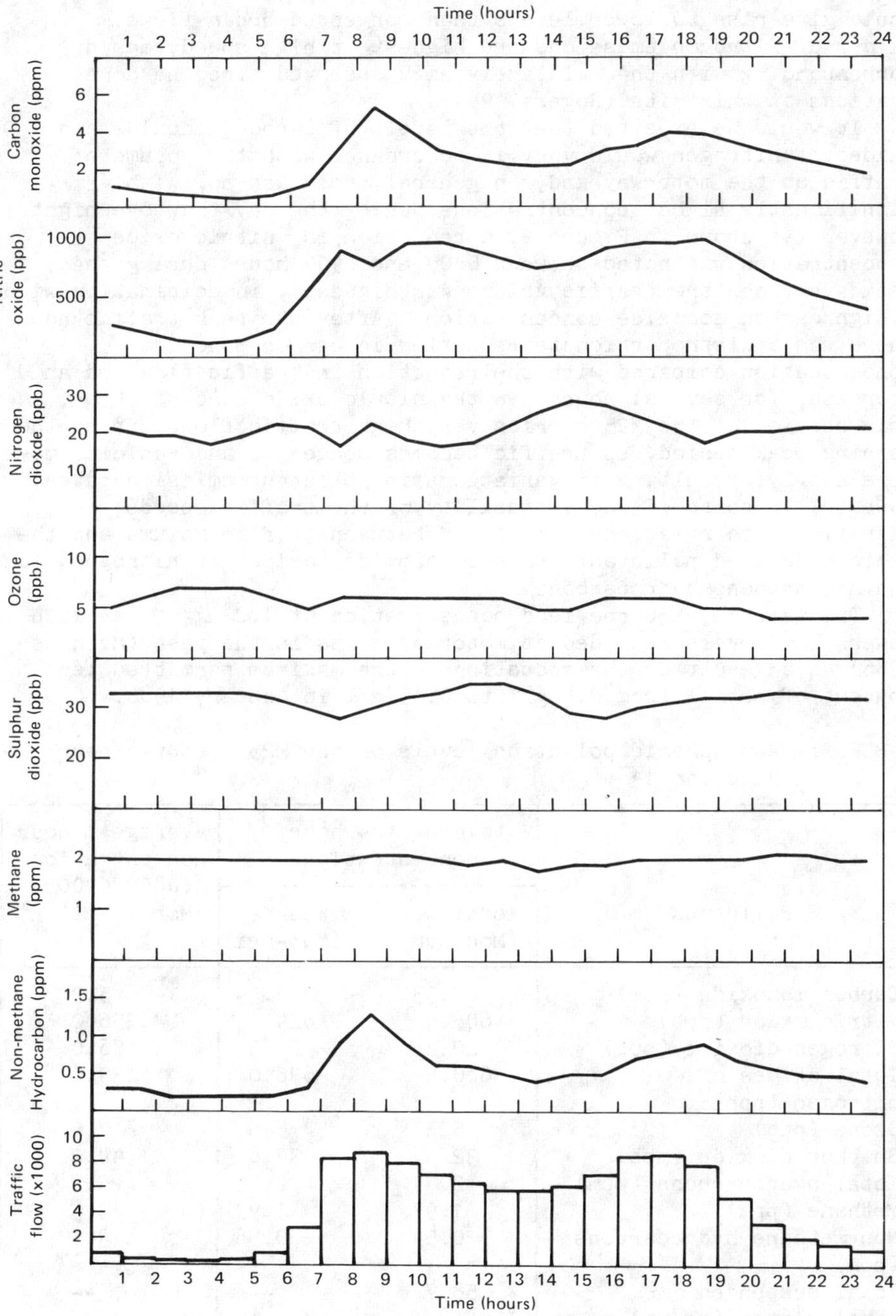

Fig. 2 Average hourly variation of pollutants and traffic on weekdays

low speed. Thus, the freely flowing, high speed motorway traffic would give rise to lower levels than congested urban flows. Oxides of nitrogen emissions are highest at high speed, again corresponding with the relatively high observed airborne concentrations at this site (Rogers 1984).

It would be expected that the levels of carbon monoxide and oxides of nitrogen would vary in accordance with the volume of traffic on the motorway and, in general, this was so, with significantly higher concentrations during the day than overnight. However, as shown in Figure 2, a reduction in nitric oxide concentration was noted between 0800 and 0900 hours during the weekdays (when the traffic volume was highest), in combination with a high carbon monoxide concentration. After the peak traffic hour there was a disproportionate reduction in carbon monoxide concentration compared with the reduction in traffic flow and an increase, for several hours, in the nitric oxide concentration. This section of the M25 carries very high traffic flows during the morning peak period, so traffic becomes congested and vehicles move more slowly, resulting in variations in pollutant emission rates. However, in spite of the variability of the traffic speeds, significant correlations were found between traffic volume and the vehicle derived pollutants carbon monoxide, oxides of nitrogen, lead and non-methane hydrocarbons.

The average airborne lead concentration of 1.2 $\mu g/m^3$ was much lower than levels recorded at motorway sites in the past (McInnes 1986) as a result of the reduction in the maximum permitted lead content of petrol from 0.4 g/l to 0.15 g/l in January 1986.

TABLE II Atmospheric pollution levels on the M25 Motorway near junction 11

Pollutant	average 24 hour concentration		average 1 hour concentration (0800-0900 Mon - Fri)
	Total week (Mon-Sun)	Weekday (Mon-Fri)	
Carbon monoxide (ppm)	2.33	2.39	5.9
Nitric oxide (ppb)	608.4	624.1	1256.0
Nitrogen dioxide (ppb)	20.2	21.3	76.0
Total oxides of nitrogen (ppb)	620.5	636.0	1239.0
Ozone (ppb)	5.3	5.5	10.0
Sulphur dioxide (ppb)	32.3	31.8	44.0
Total hydrocarbons (ppm)	2.31	2.36	5.2
Methane (ppm)	1.97	1.99	3.5
Non methane hydrocarbons (ppm)	0.54	0.56	3.5
Total suspended particulates ($\mu g/m^3$)	58.7	--	--
Particulate lead ($\mu g/m^3$)	1.18	--	--

4.2 Remote monitors

In order that the carbon monoxide concentrations from different locations on different days might be compared, the data have been considered relative to the simultaneous concentrations at the mobile laboratory (location 1). The average ratio of concentrations at each location as percentage of the concentration measured at the mobile laboratory is shown in Table III, together with the horizontal and vertical distances from the nearest point on the carriageway.

TABLE III Relative concentrations of carbon monoxide measured at the sampling locations

Location number	Distance from the road (metres) Horizontal	Vertical	Concentration relative to fixed site (percent)
1 - fixed site	5.0	1.60	100
2 - west side	5.0	0.00	171
3 - west side	10.9	3.25	114
4 - west side	17.0	7.10	83
5 - east side	5.0	0.00	112
6 - east side	9.7	2.63	91
7 - east side	14.8	5.67	75
8 - outside	5.0	0.00	39
9 - outside	7.0	0.00	80
10 - outside	18.0	7.10	37

Although locations 2 and 5 were the same distance from the carriageway as location 1, the observed concentrations were higher. This may have been partly because the intake for location 1 was 1.6 m higher than for locations 2 and 5. The high concentrations on the west side of the road (location 2) were probably the result of the dispersion characteristics of the cutting (see discussion below). Comparison between locations 2, 3 and 4 on the west side and locations 5, 6 and 7 on the east showed the levels reducing with horizontal and vertical distance from the road, with an approximate halving in concentration between the base and the top of the cutting. The concentrations behind the noise barrier at the top of the cutting on the west side (location 10) were similar to those at location 4 on the road side of the barrier (Table IV), but were a significantly lower proportion of the simultaneous levels at location 1 (Table III). Levels at locations 8 and 9, which were further along the motorway at road level outside the cutting, showed lower readings than the equivalent monitors (positions 2 and 5) inside the cutting, both in absolute terms and relative to location 1. For example, during the time when simultaneous results were obtained at locations 8 and 1 (mobile laboratory) the average concentrations outside the cutting were only 40 per cent of those found inside (Table I). Only a limited number of data were recorded for locations 8, 9 and 10 because of equipment failures.

4.3 Wind speed and wind direction

The average wind speed measured at the mobile laboratory, at a height of 4.5 m was 3.1 m/s. The predominant wind direction was 290 degrees (direction from which the wind was blowing, Figure 3a). This direction followed the orientation of the road and the cutting. Additional data from the meterological office at Heathrow Airport (measured at a height of 10 m) showed a prevailing wind direction of 200 degrees during the survey period (Figure 3b). This wind direction is perpendicular to that measured inside the cutting and illustrates the modifying effect of the cutting on the local wind conditions.

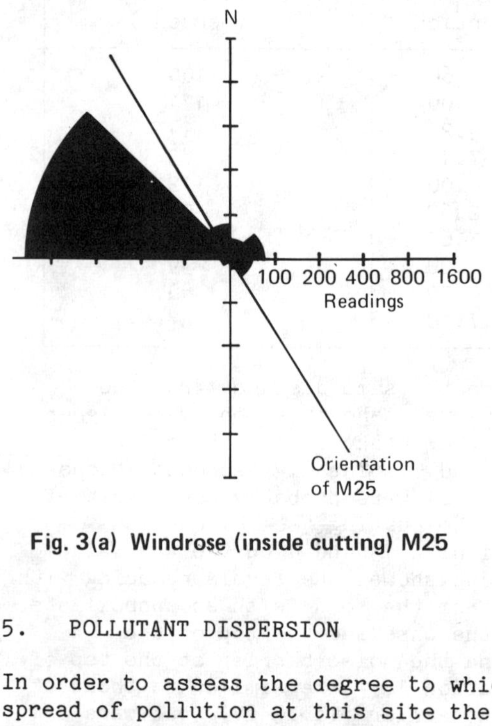

Fig. 3(a) Windrose (inside cutting) M25

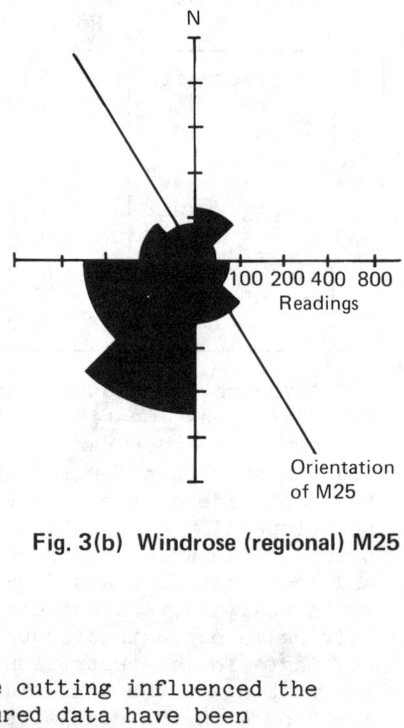

Fig. 3(b) Windrose (regional) M25

5. POLLUTANT DISPERSION

In order to assess the degree to which the cutting influenced the spread of pollution at this site the measured data have been compared with levels forecast by the current version of the computer model. This computer program is based on a Gaussian plume dispersion model that has been empirically modified for roadside dispersion of traffic emissions (Hickman et al, 1984). The method uses the detailed coordinates of a road network together with hourly traffic flows, windspeed and wind direction to calculate the carbon monoxide concentrations at a single receptor or a grid of receptor points. It has proven to be a satisfactory method in situations where the roads and receptors are all on the same level and should indicate, therefore, the concentrations likely at a level site carrying the same traffic and under the same weather conditions as found in this survey. Significant variations between the measurements and the data calculated for an open site may be attributed to the modified dispersion characteristics of this site.

5.1 Predicted concentrations

The prediction program was used to calculate carbon monoxide concentrations using the hourly traffic data and the hourly windspeed and direction data obtained at the mobile laboratory. Because of the absence of suitable equipment, it was not possible to measure traffic speeds and the values used for the calculations were estimated. For most of the time during which concentrations were measured, traffic flowed freely past the site so that the estimation of traffic speeds, based on typical values for motorways, should not lead to significant inaccuracies. It has been noted (Section 4.3) that the wind conditions measured at the mobile laboratory were different from the regional wind conditions outside the cutting. The comparison between measured and calculated values was used to identify variations in dispersion due to the geometry of the site and additional to the broad modification of the wind flow. For this reason, the data used for the calculations were those measured in the cutting.

The predicted values were averaged for each position and are shown, with the measured values, in Table IV.

TABLE IV Average predicted and measured carbon monoxide concentrations

Location Number	Average concentration		Number of hours monitored
	Measured	Predicted	
1	2.3	2.4	1285
2	3.4	2.0	127
3	2.7	2.0	314
4	2.1	1.6	285
5	3.0	3.4	118
6	2.2	3.4	111
7	1.6	3.5	83
8	1.1	2.4	56
9	2.2	1.7	44
10	2.3	1.4	23

It is not possible directly to compare the concentrations at different locations because the data (both calculated and observed) are for different periods and result, therefore, from different conditions of both traffic and wind. For example, the overall average concentrations measured at locations 1 and 10 were equal although during the 23 hours when the concentration at location 10 was measured the concentration at location 1 was substantially higher than its overall average value. During those 23 hours, the concentration at location 1 was also substantially higher than the concentration at location 10 (see Table III). The relative magnitudes of the calculated and observed data at each individual location give an indication of the effects of the cutting.

6. DISCUSSION

A number of inconsistencies can be noted between measured and predicted concentrations. The highest measured concentrations occur on the western side of the cutting while the highest predicted concentrations occur on the east side of the cutting. The east side was slightly downwind of the locally measured wind direction and the flat surface model therefore predicted the highest concentrations at that side. Through the possible interaction between the regional wind direction (Figure 3b) and the topographic structure of the cutting, the highest measured carbon monoxide concentrations in fact occurred on the west (leeward) side of the cutting.

On the western side, at ground level and midway up the embankment the measured levels exceeded the predictions but, at the top, predictions and measurements were in close agreement (Figure 4). In contrast measured and predicted concentrations were in closer agreement at the base of the eastern side of the cutting (Table IV), but the predicted concentrations for positions midway up and at the top of the cutting were significantly higher than the measurements.

Comparable results, with higher carbon monoxide concentrations and vertical concentration gradients on the leeward sides of streets than on the windward side, have been shown in field measurements and wind tunnel simulations. Figure 5 illustrates the air flow pattern postulated by other workers (see for example Stern, 1976) for similar situations. The observations made during this survey were consistent with such a flow pattern.

7. CONCLUSIONS

Significant correlations were found to exist between vehicle derived pollutants and traffic flow, with concentrations of non-vehicular pollutants remaining relatively constant throughout the survey period. Evidence of speed related effects was noted, particularly during the morning peak hour period, with a reduction in nitric oxide concentration and a corresponding increase in carbon monoxide concentration reflecting a reduction in vehicle speed.

Measurements of local wind speed and direction within the road cutting were found to be significantly different from the wind characteristics outside the cutting. Wind measurements within the cutting indicated a channelling of air along its length, at a direction approximately 90 degrees to the predominant regional wind direction.

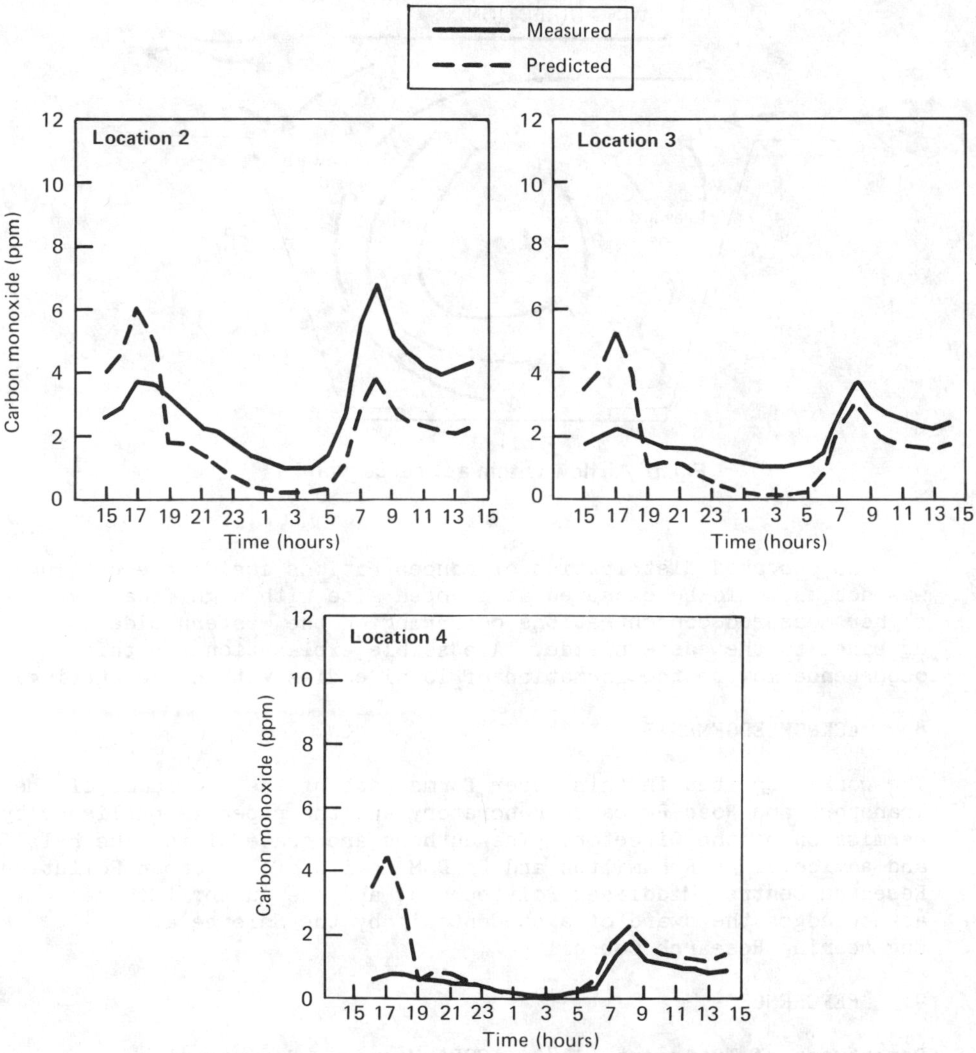

Fig. 4 A comparison of measured and predicted carbon monoxide concentrations

At road level, pollutant concentrations within the cutting appeared to be higher than would be expected at an open site. Comparisons of road level concentrations inside and outside the cutting indicated that carbon monoxide concentrations within the cutting were elevated in the order of 2.5 times. This would suggest a restriction of pollutant dispersion as a result of the cutting morphology.

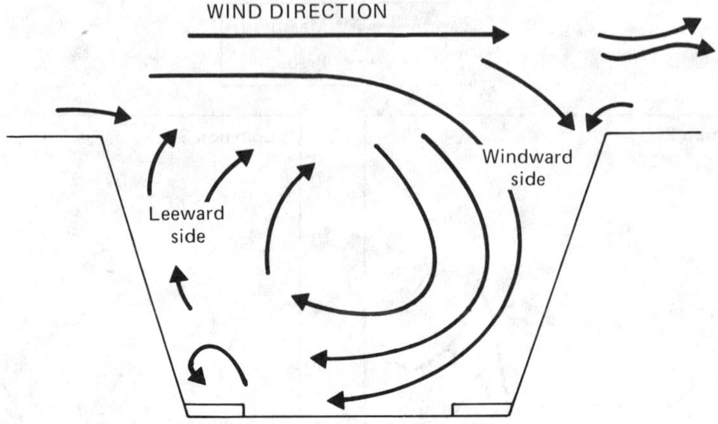

Fig. 5 Airflow within a street canyon

The observed distribution of concentrations inside the cutting was not as would be expected at an open site with significantly higher measured concentrations occurring on the western side compared to the eastern side. A possible explanation for this occurrence may be the formation of local eddies within the cutting.

8. ACKNOWLEDGEMENTS

The work described in this paper forms part of the programme of the Transport and Road Research Laboratory and the paper is published by permission of the Director. The authors are grateful for the help and advice of Dr R Hamilton and Dr D M Revitt of the Urban Pollution Research Centre, Middlesex Polytechnic, and one author (McCrae) acknowledges the award of a studentship by the Science and Engineering Research Council.

9. REFERENCES

Department of Transport (1981) COBA Manual. Department of Transport, London.

Department of Transport (1983) Manual of environmental appraisal. Department of Transport, London.

Hickman A J and D M Colwill (1982) The estimation of air pollution concentrations from road traffic. TRRL Report LR 1052, Transport and Road Research Laboratory, Crowthorne.

Hickman A J and C A Lunn (1981) Atmospheric pollution from vehicle emissions: measurements near the North Circular Road, London. TRRL Report SR 660, Transport and Road Research Laboratory, Crowthorne.

McInnes G (1986) Airborne lead concentrations and the effect of reductions in the lead content of petrol. WSL Report LR587(AP)M, Warren Spring Laboratory, Stevenage.

Rogers F S M (1984) A revised calculation of gaseous emissions from UK motor vehicles. WSL Report LR 508(AP)M, Warren Spring Laboratory, Stevenage.

Stern A C (Editor) (1976) Air Pollution Volume I: Air pollutants. Their transformation and transport. Academic Press Inc., London.

World Health Organization (1977) Environmental health criteria No 4, Oxides of nitrogen. World Health Organization, Geneva.

World Health Organization (1979) Environmental health criteria No 13 Carbon monoxide. World Health Organization, Geneva.

VOLATILIZATION OF HALOCARBONS FROM A SANITARY LANDFILL SITE

S. Haderlein, K. Pecher
Chair of Hydrology, University of Bayreuth
P. O. Box 10 12 51
D-8580 Bayreuth
West Germany

ABSTRACT. Soil gas measurements on a sanitary landfill site were carried out with a mobile sampling pipe device. After having adsorbed the halogenated organic gas constituents in the field, quantification and identification of the extracted compounds was done by GC-ECD and GC-MS in the laboratory. Several contaminated areas of the landfill site were localized by means of grid point sampling, using volatile halocarbons as tracers for industrial wastes. Shredder wastes from metal utilizing plants were found to be highly contaminated with chlorofluorocarbons (CFCs). The outgasing dynamics of CFCs from shredder wastes were estimated by several one dimensional diffusion models. A comparison of measured vertical soil gas concentration profiles with model calculations showed good agreement, using a model with two diffusion interphases - one into the atmosphere and one into the subsoil. Since the half life for CCl_2F_2 due to diffusion into the atmosphere lasts only two days, covering of the shredder wastes with compacted adsorbing layers immediately after their deposition is recommended.

1. INTRODUCTION

During the 1950's and 1960's industrial growth in Europe was rapid, and refuse from industry was often burried in uncontrolled landfill sites. Today, the location and composition of these uncontrolled sites is often difficult to determine and many of them are still unknown. By 1984, more than 30000 landfills had been detected in the FRG (1) and 1000 to 2000 of them are suspected of containing hazardous wastes in considerable amounts (2). Determination of the waste composition is essential to assess the potential toxicity of any unknown landfill site.

As volatile halocarbons are very persistent and occur in almost all industrial branches they are suitable tracers for industrial and hazardous wastes (3, 4). Due to their high vapour pressure, volatile halocarbons are easily detectable in soil air and sampling techniques have been developed to localize contaminated areas in soils and landfill sites (5, 6).

K. Grefen and J. Löbel (eds.), Environmental Meteorology, 63–73.

1.1. Investigation Area

The investigation area is an active sanitary landfill located in the bed of a former quarry to the south of Nürnberg (FRG) (Fig. 1).

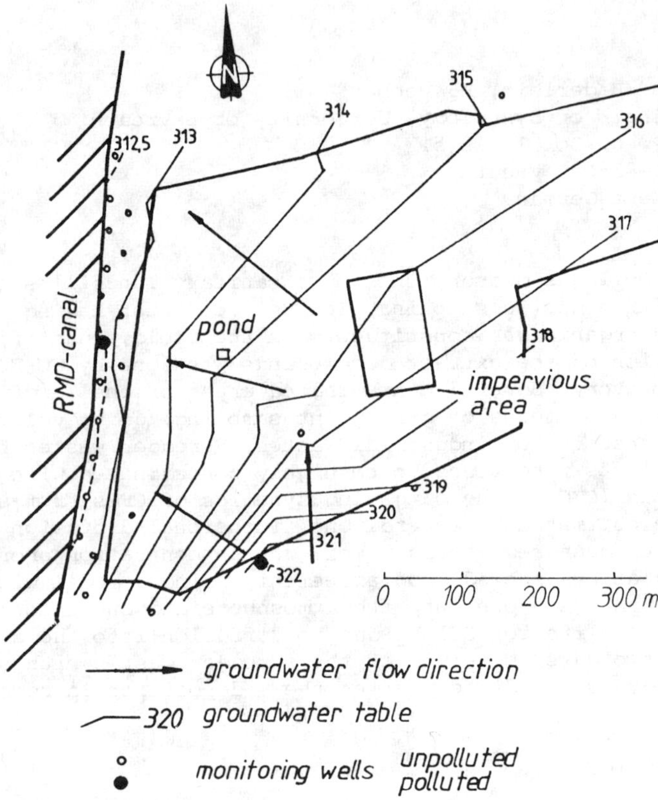

Figure 1. Hydrological map of the investigated landfill.

The landfill site covers an area of 30.000 m^2 and is subdivided into several sections containing rubble, slag from municipal waste incineration, and industrial sludge. Shredder refuse from metal and plastic foam utilizing industries is dumped together with sewage sludge and domestic wastes in an impervious area almost in the middle of the landfill. The shredder refuse consists of chopped up, non-metallic pieces of foam, tyres and other plastic materials.

2. MATERIALS AND METHODS

To detect contaminated areas, landfill gas samples were drawn by means of a mobile pipe device (Fig. 2) at grid points covering the landfill. Grid point distances were 50 m and the total number of samples taken was 120. All samples, unless otherwise stated, were taken at a depth of 1 m below the surface.

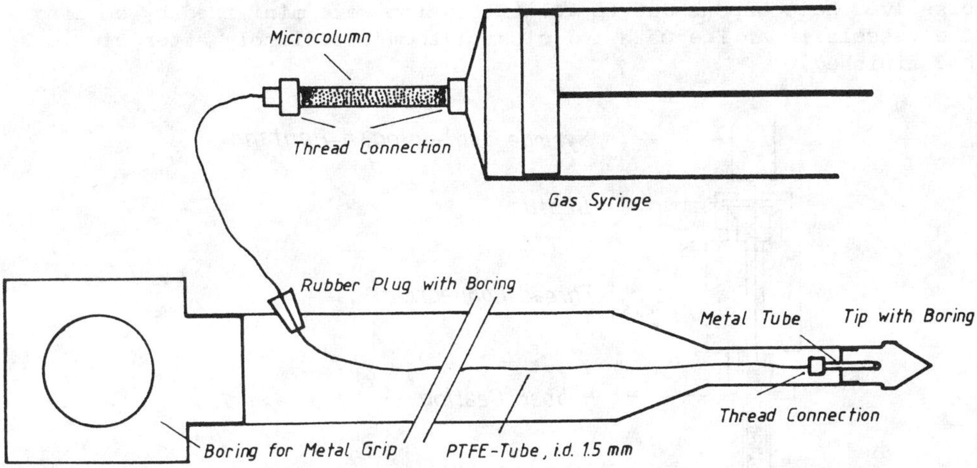

Figure 2. Sampling pipe device.

2.1. Sampling Procedure

To extract samples, a hole of about 90 cm depth was drilled into the landfill with a PÜRCKHAUER-drill. The sampling pipe, made of stainless steel, was inserted into the bore hole and driven in for further 10 cm, so that the tappered end of the pipe fitted tightly in the substrate. The tight fit prevented dilution of the sample with atmospheric gases. A microcolumn (glass, length 60 mm, i.d. 3.5 mm) packed with XAD-2 resin was fastened between the PTFE-tube and the glass syringe (Fig. 2). The resin adsorbs and concentrates the volatile halocarbons and was fixed in the microcolumn between two plugs of glass-wool.

To avoid in-situ reactions of aggressive gases like NO_x, O_2 and Cl_2 with the adsorbed substances the plugs were preconditioned in a 10 % solution of sodiumthiosulfate (7). By withdrawing the plunger of the syringe soil air was sucked through the hole in the tube tip and the PTFE-tube into the adsorber column. The dead volume of the system (2.2 ml) was neglected as the total sample volume was 200 ml. The flux rate during sampling was 70 ml min^{-1}. After having finished sampling the microcolumn was inscribed and made airtight with two silicone rubber plugs. The column was transported and stored in brown-coloured glass to avoid light induced reactions of the adsorbed substances.

2.2. Chemical Analysis

In the laboratory the microcolumns were eluted with 0.4 to 0.8 ml of n-pentane as solvent. The pentane was injected from a syringe through a septum directly onto the adsorbing resin (Fig. 3). This procedure caused the pressure in the elution device to rise and so pentane was forced to flow through the microcolumn into a glass vessel. Losses of dissolved halocarbons due to volatilization were minimized by cooling the vessels in an ice bath and closing them immediately after elution had finished.

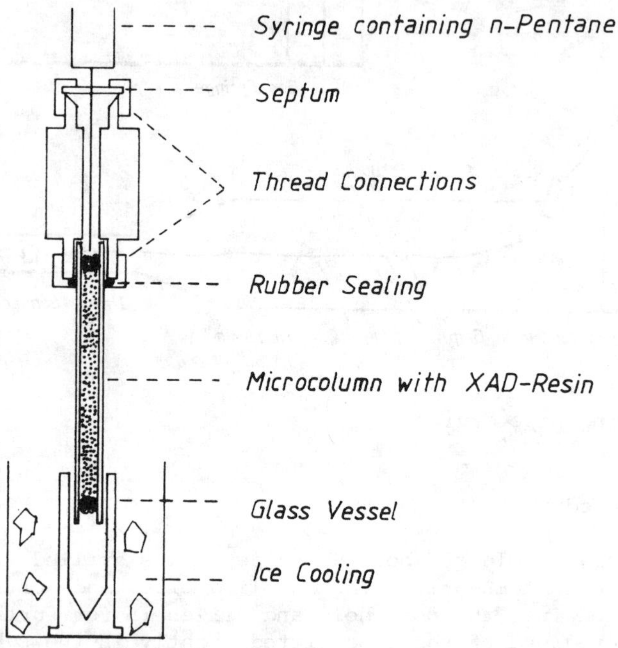

Figure 3. Elution device for microcolumns.

As most of the halocarbons are adsorbed in the top region of the micro-columns during sampling, the columns were eluted in the opposite direction. The dissolved substances were therefore not transported through the whole column during elution. Quantification and identification of the extracted compounds was done by GC-ECD and GC-MS. Table I gives the detection limits of the method.

TABLE I. Detection limits of the landfill gas sampling method (ng l^{-1} landfill gas, sampling volume = 200 ml).

CCl_2F_2	$CHCl_2F$	CCl_3F	CH_2Cl_2	$CHCl_3$	$C_2H_3Cl_3$	CCl_4	C_2HCl_3	C_2Cl_4
5	11	0.04	11	3.0*	0.1	0.06	0.2	0.1

* because of high blanks

The recoveries of the investigated compounds are near 90 % and the total error for the whole sampling and analytical procedure is not higher than 15 %. A more detailed discussion of the analytical methods is given elsewhere (8).

3. RESULTS

Figure 4 summarizes the results of the grid point sampling. The perspective column diagram shows the total concentrations of all the chlorofluorocarbons (CFCs) that were found in landfill gas.

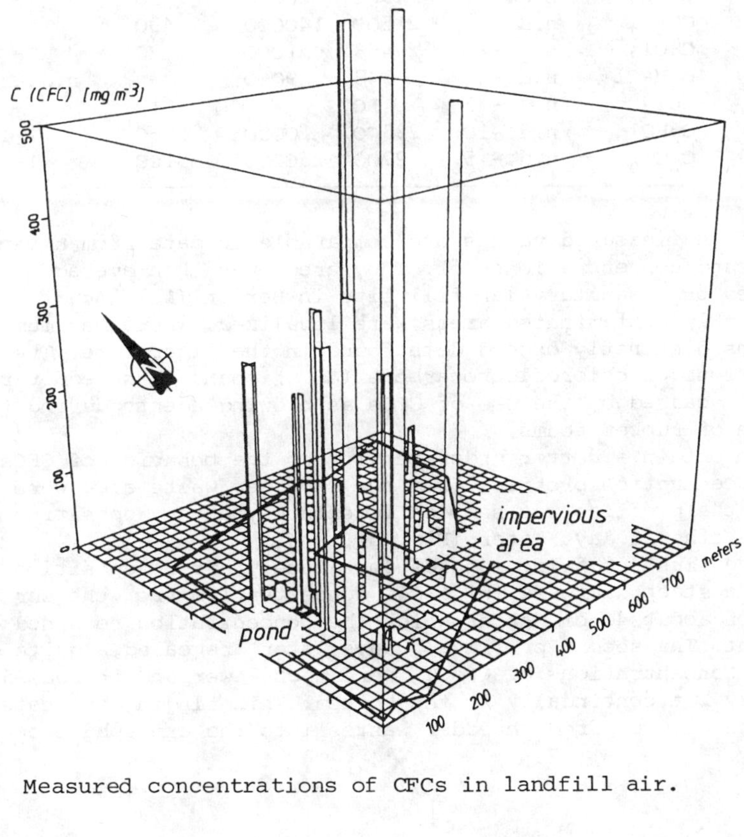

Figure 4. Measured concentrations of CFCs in landfill air.

The grid sampling pattern revealed several CFC-contaminated areas. We measured unexpectedly high concentrations in landfill gas round a small pond, where industrial sludges were dumped and on the impervious section, where the shredder wastes are located. The maximum value of CF_2Cl_2 concentration was 1.5 g m^{-3}, and was found in shredder waste landfill gas. $CFCl_3$ was found in lower concentrations, approximately 10 % of CF_2Cl_2. Results from the presented study are compared with those of other studies in table II.

TABLE II. Comparison of measured landfill gas concentrations with data from literature. Concentrations in $\mu g\ m^{-3}$ soil gas

Compound	This Study		Disposal Site (9)	Sanitary Landfill (10)	Rural Areas (5)	Urban Areas (5)
Compound	norm.	max.		mean		
CF_2Cl_2	n.d.	1469500	10000	350	-	-
$CHFCl_2$	n.d.	103300	5000	90	-	-
$CFCl_3$	n.d.-10	4500	20000	30	-	-
CH_2Cl_2	n.d.	21500	140000	490	-	-
$CHCl_3$	n.d.-10	493	< 1000	20	-	-
$C_2H_3Cl_3$	n.d.-10	348	29000	-	n.d.- 3	n.d.- 6
CCl_4	n.d.- 3	10	-	10	-	-
C_2HCl_3	n.d.-10	23300	10000	2660	1-12	8-30
C_2Cl_4	n.d.- 5	2700	250000	5100	1- 6	n.d.- 6

Most of the measured values are comparable to data from urban areas' background concentrations (5). They are lower than average values measured on a sanitary landfill site in Berlin (10). Apart from this, very highly contaminated areas were localized, where maximum concentrations distinctly exceed data found in the literature. The high concentrations of chlorofluorocarbons (CFCs) found in shredder refuse are probably caused by the use of CFCs as blowing agents during the production of rubber foams.

To obtain a better understanding of the behavior of CFCs, vertical concentration profiles from the shredder waste area were sampled. One of the profiles was sampled immediately after deposition of wastes and the other 5 days later (Fig. 5).

The first profile, sampled immediately after deposition of wastes, showed a steep increase in CF_2Cl_2 concentration from the surface to a depth of about 40 cm, after which the concentration remained nearly constant. The second profile (5 days later) revealed a quite different shape. Concentrations of CF_2Cl_2 were much lower and increased only slightly but continually to the bottom. This finding indicates, that emission of CFCs from shredder wastes into the atmosphere occurs very rapidly.

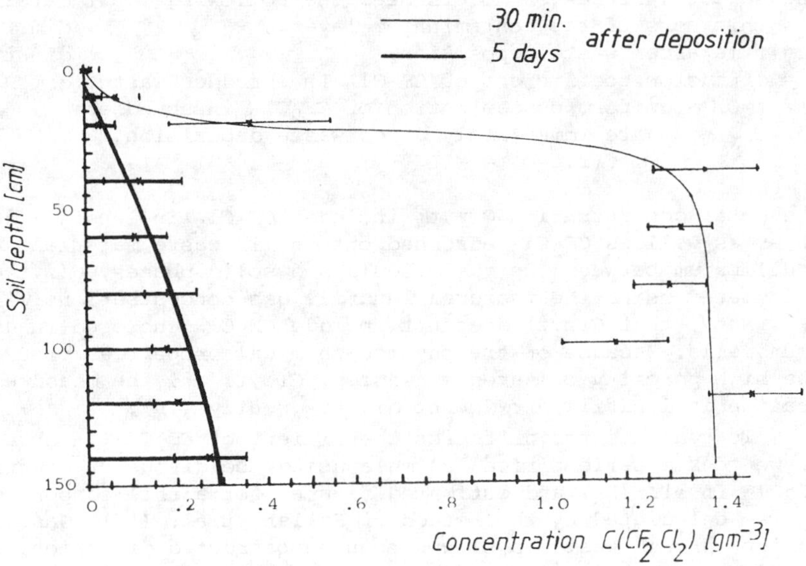

Figure 5. Measured vertical concentration profiles of CF_2Cl_2 in landfill air of shredder wastes.

3.1. Modelling of CFC Volatilization from Shredder Wastes

In order to estimate the amount and rate of outgasing of CFCs we used several one dimensional diffusion models with different boundary conditions. The models were developed by Mayer et al. (11) and are based on Fick's laws of diffusion.

Any kind of convection, e.g. due to pressure or temperature gradients within the contaminated layer, has been neglected. As this is somewhat of an oversimplification, the calculations give only a lower estimation of the rate of outgasing of CFCs (12).

All model calculations are based on the following boundary conditions:

- the concentration of CFCs immediately after waste deposition C(t=O) is constant throughout the whole contaminated layer L
- the CFC-concentration of the upper boundary layer C(z=O) remains constant over time
- concentrations of CFCs in the lower atmosphere remain zero, implying a rapid advective transport of outgasing CFCs from the boundary layer
- the diffusion coefficient D in the waste remains constant over the whole contaminated layer.

The selected diffusion models need the following input data:

L = thickness of the contaminated layer, dim L = L

t = time after waste deposition, dim t = T

D = diffusion coefficient of CF_2Cl_2 in shredder waste, dim D = L^2T^{-1}

C(z,t=O) = uniform concentration of CF_2Cl_2 in shredder waste immediately after waste deposition, dim C(z,t=O) = M L^{-3}

The total concentration C(z,t) includes CF_2Cl_2 in landfill gas and pore water as well as CF_2Cl_2 adsorbed onto solid waste material. Assuming equilibrium between the gas, liquid and solid phases, C(z,t=O) can be calculated using the measured landfill gas concentrations (13).

Note, that C(z,t) does not include CF_2Cl_2 enclosed in undamaged foam cells. Because of the physico-chemical properties of CF_2Cl_2 and the high porosity of shredder wastes, C(z,t) and the measured concentration in landfill air do not differ greatly (14).

To evaluate the diffusion coefficient of CF_2Cl_2 in shredder wastes, D, we took a semiempirical formula using the diffusion coefficient of CF_2Cl_2 in air (D_a) and estimated figures for airfilled pore volumes. D_a was calculated by the method of Fuller et al. (15). Gaseous diffusion through waste is slower than unobstructed diffusion. This reduction may involve such waste characteristics as porosity (designated by ε) and tortuosity. Porosity is defined as the fraction of open volume in cross section through the waste column. Tortuosity accounts for the fact that waste pores are not straigt channels running directly through the waste and that molecular diffusion follows channels which twist and turn. Another complicating factor is that a certain proportion of the pores (dead end pores) do not contribute to overall diffusion through the waste column.

D is calculated by (16):

$$D = D_a \left(\frac{\varepsilon_{eff}^{\frac{10}{3}}}{\varepsilon_{tot}^{2}}\right. \qquad (1)$$

ε_{eff} = airfilled pore space, dim ε_{eff} = 1

ε_{tot} = total pore space (air and waterfilled), dim ε_{tot} = 1

For model calculations, we used a value of D (CF_2Cl_2) = 0.05 cm^2s^{-1} (about 54 % of D_a).

The selected models can be used to calculate the following parameters:

- C(z,t), concentration of CF_2Cl_2 in shredder waste in the depth z and at time t, dim C(z,t) = M L^{-3}
- F(t), flux of CF_2Cl_2 through the boundary layer (z=O) soil surface /atmosphere at time t, dim F(t) = M $L^{-2}T^{-1}$

3.1.1. Model I

C(z,t) of CF_2Cl_2 is calculated by

$$C(z,t) = \frac{4\ C(z,t=0)}{\pi} \sum_{n=0}^{\infty} \frac{(-1)^n}{(2n+1)} \exp\left(-\frac{D(2n+1)^2 \pi^2 t}{4\ L^2}\right) \cdot \cos\left(\frac{(2n+1)\ \pi\ (L-z)}{2\ L}\right) \tag{2}$$

The flux into the atmosphere F(t):

$$F(t) = \frac{D \cdot C(z,t=0)}{\sqrt{\pi\ D \cdot t}} \left[1 + 2 \sum_{n=1}^{\infty} (-1)^n \exp\left(-\frac{n^2 L^2}{D \cdot t}\right)\right] \tag{3}$$

3.1.2. Model II

Compared to the model above, model II offers the advantage of introducing a second 'active' boundary layer at the bottom (z=L) of the contaminated waste layer. As contaminated wastes are deposited above layers which might have been depleted with respect to CFCs, diffusion into such a layer becomes significant. The equations are:

$$C(z,t) = \frac{C(z,t=0)}{z} \cdot 2\ \mathrm{erf}\left(\frac{z}{2\sqrt{D \cdot t}}\right) - \mathrm{erf}\left(\frac{z-L}{2\sqrt{D \cdot t}}\right) - \mathrm{erf}\left(\frac{z+L}{2\sqrt{D \cdot t}}\right) \tag{4}$$

$$F(t) = \frac{D \cdot C(z,t=0)}{\sqrt{\pi\ D \cdot t}} \cdot \left(1 - \exp\left(-\frac{L^2}{4\ D \cdot t}\right)\right) \tag{5}$$

with erf(x) being the Gauss probability function.

Figure 6 compares the measured and calculated values of CF_2Cl_2 five days after waste deposition.

Model I shows a concentration profile occuring above a completely impermeable sublayer, that differs significantly from the measured data. In contrast to model I, model II gives a good fit between measured and computed data in the upper layer. This agreement supports the hypothesis that the sublayer consists of depleted shredder wastes. Numerical integration of equation (5) yields Q(t), the total amount of evaporated CF_2Cl_2 at time t, and shows that about 2 days after waste deposition more than 50 % of the supply of CF_2Cl_2 in landfill gas is lost due to diffusion into the atmosphere.

As CF_2Cl_2 is a medium range chlorofluorocarbon with respect to molecular weight, we expect other CFCs ($CHFCl_2$, $CFCl_3$) to behave in a similar way.

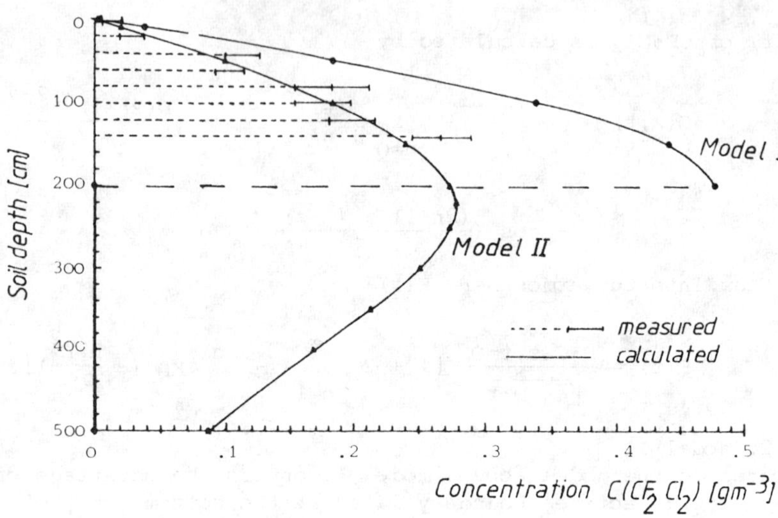

Figure 6. Comparison of measured and calculated landfill air concentration profiles five days after waste deposition.

4. DISCUSSION

The short residence times of halocarbons found in this investigation do not agree with the results of Khalil & Rasmussen (17), who found extremely long residence times for $CFCl_3$ (about 100 years) in different types of closed cell polyurethanes. However, in their experiments, they used freshly manufactured unbroken foams exposed to atmospheric pressure and room temperature. In our investigation, we attempted to determine the fate of CFCs in used rubber foams which were shredded before their deposition in a landfill site. Shredding rubber foam obviously enhances the outgasing of CFCs due to the increased number of open foam cells and increased surface area in comparison to unshredded foams. Additional factors leading to reduced residence times might be increased temperatures and squeezing of foam shreds in covered layers. These effects have not been studied in the present work. Future work will aim to estimate the total amount of CFCs in the landfill body, including CFCs in undamaged foam cells.

The results of the present study indicate that the enormous amounts of CFCs in foam cells must not longer be considered as immobile. In order to reduce the amount of CFCs entering the atmosphere which lead to ozone decomposition, a fixation in the landfill body, or at least a retardation of the outgasing process, might be achieved by covering each layer of shredder wastes with an adsorbing material immediately after deposition. A suitable and easily available material might be compacted sewage sludge.

5. ACKNOWLEDGEMENT

The authors thank Prof. Dr. R. Herrmann for helpfull advices and support of this work. Furthermore we thank Mrs. S. Koch for many valuable comments as well as Mrs. E. Misch for preparing the manuscript.

REFERENCES

(1) KERNDORFF, H., BRILL, V., SCHLEYER, R., FRIESEL, P., MILDE, G. (eds.): Erfassung grundwassergefährdender Altablagerungen - Ergebnisse hydrogeochemischer Untersuchungen. Institut für Wasser-, Boden- und Lufthygiene des Bundesgesundheitsamtes, Berlin, 1985

(2) FRANZIUS, V.: Kontaminierte Standorte - Ein internationales Problem. Symposium Kontaminierte Standorte und Gewässerschutz, Aachen 1.-3.10.1984, Umweltbundesamt, Materialien 1/85, Berlin, 1985

(3) PEARSON, C. R.: 'C_1 and C_2 Halocarbons'. In: Hutzinger, O. (ed.): The Handbook of Environmental Chemistry, Vol. 3, Part B, Springer, Berlin, 1982

(4) RUSSOW, J.: 'Fluorocarbons'. In: Hutzinger, O. (ed.): The Handbook of Environmental Chemistry, Vol. 3, Part A, Springer, Berlin, 1980

(5) NEUMAYR, V.: 'Beiträge zur Entwicklung systematischer Sanierungskonzepte bei Boden- und Grundwasserkontaminationen'. In: Aurand, K. (ed.): Bewertung chemischer Stoffe im Wasserkreislauf, Fischer, Berlin, 1981, 37-57

(6) HADERLEIN, S., PECHER, K.: Müll und Abfall (1987), in preparation

(7) BUNCH, J. E., PELLIZZARI, E. D.: J. Chrom. 186 (1979) 811-829

(8) HADERLEIN, S., PECHER, K.: Water, Air and Soil Pollution (1987), in preparation

(9) BROOKES, B. J., YOUNG, P. J.: Talanta 30 (1983) 665-676

(10) HÖFLER, F., SCHNEIDER, J., MÖCKEL, H. J.: Fresenius Z. Anal. Chem. (1987), in press

(11) MAYER, R., LETLEY, J., FARMER, W.: J. Soil Sci. Soc. Amer. Proc. 38 (1974) 563-568

(12) LYMAN, W. J.: Handbook of chemical property estimation methods, New York, 1982

(13) JURY, W. A., GROVER, R., SPENCER, W. F., FARMER, W. J.: Soil Sci. Soc. Amer. J. 44 (1980) 445-450

(14) HADERLEIN, S.: Die chemische Analyse von Deponiegasen zur Abschätzung des Gefährdungspotentials von Mülldeponien. Diplomarbeit, University of Bayreuth, 1987

(15) FULLER, E. N. et al.: Ind. Eng. Chem. 58 (1966) 1927 ff

(16) THIBODEAUX, L. J., SCOTT, H. D.: 'Air/Soil Exchange Coefficients'. In: Neely, W. B., Blau, G. E. (eds.): Environmental Exposure from Chemicals, Vol. I, Chapt. 4, Boca Raton, Florida, 1986

(17) KHALIL, M. A. K., RASMUSSEN, R. A.: Chemosphere, Vol. 16, No. 4 (1987) 759-775

MEASUREMENTS OF THE CONTRIBUTION OF A COAL-FIRED POWER PLANT TO THE OVERALL GROUNDLEVEL CONCENTRATION OF AIR POLLUTANTS BY USING THE SF_6-TRACER TECHNIQUE

H.Rötzer[+], J.Riesing[+], A.Nentwich[*], A.Szeless[*]

+ Österr. Forschungszentrum Seibersdorf Ges.m.b.H.
A-2444 Seibersdorf - Austria

* Österreichische Elektrizitätswirtschafts-AG
Am Hof 6A, A-1010 Wien - Austria

ABSTRACT. For the first time the SF_6-tracergas measuring technique was used in Austria to assess a coal-fired power plant's contribution to overall air pollution. These tracergas measurements were excellently suited to detect the plant's plume and then to measure and record it's atmospheric dispersion without interference by other emitters, even at quite large distances from the stack due to the high sensitivity of the measuring method. The most important result is the confirmation that with its scrubbed flue-gas the Dürnrohr plant's contributions to the overall measured groundlevel concentrations of air pollutants in the surroundings of the plant are quite small. Within the validity range of the calculational model computed results of plume dispersion show quite good agreement with the measured results. For specific weather conditions, such as inversions or zero wind speed the calculational model has yet to be improved to yield satisfactory agreement with the measurements. Hence, for such meteorological conditions a second measurement campaign from January to March 1987 has been executed, the results of which will be reported later.

1. INTRODUCTION

During the period of November 1985 until February 1986 measurements of the contributions of the coal-fired power plant at Dürnrohr to the overall groundlevel concentrations of air pollutants have been carried out by Österreichisches Forschungszentrum Seibersdorf (Austrian Research Centre of Seibersdorf) utilizing the SF_6-tracer technique [1,2,3,4].

The power plant at Dürnrohr is located about 40 km west of Vienna in the province of Lower Austria, in an area which is characterized by both agriculture as well as industry. The power plant comprises 2 units. Unit 1 is operated by Verbundkraft, a 100% subsidiary of Verbund-

K. Grefen and J. Löbel (eds.), Environmental Meteorology, 75–92.

gesellschaft (Federal Electricity Board). Unit 2 belongs to Niederösterreichische Elektrizitätswerke AG (Electric Utility of Lower Austria).

Although a grid of monitoring stations for air pollutants, measuring groundlevel concentrations (sulphur dioxide, nitrogen oxides, particles, ozone) is installed in the surroundings of the power station no precise statement can be made with respect to the contribution of the power plant's flue-gas plume to the overall groundlevel concentrations recorded by the environmental monitoring grid. All groundlevel concentration measurements represent superpositions of shares of various emission sources such as industrial plants, agricultural enterprises, residential housing and traffic.

Moreover, in the public's opinion it is the power station at Dürnrohr which causes the overwhelming portion if not the sole of the air pollution in that region, despite the fact that the power station, according to international standard, is equipped with highly efficient scrubbers for sulphur dioxide and nitrogen oxides as well as with electrostatic precipitators for the particles, not to speak of the 210 m high stack. Due to these environmental protection installations only relatively low contributions of the plant to the groundlevel concentrations could be expected particularly with respect to the already existing significant background concentrations in that region.

Thus for the power plant operators the advantage of measurements of the contribution of their power plant to the overall groundlevel concentrations of air pollutants by using the SF_6-tracer technique was twofold, the benefit of actually measured results and the presentation of scientifically obtained and reproducible results to the public.

In addition to the measurements model calculations of atmospheric plume dispersion were effected in order to be able to compare measured and calculationally predicted results. These model calculations of transport phenomena of air pollutants emitted by power plants are quite important, because they represent a basis to the authorities to grant operating licences to the plant owners. Moreover, those model plume dispersion calculations are also a basis of plant operating restrictions issued by the authorities when the legal limit values of groundlevel concentrations of air pollutants are exceeded in the surroundings of the plants. Experience has shown that with conservative calculational models of plume dispersion the plant owners were quite often penalized by the imposed operating restrictions. Thus calculational model improvements to match prediction and measurement should be a benefit to the plant owners.

Also it should be mentioned that parallel to the SF_6-

tracergas measurements the Österreichisches Bundesinstitut für Gesundheitswesen (Austrian Federal Institute of Public Health) has carried out transmission and groundlevel concentration measurements of the air pollutants SO_2, NO and NO_2 in the surrounding area of the power station. Whenever these measurements were executed simultaneously with the tracergas measurements they were coordinated such that measuring time and measuring area coincided. Thus, these measurements beneficially supplemented the SF_6-tracergas measurements, especially with respect to the assessment of the overall groundlevel air pollution caused by all the different pollution emitting sources in that area.

2. METHOD OF APPROACH

2.1 Purpose of Measurements

The purpose of the Dürnrohr power plant's flue-gas plume dispersion measurements using the SF_6-tracergas technique was

- measurements of plume dispersion without interference by other emitters
- assessment of the power plant's contribution to the overall environmental pollution in the surroundings of the plant
- verification of the usefulness of the SF_6-tracergas technique for large distance plume dispersion as for the first time in Austria this type of measuring technique was applied in connection with atmospheric dispersion of power plant flue-gas plumes
- comparison of predicted and measured groundlevel concentrations of air pollutants caused by power plant emissions
- proposal of calculational model improvements
- gaining additional experience in assessing the prevailing meteorological conditions in the region surrounding the power station
- putting the power plant's contribution to groundlevel air pollution into proper perspective with regard to contributions from other emission sources in that area and to inform the public, especially those living in the vicinity of the power plant.

2.2 Measuring Principle

The measuring principle of plume dispersion measurements utilizing the SF_6-tracergas technique is shown in Fig. 1 and is described as follows:

- SF_6-tracergas is added in minute quantities to the flue-gas at the bottom of the stack in order to ensure perfect mixing.

- Dilution of the SF_6-tracergas is measured at discrete measuring positions downwind of the emission source.
- The ratios of the initially injected SF_6 concentration in the flue-gas to the measured SF_6-groundlevel concentrations of the air samples taken at the selected measuring positions yield the plume diffusion characteristics.

By so measuring the SF_6-groundlevel concentrations of the air samples it is possible to assess selectively the atmospheric dispersion of a particular flue-gas plume marked by the tracergas. During all measurements meteorological data are recorded, because atmospheric dispersion of the plant's off-gas plume strongly depends on actual meteorological conditions.

It is important to point out several implicit assumptions and advantages of the SF_6-tracergas measuring method.

- SF_6 is a chemically inert gas, which does not undergo chemical changes during atmospheric dispersion
- SF_6 is non-toxic
- SF_6 does not exist in nature. Its background concentration is less than 10^{-14} cm³ SF_6/cm³ air.
- Using gas-phase chromatographs SF_6 can be detected even at very low concentrations of about 10^{-12} cm³ SF_6/cm³ air. Because of this high detection sensitivity it is possible to effect SF_6-groundlevel concentration measurements even at fairly large distances from the plant's emitting stack.
- The density of SF_6 is about 5,1 times that of air. Because of the high dilution of the tracergas in the flue-gas and during the subsequent turbulent atmospheric diffusion no separation takes place.
- As sulphur dioxide is the lead air polluting compound for all SF_6 tracergas measurements it is implicitly assumed that the diffusion process of SO_2 and SF_6 are identical. Neither tracer-measurement nor model calculation gives credit to processes such as chemical reactions, depositions etc. during transmission.

3. EXECUTION OF MEASUREMENTS

3.1 Injection of SF_6-tracergas

The sulphur-hexafluorid tracergas had a purity of 99,8%. Injection was effected at the bottom of the stack by using a thermal massflow controller. The desired flowrate of the tracergas was preselected and was kept constant during dosing to within 2% independent of temperature and pressure fluctuations. The tracergas flowrate was monitored continuously in order to assure constant injection. Originally the tracergas flowrate was chosen to be 25 l/min.

However, for the later measurements it was augmented to 50 l/min in order to increase the detection sensitivity. Due to the injection into a blower and a mixing length of about 200 m excellent mixing of the tracergas with the flue-gas could be achieved.

3.2 SF_6-concentration Measurements in Air

The measurements of the SF_6-groundlevel concentrations in air were effected by using a gas-phase chromatograph with an electron capture detector. For the first measurements the detection limit of the measuring device was 6.10^{-12} cm^3 SF_6/cm^3 air. For the later measurements the detection limit was decreased to $1,5.10^{-12}$ cm^3 SF_6/cm^3 air. The duration to analyse one air sample was approximately 2 minutes and the gas-phase chromatograph was calibrated in regular intervals with SF_6-gas-standards. The whole measuring equipment was installed in a measuring vehicle so that both the necessary mobility to follow the trajectories of the plant's plume was assured and also the transport times of those air samples collected at the fixed positioned air sample collectors could be kept short.

According to the two types of measured SF_6-groundlevel concentrations in air, namely half hour mean values and instantaneous values, two methods of air sample collection were applied. Instantaneous SF_6-groundlevel concentrations in sucked in air samples in the vicinity of the measuring vehicle were obtained in intervals of about 2 minutes while the measuring vehicle was crossing the trajectories of the plant's plume at different distances from the emission source. To measure the well known Gaussian distributions of groundlevel concentrations as a consequence of SF_6-transmission by the flue-gas it was necessary to measure along perpendicular lines to the principle wind direction. Although the measuring vehicle had to follow existing roads, which in general were not perpendicular to the principle trajectories of the plumes the SF_6-groundlevel concentration distributions could be detected in a very convincing manner. Even at large distances from the emission source groundlevel concentration distributions were measured.

Half hour mean value groundlevel concentrations were measured in air samples which were collected at 10 preselected locations covering the diffusion domain of the plant's plume, where air sample collectors have been positioned prior to the start of the tracergas measurements. Each fixed positioned air sample collector was able to measure 4 consecutive half hour mean values of SF_6-groundlevel concentrations. The start of air sample collection was triggered automatically at the preselected time.

3.3 Meteorological Data

Scientific evaluation of the measurements required complete monitoring and recording of the meteorological conditions which existed during the time intervals when the measurements were carried out. Half hour mean values of ambient air temperature, wind direction, wind speed and stability class were recorded continuously at the point of emission (roof of the vessel building) and at the 7 environmental monitoring stations in the surroundings of the power plant.

The direction of the principle dispersion trajectory of the plume was periodically determined by using a compass. The Central Bureau of Meteorology and Geodynamics supported the whole measuring campaign by supplying daily the general weather forecasts and helping to assess and analyse locally deviating meteorological conditions.

TABLE I summarizes the measurement conditions which existed during the 10 SF_6-tracergas measurements.

4. RESULTS

4.1 Measured Results

Altogether 10 plume dispersion measurements applying the SF_6-tracergas technique have been carried out at various weather conditions. TABLE I summarizes the measurement conditions during the measurements. As is evident the measurements were executed preferably for stability class 3 and 4. Only one measurement was effected at stability class 2. On three days the mean wind speeds were below 3 m/s.

TABLE II summarizes the measured results. Maximum measured values of the half hour mean values are below 10 $\mu g\ SO_2/m^3$, the highest of all values being 6,71 $\mu g\ SO_2/m^3$.

Looking at the recorded instantaneous values, all are below 5 $\mu g\ SO_2/m^3$ with the exception of measurements carried out on Jan. 29, 1986. On that day three measured instantaneous groundlevel concentrations exceeded 10 $\mu g\ SO_2/m^3$ with the maximum at 38,76 $\mu g\ SO_2/m^3$. This measured high groundlevel concentration represents a single short-time value at one point relatively close (about 4 km) to the power plant and can not be considered representative of such a high groundlevel concentration of a wider area. The 4 consecutive instantaneous measurements to the maximum one at the same point all yielded groundlevel concentrations below 0,5 $\mu g\ SO_2/m^3$.

TABLE III shows the statistics of all measured results. It contains the frequencies of measured values occuring in the various ranges of measured groundlevel concentrations. Apparently, mainly quite low SO_2 groundlevel concentrations have been measured.

In order to put the measured SO_2 groundlevel concentrations originating from the power plant into proper perspective it is obvious to compare them both with the permissible SO_2 threshold groundlevel concentration values as issued by the authorities as well as with the actually measured overall SO_2 groundlevel concentrations in the surroundings of the power station at the time of the SF_6-tracergas measurements.

According to the recommendations of the Austrian Academy of Sciences for zone I (nature protection, resort and recreation areas) [6] and also according to the second decree concerning air pollutants noxious to forests [7] the following limits of SO_2 groundlevel concentrations are valid:

half hour mean values
during the months of April to October 70 µg/m³
during the months of November to March 150 µg/m³
daily mean values
during the months of April to October 50 µg/m³
during the months of November to March 100 µg/m³

A comparison of the measured SO_2 groundlevel concentrations by the SF_6-tracer technique and the afore quoted threshold concentrations leads to the following conclusion.

- In not a single measurement any of the referenced threshold groundlevel concentrations were exceeded and all of the half hour mean values and all of the instantaneous values were very small compared to the most stringent limit for groundlevel concentrations.
- The Dürnrohr plant's contribution to air pollution in that area (called Tullnerfeld) has been very small indeed as was evident from overall SO_2 groundlevel concentration measurements automatically recorded in the 7 permanently installed environmental measuring stations in the surroundings of the plant.

4.2 Plume Dispersion Calculations

It is important to check measured results of groundlevel concentrations of air pollutants (SO_2 being the lead substance) against predicted results from model atmospheric diffusion calculations.

So far only the Austrian standard Gaussian diffusion model according to ÖNORM M 9440 (Austrian Standard Specifications) [5] has been used for numerical diffusion calculations. This diffusion model serves also as basis for regulatory purposes, despite of all its well known restrictions and limits of validity inferred by the Gaussian solution of Fick's diffusion equation.

4.3 Typical Results of the Tracer-Measurements

Rather than giving all obtained computational results in

connection with the SF_6-tracergas measurements performed at the Dürnrohr plant the measured and calculated results of three different days will be discussed briefly.

November 6, 1985

The prevailing meteorological conditions and the plant's flue-gas emission rate are given in TABLE I.

TABLE IV lists the measured (averaged half hour values) and predicted SO_2 groundlevel concentrations. For this measurement only the air sample collectors were used.

As during the measurements the wind direction fluctuated about the principle wind direction, the measured SO_2 groundlevel concentration distributions were more spread out (lower peak values, larger standard deviations) than the calculated ones.

By computing the least square sum of the logarithmic differences of measured and calculated groundlevel concentrations for various principle wind directions (all fluctuating about 280°) the best fit of the calculations to the measurements is obtained with respect to the wind direction fluctuations. TABLE V summarizes the results of the comparison of measurement and calculation. It can bee seen, that although the monitored principle wind direction was 280 degrees during the measurement period, the best calculational fit to the measurements indicated 275 degrees as principle wind direction at the very area of measurement.

Fig. 2 shows the calculated SO_2 groundlevel concentrations (lines of constant concentration) utilizing the ÖNORM M 9440 atmospheric dispersion model.

January 30, 1986

Wind was blowing from SE (120 degrees) at a mean wind speed of 5,1 m/s with stability class 4 prevailing as can be seen in TABLE I. Within the plant plume's dispersion region 109 instantanous groundlevel concentrations were measured at different measuring points by crossing the plume's trajectory at 5 different distances from the emission source using the mobil laboratory. In addition at 7 fixed positions half hour mean values were recorded 4 at each position. The measured maximum values for each of the 5 distributions are shown in TABLE VI which also includes the maximum values of groundlevel concentrations as obtained from the dispersion calculation.

As is apparent the measured groundlevel concentrations are quite low, the maximum being a half hour mean value of 3,95 µg SO_2/m^3. This value is less than 3% of the authorized limit. It is also very small compared to the overall SO_2 groundlevel concentrations of 24 to 55 µg SO_2/m^3, which were recorded at the time of measurements in the measuring area.

Fig. 3 and Fig. 4 show graphically the measured and calculated results respectively. Along the principle wind direction the measured SO_2 groundlevel concentrations, originating from the power plant are somewhat smaller than those obtained by model calculations, however of the same order of magnitude. It is valid to conclude, that for this particular meteorological situation there is quite satisfactory agreement of measurement and calculation.

Finally, it should be noticed in TABLE VI and Fig. 3, that even at a distance of 26 km the power plant's plume was clearly detectable. This proved the high sensitivity of the SF_6-tracergas technique.

February 14, 1987

According to TABLE I on that day the meteorological conditions were: principle wind direction 100 degrees, mean wind speed 6,7 m/s, stability class 3 to 4.

TABLE VII summarizes the measured and computed maximum values for each of the 5 distributions. It can be seen that the maximum instantaneous SO_2 groundlevel concentration contribution of the power plant's flue-gas was 2,3 µg SO_2/m^3 and the maximum half hour mean SO_2 groundlevel concentration contribution 1,25 µg SO_2/m^3. These values are indeed quite low compared to both the authorized limit values of SO_2 groundlevel concentrations and to the overall SO_2 pollution of 70 to 100 µg SO_2/m^3 at the time of measurement in the area of measurement.

As the stability class has been determined to be 3 to 4, model calculation for both stability classes 3 and 4 were performed. It turned out that using the calculational parameters associated with stability class 4 yielded a satisfactory agreement of measurement and calculation.

Fig. 5 and Fig. 6 show graphically the measured and calculated results. As is apparent the position of the plume close to the ground was deviated towards the north into the Danube valley by hills situated in the direct main dispersion trajectory of the plume. The dispersion of the higher portions of the plume followed the more or less straight main trajectory over the hills. Beyond the hills this portion was detected and a distinct profile measured.

5. CONCLUDING REMARKS

For the first time the SF_6-tracergas measuring technique was used in Austria to assess a power plant's contribution to overall air pollution. The Austrian Research Centre in Seibersdorf has carried out the measurements after having adapted this tracer technique to be applicable to the evaluation of medium-range air pollution transport phenomena.

It turned out that the SF_6-tracergas measurements as described in this report were excellently suited to detect the plant's plume and then to measure and record its atmospheric dispersion even at quite large distances from the stack due to the high sensitivity of the measuring method. As a further advantage this could all be done without the interference of other emitters in the plant's vicinity.

The most important result of all measurements is the confirmation that with its scrubbed flue-gas the Dürnrohr plant's contributions to the overall measured groundlevel concentrations in that area are quite small, in fact almost negligible for the so far considered well defined meteorological conditions.

In a second measurement campaign in the months of January to March 1987 special plume dispersion situation due to specific weather conditions have been investigated.

However, after the validity of SF_6-tracergas measurements in the context of measuring power plant plume dispersion phenomena has been established, it was clearly the aim of the latter measurements to put the emphasis on specific problem areas, such as

- measurements at meteorological conditions (stability classes) which so far have not been considered
- measurements at extreme weather conditions (inversions, zero wind speed situations) when model calculations fail
- measurements at very large distances from the emission source (long-range phenomena).

The results of these measurements will be subject of a later report.

TABLE I SUMMARY OF MEASUREMENT CONDITIONS DURING MEASUREMENTS

day of measurement	time of measurement	mean flue-gas emission rate at STP (m^3/h)	mean plume trajectory (%)	mean wind speed (m/sec)	stability class
6.11.1985	13.35-16.15	821.000	280	8,0	4
26.11.1985	13.00-15.00	855.000	290-340	1,5	3
29.11.1985	11.30-13.59	879.000	255	2-4	3
10.12.1985	11.33-14.49	1,038.000	230-240	5,2	4
12.12.1985	10.18-17.30	1,352.000	285	3,7	4
29. 1.1986	13.52-17.00	893.000	130	4,9	4
30. 1.1986	11.07-16.38	878.000	120	5,1	4
13. 2.1986	10.14-15.21	1,417.000	30	1,3	2-3
14. 2.1986	13.02-17.40	1,095.000	100	6,7	3-4
24. 2.1986	11.31-16.30	1,325.000	330	2,7	3-4

TABLE II SUMMARY OF MEASURED RESULTS

day of measurement	stability class	mean wind speed (m/sec)	maximum measured value ($\mu g\ SO_2/m^3$) instantaneous value	half hour mean value	SO_2-groundlevel concentration at monitoring stations ($\mu g\ SO_2/m^3$)
6.11.1985	4	8,0	4,36	3,87	0-10
10.12.1985*	4	5,2	0,46	-	0-20
12.12.1985	4	3,7	0,59	0,37	20-40
29. 1.1986	4	4,9	38,76	6,71	20-40
30. 1.1986	4	5,1	2,96	3,95	10-30
14. 2.1986	3-4	6,7	2,30	1,25	60-80
24. 2.1986	3-4	2,7	1,89	-	30-60
26.11.1985*	3	1,5	-	0,52	70-90
29.11.1985	3	2-4	-	2,73	10-30
13. 2.1986	2-3	1,3	3,16	-	80-140

* only the edge of plume has been detected

TABLE III STATISTICS OF MEASURED RESULTS

range of measured values ($\mu g\ SO_2/m^3$)	number of half hour mean values	number of instantaneous values
10-40	-	3
1-10	36	110
NG* - 1	43	158
<NG*	140	619
total	219	890

*NG is detection limit

TABLE IV CALCULATED AND MEASURED SO_2 GROUNDLEVEL CONCENTRATIONS

day: November 6, 1985

principle wind direction: 280°

measuring position	SO_2 groundlevel concentration I (mg SO_2/m^3) calculated I_c	measured I_b
1	$0{,}41.10^{-5}$	$0{,}10.10^{-3}$
2	$0{,}85.10^{-5}$	$0{,}36.10^{-4}$
3	$0{,}64.10^{-3}$	$0{,}29.10^{-3}$
4	$0{,}11.10^{-4}$	$0{,}27.10^{-4}$
5	$<10^{-8}$	$0{,}13.10^{-3}$
6	$<10^{-8}$	<NG[1]
7	$0{,}77.10^{-3}$	$0{,}19.10^{-3}$
8	$0{,}80.10^{-6}$	$0{,}15.10^{-4}$
9	$<10^{-8}$	$0{,}12.10^{-4}$
10	$<10^{-8}$	$0{,}45.10^{-4}$

1) NG is detection limit

TABLE V COMPARISON OF CALCULATED AND MEASURED RESULTS OF SO_2 GROUNDLEVEL CONCENTRATIONS

day: November 6, 1985

principle wind direction	square sum of logarithmic differences 1) 2)	agreement between calculation and measurement
(in degrees)	$\sum_{i=1}^{10} (\log I_c - \log I_m)^2$	(qualitative evaluation)
272	78,84	poor
275	44,32	good
280	62,75	fair
285	62,47	fair
290	92,60	rejected

1) implicit is the assumption of identical diffusion properties of SO_2 and SF_6

2) summing over 10 values according to 10 measuring points

TABLE VI MEASURED AND CALCULATED RESULTS OF JAN. 30, 1986

distance from emitter (m)	maximum measured instantaneous values (µg SO_2/m^3)	maximum measured half hour mean values (µg SO_2/m^3)	maximum calculated values (µg SO_2/m^3)
4.000	2,21	3,95	0,5
12.000	2,96	2,28	4,2
13.800	2,49	-	4,0
17.500	2,18	1,88	3,4
26.000	1,54	-	2,3

detection limit: 0,05 µg SO_2/m^3

TABLE VII MEASURED AND CALCULATED RESULTS OF FEB. 14, 1986

distance from emitter (m)	maximum measured instantaneous values (µg SO_2/m^3)	maximum measured half hour mean values (µg SO_2/m^3)	maximum calculated values (µg SO_2/m^3)
7.500	<0,081	-	3,5
9.000	1,66	-	4,3
13.500	1,58	1,25	5,6
20.000	2,30	-	4,8
22.500	0,59	-	4,3

detection limit: 0,081 µg SO_2/m^3

Fig. 1

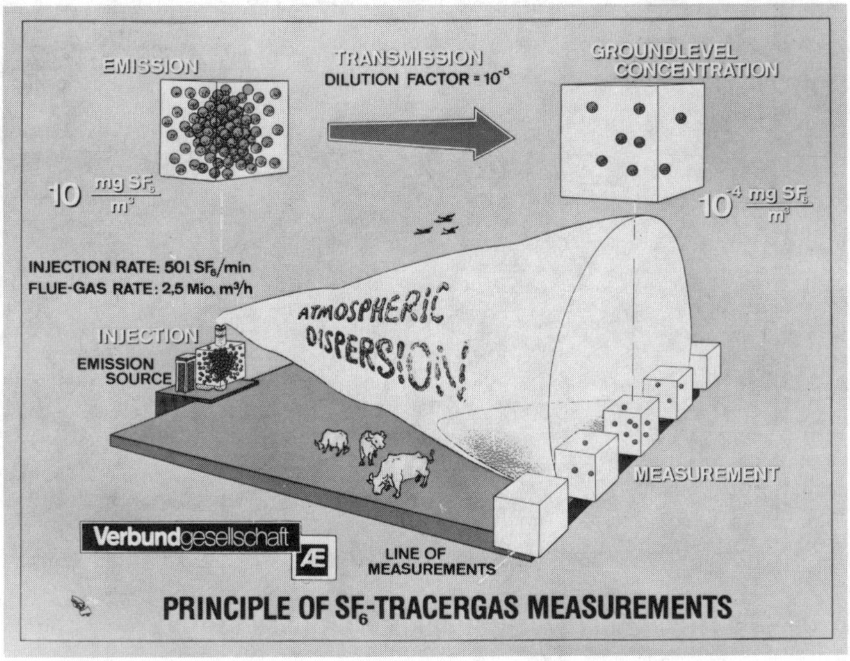
EMISSION
TRANSMISSION
DILUTION FACTOR = 10^-5
GROUNDLEVEL CONCENTRATION
10 mg SF6/m3
10^-4 mg SF6/m3
INJECTION RATE: 50 l SF6/min
FLUE-GAS RATE: 2,5 Mio. m3/h
ATMOSPHERIC DISPERSION
INJECTION
EMISSION SOURCE
MEASUREMENT
Verbundgesellschaft
LINE OF MEASUREMENTS
PRINCIPLE OF SF6-TRACERGAS MEASUREMENTS

Fig. 2

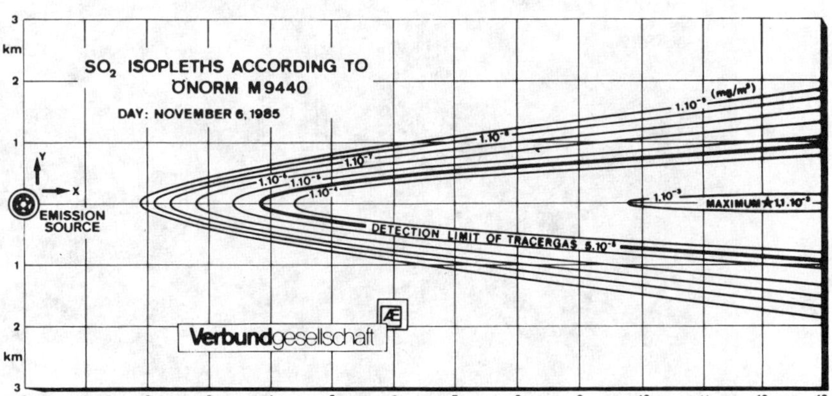
SO2 ISOPLETHS ACCORDING TO ÖNORM M9440
DAY: NOVEMBER 6, 1985
EMISSION SOURCE
DETECTION LIMIT OF TRACERGAS
Verbundgesellschaft
km

Fig. 3

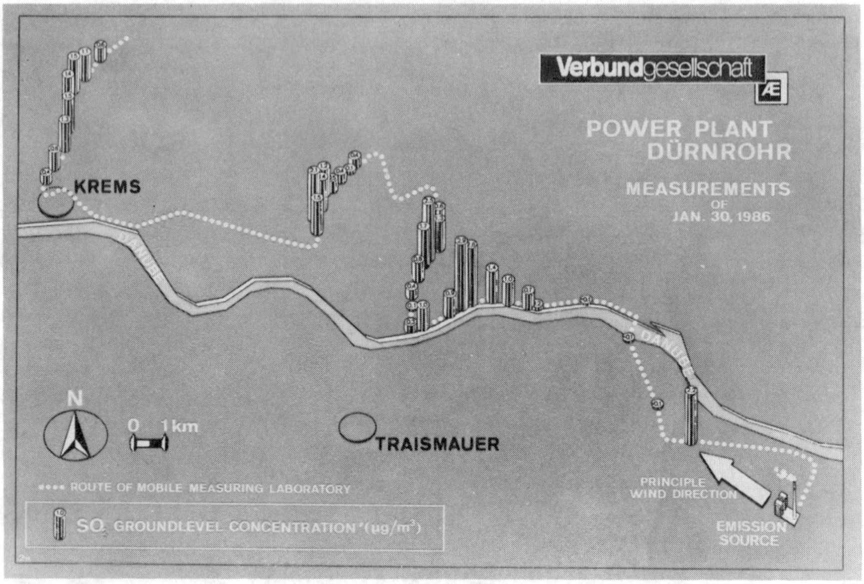
Verbundgesellschaft
POWER PLANT
DÜRNROHR
MEASUREMENTS
OF
JAN. 30, 1986
KREMS
DANUBE
N
0 1km
TRAISMAUER
ROUTE OF MOBILE MEASURING LABORATORY
PRINCIPLE
WIND DIRECTION
EMISSION
SOURCE
SO₂ GROUNDLEVEL CONCENTRATION (µg/m³)

Fig. 4

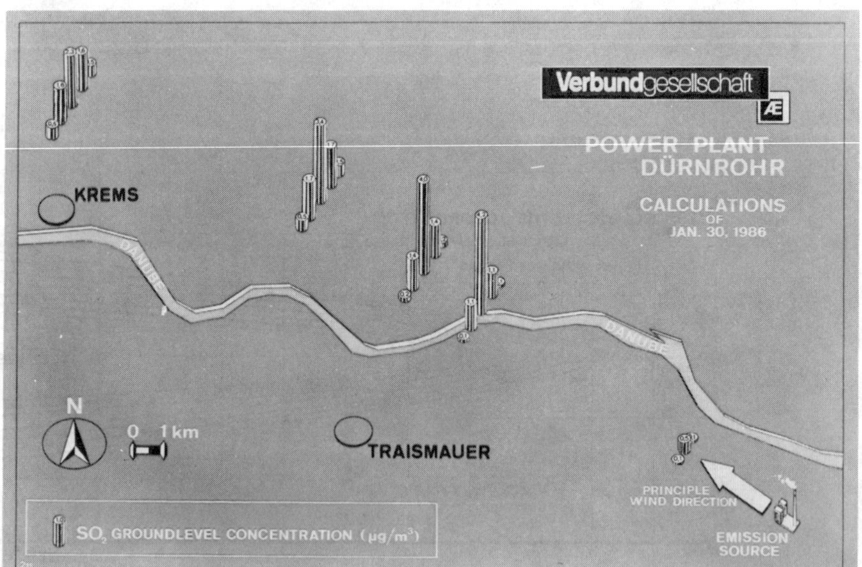
Verbundgesellschaft
POWER PLANT
DÜRNROHR
CALCULATIONS
OF
JAN. 30, 1986
KREMS
DANUBE
N
0 1km
TRAISMAUER
PRINCIPLE
WIND DIRECTION
EMISSION
SOURCE
SO₂ GROUNDLEVEL CONCENTRATION (µg/m³)

Fig. 5

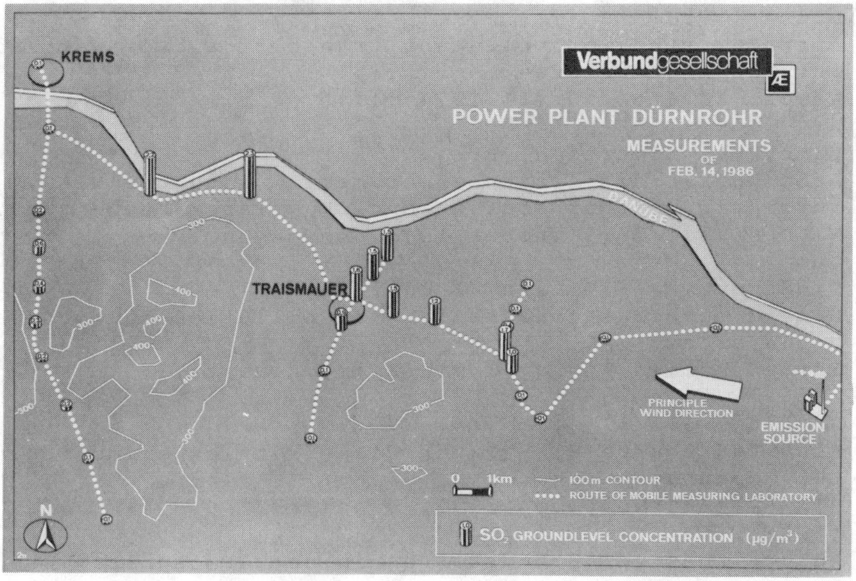
KREMS
Verbundgesellschaft
POWER PLANT DÜRNROHR
MEASUREMENTS OF FEB. 14, 1986
DANUBE
TRAISMAUER
PRINCIPLE WIND DIRECTION
EMISSION SOURCE
0 1km
100 m CONTOUR
ROUTE OF MOBILE MEASURING LABORATORY
SO_2 GROUNDLEVEL CONCENTRATION (µg/m³)
N

Fig. 6

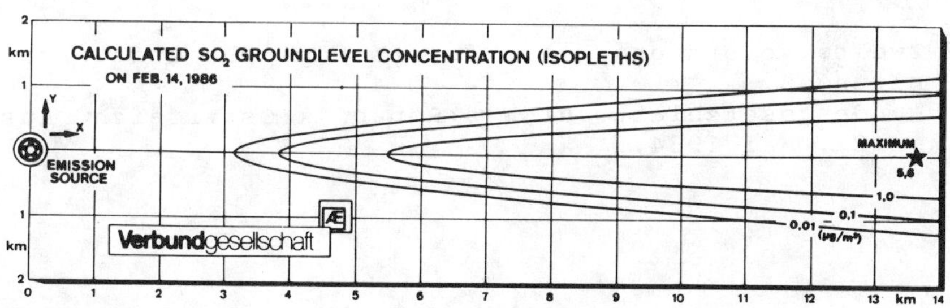
CALCULATED SO_2 GROUNDLEVEL CONCENTRATION (ISOPLETHS)
ON FEB. 14, 1986
EMISSION SOURCE
MAXIMUM 5,6
1,0
0,1
0,01 (µg/m³)
Verbundgesellschaft
km
0 1 2 3 4 5 6 7 8 9 10 11 12 13 km 14

REFERENCES

1 Drivas, P.J.; Shair, F.H.:
A tracer study of pollutant transport and dispersion in the Los Angeles area.
Atmos. Environ. 8 (11/1974) 1115.

2 Lamb, B.; Sivertsen, B.:
Atmospheric dispersion experiments using the NILU automatic weather station and SF_6 tracer techniques.
NILU Nr. 12/78, Oktober 1978, Norwegen.

3 Binenboym, J.; Gilath, I.u.a.:
Use of gaseous tracers for air pollution studies in urban areas.
Israel Atomic Energy Commission, August 1975.

4 Gryning, S.E.:
Elevated source SF_6-tracer dispersion experiments in the Copenhagen area.
RISØ-R-446, November 1981, Dänemark.

5 Ausbreitung von Schadstoffen in der Atmosphäre; Ermittlung von Schornsteinhöhen und Berechnung von Immissionskonzentrationen.
ÖNORM M 9440, 1. Juni 1982.

6 Schwefeloxide in der Atmosphäre, Luftqualitätskriterien SO_2.
Österreichische Akademie der Wissenschaften. Herausgegeben vom Bundesministerium für Gesundheit und Umweltschutz.

7 Zweite Verordnung gegen forstschädliche Luftverunreinigungen.
Bundesgesetzblatt für die Republik Österreich. Ausgegeben am 22. Mai 1984.

EVALUATION OF A MODEL PERFORMANCE WITH DATA FROM A LARGE POWER PLANT

R. Berkowicz, E. Lyck
Air Pollution Laboratory,
Danish National Agency of Environmental Protection
4000 Roskilde, Denmark

P.A. Larsen, J.S. Markvorsen
Cowiconsult, Consulting Engineers and Planners
2830 Virum, Denmark

S. Dalager, A.B. Jensen
Danish Boiler Owners Association
2860 Søborg, Denmark

ABSTRACT. Air pollution models for local-scale dispersion were developed during the years 1980-84 at the Danish Air Pollution Laboratory. The result of the project is the OML-model, a new operational air quality model. The OML-model is based on the well-known Gaussian plume formulation but the simulation of dispersion processes has been improved compared to earlier models. The improvement was obtained by describing the dispersion processes in terms of physical parameters which have been shown by recent research to be of basic importance to boundary-layer turbulence. During the period April 1984 to March 1985, an extensive dispersion experiment was conducted in the surroundings of the 1440 MW coal fired power plant, the Asnæs Power Plant, Denmark. The data set consist of one full year of 1/2-hourly measurements of SO_2-concentrations from ten monitoring stations. Meteorological measurements were made at a nearby meteorological mast. Hourly emission inventory has been performed for the power plant as well as for other major nearby sources. Results from the model evaluation are presented in the paper Detailed analyses of the OML-model performance showed that the model exhibits a heigh degree of accuracy in predicting dispersion of SO_2-pollution from the power plant. A special emphasis was put on evaluation of the physical behaviour of the model.

1. INTRODUCTION

Dispersion models are a powerfull tool for impact analyses of air pollution from large sources. Air pollution measurements, although giving very valuable informations, are usually limited to some few geographical positions only. Neither can contributions from different sources be directly evaluated from measurements alone. Using dispersion models one must, however, be sure that the model is appropriate for the particular application. The accuracy of any impact analyses will depend on the phy-

K. Grefen and J. Löbel (eds.), Environmental Meteorology, 93–114.

sical background of the model and quality of the input data, such as meteorological data and emissions. The most straightforward way to ensure the applicability of a model is to perform a model evaluation study using available measurements. The measurements used for such a study should be designed to give detailed information on model performance in variety of meteorological and emission conditions.

In this paper we describe a model evaluation study performed in connection with impact analyses of a large coal fired power plant - the Asnæs Power Plant - in Denmark. (Vestsjællands Amtskommune and Isefjordværket I/S, 1987, five parts, hereafter referred to as Part 1 to Part 5). The study covered only the sulfurdioxid (SO_2)-pollution.

The model used here is the recently developed Operational Meteorological Air Quality Model (OML). Only a short description of the main physical principles of the model is given in this paper. The emphasise is put on the strategy of the evaluation procedure as well as the conclusions that can be drawn from such a comparison.

2. THE OML-MODEL

The OML-model was developed during the years 1980-84 at the Danish Air Pollution Laboratory, National Agency of Environmental Protection in cooperation with Cowiconsult, Consulting Engineers and Planners and the Danish Boiler Owners Association.

The model is designed to predict hourly averages of air pollution from both innert and buoyant sources for distances up to 20-30 km. Computer codes (Fortran 77) are available in two major versions: OML-Point, for single point sources or several collocated sources (Olesen 1987), and OML-Multi, for multiple point and area sources (Løfstrøm, 1987). Computations are performed on a dense net of receptors (the number and positioning of receptorpoints can be freely varied) using as input time series of meteorological and emission data. The predicted time series of hourly averaged concentrations at each receptor point are used to compute long-term averages or other statistics.

2.1. The physical basis of the OML-model

The well known Gaussian plume formula is used in the model. Ground-level concentrations at a distance y from the centerline are computed by means of the expression,

$$c = (Q/\pi u \sigma_y \sigma_z)\exp(-0.5(hef/\sigma_z)^2)\exp(-0.5(y/\sigma_y)^2) \qquad (1)$$

where Q is the source strength, hef the effective height of the plume centerline above the ground, u the effective transport velocity, and σ_y and σ_z are the vertical and horizontal dispersion parameters, respectively. Reflection from the top of the boundary layer is treated in the standard way (Turner, 1964).

Though the limitations of the Gaussian approach should be recognised, it appears at present to be the only type of model capable of dealing satisfactory with different type of sources and for a wide range of atmospheric dispersion conditions. The crucial point in any Gaussian plume model is determination of the dispersion parameters σ_y and σ_z. The great majority of Gaussian models makes use of the dispersion parameters and classification methods proposed by Pasquill (1961) and later slightly modified by Gifford (1961) and Turner (1964). This method is not used in the OML-model. Instead, the dispersion parameters are calculated as contineous functions of the basic boundary-layer quantities, such as friction velocity (u_*), Monin-Obukhov length (L), convective velocity scale (w_*) and boundary-layer height (z_i). Also the dependence of the dispersion parameters on the source height is taken into account. More detailed discussion of the dispersion parameterisation procedure is given in Berkowicz et al., (1985) and Olesen (1987). The other most notable new features of the OML-mode are the following:

- Dispersion parameters depend on the plume buoyancy.
- Plume rise is modeled using the new formulas proposed by Briggs (1984) but supplemented by a number of extensions.
- In contrast to most conventional models, penetration of plume into the layer above the mixing layer is not simulated as an on/off process, instead the amount of plume penetration is taken into consideration.
- New methods are used for simulation of distortion of plume dispersion induced by tall nearby buildings. Both dilution and reduction in the effective plume height are accounted for.

2.2. The meteorological input data

Because the OML-model makes use of the basic boundary-layer parameters it is not possible to use the standard meteorological observations directly as input to the model. The boundary-layer parameters can, however, be estimated from routinely available meteorological data using methods developed for the use in connection with the OML-model (Nielsen et al., 1981; Berkowicz and Prahm, 1982b; Olesen et al., 1987). A special software-package, the "OML meteorological preprocessor", supplies the necessary meteorological input data to the dispersion model. The standard version of the preprocessor has as input hourly meteorological measurements from e.g. a synoptic surface station and twice-daily vertical profiles of temperature from e.g. a nearby radiosonde station (Olesen and Brown, 1987). Output is in this case hourly values of turbulence parameters, most essentially sensible heat flux, Monin-Obukhov length, friction velocity and mixing height. More specialised versions of the preprocessor have been designed for non-standard input data such as e.g. mast measurements (Berkowicz and Prahm, 1982a). The last method is used in connection with the Asnæs Power Plant study.

2.3. Previous tests of the OML-model

The OML-model has been evaluated using data from several different sites (Berkowicz et al., 1985). The experimental data involved measurements from large power plants in the USA and Denmark, as well as special tracer measurements. The results from the tests were used to improve the parameterisation of the model.

The new data collected in connection with the Asnæs Power Plant investigation provided the possibility to perform one of the most detailed verifications of the model.

3. THE EXPERIMENTAL SITE

The Asnæs Power Plant is situated on the Asnæs Peninsula, the west coast of Sealand (Fig. 1). The terrain is level with few, up to 60 m high hills. To the North, a 2 km wide water street (Kalundborg Fjord) separates the power plant from the Kalundborg town. To the South, there is about 3 km to the Jammerland Bay.

The power plant has five units (hereafter denoted by ASV 1-5), each with a separate stack. ASV 1-4 has a common, 64 m high boiler building with the stacks on the top of the building. The height of the four stacks above terrain is 120 m. ASV 5 is situated about 340 m West of ASV 4 and has a 220 m high stack.

About 850 meters South-East from the Asnæs Power Plant one can find another significant SO_2-source - the Statoil Refinery with three, about 90 m high stacks.

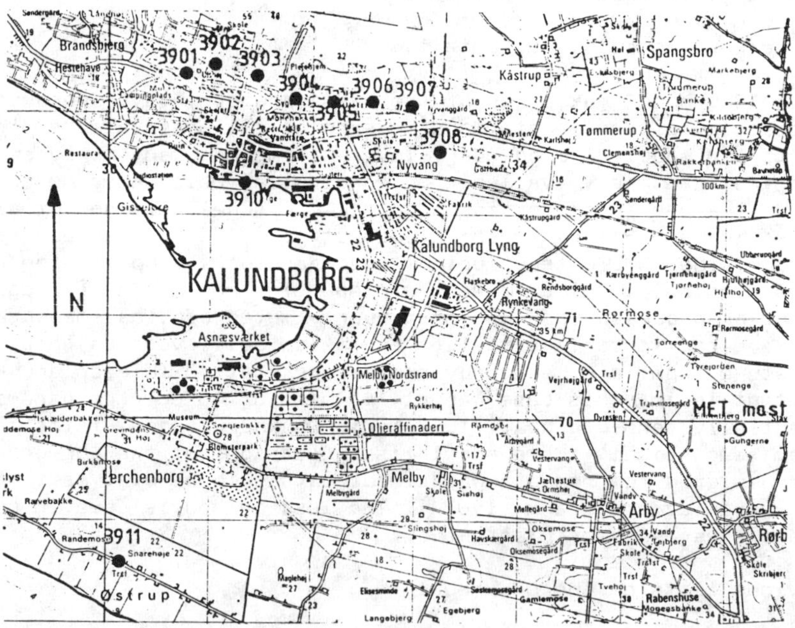

Figure 1. The experimental site. The positions of the measuring stations are indicated on the figure together with their reference numbers.

4. DATA BASE

4.1. SO_2-measurements

Continueous measurements of SO_2 were performed on 10 locations for a period of one year, from April 1984 to March 1985 (Part 3). Positions of the monitoring stations are shown in Fig. 1. Eight stations were placed in an arc, at the northern part of Kalunborg town, about 3 km from the power plant. The separation between the stations was about 400 m. One station was located at the Kalundborg harbour, about 2 km North of the plant, and another one about 2.5 km South-West of the plant. All the monitors measured 1/2-hourly averages of SO_2-concentrations. For the purpose of model testing the 1/2-hourly averages were converted to hourly averages.

4.2. Meteorological measurements

A 13 m high meteorological mast was placed about 5 km East of the Asnæs Power Plant (Part 2). The mast was operated for a period of two years, from April 84 to March 86. The measurements were designed to provide the necessary meteorological input parameters to the OML-model. Measurements included wind speed and direction at 10 m height and temperature difference between 2 and 12 m. Friction velocity, u_*, and surface heat flux, H, were estimated using the well known profile method (see e.g. Berkowicz and Prahm, 1982a).

Mixing heights were estimated using routine radio soundings performed from a station at Jægersborg, about 10 km North of Copenhagen, and about 80 km from Asnæs. The method used for estimation of the hourly values of the mixing heights is described by Olesen et al., (1987).

4.3. Emissions

A detailed emission inventory was performed for both the Asnæs Power Plant and the Statoil Refinery (Part 1). Hourly values of SO_2-emissions were estimated from the information on fuel consumption and sulphur content in the fuel. A more crude emission inventory was performed for other sources in the region. It should be noted that the sulphur emissions from the power plant and the refinery account for more than 90% of the total sulphur emission in the region.

5. THE STRATEGY OF THE MODEL EVALUATION PROCEDURE

In spite of that the OML-model had previously been tested with several experimental data, the actual project required verification of some new modelling aspects. The surroundings of the power plant (see Fig. 1) are far from being homogeneous. One could expect, that the land-water transition will influence the dispersion conditions. It should be recognised, that the OML-model is not a priori designed to treat such non-homogeneous conditions. A more detailed evaluation of the model performance in the actual conditions was thus required.

The model evaluation study was devided into 4 phases:

1. Data selection.
2. Preliminary analyses.
3. Comparison of model results with the measurements.
4. Analyses of the results.

5.1. Data selection.

The main goal of the project was to evaluate the ability of the model to describe dispersion from the Asnæs Power Plant. The main criterion for data selection was thus, that the measured SO_2-concentrations could be attributed to emissions from the power plant.

First, only the data were selected which corresponded to wind direction from the South (170° to 250°). Additionally, the highest measured concentration on the arc in Kalundborg should be higher than 50 µg/m³.

Concentration profiles along the arc in Kalundborg were evaluated and only those observations were selected, for which it was clearly seen that the plume centerline passed the chain. The criterion was that the highest concentration occured on one of the inner stations, i.e. between monitors nr. 2 and nr. 7 (see Fig. 1.). Additionally, in order to exclude observations with a significant contribution from remote sources, the concentrations measured on the station nr. 11, which was situated south-west of the power plant, should be much smaller than the concentrations measured on the arc in Kalundborg. It should be noted that significant contributions from remote sources in central and east Europe were frequently observed during the measuring campaign (Part 3). Furthermore, we have also excluded from the analyses the hours with the emissions from ASV 1-4 being smaller than the emissions from the refinery. In Fig. 2 is shown an example of a concentration profile for an hour selected according to the described procedure. Here, the highest concentration appears on the station nr. 2 and the background concentration (station nr. 11, shown here by a horizontal line) is very small.

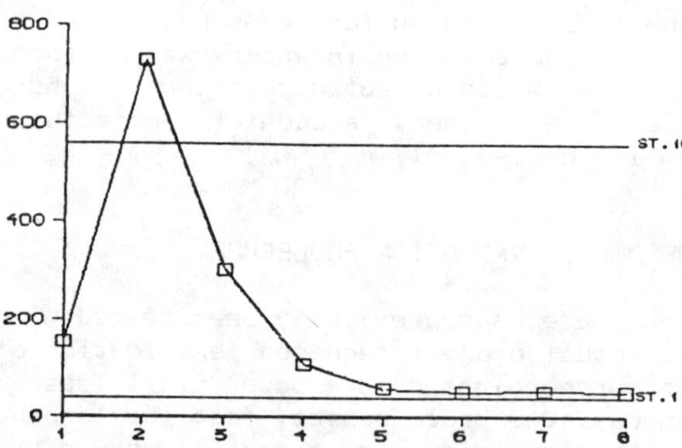

Figure 2. An example of a concentration profile from the arc in Kalundborg included in the selected data. Measurements from station nr. 10 and nr. 11 are shown by horizontal lines.

About 500 hours of SO_2-measurements were selected for the analyses.

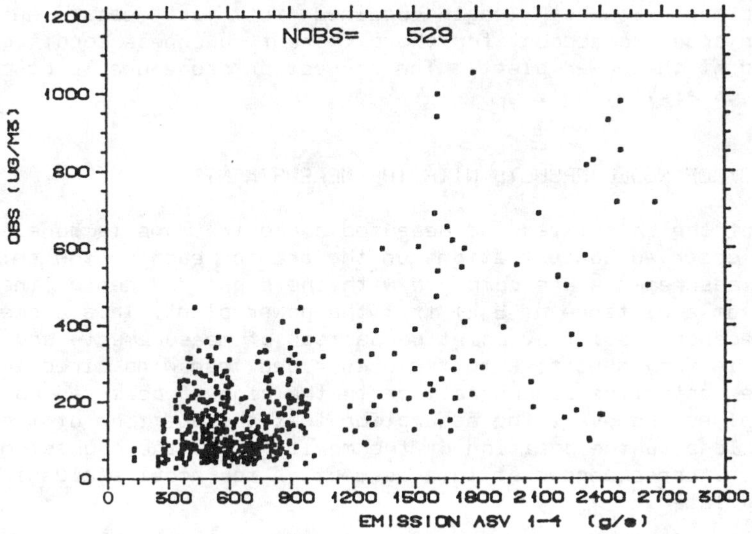

Figure 3. Dependence of the measured concentrations on emissions from ASV 1-4.

5.2. Preliminary analyses

The selected data were analysed in order to quantify the contributions from the five power plant units and the nearby refinery. In Fig. 3. the measured SO_2-concentrations are compared to emissions from ASV 1-4. Only the maximum observed concentrations for each of the selected hours are used for this comparison. As it is seen from this figure, the measured concentrations have a clear tendency to increase with increasing emissions from ASV 1-4. No such tendency could be seen with emissions from ASV 5 or the refinery. Regarding the SO_2-concentrations measured on the arc in Kalundborg (3 km from the power plant), the contributions from ASV 1-4 appear to be the most significant.

Model evaluation is thus performed using the selected 500 hours with only emissions from the Asnæs Power Plant taken into account and ASV 1-5 treated as five separate sources.

In the OML-model the dispersion parameters and the wind profile are computed making use of the roughness height at the source surroundings (Berkowicz et al., 1985). The surroundings of the Asnæs Power Plant are quite non-homogeneous, with both land, water and city areas. Because the model is not designed to treat changes in the terrain roughness, an attempt was made to find the best appropriate average roughness height. The computed concentrations are most sensitive to the roughness height for high wind speed dispersion conditions. As will be shown later, these conditions lead also to the highest ground level concentrations. From comparison of the measured concentrations with the model results for high wind speed conditions it was concluded that the approriate average value of the roughness height is 10 cm. All model calculations are subsequently performed with this value of the roughness height. Friction velocity and Monin-Obukhov length obtained from mast measurements were corrected in order to account for the different roughness conditions at the mast and at the power plant. The correction procedure is described in Gislason et al., (1985).

6. COMPARISOM OF MODEL RESULTS WITH THE MEASUREMENTS

Comparison of the calculated and measured concentrations is done only for the highest observed concentrations on the arc for each of the selected hours. The measurements are compared with the predicted centerline concentrations at a distance of 3 km from the power plant. This procedure is chosen because a point-by-point comparison of measurements and model predictions is very sensitive to the prescribed mean wind direction. Relating the centerline concentrations to the highest observed concentrations implyes, however, the assumption that the position of the centerline is close to the position of the monitor with the highest measured concentration. Consequences of this assumption for model evaluation will be discussed later.

6.1. Results.

Comparison between the predicted and the measured concentrations, using the procedure described above, is shown in Fig. 4a. Here, the observations are devided into two dispersion categories. Only the sign of the surface heat flux is used to differentiate between the to categories. The observations with positive heat flux are categorised here as unstable dispersion conditions, while all observations with a negative heat flux correspond to the class of stable dispersion conditions. This categorisation is used only for presentation of the data.

The SO_2-emissions from the Asnæs Power Plant exhibit a substantial variability, mainly due to seasonal and diurnal variations in load conditions. In order to provide a basis for the evaluation of the physics of a model, the variability in the emission must be removed from the material used for the comparison. It was done here by normalising both the measured and predicted concentrations with emissions from ASV 1-4. After normalisation the concentrations correspond to a constant emission rate of 1000 g/s. The normalisation procedure implyes the assumption that the contributions are proportional to the total emission from ASV 1-4. As one can judge from results shown in Fig. 3 this assumption appears to be reasonable. Comparison between the predicted and measured normalised concentrations is shown in Fig. 5a. The comparison is again made for the highest measured concentrations and predicted centerline concentrations.

The results shown until now were all related to measurements performed on the arc at the distance from the power plant of about 3 km. A single measuring station was situated at a distance of about 2 km from the plant (position nr. 10, the Kalundborg Harbour, see Fig. 1). For reasons discussed previously it is difficult to perform any detailed comparison of model predictions and measurements for this single position (some results are however presented in Part 4). In order to gain some information on model behaviour also at the distance of 2 km from the source, the concentrations computed at this distance were compared with

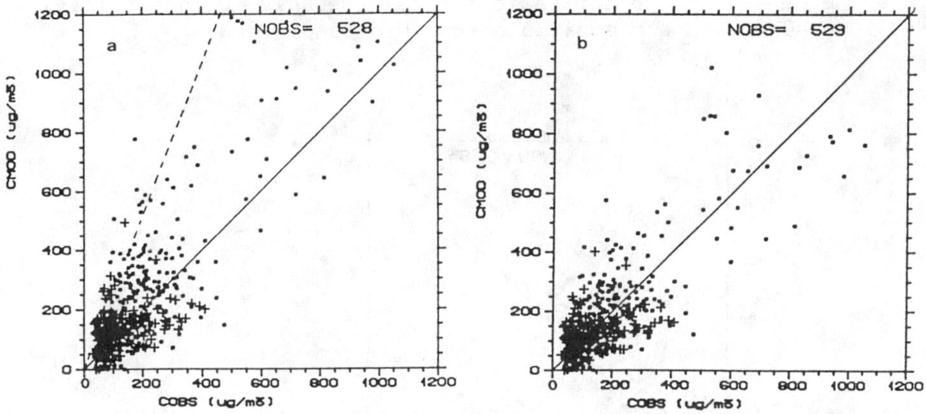

Figure 4. Comparison between the observed (COBS) and the computed (CMOD) concentrations. a) concentrations computed at the plume centerline; b) concentrations computed at a distance of 100 m from the centerline. +: unstable dispersion conditions; ▪: stable dispersion conditions.

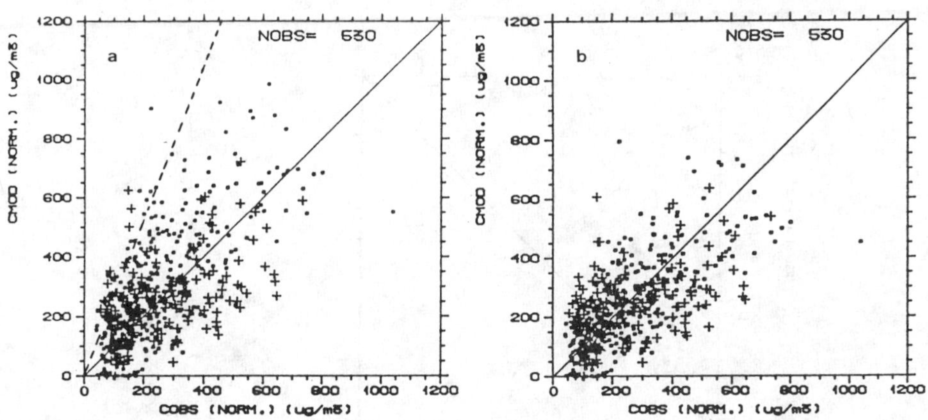

Figure 5. Comparison between the observed and computed normalised concentrations. a) concentrations computed at the plume centerline; b) concentrations computed at a distance of 100 m from the centerline.

concentrations computed for the distance of 3 km. The results are shown in Fig. 6. This figure should be compared with Fig. 7, where the measurements from the monitor at the position nr. 10 are compared with measurements from the station nr. 3 for hours where the maximum in the arc are measured at this station. As stations 3 and 10 are grossly on a straight line with the Asnæs Power Plant, it is quite probable that they measure the centerline concentrations simultaneously. Unfortunately, very few (37) measurements are available for this comparison.

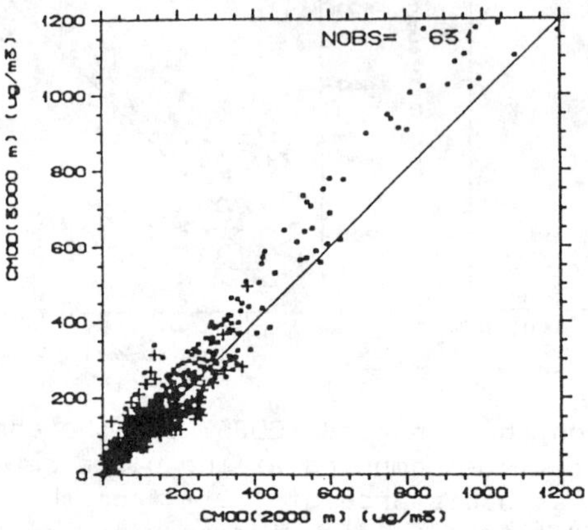

Figure 6. Comparison between concentrations computed for distances of respectively 2 and 3 km from the Asnæs Power Plant.

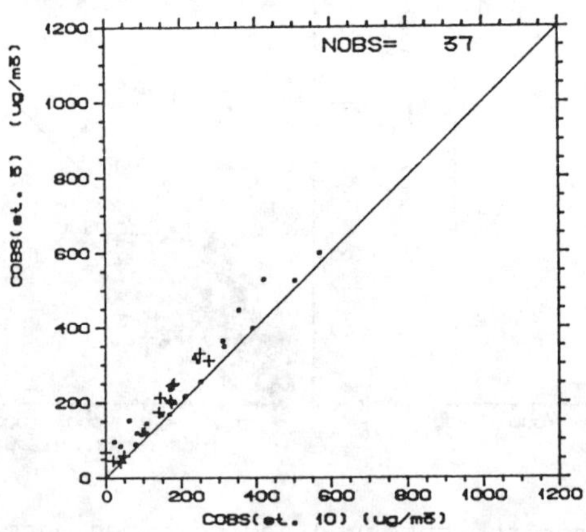

Figure 7. Comparison between concentrations measured on respectively positions nr. 3 and 10.

6.2. Discussion

Considering the results shown in previous section one can conclude that the UML-model reproduces the observed concentrations reasonably well. Some statistical measures are presented in Table I and Table II, for the non-normalised and the normalised data, respectively. Statistics is however not enough to provide a reasonable evaluation of the physical behaviour of a model. A critical evaluation of the measurements is also necessary in connection with a model testing procedure. This subject will be elaborated in more details here.

Tabel I. Statistical data for the measured and computed concentrations. All concentrations are in $\mu g/m^3$.

	All obs.		Stable		Unstable	
	C_{mod}	C_{obs}	C_{mod}	C_{obs}	C_{mod}	C_{obs}
Number	530		368		162	
Average	212	175	248	196	131	128
Standard dev.	212	157	242	177	70	82
C.V.*	1.00	0.90	0.98	0.90	0.53	0.64
Correlation	0.80		0.82		0.34	

Tabel II. Statistical data for the measured and computed normalised concentrations. All concentrations are in $\mu g/m^3$.

	All obs.		Stable		Unstable	
	C_{mod}	C_{obs}	C_{mod}	C_{obs}	C_{mod}	C_{obs}
Number	530		368		162	
Average	286	250	299	248	256	255
Standard dev.	184	158	201	157	133	159
C.V.*	0.64	0.63	0.67	0.63	0.52	0.62
Correlation	0.64		0.71		0.48	

* C.V. = Standard dev./Average

Comparisons presented in Figs. 4a and 5a are made for the computed centerline concentrations and the measured highest concentrations on the arc. However, the highest measured concentrations do not need to be identical with the centerline concentrations. How much the centerline concentration can differ from the observed maximum will depend on the distance between the plume centerline and the position of the observed maximum. As narrower the plume is with respect to the spacing between the monitoring stations as larger the difference can be. Model calculations show, that at high wind speeds the horizontal standard devition, σ_y, at the distance of 3 km from the power plant can be as small as 150 m. (see Fig. 8). Taking into account that the distance between two monitors is about 400 m, and that at the "worst case", the plume centerline can be 200 m away from the nearest monitor, it can be concluded that the concentration at the plume centerline can deviate from the measured highest concentration by as much as factor 2.5. Thus, the predicted centerline concentration can also be 2.5 times higher than the measured highest concentration. A line corresponding to this expected 2.5-times-deviation is drawn in Figs. 4a and 5a. As one can see, only very few data points are above this line. In order to clarify more closely the consequences for model evaluation of using the observed highest concentrations as a measure of the centerline concentrations we have compared the measurements with concentrations computed at a distance of 100 m from the plume centerline. The results are shown in Figs. 4b and 5b. A significantly better agreement between the predicted and the measured concentrations can be observed, especially considering the highest concentrations.

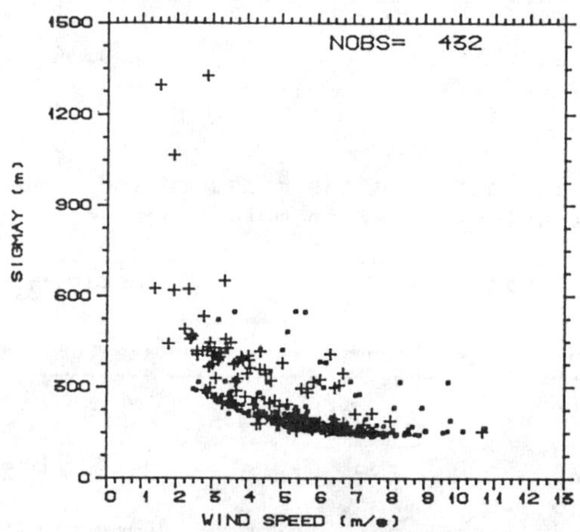

Figure 8. Dependence on wind speed of the computed horizontal standard deviation at the distance of 3 km from the source.

As mentioned before, the largest deviations between the highest observed concentrations and plume centerline concentrations are expected to occur when the plume is narrow. In Fig. 8 it is shown, that the horizontal standard deviation, σ_y, decreases with increasing wind speed. In Fig. 9a the ratio between the predicted centerline concentrations and the measured highest concentrations is shown as a function of wind speed at

10 m height. In Fig. 9b the same is shown, but for concentrations computed at a distance of 100 m from the centerline. Comparing the two pictures, one can conclude, that the tendency for overprediction seen in Fig. 9a for wind speeds above 6-7 m/s is practicaly eliminated in Fig. 9b. As will be shown later, the highest concentrations occur at high wind speeds. For model evaluation it is thus important to recognise the sensitivity of the measured and predicted concentrations towards deviations from centerline values.

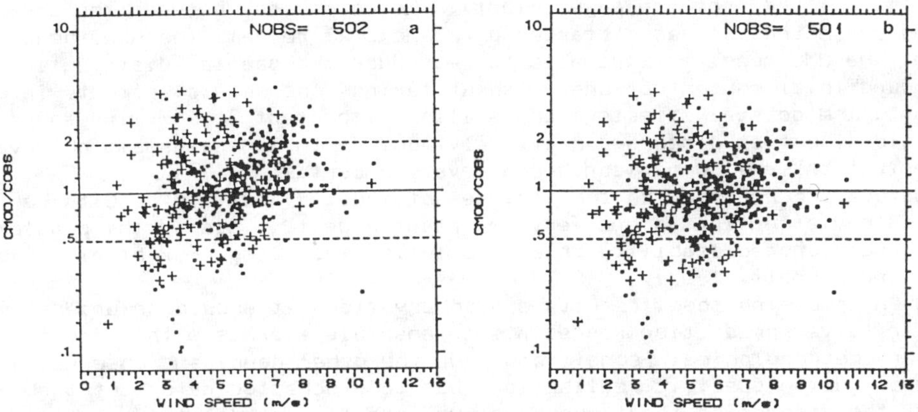

Figure 9. The ratio between the computed and the observed concentrations as a function of the wind speed. a) concentrations computed at the centerline; b) concentrations computed at a distance of 100 m from the centerline.

Considering the results shown in Fig. 9 once more, one can see, that the model underpredicts the measured concentrations somewhat for wind speeds below 3-4 m/s. There can several reasons for this behaviour. The basical physical assumptions for a stationary Gaussian model are not well founded for low wind speeds. For such a buoyant source, as the Asnæs Power Plant, the model will in general predict small ground level concentrations when the wind speed is low. The predicted concentrations are very sensitive to estimations of the plume rise. This, in turn, depends strongly on the upper layer temperature profile, especially in convective, light wind conditions (Berkowicz et al., 1985). No in site upper layer soundings were available during the measurements. The radiosoundings used for the model calculations were from a site about 80 km east of the experimental site. Uncertainety in estimations of temperature profiles and consequently erroneous evaluation of plume rise can lead to significant errors in the predicted ground-level concentrations.

Another aspect that must be considered is, that at low wind speeds, when the contribution from the Asnæs Power Plant is in general small, some other SO_2-sources may contribute significantly to the measured concentrations. As the computations presented here are performed only with emissions from the power plant, this alone will lead to systematic underpredictions.

Because the Asnæs Power Plant is situated near the coast line, one could expect that the dispersion conditions would be strongly influenced by the land-water transition effects, such as e.g. formation of an internal boundary layer (Lyons, 1975). Especially, it could be expected that concentrations at the harbour position (monitor nr. 10) should be much smaller than the inlands concentrations. The results shown in Fig. contradicts, however, this expectation. The ground-level concentrations measured at the station nr. 10 are only unsignificantly smaller than the concentrations measured at the position nr. 3, which was 1 km from the coast line (3 km from the power plant). The comparison between the predicted concentrations at distances of 2 and 3 km reveals the same behaviour. The UML-model is thus able to reproduce the spatial distribution of ground-level concentrations without taking the coastal effects into account. The coastal effects might still be important at low wind speed conditions, but as discussed previously, due to in general small observed concentrations any comparison becomes very uncertain.

Before proceeding with the analyses of model evaluation results we want to pay attention to one very important aspect. This is the problem of inherent unpredictability in air pollution modelling (Fox et al., 1983; Weber and Garrett, 1985).

When comparing model results and observations it should be understood that any given prediction represents an ensemble average with respect to similar meteorological conditions. On the other hand, an observation reflects just a specific realisation. Due to stochastic nature of turbulence, the dispersion conditions at even similar meteorological conditions will never be identical. One can thus expect, that model results will in general differ from measurements due to stochastic deviations of single realisations from an ensemble average value. Let us denote by α the ratio between the predicted and the observed concentrations,

$$\alpha = C_{mod}/C_{obs} \tag{2}$$

If the deviation between model predictions (ensemble average) and observations (single realisation) is due to the stochastic nature of turbulence one can expect that α will follow a stochastic distribution, most probable a log-normal distribution. Frequency distribution of α, defined by (2), is shown in Fig. 10. Similarity with a log-normal distribution is rather convincing. A theoretical log-normal curve has been fitted by eye to the distribution of α, as shown in Fig. 10. The fitted distribution has the following characteristics:

$$\begin{aligned} &\text{geometrical standard deviation } s_g = 1.65 \\ &\text{geometrical mean } m_g = 1.1 \end{aligned} \tag{3}$$

A cumulative distribution of α is shown in Fig. 11. From this figure one can read that 72% of all predictions lie within a factor of two with respect to measurements. It is widely recognised now that due to the aforementioned inherent unpredictability, the best one can expect from a model based on hourly averages is 80% of predictions being factor of 2 with respect to measurements (Venkatram, 1979; Weber and Garrett, 1985). Considering the results presented here, it is thus not likely that any refinement of the physics of the model will lead to a better agreement between the predictions and the measurements.

In the following we shall present an analyses of the model results and measurements with respect to the meteorological and emission conditions.

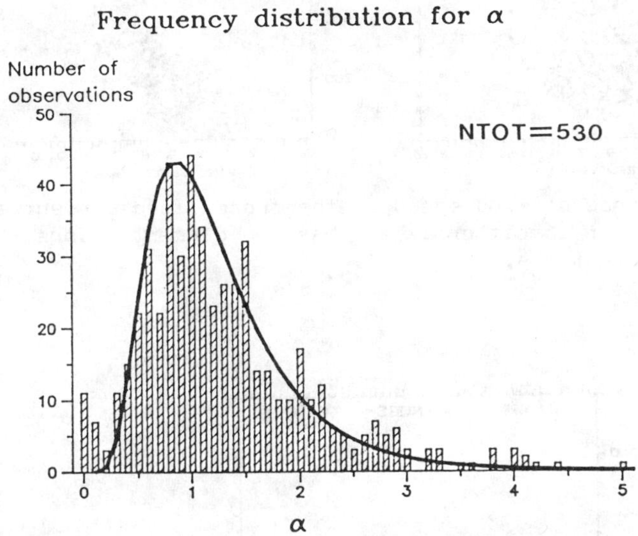

Figure 10. Frequency distribution of the ratio between the computed and the observed concentrations. A curve corresponding to a log-normal distribution is also shown.

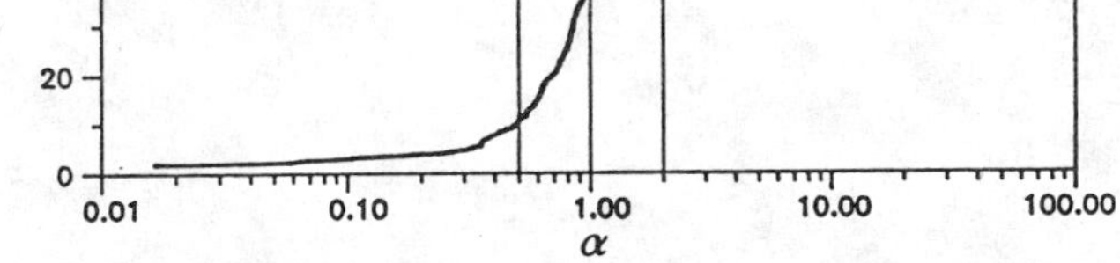

Figure 11. Accumulated distribution of the ratio between the computed and the observed concentrations.

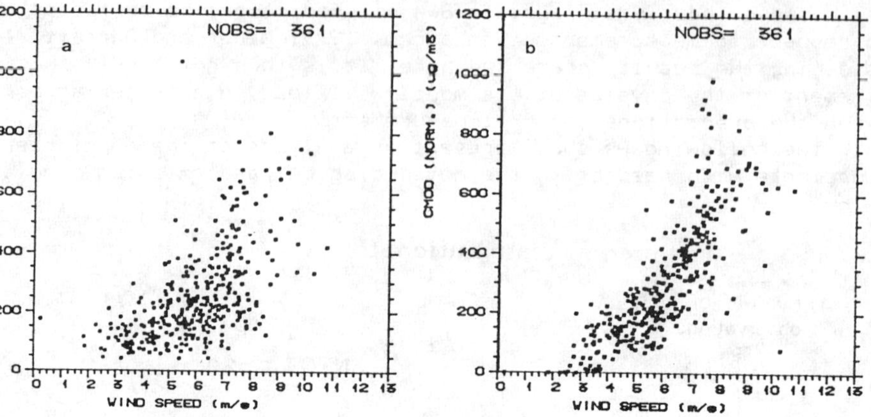

Figure 12. Dependence on wind speed of the normalised concentrations for stable dispersion conditions. a) observed concentrations; b) computed concentrations.

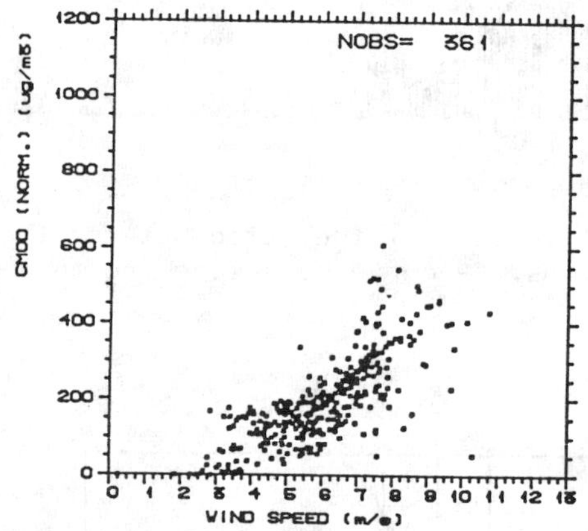

Figure 13. Dependence on wind speed of the computed normalised concentrations for stable dispersion conditions. Computations are made ignoring building effect.

7. ANALYSES OF THE RESULTS

In applying a model for assessment af air pollution it is not only important to quantify the level of pollution, but also to provide an explanation of the meteorological and emission conditions that lead to high pollution concentrations. This can be done by comparing and analysing the dependence of measurements and model results on the most important meteorological and emission parameters.

The data were for this analyses separated into two dispersion classes, the unstable and the stable, according to the sign of the surface heat flux. The dependence of the normalized concentrations on wind speed at 10 m height was evaluated.

Considering the unstable conditions, neither the measuremens nor the model results show any clear dependence on the wind speed. In connection with another model investigation, using data from an american power plant (Berkowicz et al., 1985), it was found that the highest ground-level concentrations occured during light wind convective conditions and when the plume rise was limited by a strong elevated inversion layer. It was not possible to verify this phenomena in the present investigation. The reason might be that the radiosoundings used to determine temperature profiles were not representative for the experimental site. Contributions from other sources than the Asnæs Power Plant might also mask the dependence.

The picture is totaly different for the stable dispersion conditions (Figs. 12a and 12b). Here, the concentrations (normalised) increase with increasing wind speed. Both the measured and predicted concentrations show the same tendency. Considering the measurements one must take into account the "narrow-plume-effect", which can lead to that the measured concentrations are smaller than the centerline concentrations.

The reason for the observed increase of concentrations with wind speed during stable conditions is that the vertical dispersion increases with wind speed, the horizontal dispersion decreases (see Fig. 8) and plume rise decreases too. One more effect is particularly important for the Asnæs Power Plant. This is the influence of the 64 m heigh boiler building on plume dispersion from ASV 1-4. At higher wind speeds the plumes from the four 120 m high stacks become captured in the buildings cavity wake with resulting high ground-level concentrations relatively close to the stack. In order to illustrate this effect, the dependence of the concentrations on wind speed in stable conditions is computed without taking the building effect into account. The results are shown in Fig. 13. Comparing Fig. 13 with Fig. 12 one can clearly see an evident decrease in the concentrations. The effect is most pronounced for higher wind speeds. We do remind, that all the results shown here are for the distance of 3 km from the power plant.

Emissions from a power plant like Asnæs, exhibite a significant seasonal dependence. This must be taken into account performing an impact analyses. In Fig. 14 is shown the time variation of both the measured and the predicted concentrations for all the selected hours. It is evident that there is a good agreement between the observed and predicted seasonal dependence. The highest concentrations occur in winter months (January, February), mainly because of the highest rate of power production in the winter period. However, no significant seasonal variability is observed for the normalised concentrations. It means, that at a constant level of emissions one can expect almost the same pollution impact throught the whole year.

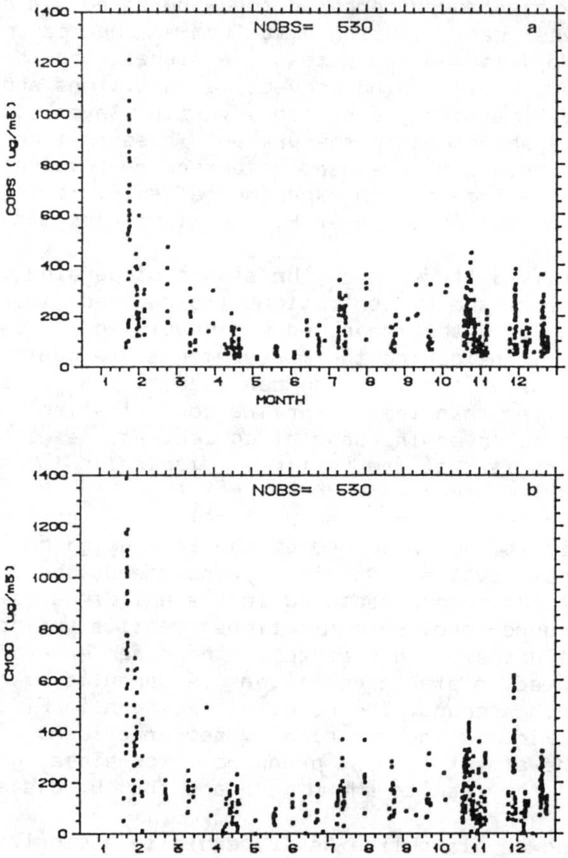

Figure 14. Seasonal dependence of concentrations. a) observed; b) computed.

The results presented here show that the OML-model is able to reproduce the observed relationships between SO_2-concentrations around the Asnæs Power Plant and the meteorological and emission conditions. After concluding so, the model was applyed for evaluation of air pollution impact of the power plant.

8. AIR QUALITY CALCULATIONS

Model calculations (Part 5) of SO_2-pollution were performed for three different scenaria:

1. Actual hourly emissions from the Asnæs Power Plant and with meteorological data for the period April 1984-March 1985.

2. Prognosticated emissions and with meteorological data for the period April 1984-March 1985.

3. Prognosticated emissions and with meteorological data for the period April 1985-March 1986.

In all the three scenaria the emissions from all the other sources than the Asnæs Power Plant were kept at the 1984/85 level. Model calculations were performed with all regional sources and subsequently with all sources minus the Asnæs Power Plant. The difference in pollution levels determined in this way was interpreted as the contribution from the Asnæs Power Plant to the SO_2-pollution in the Kalundborg area.

The air pollution measures evaluated from the model results were: 99-percentiles of hourly SO_2-concentrations on monthly basis and monthly averages.

A complete presentation of the results is given in Part 5, together with a discussion of the regulatory aspects. Here, we will only show some examples of comparison between computed and measured values of 99-percentiles.

In Fig. 15a are shown results for January 1985. The measured quantities on the 10 positions are given by stars and geographycal position is identifyed by the angle with respect to North and centrum in the Asnæs Power Plant. The computed values are from the nearest receptor circle. The distance from the power plant is given in the figure. The overall agreement between the magnitudes of the measured and the predicted 99-percentiles is very good, but the two sets of values are somwhat shifted in space. This is mainly due to the uncertainetes in predictions of plume directions. Note, that both the measurements and the model calculations show quite sharp gradients in the geographical distribution of the 99-percentiles.

An example of model results and measurements for a summer month (June 1984) is shown in Fig 15b. The agreement between the measured and the predicted 99-percentiles is very good.

The shown examples illustrate the models ability to predict the levels of pollution, but simultaneously show the difficulties of point-by-point

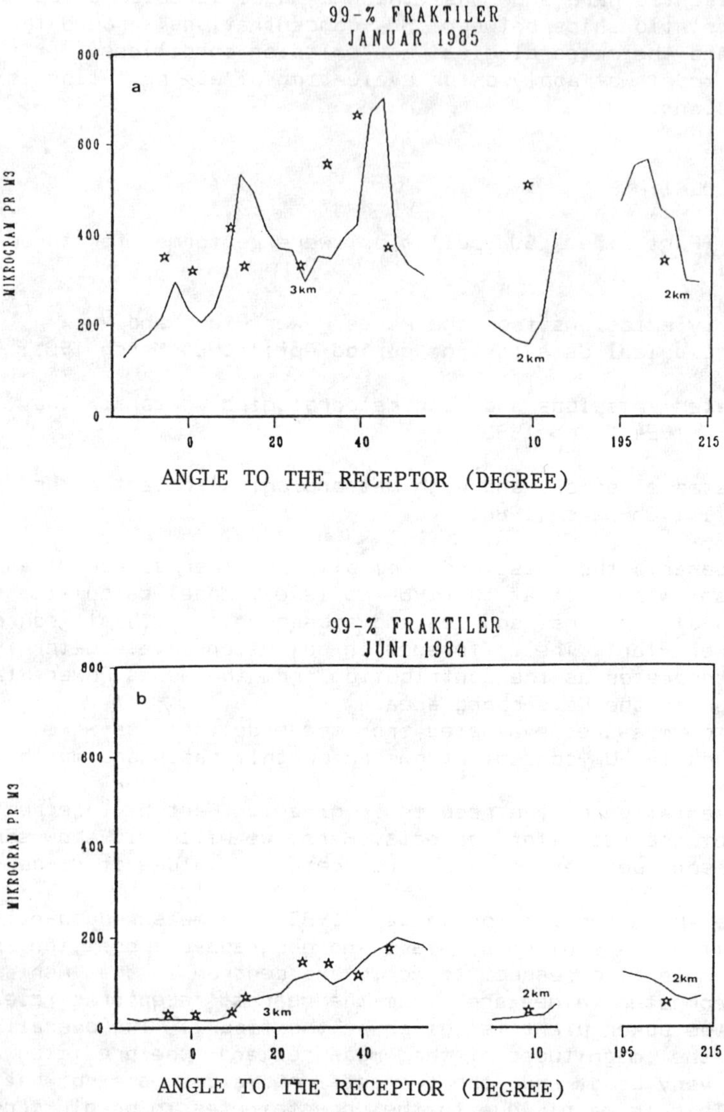

Figure 15. Computed (line) and observed (stars) 99-percentiles. a) January 1985; b) June 1984. Positions of the monitoring stations are given by the angle clockwise from North with the Power Plant as origo.

comparisons of measured and computed quantities. It should also be noted here, that the monthly averages were significantly influenced by contributions from remote sources due to long-range-transport phenomena. As an example can be mentioned that in January 1985 the contribution from remote sources was about 50% of the total measured monthly average concentration. In sommer months, the influence from the remote sources was much smaller. The highest 99-percentiles were almost totaly determined by contributions from the Asnæs Power Plant.

ACKNOWLEDGEMENTS

The Asnæs Power Plant project was jointly financed by Vestsjællands Amtskommune and Isefjordværket I/S. We appreciate the permission to use the collected data for the present publication.

REFERENCES

Berkowicz, R. and Prahm, L.P. (1982a). 'Sensible heat flux estimated from routine meteorological data by the resistance method.' J.Appl.Met., **21**, 1845-1864.

Berkowicz, R. and Prahm, L.P. (1982b). 'Evaluation of the profile method for estimation of surface fluxes of momentum and heat.' Atm.Env., **16**, 2809-2819.

Berkowicz, R., Olesen, H.R. and Torp, U. (1985). 'The Danish Gaussian air pollution model (OML): Description, test and sensitivity analysis in view of regulatory applications.' Proceedings of the 15th International Technical Meeting on Air Pollution Modelling and its Applications. St.Louis, U.S.A.

Briggs, G.R. (1984). 'Plume rise and buoyancy effects.' Atmospheric Science and Power Production. NOAA, Oak Ridge, TN.

Fox, D.G. (1984). 'Uncertainty in air quality modeling.' Bul.Am.Met.Soc., **65**, 27-36.

Gifford, F.A. (1961). 'Use of routine meteorological observations for dispersion.' Nuclear Safety 2, 47-51.

Gislason, K., Markvorsen, J.S. and Berkowicz, R. (1985). 'Test of the OML-model on data from two Danish cities.' MST LUFT-A95. Danish Air Pollution Laboratory, Risø National Laboratory, 4000 Roskilde, Denmark

Lyons, W.A. (1975). 'Turbulent diffusion and pollutant transport in shoreline environments.' Lectures on Air Pollution and Environmental Impact Analyses. American Meteorological Society, Boston, U.S.A.

Løfstrøm, P.J. (1987). User's Guide for OML-Multi. Danish Air Pollution Laboratory, Risø National Laboratory, 4000 Roskilde, Denmark

Nielson, L.B., Prahm, L.P., Berkowicz, R. and Conradsen, K. (1981). 'Net incoming radiation estimated from hourly global radiation and/or cloud observations.' J.Climat., 1, 255-272.

Olesen, H.R. (1986). User's Guide for OML-Point - an air pollution model for point sources. Danish Air Pollution Laboratory, Risø National Laboratory, 4000 Roskilde, Denmark

Olesen, H.R. and Brown, N. (1987). The OML meteorological preprocessor. Danish Air Pollution Laboratory, Risø National Laboratory, 4000 Roskilde, Denmark

Olesen, H.R., Jensen, A.B. and Brown, N. (1987). 'An operational procedure for mixing height estimation.' MST LUFT-A96. Danish Air Pollution Laboratory, Risø National Laboratory, 4000 Roskilde, Denmark

Pasquill, F. (1961). 'The estimation of the dispersion of windborne material.' Meteor. Magazine, **90**, 33-49.

Turner, D.B. (1964). 'A diffusion model for an urban area.' J.Clim.Appl.Met. **24**, 1111-1130.

Weber, A.H. and Garret, A.J. (1985). Proceedings of the DOE/AMS Air Pollution Model Evaluation Workshop. Kiawah, SC, Vol. 1. Savannah River Laboratory, Aiken, SC 29808, U.S.A.

Venkatram, A. (1979). 'The expected deviation of observed concentrations from predicted ensemble means.' Atm.Env., **13**, 1547-1549.

Vestsjællands Amtskommune and Isefjordværket I/S. (1987). Air quality study around the Asnæs Power Plant. 5 Parts and Summary Report (in Danish). Vestsjællands Amtskommune, Alléen 15, 4180 Sorø, Denmark.

BEHAVIOUR OF AIR POLLUTANTS UNDER INVERSION WEATHER CONDITIONS

G. Baumbach, K. Baumann and F. Dröscher
Institut für Verfahrenstechnik und Dampfkesselwesen der Universität Stuttgart,
Abteilung Reinhaltung der Luft
Pfaffenwaldring 23, D-7000 Stuttgart 80

ABSTRACT. In the valleys of hilly and mountainous country site relatively small emissions of pollutants can already lead to very high immission concentrations when inversion weather conditions prevail. This has been shown by the results of continuous measurements taken in two valleys in the northern Black Forest and in Schönbuch Forest.

The forests of Baden-Württemberg are beset by SO_2 on a large scale when North-East winds predominate. These long-distance transports are facilitated by atmospheric inversion layers at particular altitudes. This is examplified in a case study.

1. INTRODUCTION

The lower section of the troposphere, where emissions from vehicles, domestic heating, and industrial sources disperse in the surrounding air, is usually referred to as the mixing layer. The vertical exchange of pollutant species is restricted by barrier layers in the atmosphere. In general, these barrier layers are formed by temperature inversions in the atmosphere when the temperature increases with height. It has to be distinguished between ground inversions and elevated inversions.

Ground inversions build up as the ground surface cools due to increased radiation during clear nights. In the morning, solar irradiation leads to increasing ground temperatures resulting in the dissolution of the inversion layer proceeding from the ground. Thus, a mixing layer forms beneath the inversion layer, which acts as main barrier layer. In hilly and mountainous country site this main barrier layer is often found at altitutes well below the mountain ranges, especially in the fall and winter when solar irradiation reaches its minimum. Then, local wind systems with mountain and valley winds are virtually ineffective, resulting in a poor ventilation of the valley. In this situation the mixing volume is limited to the stagnant air mass in the valley basin by the inversion layer. Ground inversion and elevated inversion may be superimposed. In these stagnant air masses emissions from ground and medium level sources accumulate to high concentrations.

K. Grefen and J. Löbel (eds.), Environmental Meteorology, 115–124.

2. DISPERSION OF INDUSTRIAL POWER PLANT PLUME IN A BLACK FOREST VALLEY

The Forbach valley is the site of a small industrial power plant (28 MW thermal) operated continuously at constant load. The plant burns heavy oil with high sulfur content. As the SO_2 output is thus nearly constant it was possible to investigate the effects of atmospheric exchange conditions on the SO_2 dispersion as well as on the resulting SO_2 immissions in the surrounding of the factory. A valley cross-sect with four measuring stations at different altitudes was used in this study (Figure 2).

In high pressure weather situations with clear skies inversions layers build up during night hours. These prevent the vertical air circulation. In valleys this can lead to high pollutant concentrations at the adjacent slopes of the valley. Figure 1 showes all half-hour SO_2 values measured at the wooded slope of the Forbach valley in October 1985 at the medium level measuring station M2 (see figure 2). Under inversion conditions the half-hour values reach concentrations exceeding 1 mg/m^3, whereas no SO_2 is detected when the factory is shut off.

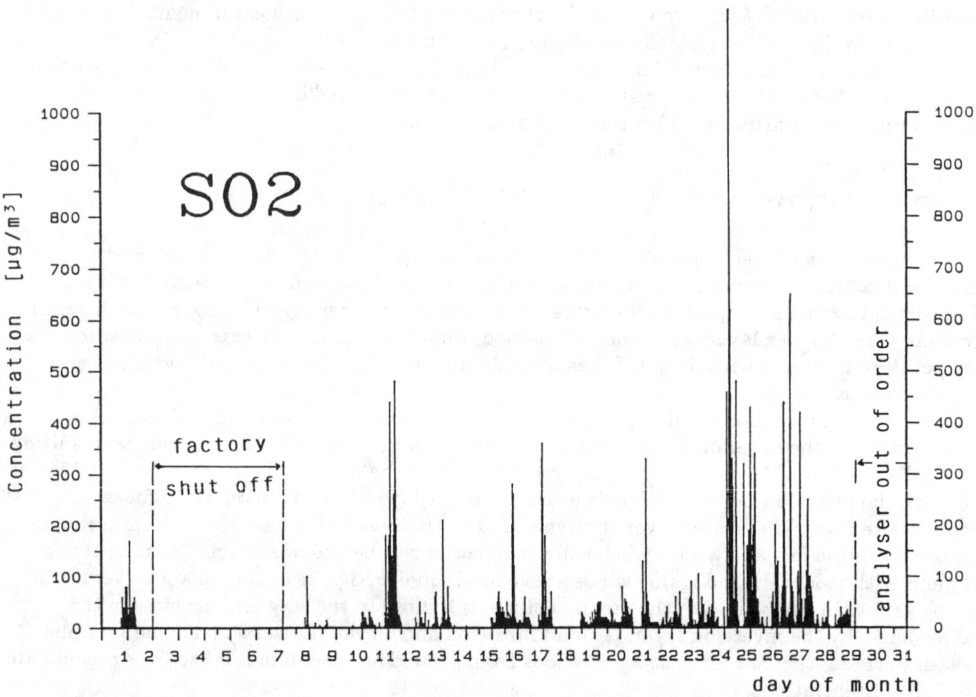

Figure 1. Half-hour SO_2 values in October 1985 at the medium level measuring station M2

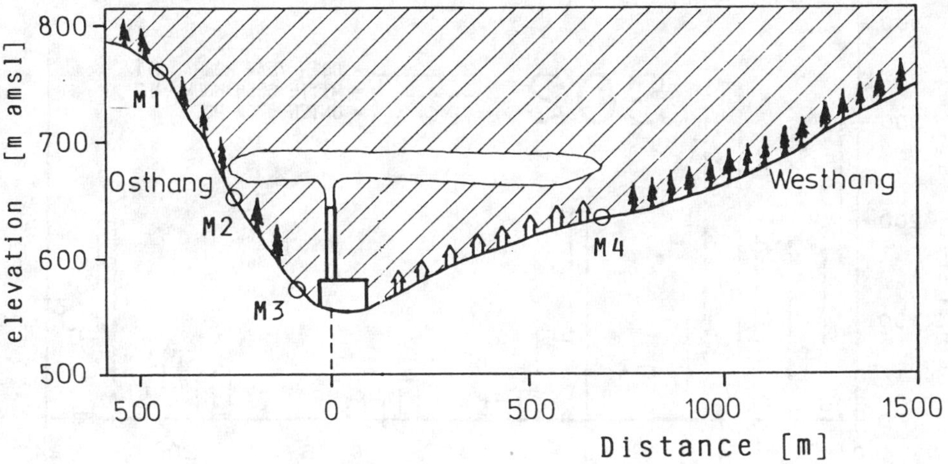

a. behaviour of the plume at night during low level inversion conditions

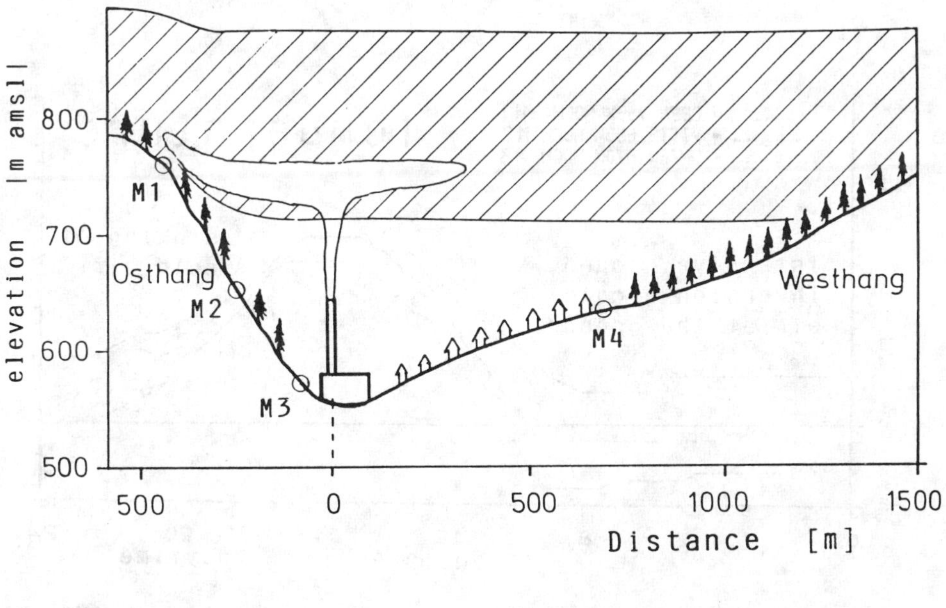

b. behaviour of the plume at the morning (8 to 10 a.m.) - the inversion base rises

Figure 2. Scheme of the behaviour of the inversion layer and the power plant plume in the Black Forest valley

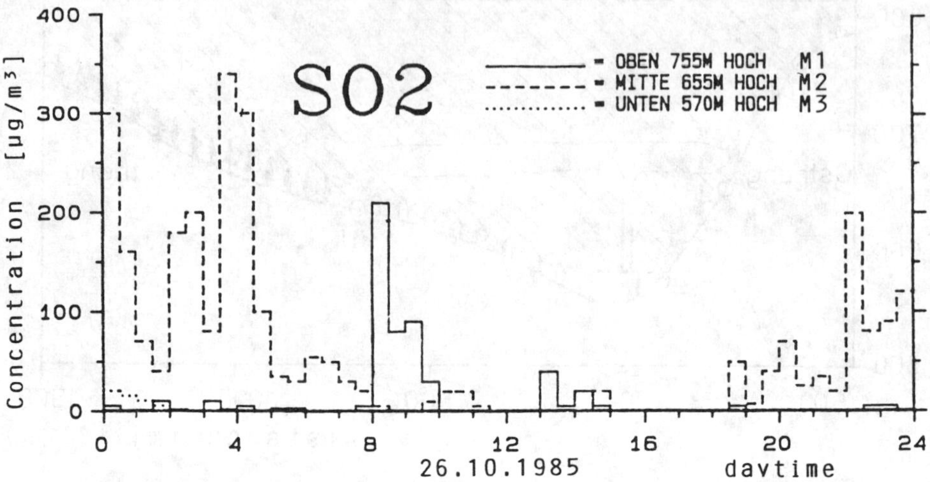

a. profiles of SO_2-concentrations

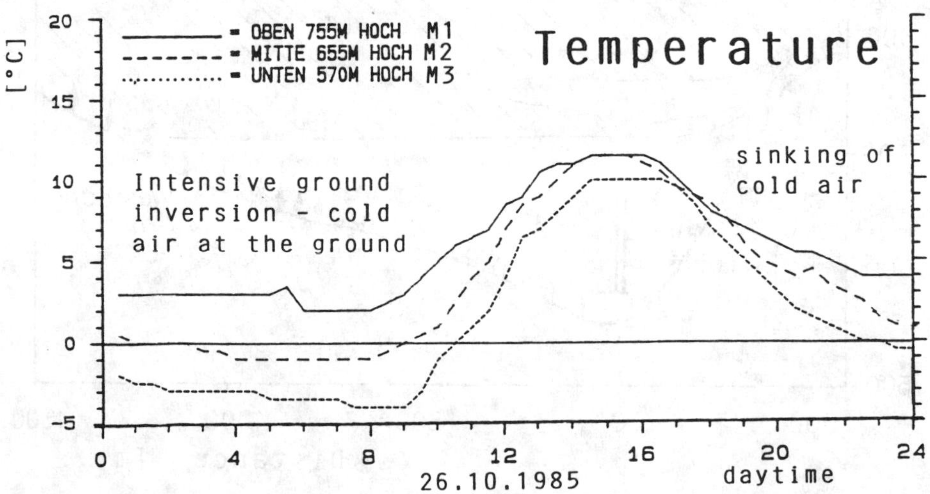

b. temperature profiles

Figure 3. SO_2 and temperature profiles at the 3 measuring stations during an examplary day

The relationships between the SO_2 concentrations and the behaviour of the inversion layers is shown schematically in figure 2 and 3.

At night (19 p.m. to 5 a.m.) during low level inversion conditions the plume of the facotry disperses at a certain hight only in horizontal directions until it reaches the valley slopes, see figure 2 a.

Then, very high SO_2-concentrations are observed at point M2 slightly above the stack exit, see figure 3a. The behaviour of the inversion layer can be determined with the temperature curves, see figure 3b.

In the morning, around 9 a.m., solar irradiation causes the inversion base to lift up from the ground since the temperature increases at the ground surface, see figure 2b. Consequently also the level at which the emissions disperse rises. The maximum SO_2-concentration is found at higher elavations at measuring point M1 on top of the hill, see figure 2b and 3a.

As the inversion builds up again in the evening the SO_2-concentrations rise again at the medium station M2, see figure 3a and b.

Concentrations of suspended particle constituents which can be attributed to the burning of heavy oil, i.e. Vanadium, Nickel, and Sulphur, at the different locations in the valley show behaviour as similar to the SO_2-concentration, see figure 4.

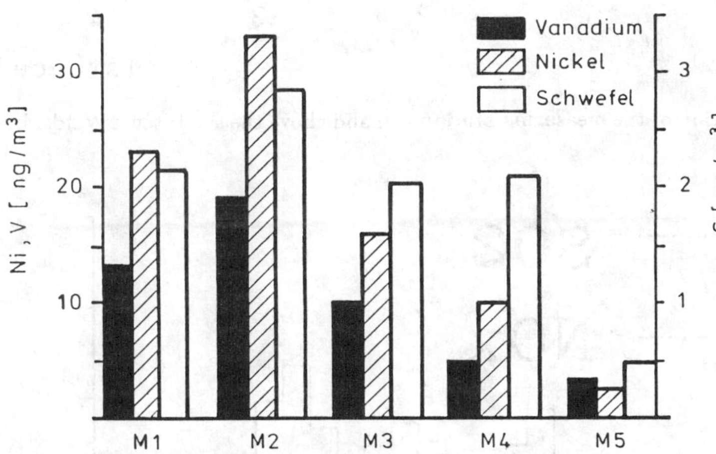

Figure 4. Concentrations of suspended particle constituents Vanadium, Nickel and Sulphur at different locations in the valley; mean values calculated for period November 15 to December 10, 1985

3. DISPERSION OF AIR POLLUTANTS EMITTED BY URBAN SOURCES IN A NARROW VALLEY

In the Aich valley in Schönbuch hills (between Stuttgart and Tübingen) pollutants are mainly emitted by cars and domestic heating. There, the vertical pollutant distribution was also studied using measuring stations at different altitudes in the valley cross-sect, see figure 5.

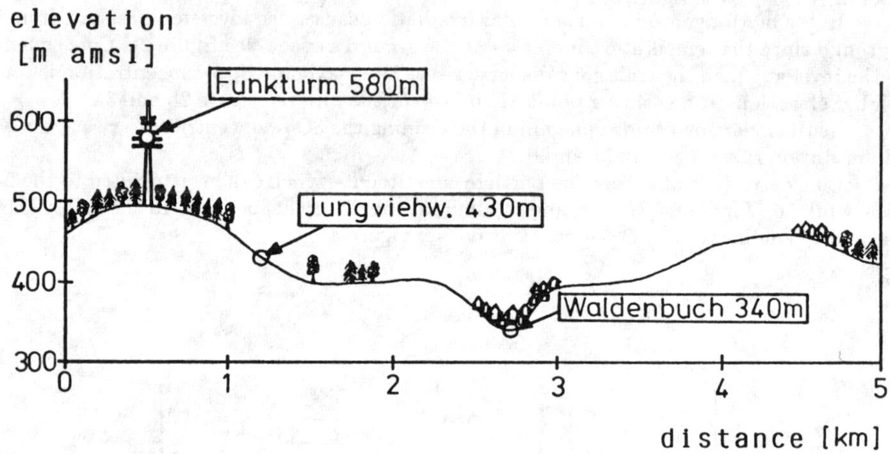

Figure 5. Location of the measuring stations in and above the small town Waldenbuch in the Aich valley

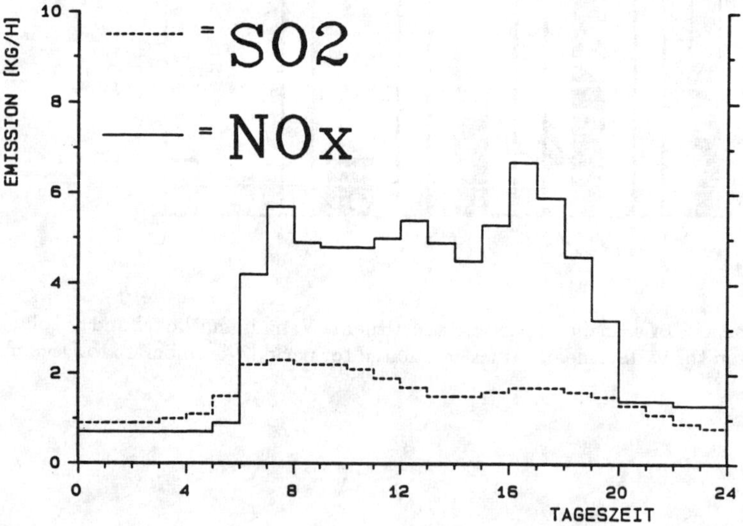

Figure 6. Average diurnal profiles of the SO_2- and NO_x-emissions in the town

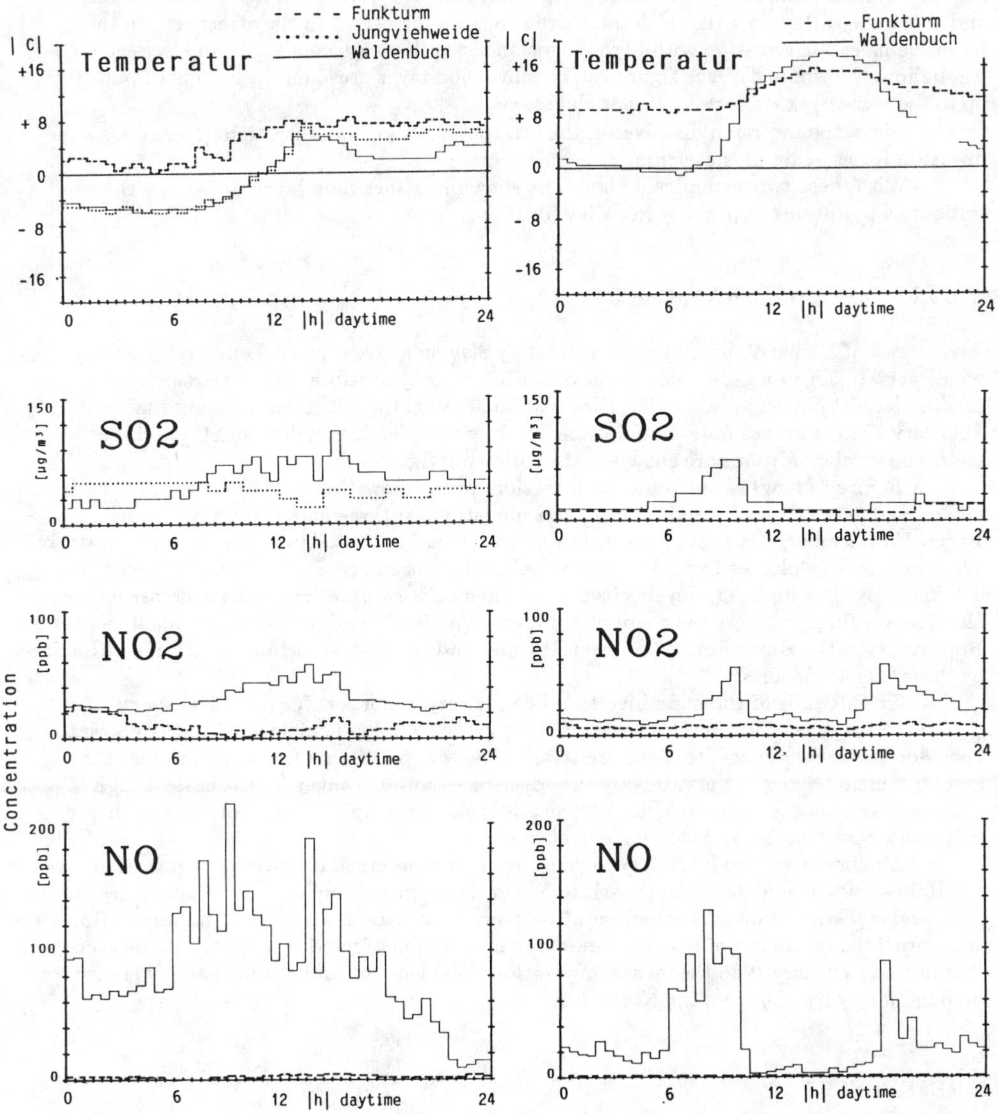

a. intesive inversion - high concentrations in valley base - very low value at higher altitudes

b. dissolving inversion in afternoon - decreasing concentrations during this time

Figure 7. Comparison of typical diurnal profiles during different inversion situations in the valley

Figure 7 presents diurnal cycles of the temperature and SO_2, NO_2 and NO concentrations in and above the valley for a day with intensive inversion dissolving in the afternoon. In the intensive inversion situation with high clouds, in the valley high concentrations persist throughout the whole day, see figure 7a. The inversion layer prevents the mixing of pollutants up to the measuring stations at higher elevations.
On clear days, the inversion dissolves in the afternoon. Then, concentrations decrease. As the inversion forms again in the evening concentrations rise.

With these two examples it should be shown how inversion layers influence the concentrations of pollutants, especially in valley locations.

4. LONG-DISTANCE TRANSPORT OF SO_2

The forests of Baden-Württemberg are beset by SO_2 on a large scale when north-easterly winds pervail. This long-distance transport of SO_2 is only possible when inversion layers predominate at particular altitudes, that is at about 1000 to 1500 m amsl. Using the month of February 1986 as an example the relationships between the distribution of SO_2 and the meteorological conditions are shown in the following figures.

Figure 8 shows the altitudes of inversion layers above Stuttgart city as well as wind vectors and the SO_2-concentrations at three measuring stations in Baden-Württemberg. The locations of these stations are indicated in figure 9. North-easterly winds advect SO_2-plumes from the liquite coal consumption centers in DDR and CSSR some 400 km away. The high SO_2-levels observed at the north-east station Aalen decrease as the cloud moves further across the country of Baden-Württemberg due to dispersion and deposition. However, even the Black Forest (between Hornisgrinde and Schauinsland) is clearly influenced by these SO_2-transports.

The altitude of the main inversion layer is very important for these transports.

In the case of low level inversions stack gases from high chimneys disperse above the inversion layer. SO_2-concentrations are small. However pollutants from sources near the ground remain locked in the valley by the adjacent mountain ranges. In the case of high altitude inversion layers (more then 2000 m amsl) there is ample mixing volume for air pollutants resulting in low SO_2-concentrations.

Maximum ground level SO_2 concentrations are observed in the case of inversion layers at the 1000 - 1500 m amsl level. In the winter when the country is snow covered, small deposition rates and moderate transport velocities allow travelling distances of several hundred kilometers. In figure 9 the frequency of SO_2 is represented as function of wind sectors. It is to be seen that during February 1986 the measuring stations Waldenbuch and Freudenstadt were influenced predominantly by SO_2 from the North-East.

5. REFERENCES

/1/ Baumbach, G.: 'Occurance of gaseous pollutants in forest stand'. In: Georgii, H.W. (ed.): 'Atmospheric pollutants in forest areas', Reidel Publ., Dordrecht, 1986, p. 177-187

/2/ Baumbach, G., K. Baumann, and F. Dröscher: 'Luftverunreinigungen in Wäldern - Ergebnisse von Schadstoffmessungen im Schönbuch und Nordschwarzwald'. IVD-Bericht 5-1987, Stuttgart University, Stuttgart 1987

/3/ Landesanstalt für Umweltschutz Baden-Württemberg, Karlsruhe: measuring data of the station Aalen, February 1986

/4/ Deutscher Wetterdienst, Weather Office Stuttgart: statistics of inversion altitudes, February 1986

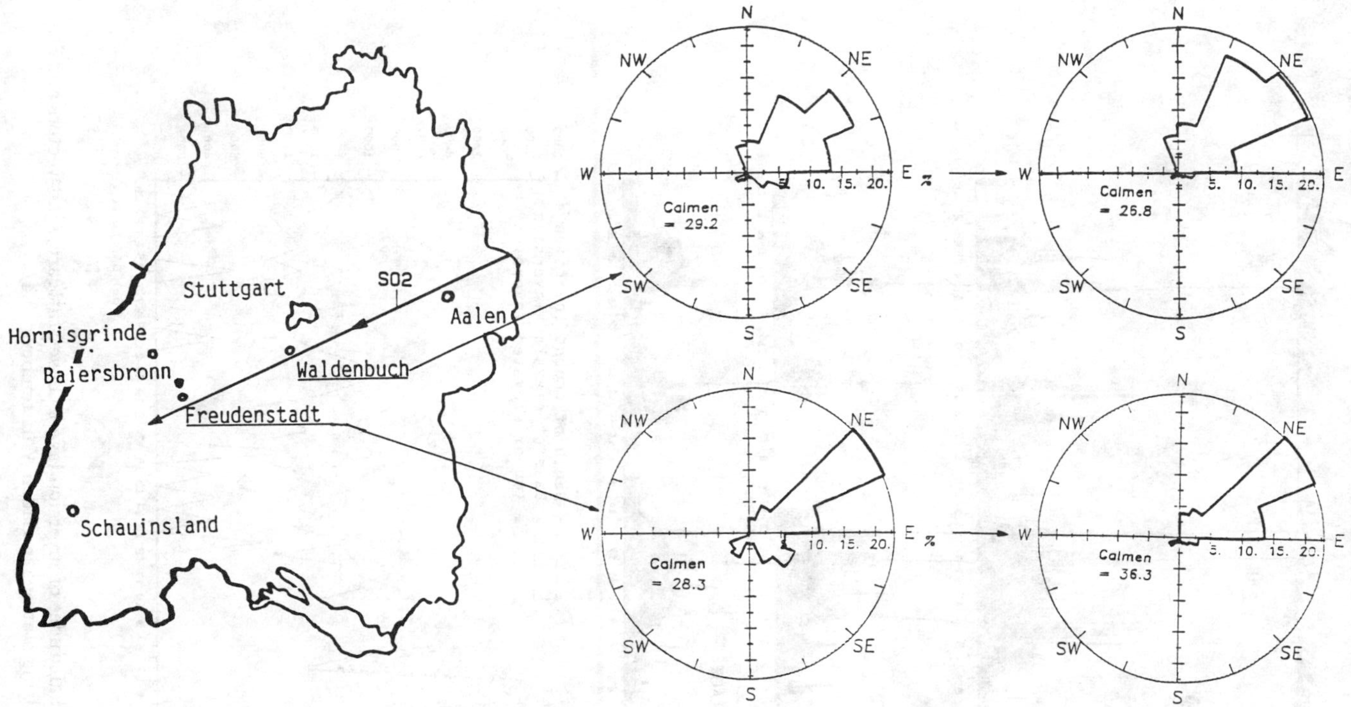

a. measuring stations in Baden-Württemberg and the parth of SO_2

b. frequency of wind direction

c. frequency of SO_2 as a function of wind sectors

Figure 9. Advection of "SO_2-clouds" from north-east to south- west across Baden-Württemberg

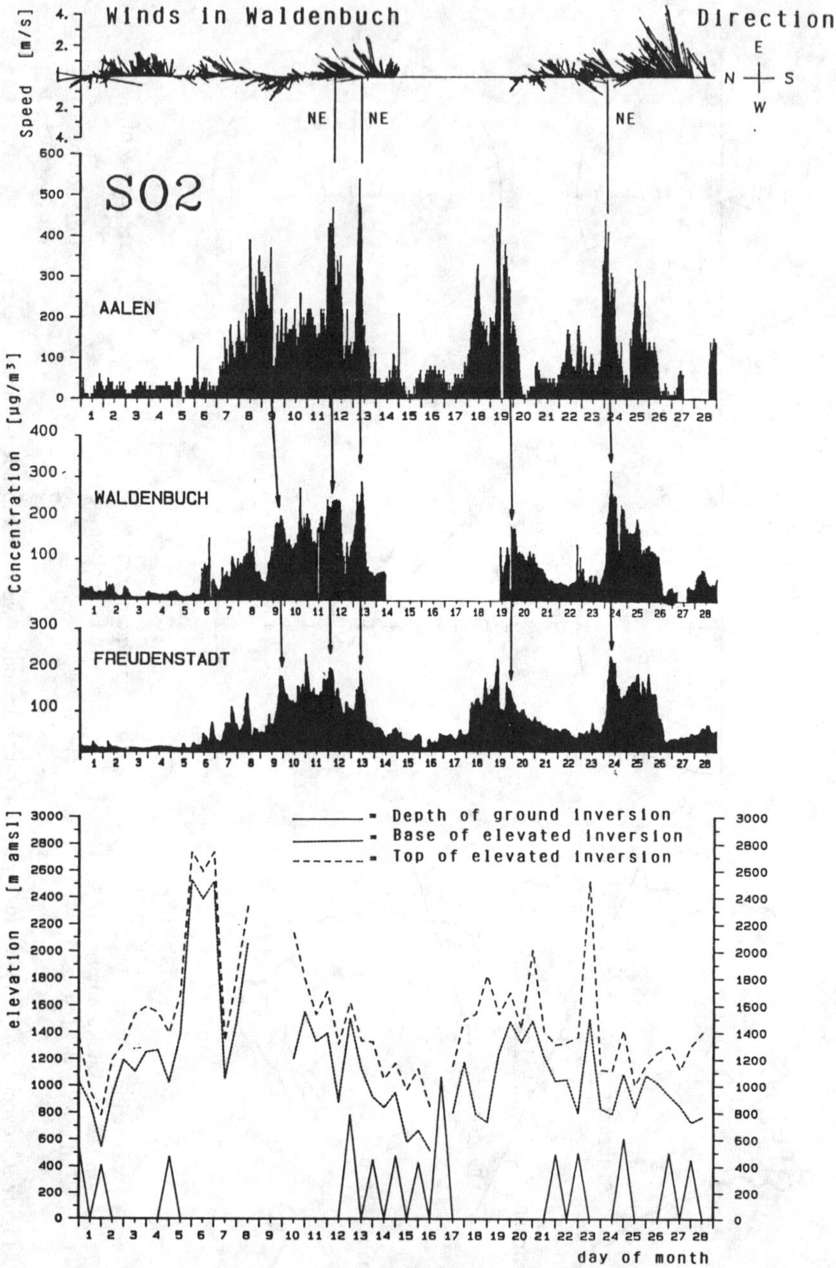

Figure 8. February 1986: altitudes of inversion layers above Stuttgart, wind-vectors and SO_2-concentrations at 3 mesuring stations in Baden-Württemberg /3,4/

AIR POLLUTANT CONCENTRATIONS AND METEOROLOGY IN TYPICAL PRE-ALPINE LANDSCAPES

T. Pelli
BMP Dr. Pelli + Co.
Postfach
8027 Zürich
Switzerland

ABSTRACT. The application of the Swiss Clean Air Act requires the use of dispersion models for air pollutants in the Prealps. This region is characterised by complex terrains with valleys, lakes and flat areas. This paper discusses the specific factors which influence the dispersion of air pollutants in the pre-alpine areas. Examples from the Lugano region (south of the Alps) and the Reppisch Valley (south of Zürich) illustrate the relevance of the asymmetry of the valley's cross-section, the effect of the combination of a valley and a lake, the importance of the accumulation process below stable layers and the occurrence of oscillating motions of air masses. The paper concludes with a brief overview of dispersion models which could be used in the pre-alpine region.

1. INTRODUCTION

On April 1, 1986, the Swiss Federal Clean Air Act became effective. This decree comprises among others emission and concentration threshold limits, regulations for licensing new industrial plants and guidelines for measures to be taken. In order to diagnose emission sources causing concentration thresholds to be exceeded at a certain site, it is indispensable to dispose of efficient models of dispersion. Major parts of populated areas in Switzerland are of pre-alpine character and show an often complex topography. Populated areas are found in valleys, at the mouth of a river and at lakesides. Local winds play a decisive role in the distribution of emitted air pollutants, especially in situations involving high concentrations.

Generally, models to be used need to comply with a high degree of versatility and accuracy, particularly since limits for concentrations have been fixed at low threshold values: For example - the annual average value for SO2 is fixed at 30 µg/m3; the 95%-percentile at 100 µg/m3.

K. Grefen and J. Löbel (eds.), Environmental Meteorology, 125–140.

2. GENERAL DEMANDS ON MODELS OF DISPERSION

Significant emissions from chimney-stacks are rare in Switzerland. Owing to the relatively high density of population, emissions at ground or roof level - resulting from traffic, heating installations and industry - become important. Often concentrations are greatest where the density of population is highest. A useful model of dispersion must therefore be capable of calculating the concentrations at the site where emissions are produced. Winds (slope-, valley-downwinds, valley-upwinds or lake-winds) will often tend to reduce the concentrations. In view of possible impairment to forests, the development of high concentrations in valleys at elevated altitudes in the Alps is of great interest. The concentrations can have their origin in the accumulation of pollutants which were brought up from the bottom of a valley with convective rising currents and which accumulated beneath a stable layer. In populated pre-alpine areas of Switzerland the influence of orographic winds is low. Table I shows the mean wind speed and the percentage of calms in Swiss towns.

Town	Height above ground m	Mean Speed m/s	Percentage of calms %	Mean Temperature ^{o}C
Lugano	30	1.5	70	11.7
Basel	16	2.1	38	9.4
Zürich	31	2.0	48	9.0
Kloten	11	2.0	49	8.5
Geneva	13	2.1	54	9.6

Tab. I: Meteorological data from Swiss towns. Calms are periods during which mean wind speeds drop below 1.5 m/sec.

A workable model should primarily be able to describe thermically generated air flow patterns. Simulations of the wind field can be based on an extensive net of meteorological readings taken at ground level - a method today practiced automatically and at relatively low cost. Conforming to the Swiss Federal Clean Air Act necessitates knowledge of mean and peak concentrations. Models used for calculation must therefore be capable of reliably estimating both.

3. DEMANDS IN PRACTICE

In practice, a model of dispersion (including the simulation of the wind field) can only be applied when its results are not too far from reality. Frequently, however, reality will divert too strongly from the ideal, limiting the practical use of pollutant dispersion models. Applying models to strongly simplified geographical formations (frequently attempted or carried out) can nevertheless help understand the physical processes involved.

In the following chapters we will discuss parameters which have a substantial influence on the formation of concentrations of pollutants and which are important in the pre-alpine area.

Geographical Orientation of Valleys

If pollutant emissions are dependent on the time of day, slope winds can have a substantial influence on the diffusion of emissions. Slope winds are in direct relation to solar radiation exposure of valley slopes, which is determinated by the geographical orientation of the valley. Therefore, the orientation of a valley has a significant effect on the diffusion of emissions, even though valley downwinds and upwinds themselves depend little on the orientation.

Cross-Section of a Valley and Asymmetries

During the rise of valley downwinds and upwinds circulative processes within the cross-section of a valley are less distinct in a narrow V-shaped valley than in a wide valley. The cross-section of a valley will also influence the limits of inversion layers, the height of anti-winds etc. Especially in the Prealps we encounter asymmetric valleys: one slope being appreciably lower or gentler than the other side. This substantially influences force and incidence of valley winds.

Incline

The incline of a valley and the height of its sides directly determine the wind force of valley winds. In valleys with gentle slopes it can happen that valley winds may change their direction several times a day in function of the solar radiation and direction and force of synoptic winds. This causes an oscillating motion of air masses and may lead to an accumulation of pollutants emitted in the valley.

Curved-Shaped Valleys

In a curviform valley not all of a slope will be either sunlit or shaded at the same time; this tends to diminish valley winds. What is more, synoptic winds will not as easily penetrate a valley of such shape.

Influence of a Lake

Local winds at a lakeside consist mainly of lake-wind, blowing from the lake to the mainland, and of land-wind, blowing in the opposite direction. If the lake lies at the foot of a mountain or at the end of a valley it will intensify circulation of air masses caused by local thermic winds.

Secluded Areas

Valley areas alone or in conjunction with a lake can have the characteristics of secluded areas (at least to some extent). During times of weak synoptic winds, accumulative processes of pollutants take place in such areas. The local winds only partly succeed in leaving the valley area and changes in wind direction bring the polluted air masses back to the site where the emissions come from.

On the basis of a few practical examples we will now illustrate the influence of the abovementioned factors on the distribution of concentrations.

4. THE AREA AROUND LUGANO AND THE CASSARATE VALLEY

4.1. Introduction

The city of Lugano is situated in a complex narrow valley on the shore of Lake Lugano (fig. 1). Since some years the concentrations of SO2, NO, NO2 and particular matters are measured continuously at a permanent measuring station in the center of the city. The measuring results, especially the SO2 concentrations, caused the administration authority of the city to carry out an extended measuring program and a detailed study concerning the source of the high local concentrations. Special importance was attached to the relationship between emissions, airflow patterns and concentrations.

From these studies (1, 2) we will draw a few examples illustrating particularities of the interrelations between emissions, concentrations, meteorology and topography in the pre-alpine area.

4.2. Influence of Valley Winds on Concentration of Pollutants

Near the mouth of the Cassarate river flowing into the Lake of Lugano an automatic meteorological measuring station is installed 30 m above the ground. Winds measured there show the typical characteristics of valley-bottom winds. Winds of more than 1.5 m/sec. blow solely in a longitudinal sense (up and down the valley). Concerned are thermic winds rarely superseded by the stronger Foehn from the north. Thermic slope and valley-winds seldom reach speeds higher than 2.5-3 m/sec. Fig. 2 shows the influence of the wind speed and temperature on the SO2 concentrations in downtown Lugano. The concentrations are distinctly lower during periods of Foehn (wind speeds greater than 3 m/sec.). Foehn is a wind coming from the Alps, bringing few pollutants. Fig. 3 shows diurnal and nocturnal variations of SO2 concentrations in downtown Lugano.

The concentrations are much smaller at night than during the day, even then when nocturnal ground inversions prevail. This can be attributed to the influence of slope downwinds and valley-downwind bringing relatively pollutant-free air into the city. Because the basic ground concentration of pollutants is of importance, differences between day and night will be less pronounced at higher temperatures.

4.3. The Influence of Topography

Fig. 4 illustrates the map of the SO2 emissions for 1984.
Fig. 5 shows the one of the SO2 concentrations (mean annual values) in Lugano for the same year. It indicates that the highest concentrations were not found downtown or in the area with the highest emission rates (fig. 1, right column). This can be attributed to the local meteorological situation: The Cassarate valley running north-south (fig. 1), the slope facing west is warmed first at dawn. The rising air masses generate slope winds directed upwards (fig. 6), air is directed from downtown Lugano towards Besso. Between 5 and 7 a.m. the heating burners of offices and apartments are put into operation, thus this air contains higher concentrations of SO2. On the other hand, the strong downward flow of air relatively poor in SO2 from the Monte Brè causes a reduction of the imissions in the center of town. When the nocturnal mountain wind (flowing north-south) has lessened near the ground, these down-flowing slope winds will hydrostatically increase the pollutant transport towards Besso.

These processes are intensified by the asymmetrical cross-section of the valley. Monte Brè's slope is steeper and higher than Besso's slope (fig. 6). Ventilation is poorer in Besso than in downtown Lugano because the valley winds downtown reach higher speeds.

4.4. Influence of Temperature Inversions and Stable Air Layers

Figure 7 shows accumulative processes at an altitude of 800 m above sea level. Nitrogen-oxides were simultaneously measured in downtown Lugano: dotted line, and at a height of 800 m at Brè: continuous line. Accumulative processes were observed in summer time during fair weather at high altitudes. Evidence of such processes was supported by simultaneous readings of hydrocarbons: benzene, toluene and m-xylene (fig. 8). The correlation between imissions of toluene and NO2 is especially high. It is possible that the high concentrations found at 800 m altitude have their origin in the city of Lugano and it is conceivable that these substances were brought up there with thermic winds and accumulated at this height due to an inversion layer. Less probable, but nevertheless possible, seems to be the assumption that the reported pollutants originate from the area of Milan and were transported in high altitude stable layers to the area of investigation.

5. THE REPPISCH VALLEY

The Reppisch Valley is situated in a pre-alpine region a few kilometers south of Zürich. The river Reppisch flows in a flat, curved valley of about 20 km length (fig. 9). The difference in altitude between the beginning and the end of the valley is 250 m. The lateral range of hills is comparatively low. Their height above the valley bottom varies between 70 and 350 m. The eastern slope is almost everywhere higher than the western one. The bottom of the valley has an average width of 500 m. Some years ago, a motorway intersection was planned roughly in the middle of the valley (as part of the southern bypass of Zürich). In order to be able to make a forecast of the concentrations to be expected from road traffic, wind readings were taken 25 m above ground level over a period of two years (3), at the site of the planned motorway intersection (fig. 9). The results of these readings illustrate that the valley is poorly ventilated. The mean wind speed is 1.4 m/sec. The narrow and curved shape of the valley slows down the penetration of synoptic winds. Due to the comparatively small differences in level and the curved shape of the valley, thermal winds can also only develop feebly. Of great significance for the forecast of the concentrations (particularly for peak values) is the fact (4) that in case of weak winds (below 1 m/sec.) frequent instabilities occur in the wind direction, which may change several times within a few hours.

Table II illustrates this process by way of hourly mean values (direction and speed) during the first fortnight of November 1984. In order to point out this fact, periods of instability have been marked with a frame. In a valley such as this, the calculation of pollutant dispersion necessitates taking the oscillating motion of the air masses and the subsequent accumulation of pollutants into account.

6. AVAILABLE MODELS OF CALCULATION

In the literature, different models have been described dealing with the simulation of wind fields and the distribution of pollutants in valleys or complex terrains. Their practical application leads only partially to the desired result: either their scope is limited to strongly simplified situations or else they can only reproduce a part of the processes.

Well represented is the family of the so-called Gaussian models (5 - 8) which are, however, not always suitable for calculations of pollutant dispersion in valleys. Their main disadvantage is that they require the wind field to be known. The wind field (or at least parts of it) is often the object of the calculations. The second important disadvantage is that they cannot reproduce an accumulation of pollutants resulting from oscillating motions of air masses.

The applications (9 - 10) of the Colorado State University Hydrostatic Mesoscale Model are interesting since it is a useful tool for describing the transition phases between valley downwind and valley upwind. We consider VALMET's model (11) to be the best currently available model. Among other things it allows reproducing fumigation processes taking place in a valley in the morning hours. Fumigation is caused by rising currents generated by solar radiation in the morning.

7. CONCLUSIONS

As shown by the examples in paragraphs 5 and 6, as well as by the comments in paragraph 7, the calculation models available at the moment are only capable of calculating the dispersion of emissions only to some extent. Models which simulate the process of dispersion during an entire daily cycle, including valley down- and upwind and transition periods, do not exist. In practice, the consideration of synoptic winds, as well as the accumulation of pollutants over a period of several days, is also of significance. As illustrated, factors such as asymmetries of the valley's cross-section, oscillation of air masses, the existence of a lake or the seclusion of the dispersion area play an important role in pre-alpine regions. Calculation models which allow taking the influence of these factors into account are practically inexistent. We are of the opinion that it should be feasible, based on a simplification of the basic equations (as e.g. in 12, 13) and extensive experimental studies, to develop models capable of calculating the expansion of pollutants in some of the abovementioned situations. The interpretation of the wind field should not be made exclusively on the basis of calculations but readings should be integrated in each case. Nowadays, the costs for such readings are reasonable.

8. REFERENCES

1. BMP Dr. Pelli + Co.: 'Le Concentrazioni di Anidride Solforosa nell' Aria della Città di Lugano', Ufficio Tecnico Comunale, Lugano, 1985.
2. BMP Dr. Pelli + Co.: 'L'Inquinamento Atmosferico a Lugano, II fase', Ufficio Tecnico Comunale, Lugano, 1986.
3. U. Steinemann: 'Windmessungen Landikon', Schindler Haerter AG, Zürich, 1986.
4. T. Pelli: 'Berechnungen für das Reppischtal mit dem geplanten Landikoner Dreieck', Schweizer Ingenieur und Architekt, 27-28/86, Zürich, 1986.
5. E.W. Burt: 'Valley Model User's Guide', EPA-450/2-77-018.
6. R.B. Harvey, Jr. and J.N. Hamawi: 'A Modification of the Gaussian Dispersion Equation to Accommodate Restricted Lateral Dispersion in Deep River Valleys', JAPCA, Vol. 36, No. 2, 1986.
7. Y. Gotaas: 'A Model of Diffusion in a Valley from a Continuous Point Source', Arch. Met. Geoph. Biokl., Ser. A, 21, 13-26, 1972.
8. H.A. Panofska and B. Prasad: 'The Effect of Meteorological Factors on Air Pollution in a Narrow Valley', Journal of Applied of Applied Meteorology, Vol. 6, June, 1967.
9. R.T. McNider and R.A. Pielke: 'Numerical Simulation of Slope and Mountain Flows', Journal of Climate and Applied Meteorology, Vol. 23, October, 1984.
10. D.C. Bader and T.B. McKee: 'Dynamical Model Simulation of the Morning Boundary Layer Development in Deep Mountain Valleys',

Journal of Climate and Applied Meteorology, Vol. 22, No. 3, 1983
11. C.D. Whiteman and K.J. Allwine: 'VALMET- A Valley Air Pollution Model', Battelle, United States Department of Energy, PNL-4728,1983.
12. N.H. Thyer: 'A Theoretical Explanation of Mountain and Valley Winds by a Numerical Method', Arch. Met. Geoph. Biokl. A15, 1966.
13. T. Yamada: 'Simulations of Nocturnal Drainage Flows by a q21 Turbulence Model', J. Atmos. Sci., 40, 1983.

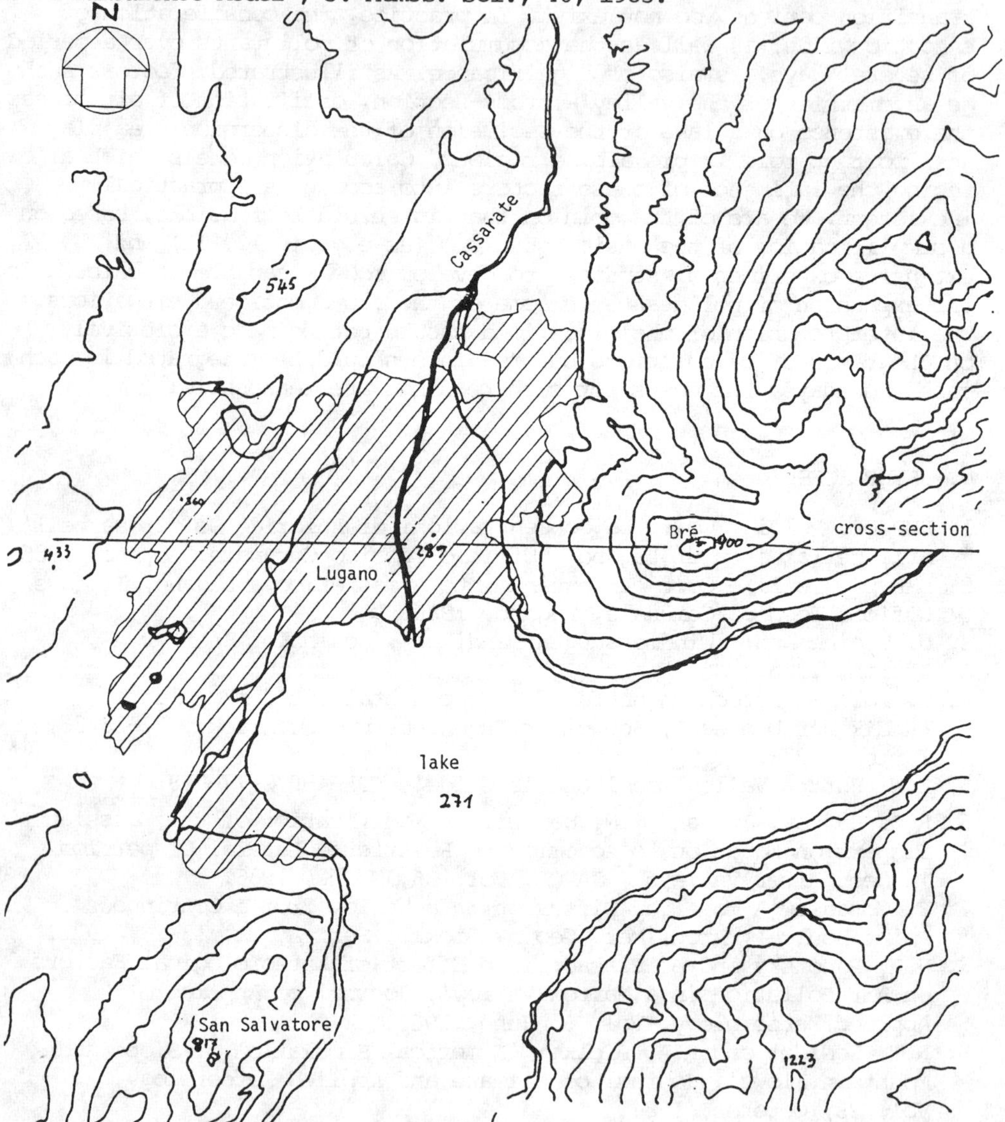

Fig. 1: Lugano and surroundings

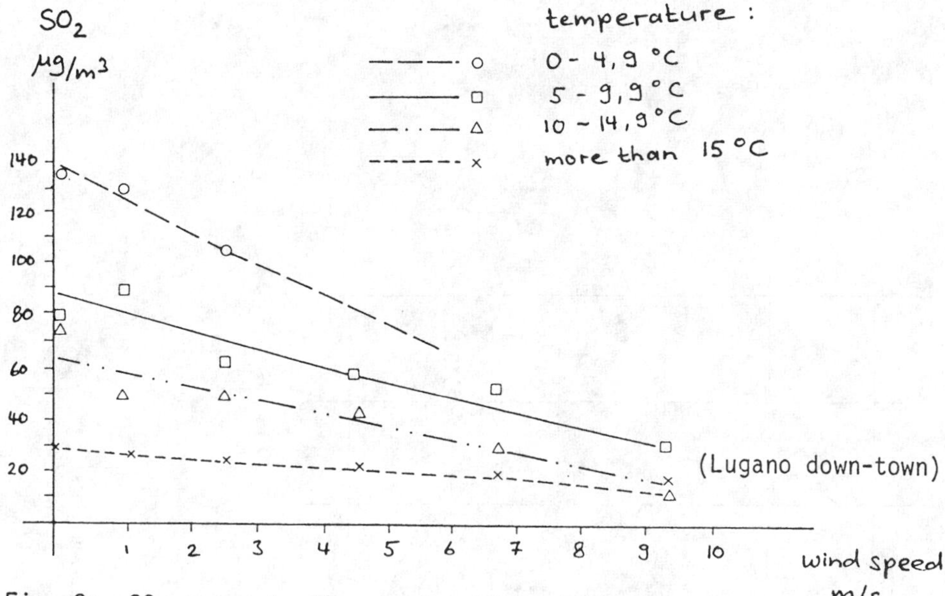

Fig. 2: SO_2-concentration in function of the wind speed

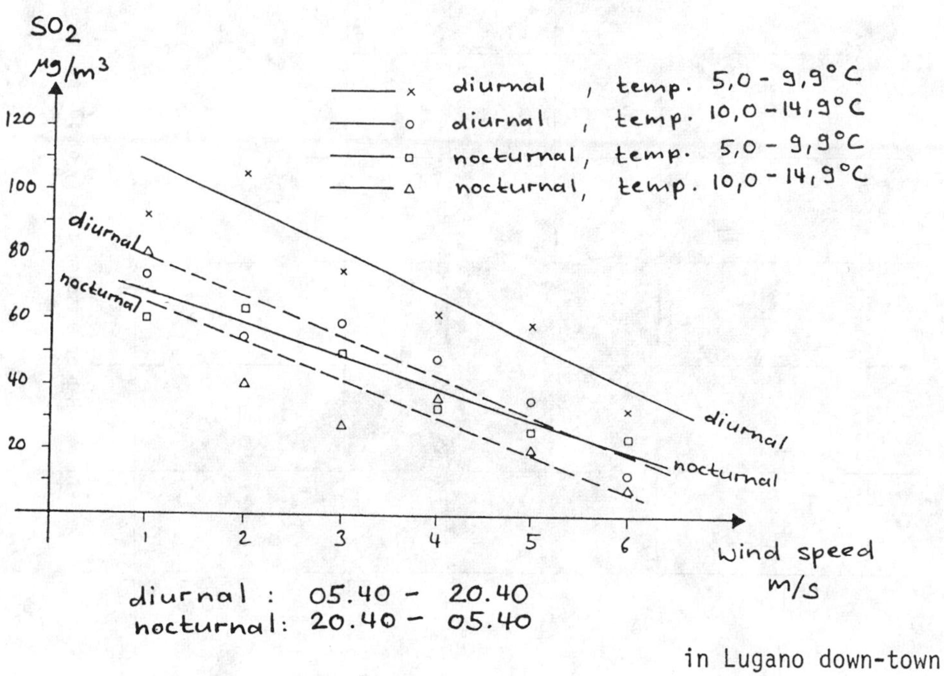

Fig. 3: Diurnal and nocturnal concentrations of SO_2

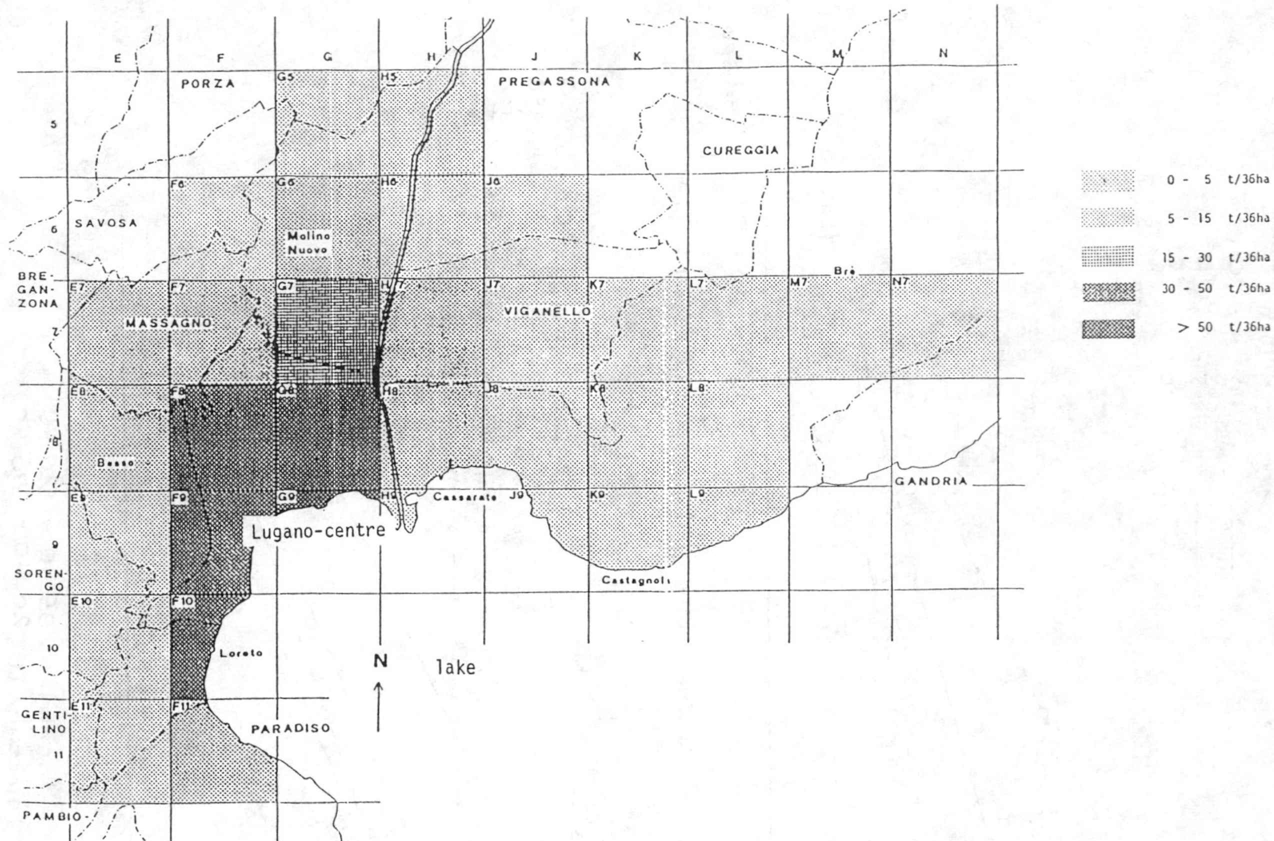

Fig. 4: Map of emission density of SO_2 at Lugano and surroundings

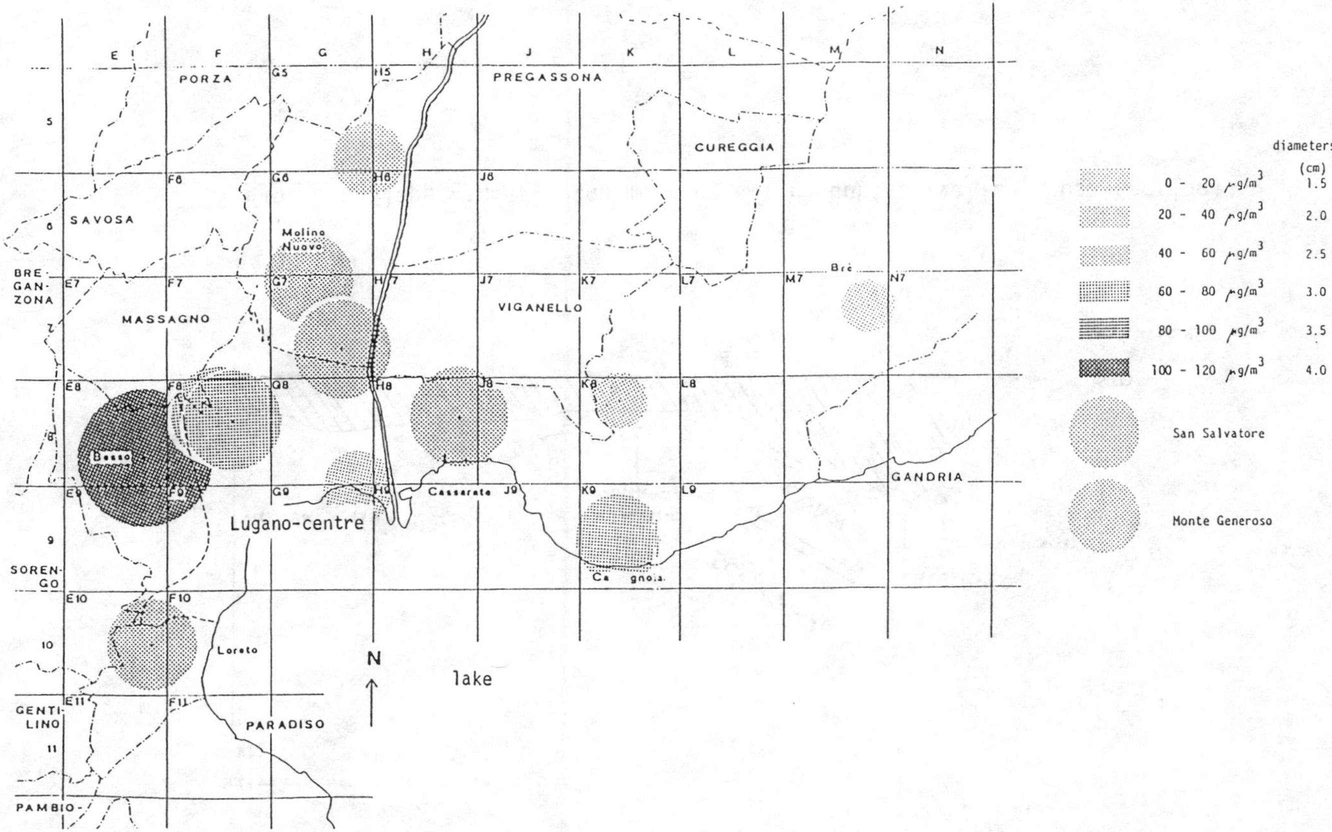

Fig. 5: Map of SO_2-concentration
(The diameters of the circles as well as the shading show the degree of the SO_2-concentration)

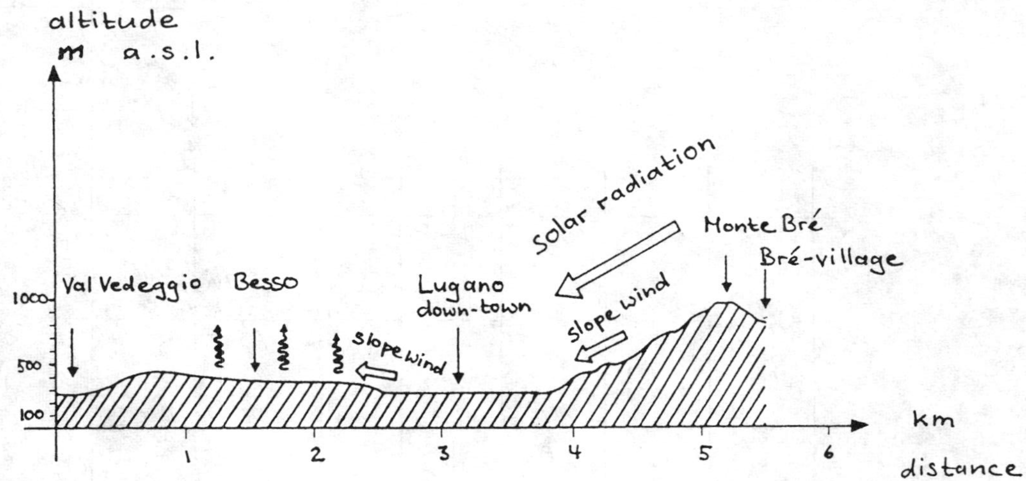

Fig. 6: Slope winds in the morning after sunrise (valley cross-section)

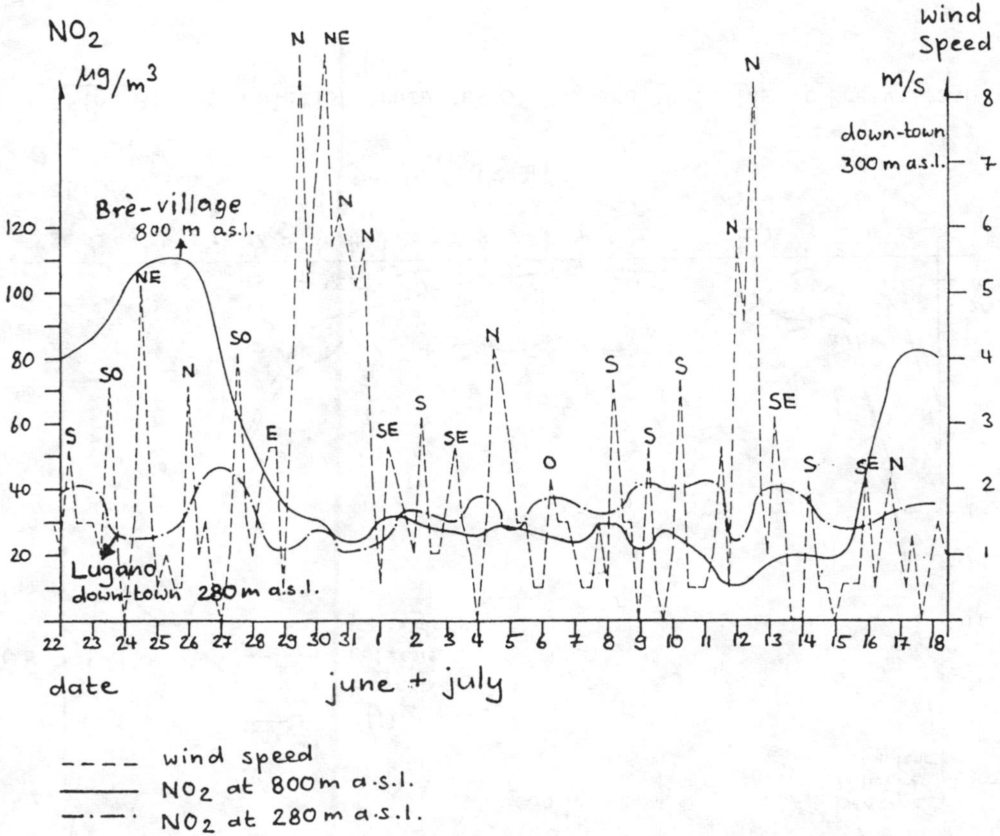

Fig. 7: Accumulation processes of NO_2 at Bré-village

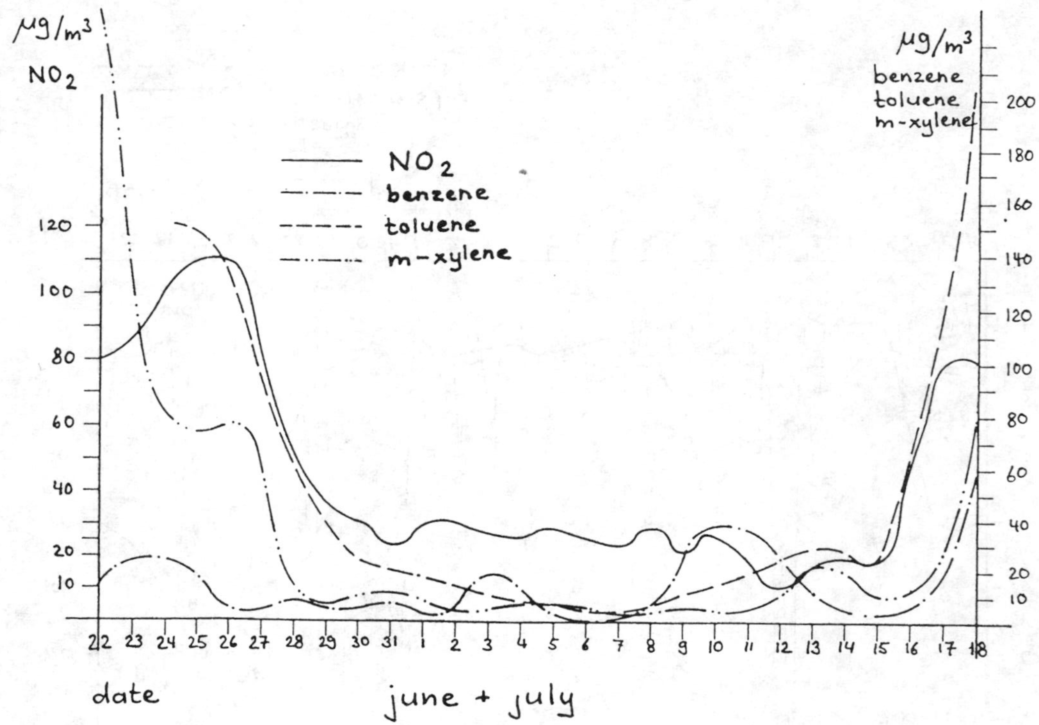

Fig. 8: Accumulation processes of NO_2 and aromatics at Bré-village

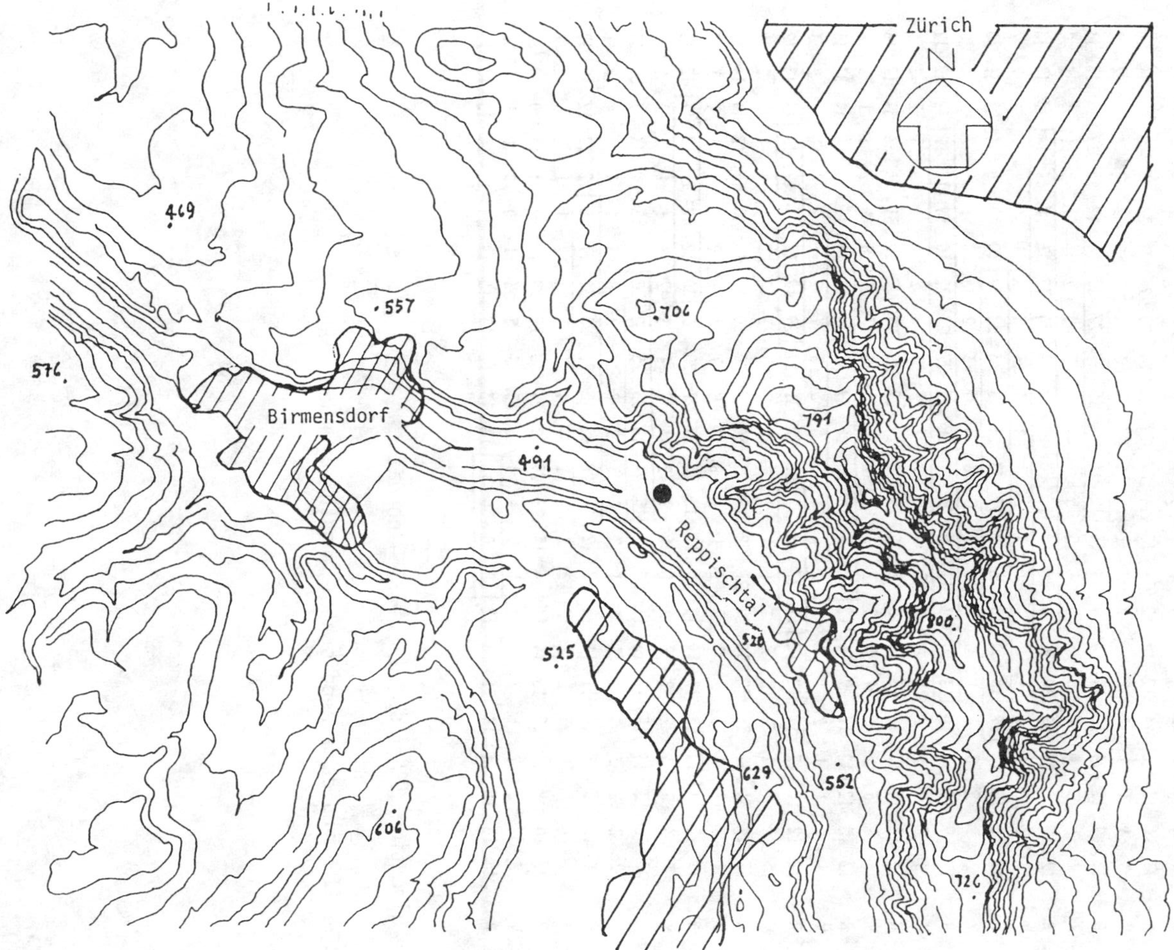

Fig. 9: The Reppischtal-valley, ● meteorological measuring station

day		0	1	2	3	4	5	6	7	8	9	10	11	12	13	14	15	16	17	18	19	20	21	22	23	24
1	R	13	13	13	13	12	12	14	-1	30	27	29	30	29	29	29	29	30	30	30	28	33	0	0	2	
	V	9	8	10	8	6	5	2	0	1	1	2	4	5	8	5	5	6	3	1	1	1	2	2	1	
2	R	13	14	13	13	14	12	10	12	14	15	14	27	29	29	29	29	28	26	33	3	18	12	11	10	
	V	3	4	4	1	4	2	1	2	2	2	1	1	3	4	3	4	2	2	2	2	1	2	2	2	
3	R	12	9	12	14	13	14	14	14	14	14	15	16	9	15	27	27	15	15	12	12	12	13	13	13	
	V	2	1	3	4	4	5	5	5	4	5	5	1	1	1	2	2	2	1	1	2	1	3	3	2	
4	R	12	12	12	29	29	30	30	30	30	9	13	30	18	12	3	30	32	6	8	14	14	13	13	13	
	V	2	2	1	4	3	4	2	1	1	3	2	4	4	4	4	3	3	1	1	2	2	4	2	2	
5	R	12	13	13	13	12	15	30	30	6	32	33	33	9	3	3	30	30	30	30	0	24	27	31	0	
	V	2	3	5	5	3	1	2	2	1	3	3	3	4	5	3	2	3	2	1	1	2	2	1	5	
6	R	12	9	12	12	10	13	10	27	12	28	30	12	18	15	27	27	27	10	9	12	12	12	12	12	
	V	3	2	2	2	2	1	2	2	2	3	5	2	2	3	3	3	2	1	1	1	3	2	2	1	
7	R	13	12	13	13	13	13	14	15	13	15	13	30	29	29	29	29	27	30	31	33	33	34	12	13	
	V	3	3	3	5	5	4	5	2	2	3	3	2	2	5	8	7	4	8	6	2	3	1	1	3	
8	R	12	13	12	12	12	9	14	12	12	14	15	30	29	29	27	30	14	14	13	14	14	14	33	13	
	V	1	1	1	1	2	1	2	1	2	2	2	5	5	5	3	7	2	2	2	2	1	1	2	3	
9	R	12	12	12	12	12	12	12	12	27	12	27	27	28	31	29	29	12	12	3	12	13	13	13	12	
	V	5	2	1	2	2	2	2	1	1	1	2	2	5	6	5	3	2	1	1	4	4	3	2	4	
10	R	13	14	31	31	31	32	31	18	27	28	29	29	30	29	30	30	15	13	13	13	12	14	14	14	
	V	4	4	2	4	3	2-	2	2	1	1	4	8	7	8	7	3	1	2	2	3	3	4	3	3	
11	R	12	13	13	13	13	13	14	13	10	13	13	15	27	27	28	2	30	12	18	27	18	17	12	18	
	V	3	3	3	3	2	2	2	2	1	1	2	2	4	5	4	3	2	1	1	1	2	2	1	1	
12	R	3	6	3	13	13	13	13	11	12	12	15	24	18	25	25	2	3	3	3	2	33	2	30	33	
	V	2	2	2	3	3	3	3	3	3	3	2	3	2	2	2	8	8	3	1	3	5	4	2	1	
13	R	9	17	5	9	10	11	11	12	13	12	15	28	15	15	21	28	31	27	8	12	12	12	8	12	
	V	1	1	1	1	3	2	2	2	2	2	2	2	3	4	3	3	2	2	1	1	2	3	2	2	
14	R	9	10	9	12	12	13	14	14	14	15	29	29	29	29	30	29	29	29	-1	-1	33	0	30	30	
	V	2	1	2	2	1	2	2	2	1	2	4	4	5	3	3	5	3	1	0	0	1	1	1	1	
15	R	18	18	15	14	14	21	18	12	18	30	30	13	13	13	12	13	15	14	14	13	14	13	13	13	

R = direction V = speed '-1' = no data

Table II Wind data Reppischtal, month november, 1984

THE DEPOSITION OF TRACE SUBSTANCES FROM THE ATMOSPHERE

J. A. Garland, K. W. Nicholson and R. G. Derwent
Environmental and Medical Sciences Division
Harwell Laboratory
Oxfordshire
OX11 ORA
England

ABSTRACT. Despite decades of study, deposition processes are incompletely understood. Wet deposition, usually studied only at ground level, results from events occurring in several kilometres of the depth of the troposphere. A full description requires three dimensional modelling of the air flows, composition and microphysics within the rain system. Future measurements by aircraft may result in much better knowledge of the vertical structure and of mechanisms. For the present, even total deposition at the surface is uncertain because of difficulties of interpolation between measurement sites.

Nucleation scavenging in clouds is probably the chief mechanism for small particle (< 3 μm) deposition, but very limited experiments show unexpectedly large rates of collection by falling drops which suggests that this process is also important. Chemical processes in cloud and rain increase incorporation of gases in precipitation by the availability of reactants.

Dry deposition of large particles is important, and is of rather uncertain, though probably minor, significance for sulphate and other small particles. For HNO_3, HCl and SO_2 dry deposition is the major removal process; it may be important in the air chemistry of other reactive gases. For HNO_3 and probably HCl removal from the air is limited only by transport to surfaces, having a time scale of only a few hours. Removal of SO_2, like O_3, is dependant on surface properties, and is an order of magnitude slower. Dry deposition in irregular terrain is uncertain; it may be augmented locally by interception of cloud and fog by hilltop vegetation.

1. INTRODUCTION

Deposition processes cleanse the atmosphere, returning material to the earth's surface. Excepting only the noble gases and the small fraction of hydrogen that escapes the earth's gravity the whole atmosphere is subject to deposition. However the term deposition is generally reserved for trace constituents of the air.

Two aspects of deposition are of vital interest in air chemistry:

K. Grefen and J. Löbel (eds.), Environmental Meteorology, 141–160.

the lifetime and hence transport range of pollutants, and the input to ecosystems i.e. the local deposition rate.

These determine the geography of the effects of air pollution and deposition occupies a central position in atmospheric chemistry. A large part of the literature in this field therefore relates more or less directly to deposition, and to review the entire field in this limited space is impracticable. Fortunately several reviews already provide a thorough description of most of the work performed in this field. Here the field is outlined only briefly and we focus on aspects of current interest and recent developments that have attracted the authors' interest.

Conventionally, wet deposition is taken to include deposition in rain and snow and dry deposition covers direct sorption of gases or particles to land surfaces (whether they are wet or dry) and also lakes and seas. Collection of fog and cloud water to vegetation (often called occult deposition) requires a treatment analogous to the dry deposition of particles and is therefore included in the dry deposition section.

2. WET DEPOSITION

2.1 General

Wet deposition has been measured for decades by deploying networks of rain collectors. However, there are a number of difficulties with this apparently straightforward method of measurement.

It is well known that dry deposition to the collector contaminates the collected precipitation (Fowler and Cape 1984). The dry deposit is not collected quantitatively (see section 3) and wet-only collectors exposed automatically, only during periods of rain, have been used in some programmes. These substantially reduce the problem, but introduce additional uncertainties as some sensors often fail to respond promptly to light rainfall, as shown by Winkler and Kroll (this volume). Their cost has prevented widespread adoption. A less tractible problem arises in interpolation between collectors. Rainfall and wet deposition vary according to the underlying surface. Near coasts, and in mountainous areas the spatial scale of variation can be too small for adequate representation in general sampling networks. Measurements of nuclear weapons fallout, (Peirson and Salmon 1959) indicate that, in a limited area, the long term average deposition rate is proportional to the annual precipitation rate, $\bar{p}$. Even so there are difficulties due to the sparsity of measurements of $\bar{p}$ in mountainous areas. Moreover, Fowler et al (1987) have shown that for some pollutants the concentration in rain, C_r, as well as $\bar{p}$ increases with height when cap clouds form over a hill top. Scavenging of the more polluted cap cloud droplets by rain falling through the cloud from above doubled the concentration in rain at the hill top (800 m) in comparison with that in the nearby valley (500 m). This finding suggests that wet deposition has often been underestimated in hilly areas, where measurements are confined to valleys.

Commonly, air and rain concentrations are measured at ground level, and results are compared or assessed using the scavenging ratio

$$W_m = \frac{\text{mass of tracer per unit mass of rain}}{\text{mass of tracer per unit mass of air}} = \frac{C_r}{C_a(0)} \quad (1)$$

Sometimes, W_v, a similar ratio based on volume concentrations is used. W_v exceeds W_m by the ratio of densities of water and air, ρ_w/ρ_a ~ 800. For modelling, the washout rate λ_w, is usually required. λ_w is the fraction of pollutant removed per second during rain. If the variation of air concentration with height is known λ_w may be deduced from W_m:

$$\lambda_w = \frac{p\, C_r\, \rho_w}{h\, C_a(0)\, \rho_a} = \frac{p\, W_m\, \rho_w}{h\, \rho_a} \quad (2)$$

$$\text{where } h = \int C_a(z)dz/C_a(0) \quad (3)$$

and p is the rate of rainfall.

In principle λ_w and W_m can be predicted from a knowledge of the processes that result in scavenging by rain and snow. In practice, field measurements often show inconsistencies with expectations, showing that understanding is unsatisfactory.

Attempts to extract λ_w from field observations suffer from inadequate knowledge of $C_a(z)$ particularly at cloud forming levels. Models often assume that pollution is confined to a boundary layer about 1 km deep, but the air at greater altitudes, in which clouds form, must have been boundary layer air. Frontal rain is common over much of Europe and forms in air that has left the boundary layer several 10^2 or 10^3 km away, but convective clouds may contain air that has ascended quite locally. Aircraft equipped for sampling air and cloud water are helping to determine the quality of air entering clouds, as well as the processes of incorporation into cloud water (ten Brink et al 1987). Because of these complications, the long history of observations of rain chemistry has not resulted in a complete evaluation of mechanisms or parameters, so there are still important uncertainties in the prediction of wet deposition.

2.2 Wet deposition of particles

Previous reviews have discussed mechanisms for incorporation of particles in cloud and rain drops (Slinn 1984, Garland 1978). Many processes may contribute for particles of various sizes, but Goldsmith et al (1963) showed that diffusio-and thermo-phoresis are unlikely to be important. For particles much smaller than 0.1 μm Brownian diffusion may result in efficient collection in cloud. Larger particles usually nucleate condensation in clouds and grow into cloud droplets. Recent aircraft observations have confirmed that essentially all the sulphate aerosol is found in cloud droplets, probably as a result of nucleation (ten Brink et al 1987). Particles of 5 μm diameter or greater are also collected efficiently by impaction with falling rain. Such large particles are expected to be removed by falling rain with a scavenging time constant of order 10^{-4} s^{-1}. Smaller particles (0.1 to 1 μm) are expected to be collected much more slowly, but the experiments of Radke et al indicated similar scavenging rates, some three orders higher than expected (see Slinn, 1984). Laboratory studies by Barlow and Latham (1983) suggest that electrostatic effects may be partly responsible.

Complete incorporation of particles into cloud water explains W_m of order ρ_a/M_w where M_w is the mass of cloud water per unit volume of air. As M_w is usually about 1-3 g m^{-3}, W_m is expected to be ~ 300 to 1000. However, this assumes that cloud water is transferred directly to rain. In temperate latitudes rain formation usually involves the distillation of water vapour from water drops to ice crystals in a process usually ascribed to Bergeron, Wegener or Findeisen. This process would much reduce the size of the cloud drops and leave the condensation nuclei unscavenged. The falling ice crystals may then collect cloud droplets and the nuclei they contain by impaction. In addition coalescence of water drops may contribute additional rain drops, and in warm clouds must be the only rainfall mechanism. The composition of the rain must depend directly on the fractions of rainwater collection due to vapour transport and impaction but these are uncertain.

Observations of various trace elements in rain and air yield values of W_m. Cawse (1974) reported a comprehensive data set. Gatz (1975) ascribed some of the variability to the effects of particle size, W_m increasing with the mass median diameter for the tracer. The range of values of W_m were consistent with the range expected on the basis of the simple arguments given above, assuming efficient incorporation into rain.

2.3 Wet deposition of gases

In cloud, there is generally ample time for the distribution of any gas between the water and gas phases to reach equilibrium. Since the liquid water content of clouds is generally of order 1 g m^{-3}, most of the gas will remain undissolved unless the solubility by volume is ~ 10^6 or more.

Below cloud the falling rain may encounter higher concentrations particularly near source areas. Peters (1983) reviewed the information required to describe uptake of gas by falling raindrops. Gas phase and liquid phase exchange processes must occur in series. For highly soluble or reactive gases the gas phase exchange, described by Frössling's equation, is limiting, while uptake of less soluble gases is controlled by diffusion and mixing in the water drop. Usually, equilibrium will be approached within the depth (~ 1 km) of the boundary layer, but uptake will be limited by solubility or the availability of reactants.

Even the most soluble inorganic vapours (HNO_3, NH_3) do not much exceed 10^3 in solubility. However, chemical reactions may result in increased uptake into the water phase. For SO_2, formation of HSO_3^- increases solubility substantially, but the reaction with O_3 will generally be limited by the resultant decrease of pH, and that with H_2O_2 by the availability of the oxidant; the latter reaction may play an important part in rain chemistry (Penkett et al 1979) but is unlikely to fix more than a small fraction of the SO_2. For this gas, the total effect of chemical processes is expected to be an increase in uptake into rain to give $W_m \sim 30$. Even so, removal in rain remains a rather inefficient process. The contamination of wet deposition by dry-deposited SO_2 is an important complication to the interpretation of field data. Fowler and Cape (1984) demonstrated that this process makes a substantial contribution to the reported sulphur deposition. This behaviour may account for some of the reported correlations between wet sulphur deposition and SO_2 concentration and leads to an overestimation of the scavenging of SO_2 by rain.

Wet deposition for some other gases is expected to be much more rapid. Dissociation of HNO_3 is practically complete, and the solubility is increased to the extent that practically all the HNO_3 in cloud will be rapidly absorbed into cloud water. Removal by falling rain will be rapid enough for a few hours of moderate rain to remove a large fraction from the boundary layer (Chang 1984, and references therein). Snow is also capable of rapid removal of HNO_3. Similar enhancement of wet deposition may apply to other strong acid gases such as HCl.

Unfortunately, the significance of contamination of rainfall collectors by dry deposition of HCl, HNO_3, NH_3 and other reactive gases has not been evaluated. This causes problems in the interpretation of observations of wet deposition and is an area requiring attention.

3. DRY DEPOSITION

3.1. General

Previous reviews (Garland 1983b, Hosker and Lindberg 1982, Chamberlain 1983, Voldner et al 1986), described the processes contributing to dry deposition in some detail. Here the principal features are summarised, and some recent concerns and developments are mentioned.

Following Chamberlain (1953), we shall describe dry deposition in terms of the deposition velocity

$$V_g = - \frac{F}{\chi(Z_r)} \qquad (4)$$

where F is the flux density of the depositing species to the surface and $\chi(Z_r)$ the concentration at some reference height, Z_r, close to the surface. Equation (4) is useful since V_g is expected to be independent of χ, in most situations of practical importance, though this has not often been demonstrated.

The appropriate deposition velocity may depend on the purpose of an investigation. For modelling the loss of material from the atmosphere, deposition to the entire surface must be included. It is then usual to define F in equation (1) as the flux to a unit area of land surface, including any vegetation, soil, buildings that may be present. However, when studying effects on vegetation or contamination of crops it may be more appropriate to specify the flux to a unit area of leaf or other surface element. In reporting, comparing and interpreting measurements it is important to specify clearly the quantities in question.

For both gases and particles, V_g is likely to depend on many variables, determined by the nature of the surface, the depositing species and also on atmospheric conditions, such as the wind speed, insolation and relative humidity. The resulting variability of V_g is sometimes overlooked, and in modelling atmospheric chemistry it is common to use a fixed value of V_g, or to accommodate only a limited number of the potential sources of variation. This may give rise to errors when changes in χ and V_g correlate, i.e. when $\overline{V_g\chi} \neq \overline{V_g}.\overline{\chi}$

Some of the variations can be predicted from a knowledge of the mechanims of deposition. The mechanistic description is facilitated by mathematically representing steps in the deposition process by separate terms. The resistance, r, equal to the inverse of deposition velocity, is often divided as follows: (Garland 1983a, Hosker and Lindberg 1982)

$$r = V_g^{-1} = r_a + r_b + r_s \qquad (5)$$

where r_a relates to turbulent transport in the air flow above and within the canopy, r_b to transport to the surface through air close to surfaces where microphysical or molecular processes, are dominant, and r_s relates to uptake at the surface. Micrometeorology provides a framework for interpreting fluxes to and from surfaces in conditions of horizontal uniformity (Thom 1975). Measurements must be made within an equilibrium layer which develops over extensive level and uniform surfaces. In the absence of large upwind disturbances to flow, a fetch of uniform surface one hundred times the measurement height is usually taken to be adequate. Thus large fields of uniform crop, forests on level ground, lakes and seas provide opportunities for studying the mechanisms of deposition from a height of a few metres.

In these circumstances r_a can be evaluated from mircrometeorological measurements (Thom 1975, Garland 1977). For gases, and often for small particles, r_b can also be estimated (Garratt and Hicks 1973, Brutsaert 1975, Slinn 1982). The surface resistance r_s usually can be evaluated only by reference to experiment. However, the evidence supports the expectation that r_s is negligible for the most reactive gases and also for particles small enough that bouncing does not occur. For some other gases the behaviour of r_s is known from measurements (see below).

This scheme allows predictions of the effects of wind speed and surface properties on the deposition of a few trace gases, and of small particles to some common, uniform, surfaces of considerable horizontal extent. Examples are SO_2 , O_3 , HNO_3 depositing to large fields of grass, cereal crops, and extensive forests and water surfaces. However, there is no formalism that permits a prediction or analysis of deposition to irregular features of the landscape, such as isolated trees, houses and hedges, or mountainous areas. Such features also disturb the flow in the boundary layer depending on their size, and there is no ready means of evaluating the consequences of that disturbance for deposition. A similar uncertainty applies at the edge of forest stands, where increased deposition may have adverse effects. These problems represent a major limitation on our knowledge of dry deposition: there have been a few relevant measurements, (eg of deposition to a hedge by Owers and Powell 1974) and in principle, the eddy correlation method using aircraft can measure deposition for the countryside as a whole (Hicks et al 1986). However no attempt has been made at a systematic investigation of the problem.

There is also no satisfactory analysis of deposition to urban areas. Some assessment of uptake by building surfaces in laboratory scale experiments has been made (Judeikis and Stewart 1976, Chamberlain et al 1984), but describing deposition to surfaces encompassing a wide range of textures, scales and geometries, including both buildings and heterogeneous vegetation, has not yet been attempted.

Measurements of dry deposition have been carried out by diverse methods. Analysis of the receiving surface gives a direct estimate of the amount deposited but is subject to difficulties when substances relevant to acid deposition are studied. Lindberg and Lovett (1983) estimated deposition of $SO_4^=$, NO_3^-, Ca^{++} and K^+ by washing leaves sampled from a canopy, and Hofken et al (1983) deduced dry deposition of $SO_4^=$, NO_3^-, NH_4^+ and Cl^- and a few metal ions by assessing the mass balance for a forest canopy. Despite the progress made by these authors, difficulties in interpretation remain due to absorption or leaching via the leaf cuticle, and the relative contributions of gases and particles for $SO_4^=$, NO_3^- and Cl^-. These difficulties can be overcome in some circumstances using a distinctive isotope, either stable, eg ^{34}S in SO_2 (Belot et al 1974), ^{15}N in NH_3 (Porter et al 1972), or radioactive ^{35}S (Garland and Branson 1977). In an alternative approach a wide variety of surrogate collectors have been adopted by various workers including filter papers (Cawse 1974), petri dishes (Lindberg and Lovett 1983), teflon plates (Davidson et al, 1985) and moss bags (Little and Wiffen 1977). Dasch (1983) found differences of about a factor of ten in quantities collected by various collectors. The collectors generally do not resemble the target surfaces either chemically or geometrically, and the relationship between the quantities collected by the two surfaces is unknown.

In a few measurements, it has been possible to expose natural surfaces, previously uncontaminated, and measure uptake directly (Little and Wiffen 1978, Cambray et al 1987). Where possible this may provide one of the most reliable methods of estimating deposition.

Micrometeorological techniques for dry deposition measurement include the gradient and eddy correlation methods. The former requires measurement of the concentration at two or more heights above an extensive, uniform surface; absorption at the surface reduces the concentration near the surface causing an increase of concentration with height. Methods for interpreting the concentration difference are well established (eg. Garland 1977), but depend on empirical flux-gradient relationships. Measurement of concentration differences of no more than a few per cent is necessary, and is often difficult to achieve. Eddy correlation methods involve measuring concentration fluctuations, c', on a time scale of about 1 second or less. Variations in the vertical component of the wind w', are measured at the same point, and the mean value of the product w'c' yields the flux directly. No empirical relationships are used in the calculation. The requirement for a rapid response sensor can be met for a few species. Both methods require measurements over extensive uniform surfaces in periods of near constant concentration and meteorological conditions. Other micrometeorological techniques (Hicks et al 1986) may enable some relaxation in instrumental requirements. The eddy accumulation technique, replaces the need in eddy correlation for fast response by the requirement to measure mean concentration with high precision. The method requires some further development (Hicks et al 1986). Another possible method depends on the proportionality between the deposition flux and the variance in concentration in a suitable region of the frequency spectrum. This method has certain advantages regarding instrumentation but has yet to be established.

All micrometeorological techniques measure the flux at some height above the surface, and assume that this is equal to the surface deposition flux. Large errors may arise as a result of divergence of the vertical flux unless the site and circumstances of measurement fulfil stringent requirements of fetch and horizontal uniformity and stationarity of concentration (Businger, 1986). These errors may be particularly significant when small deposition fluxes are being measured. Micrometeorological methods underestimate fluxes of large particles since they do not detect the sedimentation flux.

There has been considerable discussion in North America (Bromberg et al 1985) of the desirability of monitoring acid dry deposition. This implies deploying a network of collectors or instruments that continuously measure dry deposition in much the same way that rain collectors record wet deposition. At present there is no feasible method giving reliable results. The differences between various surrogate surfaces show the sensitivity of results to collector design, and it seems improbable that any collector could ever simulate a natural surface for both gases and particles. Despite the enormous amount of investigation required, it appears much more practicable to establish an understanding of the factors that determine the deposition velocity in a region and use this understanding in models, or with measured air concentrations, to calculate dry deposition.

3.2 Dry Deposition: Particles

Particle deposition from the atmosphere to vegetation canopies and other environmental surfaces is influenced by the complexity of the surface geometry and the variety of mechanisms for transport through the surface layer. For vegetation, surface elements range in size from leaf hairs to tree trunks, and an even greater range of length scales occurs in urban areas where buildings and vegetation are interspersed. Furthermore at least four mechanisms are commonly invoked to explain the variation of deposition with particle size, wind speed and surface structure (Garland 1983a, Slinn 1982).

In most atmospheric conditions, all particles smaller than about 30 μm radius are able to follow the turbulent flow in the boundary layer and are transported to the vicinity of the surface. Although the inertia of larger particles prevents their following the smallest eddies closely, these particles have substantial sedimentation velocities and are deposited chiefly by sedimentation. They usually consist of water droplets and represent the boundary between wet and dry deposition.

Very close to smooth surfaces the air is generally slow moving. Flow visualisation studies favour a picture in which turbulent sweeps bring air from the boundary layer into close contact with the surface, replacing the air previously present in a random fashion. While such processes may bring particles very near the surface some further mechanism is required to enable them to contact the surface. Particles larger than a few microns in diameter are subject to significant sedimentation and may also be carried to the surface by inertia. The numerous Aitken nuclei, smaller than about 0.1 μm are transported by Brownian motion to the surface. In an intermediate size region none of these processes operate significantly, but particles carried by the airflow close enough to graze the surface are retained in a mechanism often called interception. The precise size ranges of influence of these processes vary with wind speed and surface geometry. Dry particles larger than about 5 μm diameter may sometimes bounce at dry surfaces, but in other circumstances intermolecular forces are generally sufficient to retain the particle.

Laboratory and theoretical studies (Chamberlain 1967, Chamberlain et al 1984, Slinn 1982, Brutsaert 1975, Davidson and Friedlander 1978) have developed a consistent description of deposition. For simple surfaces it has been possible to compare in detail the results of theoretical treatments and of experiment, and although the main features are well described, there are still some areas of difficulty. Underwood (1987) modelled deposition to the simple canopy studied experimentally by Chamberlain (1967). He noted some discrepancies regarding distribution of the deposit within the canopy. However the broad features of the variation of deposition rate with size observed in this and in other wind tunnel experiments are generally well explained.

Field measurements of deposition to vegetation canopies show very wide variability (eg Sehmel 1980), and some series of measurements show unexpectably large deposition rates for 0.1 to 1 μm particles. Some of the variability can be ascribed to variation in particle size, surface structure and wind speed. Davidson et al (1982) showed that variations by about a factor of 10 in deposition velocity of lead and sulphate are to be expected between canopies of grass of similar height because of interspecies differences in morphology. A further contribution to variability due to variations in size distribution increases the total range to 20 or 30 fold. However, further contributions to variability are expected to result from random errors of measurement, and from systematic errors resulting from imperfect sites, instrumentation and technique. Many man made tracers, such as sulphate, lead and some other metallic elements are predominantly submicron in size. Observations of the amounts of these tracers that deposit on natural and artificial surfaces often result in deposition velocities several times higher than expected. Garland (1983a) assessed results of other authors (Cawse 1974, Little and Wiffen 1977, 1978) for deposition of trace elements on filter paper, and of lead on grass and found that the deposition velocity was up to an order of magnitude larger than expected. Höfken et al (1983) also found high deposition velocities to leaves. On the other hand the Chernobyl accident provided a rare opportunity to measure deposition of a tracer to natural surfaces with no possible interferences. V_g for grass in S. England was about 0.5 mm s^{-1}, consistent with the expected value (Cambray et al 1987).

Micrometeorological methods have also been used for the measurement of the dry deposition of small particles. Results have sometimes been consistent with expectation. For example the concentration gradient method indicated low deposition velocities for Aitken nuclei of 0.05 to 0.2 μm diameter (Garland and Cox 1982), and for sulphate particles (Nicholson and Davies 1987). However other measurements have yielded values substantially larger than expected. It seems probable that some apparent disagreements are due to unexpectedly wide size distributions and others to differences between the flux at the effective measurement height and that at the surface. For small particles, with a small expected deposition velocity, the requirements of fetch and of horizontal uniformity and invariance with time of the concentration are particularly strict.

Sievering (1983) and Wesely et al (1985) have commented that the deposition velocity of small particles to grass and short crops is substantially higher in daytime conditions, when the heating of the surface by the sun results in instability in the air near the surface, than in thermally stable conditions at night. The difference exceeds the change in r_a (equation 5) attributable to the variation in stability and the consequent change in windspeed, and implies that r_b can be substantially reduced in unstable conditions. No mechanism is known to explain this change (Wesely et al 1985). Some effect of windspeed on r_b is to be expected (eg. see Garland 1977) but the observed change far exceeds this. Slinn (1983) discussed the effect of stability at smooth surfaces, rejecting the generally accepted view that the effect of stability on exchange processes vanishes close to the surface. It seems improbable that the change in structure of the flow within ~ 1 cm of the smooth surface postulated by Slinn is relevant for deposition to crops, and the apparent dependence on stability is yet to be explained. It is not clear whether an effect of instability can explain the unexpectedly large values of V_g.

These uncertainties in the deposition of particles in this limited but important size range will continue unless these discrepancies can be explained. Unfortunately, there have been few measurements in which the size distribution of the particles and the morphology of the surface were well documented, so that models of the deposition process could be validated, and the conditions of measurements are specified thoroughly so that possible errors of measurement could be estimated. Particulate species of environmental significance are generally polydisperse. Rapid deposition of the largest particles is expected to occur close to the source and only the slowly depositing small particles may travel significant distances. The rate of growth of the remaining particles by coagulation or condensation processes may then have an important effect on the rate of deposition. Although models for the development of the size distribution are available, there have been no attempts, to the authors' knowledge, to use them to describe the expected variation of deposition rate with distance from the source.

3.3 Dry Deposition: Gases

Sulphur dioxide and ozone have received considerable attention, and the general features of their deposition have been appreciated for some time (eg. Garland 1977, Wesely 1983). There is little to add to the summary presented in a previous review (Garland 1983b).

Both gases are taken up by vegetation. Diffusion through stomata appears to be important, so that the surface resistance, r_s, varies according to the physiological activity of the vegetation and is often similar for the two gases. There are additional sinks in vegetation canopies; for SO_2, reactions in dew forming on the leaf surfaces appear important. Baldocchi et al (1987) have shown how a model may be constructed using meteorological variables to predict the variation of deposition velocity of such gases. Typical deposition velocities to plant canopies are about 8 mm s^{-1} in daytime, but much smaller at night. Diffusion through the crop to the soil surface, and deposition there, is also important, but here the gases differ. SO_2 is more readily absorbed by damp and more alkaline soils. Ozone reacts more readily at a dry soil surface. The behaviour at the surface of the sea shows a corresponding difference: while SO_2 shows little r_s, O_3 experiences a large resistance and V_g for this gas is an order of magnitude smaller.

Studies of these gases have highlighted the diurnal variation of deposition rate. Over land, both V_g and χ are usually lower at night as a consequence of atmospheric stability and suppressed mixing near the ground, decreased wind speed, and closed stomata of vegetation. Similar changes probably apply to several other depositing gases.

The deposition of nitrogen oxides is more complex and is not yet well resolved. NO_2 shows stomatal control in field experiments at higher concentrations as demonstrated by Grennfelt et al (1983) and Johansson (1987). NO deposition occurs much more slowly. The latter reference also shows that there is an additional internal resistance to NO_2 and that uptake ceases or reverses when concentrations fall to about 2 ppb. The behaviour of NO_x (= NO + NO_2) is further complicated by the possibility of emission from soil and vegetation. Galbally and Roy (1978) observed emission of NO from soil. Parrish et al (1987) measured upward fluxes as large as 21 ng m^{-2} s^{-1} of N as NO from grassland in Colorado stimulated by rain. Höfken et al (1986) and Dollard et al (1987a) measured both NO and NO_2 fluxes and apparently saw both emission and uptake of NO_x on different occassions at the same sites.

One is led to speculate that there is a compensation point, ie. an atmospheric concentration below which emission occurs and above which there is deposition for NO if not for NO_2 as well. The compensation point would depend on the physical, chemical and microbial status of the soil. In any case Johansson (1987) concluded that, at the low concentrations that occur in rural air in Sweden, the deposition velocity is practically zero, while at higher concentrations in England and America Delany and Davies (1983) and Wesely et al (1982) observed substantial uptake at rates equivalent to deposition velocities of ~ 5 mm s^{-1} for NO_x. However Dollard et al (1987a) saw little net flux although the concentrations they experienced were greater than Johanssons's. The outcome of recent studies is less ambivalent for the more reactive HNO_3. Huebert (1983), Dollard et al (1987b) and Höfken et al (1986) measured fluxes by the gradient method. No emission was observed, and deposition velocities were about 0.2 m s^{-1} to wheat and higher than for any gas discussed above. (Dollard et al noted similar results for HCl). The data from all investigations mentioned are consistent with negligible surface resistance. Meixner et al (1987) observed deposition velocities of order 0.2 m s^{-1} to forest at rather modest windspeeds; for higher winds even higher V_g to forest should be anticipated.

Ammonia fluxes have been observed by Denmead et al (1974) Dollard et al (1987a) and Duyzer et al (1987). Persistent emission was observed from grazed pasture and in ripe wheat crops. If maintained throughout the year the emissions would represent a loss of order 100 kg ha^{-1}, a substantial fraction of the nitrogen budget. By contrast Hove et al (1987) have shown uptake by leaves with evidence of stomatal control, and Duyzer et al found deposition at environmental concentrations to heather moorland and a pine plantation. The mean V_g was 1.9 cm s^{-1}. Surface wetness reduced r_s to zero. Perhaps the contrasting results are explained by differences in soil conditions, and Duyzer's results may be indicative of ammonia inputs of order 20 kg N $ha^{-1}a^{-1}$ by deposition to areas with acid soils in Central Europe.

A picture of some complexity and much uncertainty is emerging for the nitrogen compounds. Most measurements relate to agricultural land and there HNO_3 may make the only significant input. For poor land both NH_3 and HNO_3 can make important contributions to deposition. A comparable input for NO_x may occur if the soil conditions suppress emission and gas concentrations are of order 10 ppb. For forest, HNO_3 deposition is likely to be increased a few times, and this may be the dominant flux. NH_3 is converted to nitrate in soil and so is assumed to have an equivalent acidifying effect.

Further interest may focus on oxidising species. Recently Dollard et al (1987a) have indicated that H_2O_2 deposits with a V_g of about 1 cm s^{-1}. More reactive species, such as free radicals may perhaps be removed with V_g similar to that of HNO_3, ie about 20 cm s^{-1}. In the daytime boundary layer (h~2 km) this is equivalent to a lifetime of order 3 hours, and deposition has the potential to influence the concentration in the boundary layer of any reactive species expected to have a chemical lifetime of this order or longer. There may be a need to investigate the dry deposition of a number of reactive species that fit this criterion.

NH_3 and HNO_3 or HCl undergo reversible reactions in the air, as do also NO and O_3. Removal of one of each pair of reactants by deposition may disturb an equilibrium in the surface layers. Where measurements are made by a micrometeorological technique, reactions occurring in the air below the level of measurement may result in a difference between the dry deposition flux and that observed. Such difficulties can be overcome if composite species are measured instead of single gases (van Aalst 1983). Thus, NO_x (= NO + NO_2), NH_3^T (= NH_3 + NH_4^+), oxidant (= O_3 + NO_2), nitrate (= HNO_3 + NO_3^-) are conserved in rapid airborne reactions, and are not subject to the difficulty described. It has yet to be determined whether the reactions occurring below a typical measurement height of a few metres are sufficient to cause appreciable errors.

3.4 Deposition in Cloud and Fog Water

'Occult precipitation' ie, cloud water collected by vegetation but not recorded by conventional rain gauges, has been recognised for decades as a significant hydrological input in some areas (see references in Lovett and Reiners 1983). Inclusion of airborne particles in cloud droplets (of order 20 μm) greatly enhances deposition so that deposition velocities of 5 to 50 cm s^{-1} may apply. The concentration of trace species in cloud water is substantially greater than in rain so that a minor increase in precipitation due to cloud water impaction may result in a large increase in the deposition of trace substances.

The importance of cloud water deposition of acidity was demonstrated by Lovett et al (1982) and Dollard et al (1983). Lovett estimated the deposition of sulphate to a forest at 1200 m altitude in New Hampshire to be about twice that in rain. Waldman et al (1985) showed that a forest 700 m above sea level near the Californian coast received acid deposition of comparable magnitudes in cloud deposition and in rain, and Dollard found that sulphate deposition in cloud water added 20 percent to that deposited in rain to a hilltop moor in England. The difference in relative significance is attributable largely to varying frequencies of cloud in contact with the surface and the greater rate of impaction onto forest. At higher altitudes low temperatures may cause rime accretion: this may contribute a substantial fraction to the snow pack and over 90% of the deposition (Borys et al 1983).

It has been speculated that the high concentration of cloud water may result in direct damage to vegetation. Waldman et al (1985) collected cloud water samples with pH substantially below 3, a range where direct foliar injury has been observed.

4. CONCLUSIONS

The previous sections show areas of knowledge and of uncertainty in both the wet and dry depositon of particles, from the atmosphere. For some gases rather more confidence can be placed in our understanding the physical and chemical aspects of deposition but the precision of estimates of overall rates is limited by inadequacies in the description of the occurrences of rain and cloud, transport above the boundary layer, and exchange processes at rough, irregular terrain.

However, some generalisation regarding the time scale for deposition can be attempted. For particles (assuming efficient incorporation into rain) and for the most soluble gases, it is likely that rain can clean the atmosphere rather quickly. An upper limit to the significance of wet deposition is estimated by assuming that falling rain immediately reduces the concentration to zero. The mean concentrations of a species generated or emitted at a rate equivalent to a concentration increase of q (ppb h^{-1}) will be

$$\bar{c}_{dry} = \frac{t_d\, q}{2} \quad \text{and} \quad \bar{c}_{wet} = 0 \tag{6}$$

during dry and wet periods of duration t_d and t_w respectively.

Then the long term mean concentration will be

$$\bar{c} = \frac{t_d\, q}{2} \quad \frac{t_d}{t_d + t_w} \tag{7}$$

The mean removal rate must equal the production rate, q, and so the mean removal coefficient for wet deposition becomes

$$\Lambda_w = \frac{\text{mean removal rate}}{\text{mean concentration}} = \frac{q}{\bar{c}} = \frac{2\ (t_d + t_w)}{t_d^2} \tag{8}$$

Similar arguments have been developed in detail by Rodhe and Grandell (1981). Over Europe dry and wet periods are estimated to have lengths of ~ 40 to 70 h and 6 to 12 h respectively, in a lagrangian frame, and we therefore estimate an upper limit on Λ_w of 0.03 to 0.06 h^{-1} or 10^{-5} to 2×10^{-5} s^{-1}. For a global estimate this removal coefficient would require scaling down by the lower global frequency of rain events, taking into account the drier land and sea areas, and the incomplete removal that occurs in rain events for real pollutants. Even for water vapour, its turnover time in the troposphere (~ 10 days) implies a value of Λ_w of ~ 10^{-6} s^{-1}. It is improbable that even the most soluble gases will have a higher Λ_w. The difference between this and the previous estimate may relate to the fraction of water or gas that escapes deposition within an area of rainfall.

For dry deposition, an estimate of the removal time constant, Λ_d, is obtained from the ratio V_g/h where h is the mixing height, defined in eq 3 for the tracer.

Unfortunately h is not well known. For short lived species ($V_g \gtrsim 10^{-2}$ m s^{-1}) a tracer with a mean surface source should be chiefly confined to the mixing layer, with h ~ 1 km. Then $\Lambda_d \sim 10^{-5}$ s^{-1}, equivalent to a residence time of order 1 day. Species with slower removal rates may have time to become dispersed through a greater depth of the troposphere. Less reactive gases and small particles have $V_g \lesssim 10^{-3}$ m s^{-1}, h may be $\gtrsim$ 3 km so that $\Lambda_d \simeq 10^{-7}$ s^{-1}. Thus for small particles wet deposition, with $\Lambda_d \sim 10^{-6}$ is the more efficient mechanism.

This was born out by the behaviour of the finely divided ^{137}Cs dispersed in the atmosphere by the Chernobyl accident (Cambray et al 1987). Measurements of deposition where rain did not occur during the initial passage of the plume showed $V_g \approx 0.5$ mm s^{-1} . Where rain fell, deposition was two or three orders greater. The rate of loss from the atmosphere followed a 10 day retention time, and as dry deposition could not support such rapid deposition, Λ_w must have been about 10^{-6} s^{-1}.

In contrast, for the most reactive gases, such as HNO_3, $V_g \sim 10^{-1}$ m s^{-1}, dry deposition will give $\Lambda_d \sim 10^{-4}$ s^{-1}, and the very short residence time (~ 3 h) will result in only a minor fraction surviving long enough to be removed in rain. Although such a gas will be washed out rapidly during rainfall, in the long term dry deposition must be expected to dominate.

In addition, of SO_2 emitted to the atmosphere about one half is removed by dry deposition (Garland 1978). For this gas conversion to sulphate in the atmosphere is the chief competing process; the sulphate produced is largely deposited in rain and direct washout of SO_2 is of minor importance. These ideas successfully describe the behaviour of sulphur in the atmosphere as observed in the extensive EMEP programme (Eliassen and Saltbones 1983). While these considerations indicate the lifetime before deposition of species of interest, more local effects control the spatial distribution of deposition. Variations in rainfall are important in hilly areas and may cause both regional and local increases in deposition where orography enhances rainfall. The increased concentration in the presence of cap clouds is relevant here. Hill tops may also intercept cloud water and suffer additional deposition. In total the deposition to hill tops may be several times higher than at nearby valley sites. Dry deposition to forest may also be enhanced by the greater surface roughness. For SO_2, NO_x and O_3, stomatal control limits or negates the increase, but the highly reactive gases such as HNO_3 which experience little or no surface resistance may deposit a few times more rapidly to forest than to surrounding grassland. Thus the effects of air pollution may be increased at just those locations where poor soils or marginal growth conditions often make ecosystems particularly sensitive.

References

Baldocchi, D.D., Hicks, B.B. and Camara, P. (1987) 'A canopy stomatal resistance model for gaseous deposition to vegetated surfaces'. Atmospheric Environment 21 91-101.

Barlow, A.K. and Latham, J. (1983) 'A laboratory study of the scavenging of sub-micron aerosol by charged raindrops'. Quart J Roy Met Soc **109** 763-770.

Belot, Y., Bourreau, J.C., Dubois, M.L. and Pauly, L.S. (1974) 'Mesure de la vitesse de captation du dioxide de soufre sur les feuilles des plant au moyen du soufre -34'. IAEA-SM-191-18, IAEA, Vienna.

Borys, R.D., Demott, P.J., Hindman, E.E. and Feng, D. (1983) 'The significance of snow crystal and mountain surface riming on the removal of atmospheric trace constituents from cold clouds'. Precipitation Scavenging, Dry Deposition and Resuspension. Pruppacher, H.R., Semonin, R.G. and Slinn, W.G.N. (eds) Elsevier, New York 181-188.

Brink, H.M. ten, Schwartz, S.E. and Daum, P.H. (1987) Efficient scavenging of aerosol sulfate by liquid water clouds. Atmospheric Environment **21** 2035-2052.

Bromberg, S.M., Hicks, B.B. and Durham, J.L. (1986) 'Monitoring acid deposition'. Aerosols; Research, Risk assessment and Control Strategies. Lee, S.D., Scheider, T., Grant, L.D. and Verkerk, P.J. (eds) Lewis Publishers Inc. Michigan. 239-245.

Brutsaert, W. (1975) 'A theory for local evaporation (or heat transfer) from rough and smooth surfaces at ground level'. Water Resources Research **11** 543-550.

Businger, J.A. (1986) 'Evaluation of the accuracy with which dry depositon can be measured with current micrometeorological techniques'. J. Climate and Applied Meteorology **25** 1100-1124.

Cambray, R.S., Cawse, P.A., Garland, J.A., Gibson, J.A.B., Johnson, P., Lewis, G.N.J., Newton, D., Salmon, L. and Wade, B.O. (1987) 'Observations on radioactivity from the Chernobyl accident. Nuclear Energy **26** 77-101.

Cawse, P.A. (1974) 'A survey of atmospheric trace elements in the UK' (1972-1973). AERE-R7669 HMSO London.

Chamberlain, A.C. (1953) 'Aspects of travel and deposition of aerosols and vapour clouds'. AERE HP/R1261 Harwell Laboratory, UKAEA.

Chamberlain, A.C. (1967) 'Deposition of lycopodium spores and other small particles to rough surfaces'. Proc. R. Soc. Lond **A 296** 45-70.

Chamberlain, A.C. (1983) 'Deposition and resuspension'. Precipitation Scavenging, Dry Deposition and Resuspension . Pruppacher, H.R., Semonin, R.G. and Slinn, W.G.N. (eds) Elsevier New York 731-751.

Chamberlain, A.C., Garland, J.A. and Wells, A.C. (1984) 'Transport of gases and particles to surfaces with widely spaced roughness elements'. Boundary-Layer Meteorology **29** 343-360.

Chang, T.Y. (1984) 'Rain and snow scavenging of HNO_3 vapour in the atmosphere'. Atmospheric Environment **18** 191-197.

Dasch, J.M. (1983) 'A comparison of surrogate surfaces for dry deposition collection'. Precipitation Scavenging Dry Deposition and Resuspension. Pruppacher, H.R., Semonin, R.G. and Slinn, W.G.N. (eds) Elsevier New York 883-900.

Davidson, C.I., Lindberg, S.E., Schmidt, J.A., Cartwright, L.G. and Landis, L.R. (1985) 'Dry deposition of sulphate onto surrogate surfaces'. J. Geophys. Res. **90** 2123-2130.

Davidson, C.I. and Friedlander, S.K. (1978) 'A filtration model for aerosol dry deposition: application to trace metal deposition from the atmosphere'. J. Geophys. Res. **83** 2343-2352.

Davidson, C.I., Miller, J.M. and Pleskow, M.A. (1982) 'The influence of surface structure on predicted particle dry deposition to natural grass canopies'. Water Air and Soil Pollution **18** 25-43.

Delany, A.C. and Davies, T.D. (1983) 'Dry deposition of NO_x to grass in rural East Anglia'. Atmospheric Environment **17** 1391-1394.

Denmead, T.O., Simpson, J.R. and Freney, J.R. (1974) 'Ammonia flux onto the atmosphere from a grazed pasture'. Science **185** 609-610.

Dollard, G.J., Unsworth, M.H. and Harvey M.J. (1983) 'Pollutant transfer in upland regions by occult precipitation'. Nature **302** 241-243.

Dollard, G.J., Davies, T.J. and Lindstrom, J.P.C. (1987a) 'Measurements of the dry deposition rates of some trace gas species'. Physico Chemical Behaviour of Atmospheric Pollutants. Reidel, Dordrecht.

Dollard, G.J., Atkins, D.H.F., Davies, T.J. and Healy, C. (1987b) 'Concentrations and dry deposition velocities of nitric acid'. Nature **326** 481-483.

Duyzer, J.H., Bouman, A.M.M., van Aalst, R.M. and Diederen, H.S.M.A. (1987) 'Assessment of dry deposition fluxes of NH_3 and NH_4^+ over natural terrains'. TNO report P87/034 TNO, The Hague.

Eliassen, A. and Saltbones, J. (1983) 'Modelling of long range transport of sulphur over Europe : a two year model run and some model experiments'. Atmospheric Environment **17** 1457-1473.

Fowler, D. and Cape, J.N. (1984) 'The contamination of rain samples by dry deposition on rain collectors'. Atmospheric Environment **18** 183-189.

Fowler, D., Cape, J.N., Leith, I.D., Choularton, J.W., Gay, M.J. and Jones, A. (1987) 'The influence of altitude on rainfall composition'. Atmospheric Environment, to be published.

Galbally, I.E. and Roy, C.R. (1978) 'Loss of fixed nitrogen from soils by NO exhalation'. Nature **275** 734-5.

Garland, J.A. (1977) 'The dry deposition of sulphur dioxide to land and water surfaces'. Proc. Roy. Soc. London **A354** 245-268.

Garland, J.A. and Branson, J.R. (1977) 'The deposition of sulphur dioxide to pine forest assessed by a radioactive method'. Tellus **29** 445-454.

Garland, J.A. (1978) 'Dry and Wet removal of sulphur from the atmosphere'. Atmospheric Environment **21** 349-362.

Garland, J.A. and Cox, L.C. (1982) 'Deposition of small particles to grass'. Atmospheric Environment **16** 2699-2702.

Garland, J.A. (1983a) 'Dry deposition of small particles to grass in field conditions'. Precipitation Scavenging Dry Deposition and Resuspension. Pruppacher, H.R., Semonin, R.G. and Slinn, W.G.N. (eds). Elsevier New York 849-854.

Garland, J.A. (1983b) 'Principles of dry deposition: application to acidic species and ozone'. Saure Niederschläge, VDI-Berichte **500**, 83-95.

Garratt, J.R. and Hicks, B.B. (1973) 'Momentum heat and water vapour transport to and from natural and artificial surfaces'. Quart J.R. Met. Soc. **99** (1973) 680-687.

Gatz, D.F. (1975) 'Wet deposition estimation using scavenging ratios'. J. Great Lakes Res. **2** Suppl. 1. 21-32.

Goldsmith, R., Dellafield, H.J. and Cox, L.C. (1963) 'The role of diffusiophosesis in the scavenging of radioactive particles from the atmosphere'. Quart. J. Roy Met Soc **89** 43-61.

Grennfelt, P., Bengtson, C. and Skärby, L. (1983) 'Dry deposition of nitrogen dioxide to scots pine needles'. Precipitation Scavenging, Dry Deposition and Resuspension. Pruppacher, H.R., Semonin, R.G. and Slinn, W.G.N. (eds). Elsevier New York 753-761.

Hicks, B.B., Wesely, M.L., Lindberg, S.E. and Bromberg, S.M. (1986) 'Proceedings of the NAPAP Workshop on Dry Deposition'. NDAA/ATDD, Oak Ridge.

Höfken, K.D., Meixner, F.X. and Ehhalt, D.H. (1983) 'Deposition of atmospheric trace constituents onto different natural surfaces'. Precipitation Scavenging Dry Deposition and Resuspension. Pruppacher, H.R., Semonin, R.G. and Slinn, W.G.N. (eds). Elsevier New York 825-836.

Höfken, K.D., Meixner, F., Miller, K.P. and Ehhalt, D.H. (1986) 'Untersuchungen zur trockenen Deposition und Emission von atmosphärischem NO, NO_2 und HNO_3 on naturlichen Oberflächen'. Jul 2054, KfA Jülich.

Hosker, R.P. and Lindberg, S.E. (1982) 'Atmospheric deposition and plant assimilation of gases and particles'. Atmospheric Environment 16 889-910.

Hove, L.W.A. van, Koops, A.J., Adema, E.H., Vredenberg, W.J. and Pieters, G.A. (1987) 'Analysis of the uptake of atmospheric ammonia by leaves of phaseolas vulgavis, L'. Atmospheric Environment **21** 1759-1764.

Huebert, B.J. (1983) 'Measurements of the dry deposition flux of nitric acid vapor to grasslands and forest'. Precipitation Scavenging, Dry Deposition and Resuspension. Pruppacher, H.R., Semonin, R.G. and Slinn, W.G.N. (eds). Elsevier New York 785-792.

Johansson, C. (1987) 'Pine forest: A negligible sink for atmosphere NO_x in rural Sweden'. Tellus: to be published.

Judeikis, H.S., and Stewart, T.B. (1976) 'Laboratory measurements of SO_2 deposition velocities on selected building materials and soils'. Atmospheric Environment **10**, 769-776.

Lindberg, S.E. and Lovett, G.M. (1983) 'Application of surrogate surfaces and leaf extraction methods to estimation of dry deposition'.

Precipitation Scavenging Dry Deposition and Resuspension. Pruppacher, H.R., Semonin, R.G. and Slinn, W.G.N. (eds). Elsevier New York 837-848.

Little, P. and Wiffen, R.D. (1977) 'Emission and deposition of petrol engine exhaust Pb-I deposition of exhaust lead to plant and soil surfaces'. Atmospheric Environment **11** 437-447.

Little, P. and Wiffen, R.D. (1978) 'Emission deposition of petrol engine exhaust Pb-II. Airborne concentration particle size and deposition of lead near motorways'. Atmospheric Environment **12** 1331-1341.

Lovett, G.M. and Reiners, W.A. (1983) 'Cloud water: an important vector of atmospheric deposition'. Precipitation Scavenging, Dry Deposition and Resuspension. Pruppacher, H.R., Semonin, R.G. and Slinn, W.G.N. (eds) Elsevier, New York 171-179.

Lovett, G.M., Reiners, W.A. and Olson, R.K. (1982) 'Cloud droplet deposition in subalpine balsam fir forest'. Science **218** 1303-1304.

Meixner, F.X., Franken, H.H., Duyzer, J.H. and Van Aalst, R.M. (1987) 'Dry deposition of gaseous HNO_3 to a pine forest'. 16th International Meeting on Air Pollution and its Applications. Lindau April 1987.

Nicholson, K.W. and Davies, T.D. (1987) 'Field measurements of the dry deposition of particulate sulphate'. Atmospheric Environment **21** 1561-1571

Owers M.J. and Powell, A.W. (1974) 'Deposition velocity of SO_2 on land and water surfaces using a ^{35}S tracer method'. Atmospheric Environment **8** 63-68.

Parrish, D.D., Williams, E.J., Fahey D.W., Liu S.C. and Fehsenfeld, F.C. (1987) 'Measurements of nitrogen oxide fluxes from soils: intercomparison of enclosure and gradient measurement techniques'. J. Geophys Res. **92** 2165-2171.

Peirson, D.H. and Salmon, L. (1959) 'Gamma radiation from deposited fallout'. Nature **184** 1678-1679.

Penkett, S.A., Jones, B.M.R. and Eggleton, A.E.J. (1979) 'The importance of atmospheric ozone and hydrogen peroxide in oxidising sulphur dioxide in cloud and rain water'. Atmospheric Environment **13** 123-127.

Peters, L.K. (1983) 'Gases and their precipitation scavenging in the marine atmosphere'. Air Sea Exchange of Gases and Particles. Liss, P.S. and Slinn, W,G,N, (eds) Reidel, Dordrecht, 299-405.

Porter, L.K., Viets, F.G. and Hutchinson, G.L. (1972) 'Air containing nitrogen-15 ammonia: foliar absorption by corn seedlings'. Science **175** 759-761.

Rodhe, H. and Grandell, J. (1981) 'Estimates of characteristic times for precipitation scavenging'. J. Atmospheric Sci. **38** 370-386.

Sehmel, G.A. (1980) 'Particle and gas dry deposition, a review'. Atmospheric Environment **14** 983-1011.

Sievering, H. (1983) 'Eddy flux and profile measurements of small-particle dry-deposition velocity at the Boulder Atmospheric Observatory'. Precipitation Scavenging, Dry Deposition and Resuspension. Pruppacer, H.R., Semonin, R.G. and Slinn, W.G.N. (eds) Elsevier, New York 963-978.

Slinn, W.G.N. (1982) 'Predictions for particle deposition to vegetative canopies'. Atmospheric Environment **16** 1785-1794.

Slinn, W.G.N. (1983) 'A potpourri of deposition and resuspension questions'. Precipitation Scavenging, Dry Deposition and Resuspension. Pruppacher, H.R., Semonin, R.G. and Slinn W.G.N. (eds) Elsevier, New York 1361-1416.

Slinn, W.G.N. (1984) 'Precipitation Scavenging'. Atmospheric Science and Power Production. Randerson (ed) DOE/TIC-27601, National Technical Information Service, Springfield, Virginia.

Thom, A.S. (1975) 'Momentum, mass and heat exchange of plant communities'. Vegetation and the Atmosphere (Ed. J.L. Montieth). Academic Press 57-109.

Underwood, B.Y. (1987) 'Dry deposition to a uniform canopy: evaluation of a first-order closure mathematical model'. Atmospheric Environment **27** 1573-1585.

van Aalst, R.M. (1983) 'Dry deposition of acid precursors in the Netherlands'. Saure Niederschläge VDI-Berichte **500** 97-102.

Voldner, E.C., Barrie, L.A. and Sirois, A. (1986) 'A Literature review of dry deposition of oxides of suphur and nitrogen with emphasis on long range transport modelling in North America'. Atmospheric Environment **20** 2101-2123.

Waldman, J.M., Murgen, J.W., Jacob, D.J. and Hoffmann, M.R. (1985) 'Chemical characterisation of stratus cloudwater and its role for pollutant deposition in a Los Angeles pine forest'. Tellus **37B** 91-108.

Wesely, M.L. (1983) 'Turbulent transport of ozone to surfaces common in the eastern half of the United States'. Trace Atmospheric Constituents. S.E. Schwartz (ed) John Wiley & Sons, New York.

Wesely, M.L., Eastman, J.A., Stedman, D.H. and Valvac, E.D. (1982) 'An eddy correlation measurement of NO_2 flux to vegetation and comparison to O_3'. Atmospheric Environment **16** 815-820.

Wesely, M.L., Cook, D.R. and Hart, R.L. (1985) 'Measurements and parameterisation of particulate sulphur dry deposition over grass'. J. Geophys Res. **90** 2131-2143.

A NUMERICAL METHOD FOR DETERMINING THE DRY DEPOSITION OF ATMOSPHERIC TRACE CONSTITUENTS

Gerhard Kramm
Industrieanlagen-Betriebsgesellschaft mbH
P. O. Box
8012 Ottobrunn
Federal Republic of Germany

ABSTRACT. A diagnostic deposition model (DIADEM) based on the generally accepted micrometeorological laws of the transfer of momentum, sensible heat and matter near the earth's surface will be presented.

The model evaluates deposition fluxes and velocities of long-lived atmospheric trace gases by means of a detailed determination of the micrometeorological quantities such as the friction velocity u_*, the temperature and humidity scales Θ_* and q_*, the roughness length z_o and the zero-displacement d. The parameterization of the fluxes is founded on the flux-gradient relationships in the turbulent region of the surface layer and the sublayer Stanton number as well as the Reynolds' analogy between concentration, temperature and wind velocity distributions in the underlying sublayer. The model requires only the vertical profile data of wind velocity, dry and wet bulb temperatures and trace gas concentrations from the turbulent part of the surface layer.

DIADEM has been applied to vertical profile data collected in field experiments such as the GREIV I 1974 project and the Great Plains Turbulence Project.

In order to illustrate in which way the model can also be used to evaluate deposition fluxes and velocities of reactive trace gases, the method has been applied to observed concentrations of NO, NO_2 and O_3.

1. INTRODUCTION

Our knowledge about trace constituents in polluted air reveals that the effect of dry deposition plays an important role in the mass budget of many gaseous species. The essential natural influences on dry deposition of trace gases are the micrometeorological conditions in the atmospheric surface layer and the capability of vegetation and the soil to take up matter. The individual mass sorption behaviour of the ground in combination with the special mass and energy transfer in the surface layer, characterized to a large degree by the mechanical and thermal inter-

K. Grefen and J. Löbel (eds.), Environmental Meteorology, 161–181.

actions between the atmosphere and the ground (vegetation and soil), provide the governing cause-effect relationship for the phenomenon of dry gas deposition.

2. THEORETICAL BACKGROUND

2.1. Preliminary Remarks

In order to evaluate the uptake matter we have to assume that the sorption flux S_i is equal to the observed atmospheric deposition flux F_i, i. e.

$$S_i - F_i = 0 \quad . \tag{1}$$

The sorption flux S_i, which strongly depends on the surface concentration $c_{s,i}$ (expressed in kg m^{-3} or ppm) and the physical and biochemical properties of vegetation and the soil, may be parameterized by an empirical approach (cf. Galbally, 1971)

$$S_i = - v_{s,i}\, c_{s,i} \quad , \tag{2}$$

where $v_{s,i}$ is the sorption velocity. The reciprocal expression of $v_{s,i}$ is the so-called sorption resistance $r_{s,i}$. $v_{s,i}$ and $r_{s,i}$, respectively, may be described as functions of the biochemical and physical processes taking place in the soil or within the plants. However, such a description proves to be particularly difficult because most of these processes cannot be formulated mathematically in an acceptable manner. Therefore, the direct determination of $v_{s,i}$ and $r_{s,i}$, respectively, is not to be taken into consideration.

In analogy to the semi-theoretical mixing laws of the micrometeorology, the deposition flux F_i can be described by the flux-gradient relationship (see e.g. Chamberlain, 1968; Businger, 1973)

$$F_i = - (D_i + K_i) \frac{\partial c_i}{\partial z} \quad , \tag{3}$$

where c_i is the concentration (also expressed in kg m^{-3} or ppm) and z the height above ground. In Eq. (3), the molecular diffusivity D_i is a substance-specific quantity, whereas the eddy diffusivity K_i essentially depends on the dynamic boundary layer conditions and, therefore, strongly on height. One assumes that this dependence is satisfactorily described by setting K_i equal to K_h, the eddy diffusivity of heat. This generally accepted statement is based on the empirical result that turbulent transports of sensible heat and water vapour are similar phenomena (e.g. Swinbank and Dyer, 1967; Dyer and Hicks, 1970), which are adopted to the other trace gases as well (see also Businger, 1986). The following expressions are used for K_h:

$$K_h = \begin{cases} K_m & \text{for Ri} \geq 0 \\ K_m\,(1 - 16\,\text{Ri})^{1/4} & \text{for Ri} < 0 \end{cases}, \tag{4}$$

where K_m, the eddy diffusivity for momentum, is defined by

$$K_m = (\kappa\,(z - d))^2 \frac{\partial U}{\partial z} \begin{cases} (1 - 5\,\text{Ri})^2 & \text{for } 0 \leq \text{Ri} < \text{Ri}_{cr} \\ (1 - 16\,\text{Ri})^{1/2} & \text{for Ri} < 0 \end{cases} \tag{5}$$

$$U = (u^2 + v^2)^{1/2}$$

and Ri, the Richardson number, by

$$\text{Ri} = \frac{g}{\Theta_m} \left(\frac{\partial \Theta}{\partial z} + 0.608\ \Theta_m \frac{\partial q}{\partial z}\right) \left(\frac{\partial \mathbf{v}}{\partial z}\right)^{-2} \tag{6}$$

Ri_{cr} = 0.2 is the critical Richardson number, κ = 0.4 the von Karman constant and d the so-called zero-displacement. u and v are the components of the horizontal wind vector **v**. The relations for K_m and K_h were derived from the empirical results found by Dyer and Hicks (1970) for unstable conditions and by Webb (1970) for stable conditions.

2.2. Deposition Rates and Velocities of Trace Gases without Chemical Transformations

The most practicable theoretical approach to develop a calculation scheme for dry deposition in the absence of chemical reactions is a full set of constant flux conditions for the surface layer. These model formulations imply that the transfer rates passing through the thin viscous sublayer directly above ground surface are equal to the fluxes observed in the turbulent part of the surface boundary layer.

Integrating Eq. (3) over the constant flux layer $0 \leq z \leq z_h$ provides

$$F_i = - r_a^{-1} (c_{h,i} - c_{s,i}) = - u_* c_{*,i} = \text{const.}, \tag{7}$$

where z_h is the top of the surface layer, $c_{h,i}$ the concentration at z_h and r_a the integral

$$r_a = \int_0^{z_h} (D_i + K_h)^{-1}\, dz \tag{8}$$

defining (quite analogously to Ohm's law) a bulk resistance of the air against the mass transfer (denoted as the total aerodynamic resistance). It should be pointed out that the determination of the deposition flux with Eq. (7) is only admissible when F_i is independent of height.

Taking the budget equation (1) into account, we can derive two results:

Firstly, the deposition flux can be parameterized by

$$F_i = - \frac{c_{h,i}}{r_a + r_{s,i}} = - \tilde{v}_{d,i} \; c_{h,i} \quad , \tag{9}$$

where $\tilde{v}_{d,i} = (r_a + r_{s,i})^{-1}$ is the special deposition velocity at the surface, which not only depends on the transfer ability of the atmosphere but also on the sorption properties of vegetation and the soil.

Secondly, the sorption resistance

$$r_{s,i} = - (\frac{c_{h,i}}{F_i} + r_a) \tag{10}$$

can be determined indirectly.

Since the transfer conditions of the fully turbulent regime differ considerably from the transfer properties of the molecular-turbulent sublayer it is advantageous and advisable to formulate an approach for the total aerodynamic resistance in Eq. (8) by setting $r_a = r_t + r_{mt}$, where

$$r_t = \int_{\delta}^{z_h} K_h^{-1} \, dz \tag{11}$$

denotes a purely turbulent resistance and

$$r_{mt} = \int_{0}^{\delta} (D_{c,i} + K_h)^{-1} \, dz \tag{12}$$

a molecular-turbulent resistance.

As to δ, the lower boundary of the fully turbulent regime, it is most useful to assume the first observation level above the characteristic height of the roughness elements z_r.

In order to solve the r_t-integral in Eq. (10), the eddy diffusivity relations (4) and (5) and the constant flux assumptions for the turbulent fluxes of momentum (τ_o), sensible heat (H_o) and water vapour (Q_o), i.e.

$$\tau_o = \rho_L \, K_m \frac{\partial U}{\partial z} = \rho_L \, u_*^2 = \text{const.} \tag{13}$$

$$\begin{Bmatrix} H_o \\ \\ Q_o \end{Bmatrix} = - \rho_L K_h \begin{Bmatrix} c_{p,o} \dfrac{\partial \Theta}{\partial z} \\ \\ \dfrac{\partial q}{\partial z} \end{Bmatrix} = - \rho_L u_* \begin{Bmatrix} c_{p,o} \Theta_* \\ \\ q_* \end{Bmatrix} = \text{const.} \quad , \tag{14}$$

are taken into consideration. So, integration of Eq. (11) yields the relation

$$r_t = \frac{1}{u_* \kappa} \left(\ln \frac{z_h - d}{\delta - d} - \Psi_h \right) \quad , \tag{15}$$

which includes the dimensionless integral stability function for heat and matter

$$\Psi_h = \begin{cases} -\dfrac{5}{L}(z_h - \delta) & \text{for } L > 0 \text{ (stable)} \\ 0 & \text{for } L \to \infty \text{ (neutral)} \\ 2 \ln \dfrac{1 + y_h^2}{1 + y_\delta^2} & \text{for } L < 0 \text{ (unstable)} \end{cases} \tag{16}$$

with

$$y_\delta = (1 - \frac{16}{L}(\delta - d))^{1/4} \quad \text{and} \quad y_h = (1 - \frac{16}{L}(z_h - d))^{1/4} \quad , \tag{17}$$

the reciprocal expressions of the dimensionless wind gradients in the unstable case at the two heights δ and z_h, and

$$L = \frac{u_*^2}{\kappa \frac{g}{\Theta_m} (\Theta_* + 0.608\, \Theta_m q_*)} \quad , \tag{18}$$

the Monin-Obukhov stability length.

The determination of the height invariant scale parameters u_* (friction velocity), Θ_* (heat flux temperature), q_* (water vapour flux concentration) and $c_{*,i}$ (trace gas flux concentration) is based on the relations

$$u_* = \frac{\kappa}{\ln \frac{z_h - d}{\delta - d} - \Psi_m} (U_h - U_\delta) \tag{19}$$

$$\begin{Bmatrix} \Theta_* \\ q_* \\ c_{*,i} \end{Bmatrix} = \frac{1}{u_* \, r_t} \begin{Bmatrix} \Theta_h - \Theta_\delta \\ q_h - q_\delta \\ c_{h,i} - c_{\delta,i} \end{Bmatrix} , \tag{20}$$

where the dimensionless integral stability function for momentum is given by

$$\Psi_m = \begin{cases} \Psi_h & \text{for } L > 0 \\ 0 & \text{for } L \to \infty \\ 2 \ln \dfrac{1 + y_h}{1 + y_\delta} + \ln \dfrac{1 + y_h^2}{1 + y_\delta^2} - 2 \arctan \dfrac{y_h - y_\delta}{1 + y_h \, y_\delta} & \text{for } L < 0 \end{cases} \tag{21}$$

Eqs. (15) to (21) describe a closed system to determine r_t and the fluxes $F_i = - u_* \, c_{*,i}$ as functions of wind velocity (U), potential temperature (Θ), specific humidity (q) and trace gas concentrations (c_i) at the two heights δ and z_h.

The r_{mt}-integral in Eq. (12) was solved approximately for smooth surfaces by Reichardt (1951) (see also Hasse and Liss, 1980). For rough surfaces (vegetations etc.), however, there is no customary way to solve this integral in an elementary manner. Therefore, the transfer of sensible heat and matter between the earth's surface and δ has to be described by empirical transition formulas

$$\begin{Bmatrix} H_o \\ Q_o \end{Bmatrix} = - \rho_L \, \zeta \, U_\delta \begin{Bmatrix} c_{h,o} \, (\Theta_\delta - T_s) \\ q_\delta - q_s \end{Bmatrix} = \text{const.} \quad , \tag{22}$$

$$F_i = - \zeta \, U_\delta \, (c_{\delta,i} - c_{s,i}) \quad . \tag{23}$$

Analogous to the local drag coefficient $\zeta_d = (u_*/U_\delta)^2$, ζ represents dimensionless transfer coefficients, so-called Stanton numbers, for heat and matter, which are given by

$$\zeta = \frac{1}{\dfrac{U_\delta}{u_*} \left(\dfrac{U_\delta}{u_*} + B^{-1} \right)} \tag{24}$$

B is the so-called sublayer Stanton number (see e. g. Owen and Thomson, 1963; Chamberlain 1968). It is a measure of the difference in the transfer rates of momentum and sensible heat as well as matter to and from the rough surface.

As the heat and matter transfer is in the viscous sublayer of the molecular-turbulent structure, the sublayer Stanton numbers also have to be functions of the special Reynolds number $Re_* = u_* z_r/\nu$ (ν = kinematic viscosity) and the Prandtl number $Pr = \nu/\alpha$ (α = thermal diffusivity), respectively, of Re_* and the Schmidt number $Sc_i = \nu/D_i$. Such functions of the structure

$$B = b^{-1} Re_*^{-m} \begin{cases} Pr^{-n} & \text{for heat} \\ Sc_i^{-n} & \text{for water vapour and trace gases} \end{cases} \tag{25}$$

were derived by Owen and Thomson (1963) for the heat transfer and by Chamberlain (1968) for the thorium-B and water vapour transfer from wind tunnel investigations, where Chamberlain obtained a similar result for the mass transfer in the range $0.33 < Re_* < 33$ as did Owen and Thomson for the heat transfer.

The quantity z_r may be considered as the characteristic height of the surface roughness elements. Such a quantity is Nikuradse's (1933) equivalent sand roughness which is approximately equal to 30 z_o, with z_o being the roughness length. It was used by Owen and Thomson as well as by Chamberlain. In contrast to these, the relation $z_r = z_o + d$ is to be suggested here because the mean wind velocity becomes zero at this height.

Comparing Eq.(23) with Eq. (7) provides for the molecular-turbulent resistance

$$r_{mt} = \frac{1}{\zeta U_\delta} = \frac{1}{u_*}\left(\frac{U_\delta}{u_*} + B^{-1}\right) \quad . \tag{26}$$

Thus, we obtain for the total aerodynamic resistance

$$r_a = r_{mt} + r_t = \frac{1}{u_*}\left(\frac{U_\delta}{u_*} + B^{-1} + \frac{1}{\kappa}\left(\ln\frac{z_h - d}{\delta - d} - \Psi_h\right)\right) \quad . \tag{27}$$

As F_i may be determined only from the observation data of the fully turbulent part of the Prandtl layer, the surface concentration $c_{s,i}$ and the sorption resistance $r_{s,i}$ can be calculated.

2.3. The Calculation Scheme for the Determination of the Surface Layer Parameters

Deposition fluxes and transfer resistance can only be determined if the turbulent fluxes of momentum, sensible heat and water vapour are known. It has been shown earlier that in any diabatic case the knowlegde of the mean values of wind velocity, temperatur and humidity collected at the two heights δ and z_h is sufficient (at least essentially) to determine these turbulent fluxes. However, the zero-displacement d and the roughness length z_o as well as the reference values Θ_r, q_r and $c_{r,i}$ have to be known.

Micrometeorological profile measurements by instrumented masts or towers, however, are generally carried out in such a manner that the mean values of the field quantities are observed at more than two levels. Thus, the turbulent fluxes may be determined with some redundancy. Nevertheless, such profile measurements are necessary to compute more appropriate mean values of the turbulent fluxes and especially to evaluate z_o and d as well as the required reference values $\Theta_r = \Theta(z_o+d)$, $q_r = q(z_o+d)$ and $c_{r,i} = c_i(z_o+d)$ by least squares fitting.

In order to be able to establish the best approximations for d and z_o and the relevant reference values Θ_r and q_r the following expressions have to be minimized:

$$\sum_{i=1}^{N} (U_{M,i} - U_{C,i}((z_o,d)^T))^2 \tag{28}$$

$$\sum_{i=1}^{N} (\Theta_{M,i} - \Theta_{C,i}(\Theta_r))^2 \tag{29}$$

$$\sum_{i=1}^{N} (q_{M,i} - q_{C,i}(q_r))^2 \tag{30}$$

$N \geq 3$ (number of observation levels)

Here $U_{M,i}$ is the wind velocity observed at the height z_i; the corresponding values $\Theta_{M,i}$ and $q_{M,i}$ for the potential temperature and the specific humidity can be established from measurements of the dry and wet bulb temperatures at the same height. The profile functions

$$U_{C,i}((z_o,d)^T) = \frac{u_*}{\kappa}\left(\ln \frac{z_i - d}{z_o} - \Psi_{m,i}\right) \tag{31}$$

$$\begin{Bmatrix} \Theta_{C,i}(\Theta_r) \\ q_{C,i}(q_r) \end{Bmatrix} = \begin{Bmatrix} \Theta_r \\ q_r \end{Bmatrix} + \frac{1}{\kappa}\begin{Bmatrix} \Theta_* \\ q_* \end{Bmatrix}\left(\ln \frac{z_i - d}{z_o} - \Psi_{h,i}\right) \tag{32}$$

can be derived from Eqs. (15) to (21), where δ and z_h are replaced by z_o + d and z_i and the relevant profile values are to be considered accordingly. Appropriate values for the scaling parameters u_*, Θ_* and q_* are provided by the arithmetic averages

$$\begin{Bmatrix} u_* \\ \Theta_* \\ q_* \end{Bmatrix} = \frac{1}{N-1} \sum_{i=1}^{N-1} \begin{Bmatrix} u_{*,i} \\ \Theta_{*,i} \\ q_{*,i} \end{Bmatrix} , \tag{33}$$

which are also used for the calculation of an appropriate value of the Monin-Obukhov stability length (analogous to Eq. (18)). $u_{*,i}$, $\Theta_{*,i}$ and $q_{*,i}$ can be calculated simultaneously from the profile data collected at the adjacent observation levels z_i and z_{i+1} with the integral bulk formulations of the previous chapter. In any unstable case the determination of $u_{*,i}$, $\Theta_{*,i}$ and $q_{*,i}$ has to be carried out by a method of successive approximations where the equation set (15) to (21) serves as the iteration prescription. In contrast to this, there are analytical solutions for neutral and stable conditions (cf. Businger, 1973)

$$\begin{Bmatrix} u_{*,i} \\ \Theta_{*,i} \\ q_{*,i} \end{Bmatrix} = \beta \begin{Bmatrix} U_{M,i+1} - U_{M,i} \\ \Theta_{M,i+1} - \Theta_{M,i} \\ q_{M,i+1} - q_{M,i} \end{Bmatrix} \tag{34}$$

with

$$\beta = \frac{\kappa}{\ln \frac{z_{i+1} - d}{z_i - d}} (1 - 5\, Ri_b) \tag{35}$$

and

$$Ri_b = \frac{g}{\Theta_m} \frac{\Theta_{M,i+1} - \Theta_{M,i} + 0.608\, \Theta_m (q_{M,i+1} - q_{M,i})}{(U_{M,i+1} - U_{M,i})^2} (z_{i+1} - z_i) , \tag{36}$$

the so-called bulk Richardson number.

Solving the linear least squares problems according to expressions (29) and (30) leads to the minimization conditions

$$\begin{Bmatrix} \Theta_r \\ q_r \end{Bmatrix} = \frac{1}{N} \sum_{i=1}^{N} \begin{Bmatrix} \Theta_{M,i} \\ q_{M,i} \end{Bmatrix} - \frac{1}{\kappa} \begin{Bmatrix} \Theta_* \\ q_* \end{Bmatrix} \sum_{i=1}^{N} \left(\ln \frac{z_i - d}{z_o} - \Psi_{h,i}\right) . \tag{37}$$

The non-linear least squares problem in accordance with expression (28) can only be solved by a method of successive approximations. Such a procedure may be derived from the approximation of the non-linear problem by series of linear least squares problems, i. e., if z_o and d are approximations for the optimum solution, then the solutions $\tilde{z}_o$ and $\tilde{d}$ of the linear least squares problem

$$\sum_{i=1}^{N} \left(U_{M,i} - U_{C,i}((z_o,d)^T) - \frac{\partial U_{C,i}}{\partial z_o}(\tilde{z}_o - z_o) - \frac{\partial U_{C,i}}{\partial d}(\tilde{d} - d)\right)^2 = \min \tag{38}$$

will in general be better solutions of the non-linear least squares problem than z_o and d. $\partial U_{C,i}/\partial z_o$ and $\partial U_{C,i}/\partial d$ can be approximated by the following expressions:

$$\frac{\partial U_{C,i}}{\partial z_o} = - \frac{u_*}{\kappa\, z_o}\, \phi_{m,o} \tag{39}$$

$$\frac{\partial U_{C,i}}{\partial d} = - \frac{u_*}{\kappa\,(z_i - d)}\, \phi_{m,i} + \frac{1}{\kappa}\left(\ln \frac{z_i - d}{z_o} - \Psi_{m,i}\right) \frac{\partial u_*}{\partial d} \tag{40}$$

with

$$\phi_{m,o} \cong 1 \tag{41}$$

$$\phi_{m,i} = \begin{cases} 1 + \frac{5}{L}(z_i - d) & \text{for } 0 < Ri < Ri_{cr} \\ 1 & \text{for } Ri = 0 \\ {y_i}^{-1} & \text{for } Ri < 0 \end{cases} \tag{42}$$

and

$$\frac{\partial u_*}{\partial d} = \frac{1}{N-1} \sum_{i=1}^{N-1} \begin{cases} \dfrac{\dfrac{1}{z_{i+1} - d} - \dfrac{1}{z_i - d}}{\ln \dfrac{z_{i+1} - d}{z_i - d}}\, u_{*,i} & \text{for } 0 < Ri < Ri_{cr} \\[2ex] \dfrac{\dfrac{{y_{i+1}}^{-1}}{z_{i+1} - d} - \dfrac{{y_i}^{-1}}{z_i - d}}{\ln \dfrac{z_{i+1} - d}{z_i - d} - \Psi_{m,i}}\, u_{*,i} & \text{for } Ri < 0 \end{cases} \tag{43}$$

y_i and y_{i+1} are to be determined in the sense of Eq. (17), with L_i for the layer $z_i \leq z \leq z_{i+1}$ being evaluated from $u_{*,i}$, $\Theta_{*,i}$ and $q_{*,i}$, in accordance with Eq. (18).

The iteration procedure may be ceased if, after the kth iteration step, the desired degree of accuracy

$$| z_o^{(k+1)} - z_o^{(k)} | < 0.1 \text{ mm} \quad \text{and} \quad | d^{(k+1)} - d^{(k)} | < 0.1 \text{ mm}$$

is achieved. When the micrometeorological parameters z_o, d, u_*, Θ_* and q_*, Θ_r and q_r have been obtained, $c_{*,i}$ and $c_{r,i}$ can be determined analogously.

3. MODEL TESTS

Although the vertical profiles of wind velocity, temperature and humidity and trace gas concentrations are necessary to determine the deposition fluxes and velocities, only a few profile data sets exist, which make it possible to test numerical methods of this kind. Test calculations with the micrometeorological part of the deposition model were carried out with observation data from the GREIV I experiment (see Beyer and Roth, 1976). This experiment took place over a flat site, covered with winter barley (about 25 cm high) and rape (50 - 75 cm high), near Meppen/Emsland in northern Germany in April 1974. The simultaneously measured 30-min averages of wind velocity, dry and wet bulb temperatures of the working groups from Universities of Kiel (April 20 - 24, 1974) and Munich (April 24 - 27, 1974) were used for the calculations of the turbulent fluxes of momentum, sensible heat and water vapour. The observations of the Kiel group were carried out at heights of 0.50, 1.26, 3.18 and 8.00 meters and those of the Munich group at heights of 0.50, 1.00, 2.00, 4.00, 8.00 and 16.00 meters. In addition, the 30-min averages of the surface temperature (infrared temperature measurements by the University of Hannover working group) were used for calculating sublayer Stanton numbers and the 30-min run data of directly determined fluxes of momentum, sensible and latent heat flux (ultrasonic anemometer-thermometer and Lyman alpha hygrometer measurements by the University of Mainz working group) for comparison purposes. The flux measurements of the Mainz group were carried out in the vicinity of the instrumented mast of the Kiel group. The height of the observation level came up to 2.0 meters.

109 profile data sets of the Kiel group were taken into account, 77 of which were suitable for evaluation. The calculations made on the basis of 110 profile data sets of the Munich group were carried out for five (without the 16.00 m level) and six observation levels, 73 and 69 data sets, respectively, were appropriate to be evaluated. Since the calculations for six levels provided "errors" which were a little bit greater than those for five levels, their results were no longer taken into consideration.

If $z_o + d > z_1$ or more than 40 iteration steps had been required to determine z_o and d, the profile data sets were generally rejected by the computer program. Such criteria occurred for profile data which were collected under strongly stable conditions with low wind velocities and temperature inversions or in the transition phase between lapse and inversion conditions, where steady-state conditions must not be expected (see also Stearns, 1970).

The examples of various stability conditions in Fig. 1 show, that the least squares fits calculated coincide very accurately with the values observed. Clear consistency also exists between the directly measured fluxes of momentum, sensible and latent heat and the calculated fluxes (see Fig. 2).

Deposition fluxes and velocities of NO, NO_2 and ozone were calculated on the basis of vertical profiles observed by a working group of the Institute of Atmospheric Chemistry III at the Nuclear Research Unit at Jülich in western Germany (cf. Höfken et al., 1986) and by several working groups involved in the Great Plains Turbulence Project (cf. Lettau and Davidson, 1957).

The measurement results of the Jülich working group were obtained in September 1984 at heights of 0.40, 2.60 and 17.20 meters over even terrain near Jülich, the measurements being averaged over 5 minutes. During the observation phase the height of vegetation (turnips) in the vicinity of the measurement area came up to about 25 cm.

The Great Plains Turbulence Project took place over even terrain near O'Neill, Nebraska, USA, in the summer of 1953. The required wind (60-min averages), temperature and humidity data (5-min averages) for the calculation of ozone deposition were observed at heights of 0.40, 0.80, 1.60, 3.20 and 6.40 meters by the working group from the Johns Hopkins University, and the ozone values (60-min averages) were registered at heights of 0.40, 1.60 and 6.25 meters by the working group from the University of New Mexico .

Results of the deposition calculations for fluxes independent of height are shown in Figures 3 to 6. The calculated least squares fits match the observed values very well. Thus, we can assume that the deposition fluxes $F_i = - u_* c_{*,i}$ are evaluated with a sufficient degree of accuracy. Particularly noteworthy are the results presented in Figures 4 and 5. They show that, at some times, the ground (vegetation and soil) can be regarded as a trace gas source.

The results presented in Fig. 7 were calculated for fluxes of NO, NO_2 and ozone including the reactions among these trace gases. These reactions can be summarized by

$$NO_2 + h\nu \rightarrow NO + O_3 \; , \; k_1 = \text{variable}\,(\text{min}^{-1}) \qquad (44)$$

$$NO + O_3 \rightarrow NO_2 + O_2 \; , \; k_2 = 3{,}100 \exp(1{,}450\; T^{-1})\; (\text{ppm}^{-1}\text{min}^{-1}) \; , \qquad (45)$$

The results are based on the same observations as those evaluated for fluxes invariant with height (shown in Fig. 6). For the calculations, the fluxes of the "trace gases" $c_1 = [NO] + [NO_2]$, $c_2 = [NO_2]+[O_3]$ and

$c_3 = [NO] - [O_3]$ considered as constant in height (cf. Lenschow, 1982) were determined with the procedure described earlier. In addition, a system of ordinary differential equations, for example, for the ozone concentration

$$\frac{\partial}{\partial z} F_{O_3} = k_1 c_2 - (k_1 + k_2 c_3) [O_3] - k_2 [O_3]^2 \tag{46}$$

$$\frac{\partial}{\partial z} [O_3] = - \frac{F_{O_3}}{K_h} \tag{47}$$

was necessary to determine the deposition fluxes of NO, NO_2 and ozone. It can be seen that, despite the low value of the photolysis frequency of $k_1 = 0.087 \ min^{-1}$ (degree of cloud coverage = 8/8), the chemical transformations cause different flux divergences and thus height-dependent fluxes which are also reflected in the changed concentration profiles. It is clear that the results of the differing flux calculations match only in one level (z = 4.13 m). The neglect of chemical processes is, therefore, justified only in particular instances. The question as to how far the results presented in Figures 3 to 6 are influenced by chemical reactions cannot be answered on the basis of the available data.

The model tests discussed earlier show that the calculated concentration profiles match very well with the observed profile data. It can, therefore, be assumed that the described determination of the deposition (or exhalation) of trace gases in the surface boundary layer leads, in terms of the model, to qualitatively and quantitatively reasonable results. However, the verification of the overall physics of the model is necessary to be able to guarantee the practical suitability of the deposition model. This would presuppose that directly measured trace gas fluxes can be used for the purpose of comparison. Such trace gas flux measurements were not carried out within the framework of the field experiments used earlier, so that a check of the calculated fluxes is not possible at present. There is an urgent requirement of future field experiments that, apart from the vertical profiles of wind velocity, temperature, humidity and trace gas concentrations, also the relevant deposition fluxes and soil source strengths will be determined directly and that in such a way that the influence of the chemical reactions on the deposition fluxes can be estimated.

4. ACKNOWLEDGMENT

I should like to express my thanks to the Deutsche Forschungsgemeinschaft (SFB 73 "Atmosphärische Spurenstoffe" - Projekt F9) and the Bundesminister für Forschung und Technologie (Förderschwerpunkt "Physikalisch-chemische Prozesse in der Atmosphäre" - Förderkennzeichen: 07VTD01 1) for their support of this study.

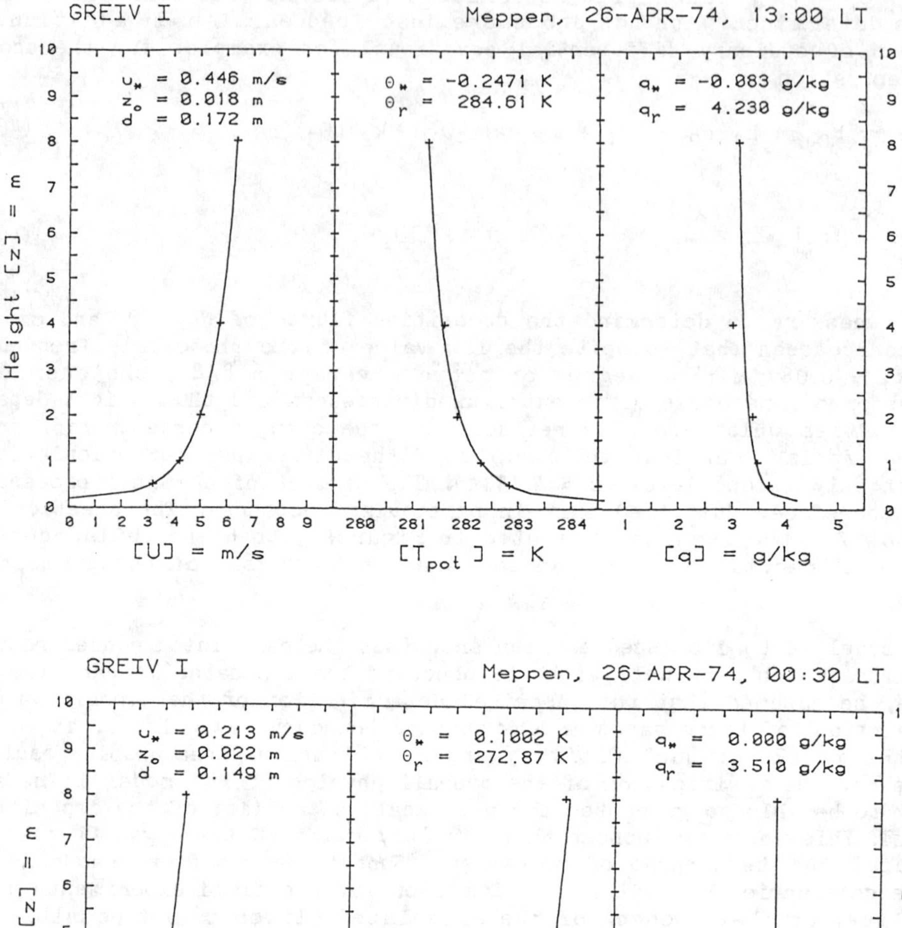

Figure 1. Typical vertical profiles of wind velocity (U), potential temperature (T_{pot}) and specific humidity (q) calculated for stable ($\Theta_* > 0$) and unstable conditions ($\Theta_* < 0$). The crosses represent the observed values and the solid lines the calculated profiles.

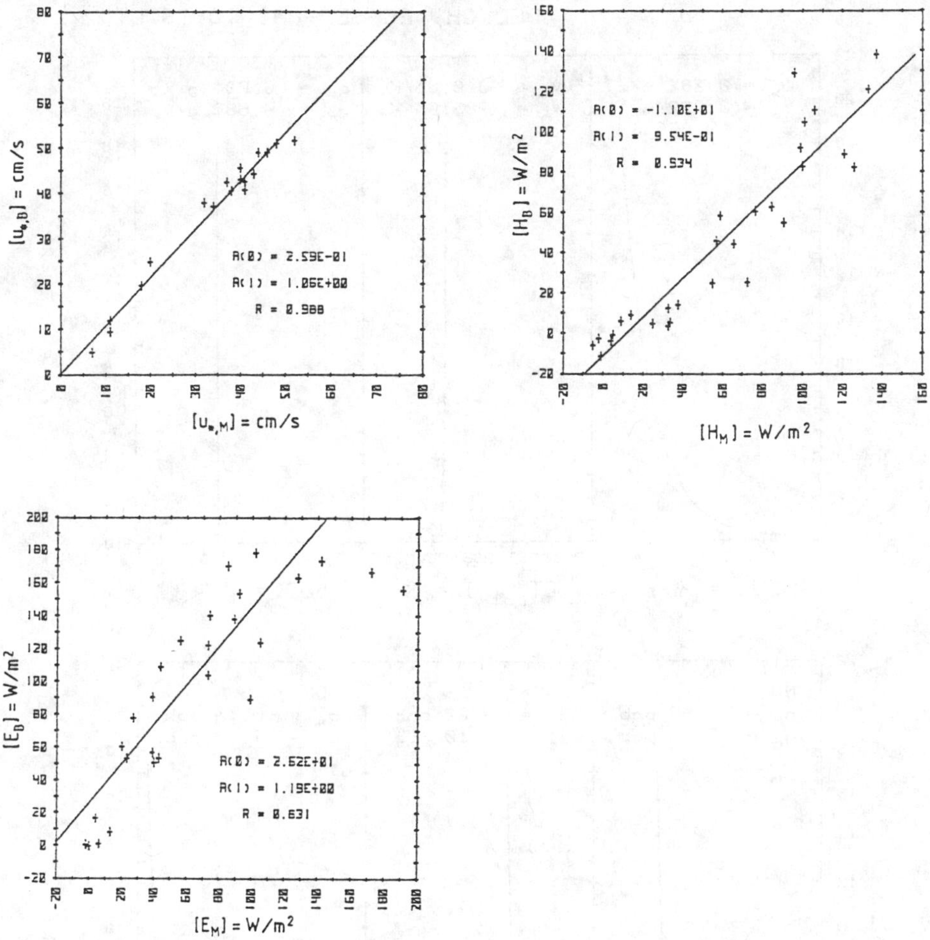

Figure 2. Calculated fluxes (index B) of momentum (u_*), sensible (H) and latent heat (E) vs. directly determined fluxes (index M). A(0) and A(1) are the coefficients of the least squares fit, and R is the correlation coefficient.

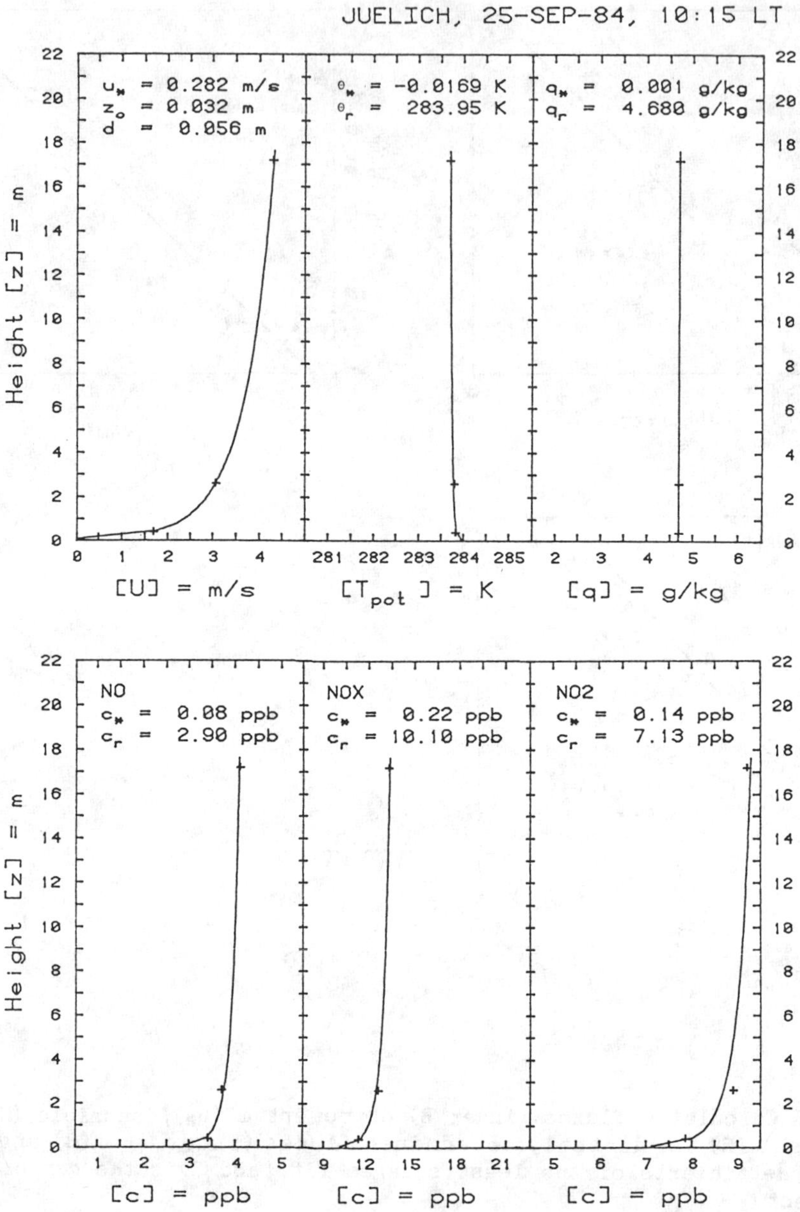

Figure 3. Typical vertical profiles of wind velocity (U), potential temperature (T_{pot}) and specific humidity (q) as well as NO, NO_2 and NO_x calculated for unstable conditions ($\Theta_* < 0$). The crosses represent the observed values and the solid lines the calculated profiles.

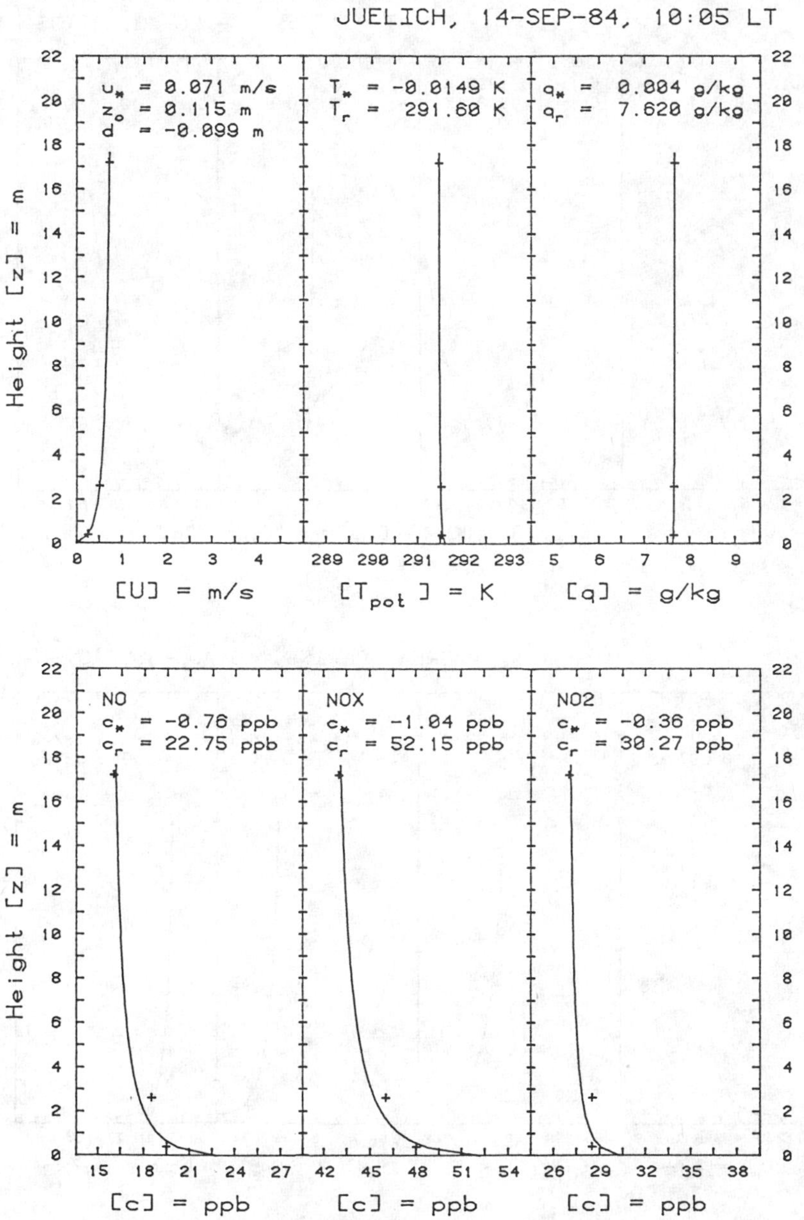

Figure 4. Typical vertical profiles of wind velocity (U), potential temperature (T_{pot}) and specific humidity (q) as well as NO, NO_2 and NO_x calculated for unstable conditions ($\Theta_* < 0$). The crosses represent the observed values and the solid lines the calculated profiles.

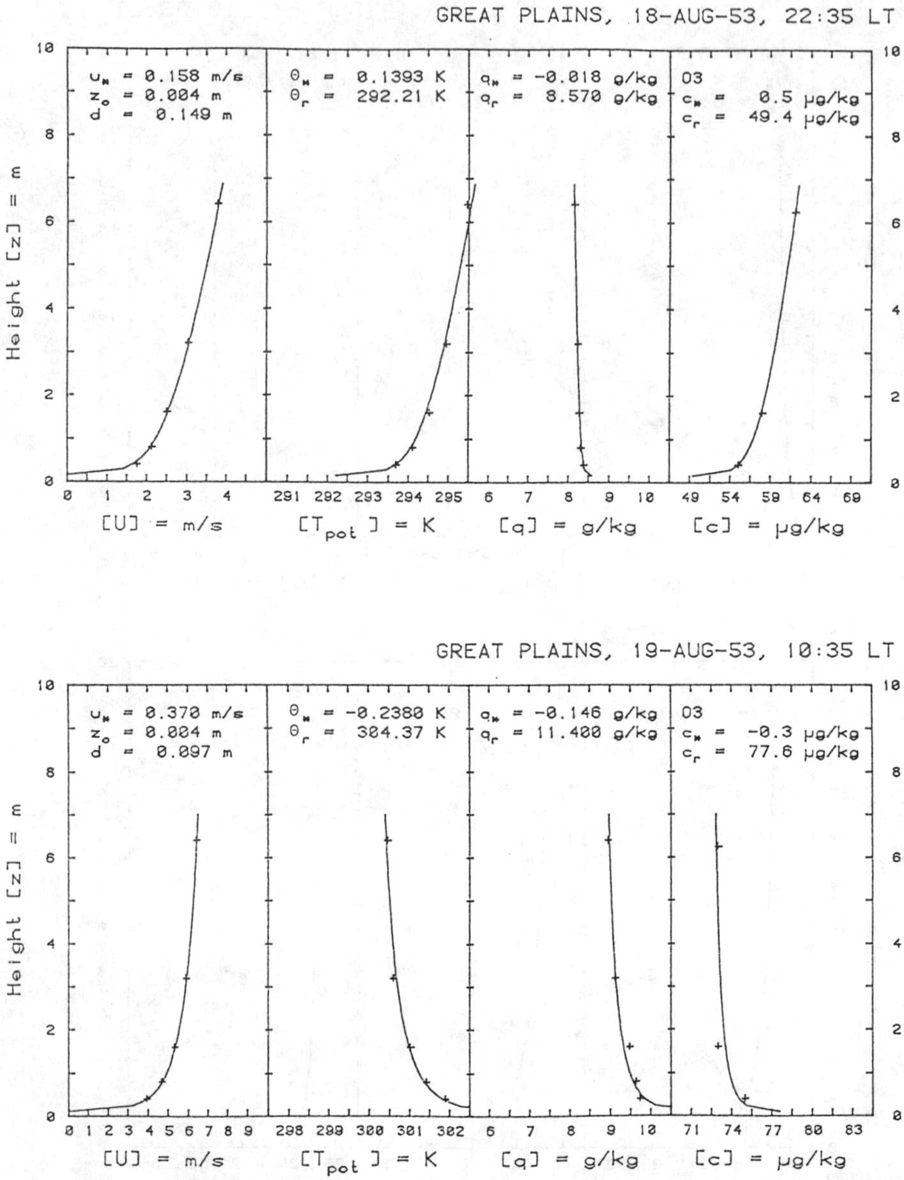

Figure 5. Typical vertical profiles of wind velocity (U), potential temperature (T_{pot}), specific humidity (q) and ozone (O_3) calculated for stable ($\Theta_* > 0$) and unstable conditions ($\Theta_* < 0$). The crosses represent the observed values and the solid lines the calculated profiles.

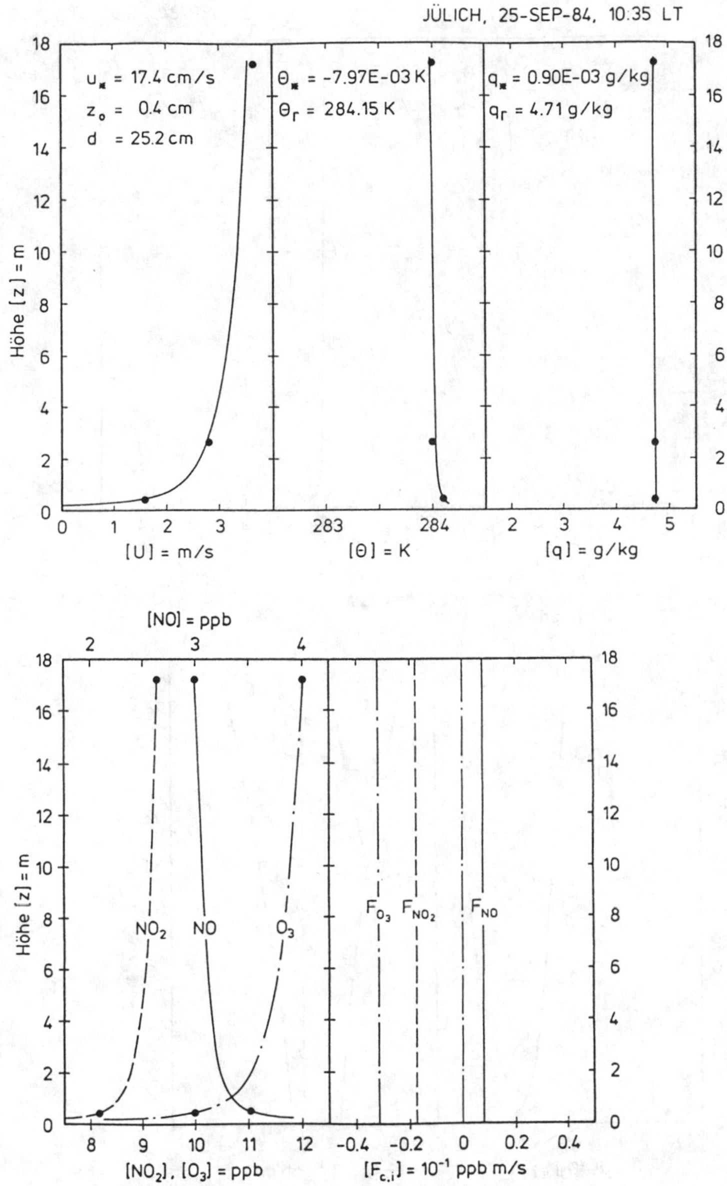

Figure 6. Typical vertical profiles of wind velocity (U), potential temperature (Θ) and specific humidity (q) as well as NO, NO_2 and O_3 calculated without chemical transformations ($F_{c,i}$ = const.). The dots represent the observed values and the lines the calculated profiles.

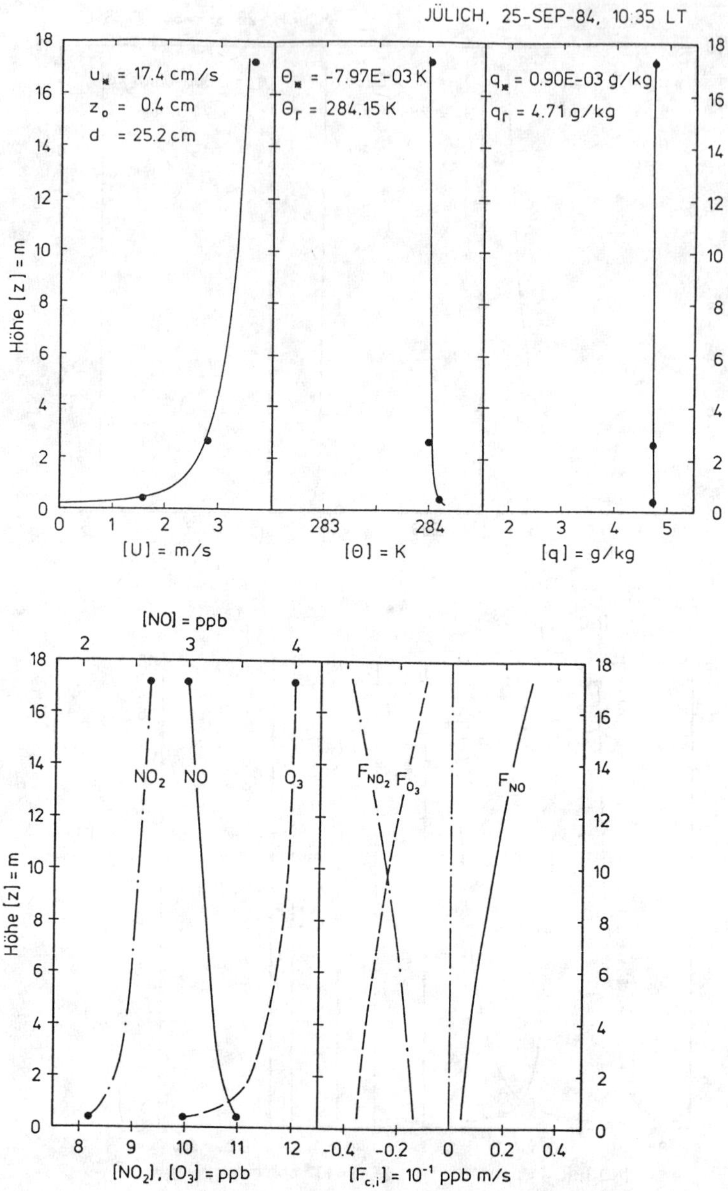

Figure 7. Typical vertical profiles of wind velocity (U), potential temperature (Θ) and specific humidity (q) as well as NO, NO_2 and O_3 calculated with chemical transformations ($F_{c,i} \neq$ const.). The dots represent the observed values and the lines the calculated profiles.

5. REFERENCES

Beyer R, Roth R (1976): GREIV I 1974-Meßdaten. Berichte des Inst f Meteor und Klimat der TU Hannover **16**

Businger JA (1973): Turbulent transfer in the atmospheric surface layer. In: Haugen DA (Ed.) Workshop on Micrometeorology. AMS, p 392

Businger JA (1986): Evaluation of the accuracy with which dry deposition can be measured with current micrometeorological techniques. J Appl Meteor **25**: 1100-1124

Chamberlain AC (1968): Transport of gases to and from surfaces with bluff and wave-like roughness elements. Quart J R Met Soc **94**: 318-332

Dyer AJ, Hicks BB (1970): Flux-gradient relationships in the constant flux layer. Quart J R Met Soc **96**:715-721

Galbally IE (1971): Ozone profiles and ozone fluxes in the atmospheric surface layer. Quart J R Met Soc **97**:18-29

Hasse L, Liss PS (1980): Gas exchange across the air-sea interface. Tellus **32**:470-481

Höfken KD, Meixner F, Müller KP, Ehhalt DH (1986): Untersuchungen zur trockenen Deposition und Emission von atmosphärischem NO, NO_2 und HNO_3 an natürlichen Oberflächen. Berichte der Kfa Jülich, Nr. **2054**

Lenschow DH (1982): Reactive trace species in the boundary layer from a micrometeorological perspective. J Meteor Soc Japan **60**(1):472-480

Lettau HH, Davidson B (1957): Exploring the atmosphere's first mile, Vol 2. Pergamon Press, New York, p201

Nikuradse J (1933): Strömungsgesetze in rauhen Rohren. Forsch Arb Ing-Wes **361**

Owen PR, Thomson WR (1963): Heat transfer across rough surfaces. J Fluid Mech **15**:321-334

Reichardt H (1951): Vollständige Darstellung der turbulenten Geschwindigkeitsverteilung in glatten Rohren. Z angew Math Mech **31**:208-219

Stearns CR (1970): Determining surface roughness and displacement height. Boundary-Layer Meteorology **1**:102-111

Swinbank WC, Dyer AJ (1967): An experimental study in micro-meteorology. Quart J R Met Soc **93**:494-500

Webb EK (1970): Profile relationships: The log-linear range, and extension to strong stability. Quart J R Met Soc **96**:67-90

The Importance of Heterogeneous Reactions for the Sulfate Formation in Plumes

S. Jordan and K. Nester
Kernforschungszentrum Karlsruhe GmbH
Laboratorium für Aerosolphysik und Filtertechnik/
Institut für Meteorologie und Klimaforschung
Postfach 3640, D-7500 Karlsruhe

ABSTRACT

Field measurements have shown that gas-to-particle conversion as well as heterogeneous reactions can be an important source of secondary sulfate aerosol in power plant plumes. Especially for high humidity conditions when plumes are mixed with clouds the measured rate of SO_2 removal in the plume can exceed 5-7% which can only be explained by heterogeneous SO_2-oxidation on Water containing fly ash particles or by H_2O_2 reactions.

A three-dimensional stationary numerical model is used to simulate the process of sulfate particle formation by heterogeneous reactions in humid atmosphere. This formation process on wet fly ash particles is parameterized and based on data of experimental studies in a 4 m^3 smog chamber. The sulfate formation on particles is limited by the sulfate formation itself, the pH-value and the relative humidity. Calculations are performed for plumes of coal fired power plants under different metereological conditions.

1. INTRODUCTION

Models which describe the chemical dispersion and transformation of the emitted products in the atmosphere can give an quantitative estimate of the environmental burden of emissions from industrial plants and power plants. Sponsored by the Umweltbundesamt a dispersion model was developed at KfK, which describes the dynamic and chemical effects caused by the interaction of the plumes from a power plant and a cooling tower. Necessary for the application of this dispersion model is the availability of formulas which describe the transformation processes of the emission products in the atmosphere. These physical-chemical moduls were derived from laboratory experiments *), at which the microphysical chemical conditions in the atmosphere were simulated. These experiments were performed to investigate the SO_2-oxidation process on particles and in droplets in power plant plumes. The aim of the laboratory experiments was to determine the oxidation rates under variable atmospheric parameters and the maximum formed sulfate in the aerosol system respectively the droplet system.

---

*) sponsored by the Umweltbundesamt and Projekt Europäisches Forschungszentrum für Maßnahmen zur Luftreinhaltung

K. Grefen and J. Löbel (eds.), Environmental Meteorology, 183–191.

2. SMOG CHAMBER EXPERIMENTS

Heterogeneous trace gas reactions on particles were investigated using different chemical reactors: The reactions on airborne particles were investigated in a smog chamber. Under special conditions and for the reactions on solid particles a fixed-bed reactor was used . The reaction in droplets was investigated by using a stirred vessel reactor [1]. Different kinds of particles and gases were introduced into the reactors. Most experiments were performed using industrial particles or fly ash particles sampled from the off-gas of power plants behind the electro-static precipitator [2]. After the physical and chemical characterization of the properties of the particles, the aerosol was injected together with the reacting gas into the chemical reactor. The chemical reaction in the reactor was followed by the analysis of the reaction products as a function of the reaction time. From these data rate laws were derived and introduced into a physico-chemical model of SO_2-reactions on aerosols [3]. Fig. 1 shows a typical sulfate formation process on fly ash. In this figure the formation of sulfate per gram aerosol is shown with increasing reaction time. The heterogeneous sulfate formation on particles is a saturation process. This process ends after 60 to 150 minutes. The process depends strongly on the relative humidity. For high humidities fast reaction and a high sulfate formation is observed. The main experimental results can be summarized for solid fly ash particles as follows:The oxidation rate and the maximum amount of sulfate formed on the particles depends on: Relative humidity, SO_2-concentration and the pH of the soluble portion of the particles.

Fig. 1

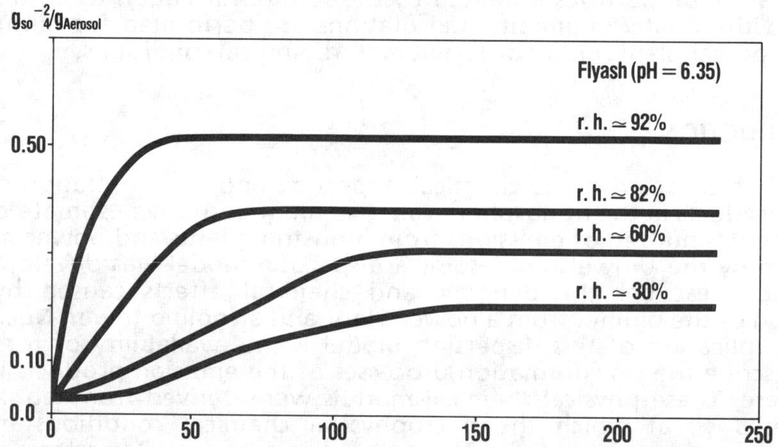

Sulfate formation on fly ash by catalytic reaction

The sulfate formation in droplets is limited to pH-values higher than 1.5 to 0.1. In droplets the sulfate formation rate and the maximum sulfate formed in the droplet depends on the water content of the particles (m_w), the SO_2-concentration and the pH-value. Only the sulfate formation rate in droplets increases with increasing concentration of catalytic ions in the water content of the particles.

Fig. 2

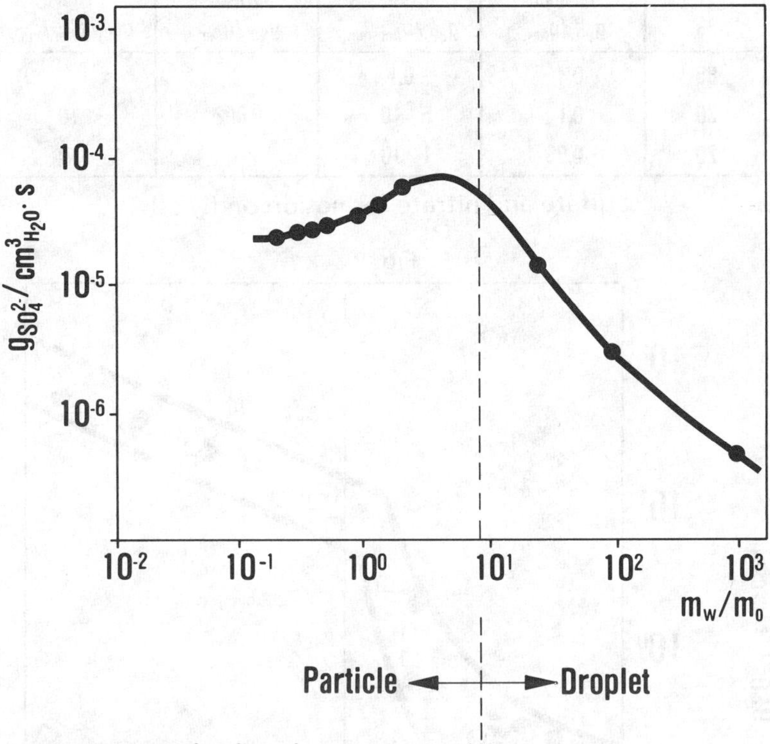

Reaction rates for fly ash particles
m_W = mass water
m_O = dry particle mass

Fig. 2 gives an overview on the measured sulfate formation rates on fly ash particles depending on the water content of particles. It is assumed that a particle becomes a droplet, when the mass of water is ten times the mass of the original particle. The formation rate increases with increasing relative humidity, reaches a maximum at about 90% relative humidity and decreases with higher water content of the particles by the dilution effect of the catalytic ions in the droplet. Fig. 3 gives an overview on the maximum sulfate formed on fly ash particles. The higher the water content of the particles is, the higher is the mass of sulfate formed on the fly ash particles.

Nitric oxides in the flue gas have only a limited influence on the sulfate formation on the fly ash particles (see Table 1). The nitrate on the particles is less than 1/10 of the formed sulfate and the rate of the nitrate formation is much slower than that of the sulfate formation.

From the simulation experiments parametric equations can be derived for the heterogeneous oxidation on particles and in droplets.

Table 1

r.h. %	Max. Sulfate formed $g_{SO_4}/g_{Aer.}$	Formation rate $g_{SO_4}/g_{Aer.\text{-}min.}$	Max. Nitrate formed $g_{NO_3}/g_{Aer.}$	Formation rate $g_{NO_3}/g_{Aer.\text{-}min.}$
85	0,4	0,1		
30	0,1	$5 \cdot 10^{-3}$	0,002	$< 10^{-5}$
20	0,05	$1 \cdot 10^{-3}$		

Sulfate and nitrate formation on fly ash

Fig. 3

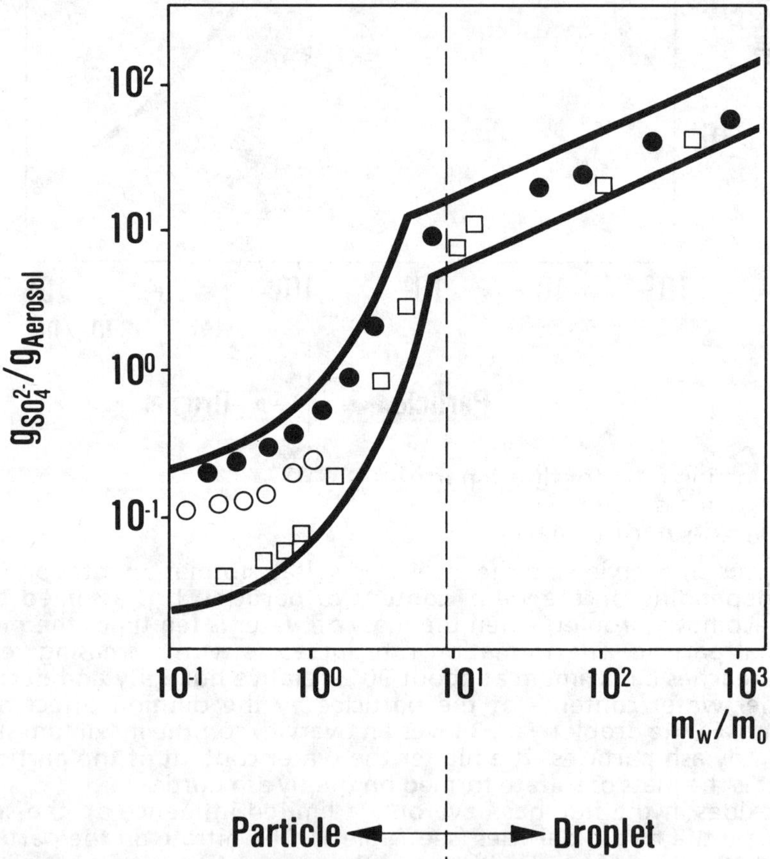

Maximum sulfate formed on fly ash particles
m_w = mass water on particle
m_o = mass of dry particle

Heterogeneous oxidation on particles:

$$\frac{d}{dt}\left(\frac{m_{SO_4^{2-}}}{m_A}\right) = a_o\left(1 - \frac{m_{SO_4^{2-}}}{m^{\infty}_{SO_4^{2-}}}\right)$$

$$a_o = A\frac{m_w}{m_A}(p_{SO_2})^{\beta}; \qquad 0.4<\beta<1$$

with $m_{SO_4^{2-}}$ = mass of sulfate
m_A = Mass of dry aerosol
$m_{SO_4^{2-}}$ = maximum formed sulfate
P_{SO_2} = partial pressure of SO_2
m_w = water content of particle

Heterogeneous oxidation in droplets:

$$\frac{d}{dt}[SO_4^{2-}]_t^w = K_t([SO_4^{2-}]_t^w)^{\varepsilon}\cdot([Fe(III)]^w)^{\lambda}\cdot[H^+]^{\phi}\cdot[HSO_3^-]^w$$

$\varepsilon = 1.16$; $\lambda = 0.61$; $\phi = -1.11$, $\omega = -0.43$

with $[SO_4^{2-}]_t^w$ = sulfate concentration in droplets
[Fe(III)] = iron (III) concentration
$[HSO_3^-]$ = HSO_3-ion concentration
k_t = rate constant

Because the iron content of the soluble portions of the fly ash particles was at least a factor of ten higher than other catalytic elements, the heterogeneous oxidation of SO_2 in droplets was derived for iron 3-ions.

To compare the calculated sulfate formation in plumes by heterogeneous reactions with the sulfate formation by homogeneous oxidation processes in the plume, a simple model for homogeneous oxidation of SO_2 by OH-radicals was used. The concentration of OH-radicals in the off-gas is determined by different coupled reactions with water-steam, carbon-hydroxides, ozone, NO_x and SO_2. Important for the oxidation process is a production of OH in the plume. The calculation of the homogeneous OH-oxidation uses a simplified scheme (compare [5]), which bases on the model introduced by Cocks and Fletcher [4]. Reaction rates were used valid for solar radiation on August 1st, 50° North and cloudless skies.

3. THE PLUME MODEL

For the computation of the interaction between the plumes of cooling tower and stack the three-dimensional model WALKÜRE [5] was developed. It relies on the following equations:

- The Navier-Stokes equation of motion,
- the continuity equation,
- the first theorem of thermodynamics,

- the conservation equation for water vapor and droplets,
- the equation for the turbulent kinetic energy and its dissipation.

The additional equation to simulate the sulfate particle formation by homogeneous and heterogeneous reactions in a humid atmosphere are:
- the diffusion equation for SO_2,
- the conservation equation for non-reactive gases and particles,
- the production equation of SO_4^{2-} by homogeneous oxidation processes,
- the production equation of SO_4^{2-} by heterogeneous oxidation processes.

This gives a set of eleven partial differential equations. The following essential assumptions are made to solve these equations:
- Steady-state conditions,
- the turbulent exchange in the direction of transport is neglected in favour of the horizontal advection,
- changes in density are taken into account via the Boussinesq approximation, otherwise the medium is considered to be incompressible.

These equations are converted into difference equations and solved numerically in the direction of transport. The numerical procedures used are the iterative method developed by Stone [6] and the cyclic reduction [7]. With the choice of appropriate initial conditions the model calculates single-plume development and mixing of several plumes from stack and cooling towers.

The initial rise of stack plumes is not treated in the WALKÜRE-model. Therefore the stack plume is described by an integral model up to a distance of some deca meters. Beyond this distance the conditions in the stack plume allow to employ the WALKÜRE model. The result of the integral model is used as initial condition for the WALKÜRE model.

4. CALCULATIONS ON THE SULFATE FORMATION IN PLUMES

For the calculation two weather categories are considered (table 2):
Category 1 has a high relative humidity (85 %) and has low solar radiation. The stratification of the atmosphere is neutral.

Table 2

Weather Category	Relative Humidity	Solar Radiation	Stratification of Atmosphere
I	high (85 %)	low	neutral
II	low (55 %)	high	slightly unstable

Weather Categories

Weather category 2 has low relative humidity (55 %), solar radiation is strong and the stratification of the atmosphere is slightly unstable.

The following figures 4 and 5 consider only the sulfate formation in particles by heterogeneous reactions. Fig. 4 shows high formation rates of sulfate close to the source. This is caused by the high relative humidity and high SO_2-concentration. With increasing distance from the stack the SO_2-concentration decreases. After 500 m distance from the stack the high formation rate caused by the high relative humidity in the plume is compensated by the dilution effect. The sulfate formation rate decreases after 3000 m because the sulfate formation process is already saturated. In comparison a wet desulfurized plume shows a slower decrease of the sulfate formation rate caused by the higher relative humidity in the plume. More sulfate is formed up to 1500 m. The sulfate content of particles is closer to the saturation value. That causes a faster decrease of the formation rate of sulfate with increasing distance.

Fig. 4

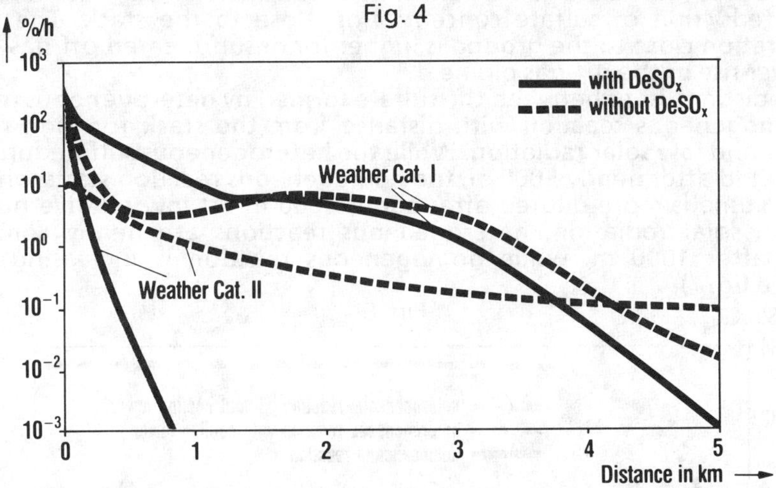

Sulfate formation rate in the power plant plume by heterogeneous reactions

Fig. 5

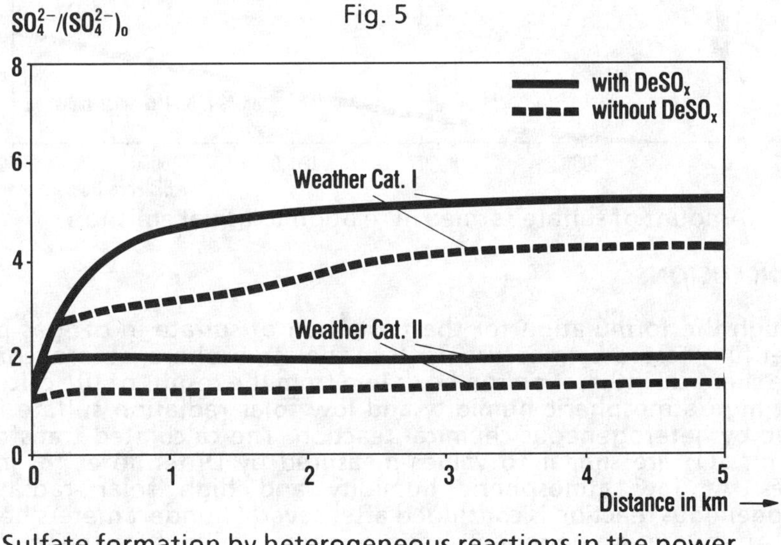

Sulfate formation by heterogeneous reactions in the power plant plume

For weather category 2 (dry weather) the differences in the formation rate are more characteristic . Without desulfurization the sulfate formation rate decreases from 10 %/h close to the stack to 1%/h in a distance of about 1000 m.With desulfurization the rate decreases much faster from relatively high values to near zero in a distance of 1000 m. Fig. 5 shows the relative increase of the sulfate content in the off-gas plume. The calculations show that in wet desulfurizated offgas the sulfate formation is higher than in not desulfurizated gas. Desulfurization brings reduction of SO_2 -concentration but no reduction of sulfate concentration close to the stack. The sulfate concentration close to the ground is higher for desulfurizated off-gas caused by a lower rise of the off-gas plume.

Fig. 6 distinguishes between the sulfate formed by heterogeneous reaction and homogeneous reaction with distance from the stack for high relative humidity and low solar radiation. While the heterogeneous sulfate formation is completed after nearly 5000 m, the homogeneous reaction starts when the off-gas plume is more diluted after about 3000 m. At low relative humidity and high solar radiation, heterogeneous reactions are nearly completed already after 1000 m, while homogeneous reaction is increasing slowly (compare fig. 7).

Fig. 6

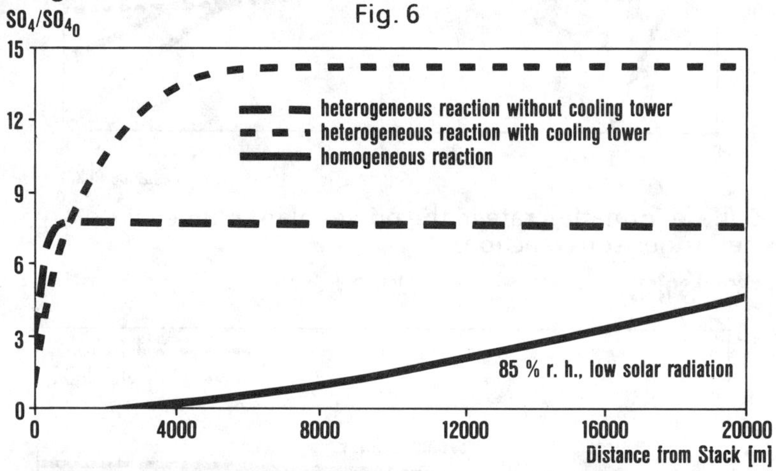

Amount of sulfate formed in relation to initial emission

5. CONCLUSIONS

Although the formulation for the formation of sulfate in off-gas plumes is not yet fully tested by measurements in the atmosphere, several conclusions on the chemical interaction can be drawn from the results of the calculations.

At high atmospheric humidity and low solar radiation sulfate is mainly formed by heterogeneous chemical reaction. The calculated transformation rates for SO_2 are similar to values measured by Dittenhöfer [8] in off-gas plumes. At low atmospheric humidity and high solar radiation the heterogeneous reaction is concluded after several hundert meters behind the stack. Afterwards homogeneous reactions are dominant for the formation of sulfate in the plume. The application of wet desulfurization decreases the

Fig. 7

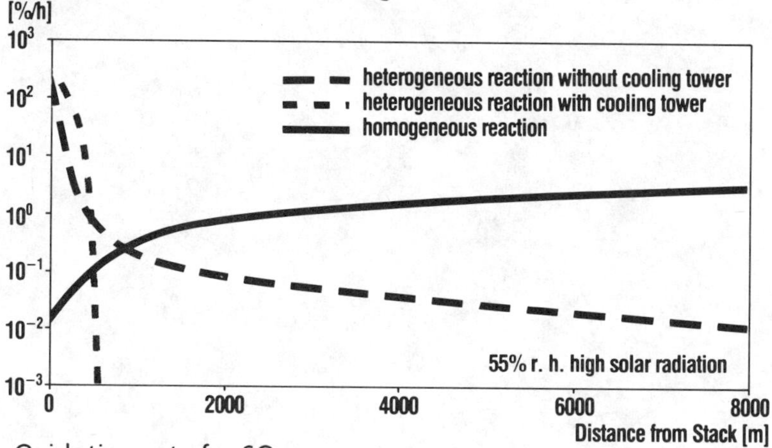

Oxidation rate for SO_2

SO_2-content in the off-gas plume, but increases the sulfate content by increased heterogeneous SO_2-reactions. The reduction of the plume rise increases the sulfate concentration near the ground for wet desulfurizated off-gas. To have an advantage from the desulfurization process also for the closer environment of the power plant, the plume rise must be increased. This can be attained by introducing the off-gas into a cooling tower. The disadvantage of this method are droplets in the cooling tower plume with pH-values between 2.5 and 3.5. Therefore the plume should not reach the ground.

REFERENCES

[1] R. Dlugi, 'SO_2-Oxidation in aerosol-particles and droplets', *J. Aerosol Sc.* **14** (1988), 292-297

[2] R. Dlugi, S. Jordan, E. Lindemann, 'Experimentelle Untersuchungen zur Entstehung von schwefelhaltigen Aerosolen unter atmosphärischen Bedingungen', KfK 3187, 1981

[3] R. Dlugi, S. Jordan, S. Manegold, 'Chemische Reaktionen und Aerosolverhalten in Rauchfahnen mit kondensierender Atmosphäre', KfK-Bericht 3817, 1985

[4] A.T. Cooks, J.S. Fletcher, 'A model of gas phase chemical reactions of power plant plume constituents', CERI-report, RD/L/R 1999, 1979

[5] K. Nester, H. Verenkotte, 'Dynamische und chemische Wechselwirkungen zwischen Abgas und Kühlturmfahnen', KfK-Bericht 3955, 1985

[6] H.L. Stone, 'Iterative Solution of implicit approximations of multidimensional partial differential equations', SIAM, *Journal of Numerical Analysis*, Vol. **5**, 1968, p. 530-558

[7] U. Schumann, 'Über die direkte Lösung der diskretisierten Poisson-Gleichung mittels zyklischer Reduktion'. Ext. Bericht, KfK-Ext. 8/75-6, 1975.

[8] Dittenhofer A.C., 'The chemical transformation of sulfur-dioxide to sulfate in the plume of the cool burning Keystone power plants', Thesis Penn. State University, (1979)

COMPARISON OF VARIOUS PRECIPITATION GAUGES AND SENSORS FOR THE DETERMINATION OF WET DEPOSITION

P. Winkler, S. Jobst
Deutscher Wetterdienst
Meteorologisches Observatorium Hamburg
Frahmredder 95
D-2000 Hamburg 65

ABSTRACT. Basic analysis of the precipiation process has shown that an intensity of 0.1 mm/h seems to be a reasonable upper threshold for activation or rain sensors. Determination of the response of sensors as a function of intensity has shown that some of the presently available sensors met this threshold, while others do not. Wet-only collectors with less sensitive sensors show for the most major ions lower concentrations than collectors with highly sensitive sensors. Obviously those gauges do not collect the first highly concentrated rain drops of precipitation events. Another problem arises from different funnel shapes of the various devices which cause increasing wetting capacity with increasing funnel surface related to the funnel opening area. This wetting capacity causes a memory effect which may be responsible for a more than 20 % higher trace substance content in a sample. Other problems with available devices exist in unreliable trace element concentrations because some of the collectors contaminate the samples with heavy metals.

1. INTRODUCTION

This paper presents preliminary results of an intercomparison of wet-only precipitation collectors and rain sensors, presently conducted at the Meteorological Observatory Hamburg in cooperation with the Institute for Inorganic and Applied Chemistry of the University of Hamburg. This intercomparison was started in the middle of 1986 and continues to the end of 1988. In this intercomparison 11 different collector types are taking part, 9 of them in two identical exemplars.

Since the evaluation of the data is not complete, some basic considerations concerning the collection of precipitation as well as parameters which influence the

K. Grefen and J. Löbel (eds.), Environmental Meteorology, 193–200.

collection process are addressed. Because the collectors were designed for the collection of rain we confine to this precipitation type. The basic question, when we investigate wet deposition, is "What is rain ?" The meteorological glossary defines rain as precipitation in the form of liquid water drops which have diameters greater than 0.5 mm. If the droplets are smaller, we distinguish drizzle. Drizzle droplets are displaced horizontally by the wind. The intensity of the rainfall can vary from very light, that means scattered drops which do not completely wet an exposed surface regardless of duration, over light, which means that the rainfall rate varies between a trace and an intensity up to 0.25 mm/h, to moderate and heavy. Here we have a problem similar to sunshine where we also know all transitions between blue sky over slight cirrus clouds which scarcely weaken the radiation, to dense cirrus, where only the disk of the sun can be seen but no shadows develop any more. In order to distinguish between sunshine or not it was necessary to define a threshold radiation intensity.

A similar problem arises when we try to define the conditions for wet deposition. Two basic questions can be formulated in this context:

1) Which precipitation amount per event contributes which percentage to total deposition ?
2) Is the precipitation amount falling at a certain intensity distributed in such a manner that it is possible to define a threshold intensity at which deposition gauges should start to collect precipitation in order to achieve comparable results ?

2. STRUCTURE OF PRECIPIATION

To answer these questions measurements of the precipitation distribution (Riedl, 1986) have been evaluated. Fig.1 shows the cumulative precipitation amount as function of intensity for the station Hohenpeissenberg in southern Germany. This curve shows that about 98% of the precipitation amount are falling at intensities greater than 0.1 mm/h and below this intensity only 2%. A precipitation sensor therefore should be able to recognize an intensity of at least 0.1 mm/h. From this curve we also see that the definition of a threshold intensity is reasonable.

It must be emphasized however, that this curve is valid for a typical climatic regime. In another region, for example in England where drizzle is more frequent, the intensity range between 0.1 and 1 mm/h may contribute a higher fraction to the total amount than given by the curve in fig. 1. On the other hand in the tropics with mostly heavy rains, slight intensities might contribute only a minor fraction to total deposition.

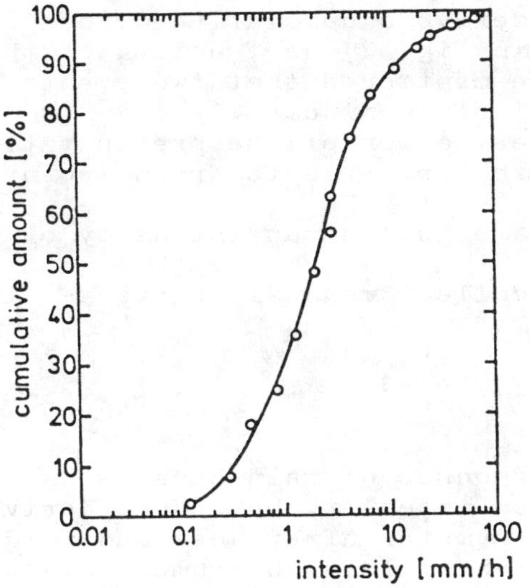

Figure 1:
Cumulative amount of precipitation as function of precipitation intensity. (Hohenpeissenberg 1972-1984, Riedl, 1987).

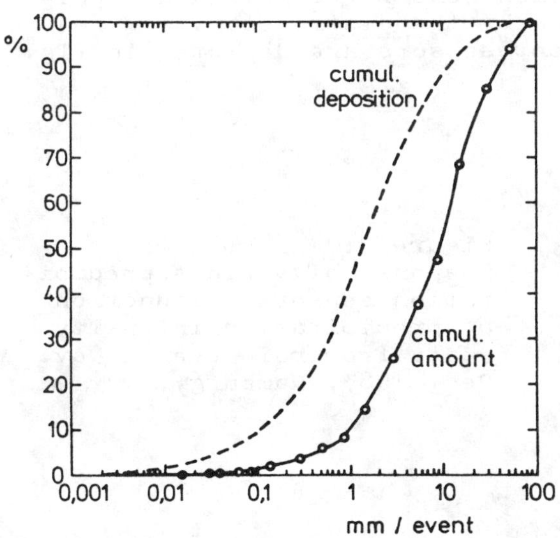

Figure 2:
Cumulative amount of precipitation and cumulative deposition of an arbitrary trace substance as function of the productiveness per event. (Data of Hohenpeissenberg 1980-1986, Riedl, 1987).

Another aspect can be derived from the cumulative amount as function of the productiveness of rain events (fig.2). Again only 1-2 % of the total precipitation amount is connected with events of a productiveness of 0.1 mm/event (solid curve). It is known from literature (e.g. Junge

1963, Georgii 1965) that events with little productiveness carry higher concentrations of trace substances. Assuming average conditions, we can derive a cumulative deposition distribution (dashed curve in fig. 2) as function of the productiveness. It should be mentioned that two events are separated by a time interval of 10 minutes.

From the above characterization of the precipitation structure we can formulate the criteria for rain sensors and collectors:

- are sensors able to give a signal at an intensity of 0.1 mm/h, and
- if so, is it possible to collect and analyse events with a productiveness of 0.1 mm.

3. RESPONSE OF RAIN SENSORS

From our field study the responses of rain sensors as a function of the precipitation intensity have been derived. We used the output of a distrometer after Joss and Waldvogel (1967) as reference signal. This instrument gives the raindrop size distribution every minute. The intensity is derived by integration of the size distribution. To characterize the sensors it has been determined how frequent a certain precipitation sensor recognized a precipitation intensity as given by the distrometer. The response curves of 6 different sensors are depicted in fig.3.

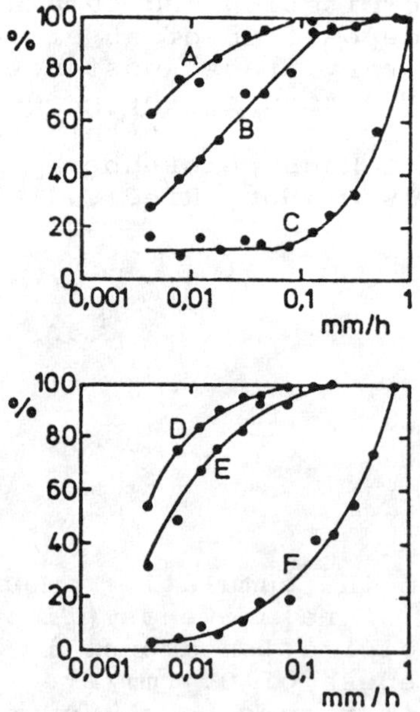

Figure 3:
Response of various precipitation sensors as function of precipitation intensity (data from rain events Nov.-Dec. 1987, Hamburg).

It can be seen that the sensors A,B,D,E met the threshold intensity of 0.1 mm/h satisfactorily, but that the sensors C,F do not react good enough. It will be shown later, that differences in sensor response are reflected in the trace substance content of the samples of wet only collectors.

4. RESULTS FROM RAIN COLLECTORS

In the measuring field 20 collectors of 11 designs have been installed. The collection bottles were changed at working days and analyzed for amount, el. conductivity, pH, anions ($SO_4^=$, NO_3^-, Cl^-), cations (NH_4^+, Na^+, K^+, Ca^{++}, Mg^{++}), and trace elements Fe, Cu, Cd, Pb, Al. For most of these parameters deviations between the various samplers scatter in most cases within $\pm$ 20% exeptionally up to 30%. For several samplers contamination due to use of improper materials has been identified. This causes even larger deviations. It is not the focus of this paper to discuss these problems here in detail because the measuring program has not been finished yet. Instead, it seems of interest to discuss more general aspects.

In the following table the mean values of daily deviations from the average in percent of the various collectors for nitrate are shown:

sampler	1	11	2	12	3	13	4	14	5	15
NO_3(%)	92	91	83	85	87	97	108	105	91	90
sampler	6	16	7	17	8	18	9	19		
NO_3(%)	101	104	103	130	92	85	90	92		

For other ions the scatter is similar and the main question arises if the scatter arises from random influences or if design or constructional aspects can be made responsible. And this arises the question of central importance: Which is the true reference value ?

In a first step we have devided the instrument pool in two groups: those with good and those with poor sensitivity of sensors. It turned out that the instruments with low sensitive sensors give on the average about 10% lower concentrations for many trace substances. We can conclude from this result that instruments with poorly reacting sensors obviously frequently fail in collecting the first highly concentrated rain drops of an event.

Another aspect of general interest concernces the influence of the wetting capacity. One must ask what happens with an event of little productiveness. Some of the precipitation water will adhere to the funnel walls after the end of an event and evaporate there so that the trace substances stick at the walls. A follow-up event will redissolve the material and wash it into the collection bottle. Since the funnels of the various instruments are differently shaped the wetting capacity varies by more than a factor of 3, even if the total funnel surface is normalized to the opening area. In terms of precipitation depth the amount of water adhering to the walls corresponds to 0.07 to 0.24 mm. It can be shown that the precipitation amount as determined with instruments of non-standardized funnel shape becomes systematically lower with increasing wetting capacity.

This water remaining and evaporating in the funnel and releasing its trace substances there causes a memory effect. An experiment, where artifical rain water (el. conductivity 128 µS/cm) was applied and evaporated, was conducted in order to quantify this memory effect. A follow-up event with destilled water corresponding to a rain of a productiveness of 1 mm depth confirmed that increasing wetting capacity increased the memory effect. The rinsing water showed conductivities up to 16 % of the previous event. An event of smaller productiveness would have produced a higher memory effect.

From precipitation statistics it is known that more than 80 % of all event have a productiveness of less than 0.1 mm (fig. 4). Although the precipitation water of these events remains almost completely in the funnels it will not evaporate completely until the next event in all cases, however, it will cause a memory effect in numerous

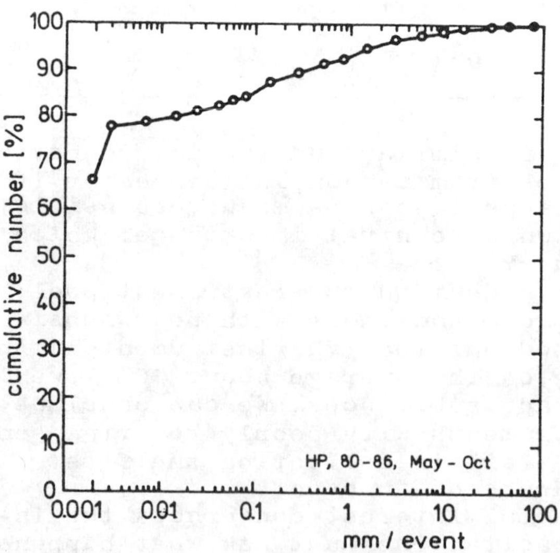

Figure 4:
Cumulative number frequencies of precipitation events as function of the productiveness per event. (Data of Hohenpeissenberg 1980-1986, Riedl, 1987).

cases, especially in those where the follow-up event in turn has a small productiveness. In order to harmonize and standardize the technique of measurement for wet deposition the above described results strongly support the postulation that funnels of different collection devices are comparable in their ratio total funnel surface to funnel opening area, and shaped in such a way that the wetting capacity becomes minimal and of comparable magnitude.

Another consequence of these findings concerns the interpretation of deposition measurements. Caution should be taken especially in those cases when the precipitation amount is small. Although the precipitation amount collected may be nearly correct, the concentration may be wrong. On the other side the deposition may be nearly correct if the sensor has a high sensitivity.

5. CONCLUSIONS

During a still ongoing investigation of rain sensors and rain collectors the following general results have been obtained:

- The structure of precipitation seems to allow the definition of a threshold intensity of about 0.1 mm/h above which 98 % of the total water amount are deposited.
- Precipitation sensors are presently not comparable with respect to their response to slight precipitation intensities or to a certain threshold intensity.
- Precipitation collectors may indicate the trace substance content about 10 % to low if their sensors are of poor sensitivity.
- Precipitation collectors show a wetting capacity depending on the ratio of total funnel surface to funnel opening area.
- Depending on the wetting capacity a memory effect is caused, which tends the trace substance content to be more or less increased. With increasing wetting capacity the precipitation amount is indicated to low.

There remain some open questions which have to be investigated in future: Shall the funnels be rinsed regularly with destilled water in order to avoid the memory effect ? Are collectors using funnels designed optimally or do bucket collectors give better results ? How should sensors be constructed in order to respond to a certain threshold intensity ? Which are the pathways by which samples are contaminated by use of improper materials ? Results on these problems will be published elsewhere.

ACKNOWLEDGEMENT

This work was supported by the Ministery of Research and Technology (grant 07431073). The supply with data of the Hohenpeissenberg by J. Riedl is greatfully acknowledged.

LITERATURE

Georgii, H.W.: Untersuchungen über Ausregnen und Auswaschen atmosphärischer Spurenstoffe durch Wolken und Niederschlag. Berichte DWD, Bd.14 Nr. 100 (1965).

Joss, J., A. Waldvogel: Ein Spektrograph für Niederschlagstropfen mit automatischer Auswertung. PAGEOPH 68 (1967) 240-246.

Junge, Ch.: Air chemistry and radioactivity. Academic Press, New York 1963.

Riedl, J.: Private communication 1987.

EXPERIMENTAL INVESTIGATIONS ON THE DEPOSITION OF TRACE ELEMENTS IN FOREST AREAS

S. Grosch and G. Schmitt
Institute for Meteorology and Geophysics
University of Frankfurt
Feldbergstraße 47
6000 - Frankfurt/Main
Federal Republic of Germany

ABSTRACT Detailed deposition measurements aimed to determine the efficiency of a canopy for dry deposition were carried out within a forest area located about 25 km NW of the city of Frankfurt since June 1985. Continuous wet-only sampling was applied below the canopy of a spruce stand and on a nearby clearing. In 1987 the sampling site was extended, now providing sampling facilities below deciduous trees (beech/oak). Dry deposition rates were derived by using a flux balance concept mainly based on the different element concentrations in throughfall compared to precipitation. Data about aerosol and trace gas concentrations measured throughout the same time allowed to estimate deposition velocities (v_d) for the aerosol components Fe, Pb, Cd, $SO_4^=$, NO_3^- and Cl^- as well as the gases SO_2 and NO_2 valid for a spruce canopy. The results range from 1.26 cm/s to 2.08 cm/s for aerosol particles whereas lower figures are estimated for SO_2 (0.85 cm/s) and NO_2 (0.48 cm/s). With a fraction of about 70-80%, dry deposition shows up as the most important flux of the total atmospheric input. Fog has to be considered as an important deposition pathway. Under certain conditions the interception of fog droplets within a canopy can lead to measurable amounts of throughfall. Preliminary measurements enables differentiated estimations about the interception deposition caused by fog.

1. INTRODUCTION

Chemical compounds bound on aerosol particles and gases once released into the atmosphere from natural or antropogenic sources, will be removed again by wet and dry deposition. Due to the still apparent problems of forest decline (BMELF, 1986) and the impact of air pollution, the determination of deposition rates of atmospheric constituents especially in forest areas has become a subject of particular interest.
Whereas the sampling methods for the wet deposition are fairly well established, the direct estimation of the dry deposition appears to be

K. Grefen and J. Löbel (eds.), Environmental Meteorology, 201–216.

much more complicated. Besides the meteorological conditions within the atmospheric boundary layer, the specific surface properties highly influence the efficiency of dry deposition processes (Sehmel, 1980).
As a third deposition pathway of particular importance for high elavated forest regions, fog has to be considered . Because of the enhanced element concentrations in fogwater (Schmitt, 1987; Winkler, 1986) the impaction of fog droplets within a canopy can lead to high amounts of additional deposition.
In former investigations wet and dry deposition rates in different forest areas were determined, the latter using a sampling device based on the "Bergerhoff method" (Georgii et al., 1986; Grosch and Schmitt, 1984). Due to the different surface properties, results obtained from this method cannot simply be transfered to a natural surface. As an approach to determine the dry deposition within a canopy, a proper flux balance has to be established . Therefore simultaneous sampling of throughfall and of unaffected rainwater on a nearby clearing has been carried out since summer 1985. From additional air concentration measurements approximate mean deposition velocities can be estimated .
The interception of fogwater will contribute to the total deposition in forest areas with suitable conditions (frequency, density and duration of fog events). Extensive investigations on fogwater chemistry showed an extreme enhancement of concentrations compared to rainwater (Schmitt, 1987).

2. MATERIALS AND METHODS

2.1. Measuring site and instrumentation

The sampling site "Königstein" is located about 25 km NW of the city of Frankfurt at an elevation of 520 m asl.. This station at the same time is part of another measuring network. Detailed information about the additional measurements and the instrumentation can be taken from Hess. Landesanst. f. Umwelt (1984). The station basicly includes a measuring site within a spruce-stand (Picea abies) and one on a nearby clearing as reference (distance: about 250m). Measurements of aerosol particles (sampling period: 14 days) and trace gases (continuously: SO_2, NO, NO_2, O_3) are applied. On the clearing the wet deposition is sampled by means of a wet and dry deposition sampler. For further information refer to Georgii et al. (1984); a detailed description of the sampling device is given by Rohbock and Georgii (1982).
For the investigation of throughfall, a system of eight wet-only samplers has been spread out on an area of approximately 30 40 m. Fig. 1 shows a sketch of this sampler. All samplers are governed by one rainsensor. The samples are taken on a daily basis. Continuous and simultaneous sampling has been applied since June 1985. In May 1987 an additional site with comparable characteristics (number of samplers, area, distance to the clearing) within a beech/oak-stand (Fagus silv./Quercus robur) was build up.

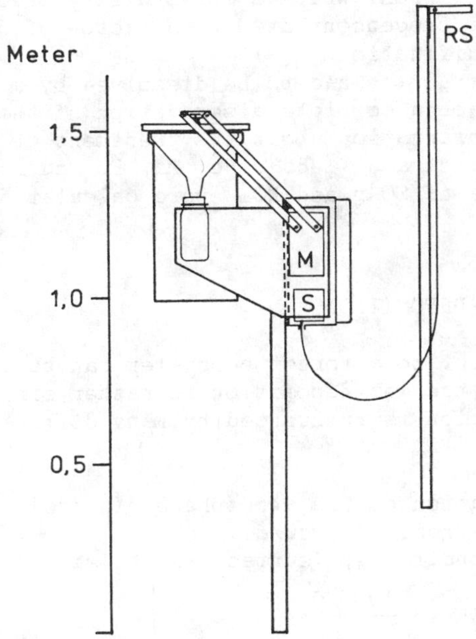

Fig. 1: Wet-only sampler. RS: rainsensor, S: electronical steering
M: motor

To estimate the amount of percolating fogwater, simple systems of samplers (just a funnel and sampling bottle) were located underneath spruce trees in a height of 1.50 m. One sampler was located on a forest-aisle as a reference to make sure, that the samples are not influenced from sedimenting droplets. These investigation were carried out at a mountain station with high frequency of fog events. The station "Kleiner Feldberg" is located about 30 km NW of the city of Frankfurt at an altitude of 805 m asl..

2.2. Sample treatment and chemical analysis

During the first six months all throughfall samples were analysed separately. Later on, after determining the individual volumes (by weight), aliquot parts of the eight samples were combined for a mean, volume-weighted sample.

All precipitation samples were first analysed for pH (glass electrode, Ingold 405-88-1m) and total conductivity (electrode: WTW LTA 100). Next a vacuum filtration was applied (filter: Satorius 11306, celluloseacetat, 0.45 µm poresize) thus seperating a "soluble" and "insoluble" fraction. The anions $SO_4^{=}$, NO_3^- and Cl^- in the soluble fraction were analysed with Ioncromatography. For other analysis the soluble part of throughfall

samples were acidified (1% HNO_3) whereas for ordinary precipitation samples this fraction was preconcentrated by a Factor of 5 by heating and evaporating prior to acidification.
The filter as insoluble fraction had to be dissolved by means of HNO_3 and additional heating to reach a complete disgradation of the filter material. For detailed information about the treatment of the insoluble fraction refer to Georgii et al. (1986). Fe, Mn, Pb and Cd were analysed by means of AAS. All data of $SO_4^=$ and NO_3^- were calculated as S and N respecively.

3. ESTIMATION OF THE DRY DEPOSITION

The total atmospheric input to a forest ecosystem can be characterized as shown in fig. 2. Whereas the wet deposition is rather simple to describe, the amount of dry deposition is influenced by many different parameters to be summerized as:

a) thermal stratification of the atmosphere; (dT/dz)
b) conditions of airstream; (u, du/dz)
c) special surface conditions; (surface roughness)

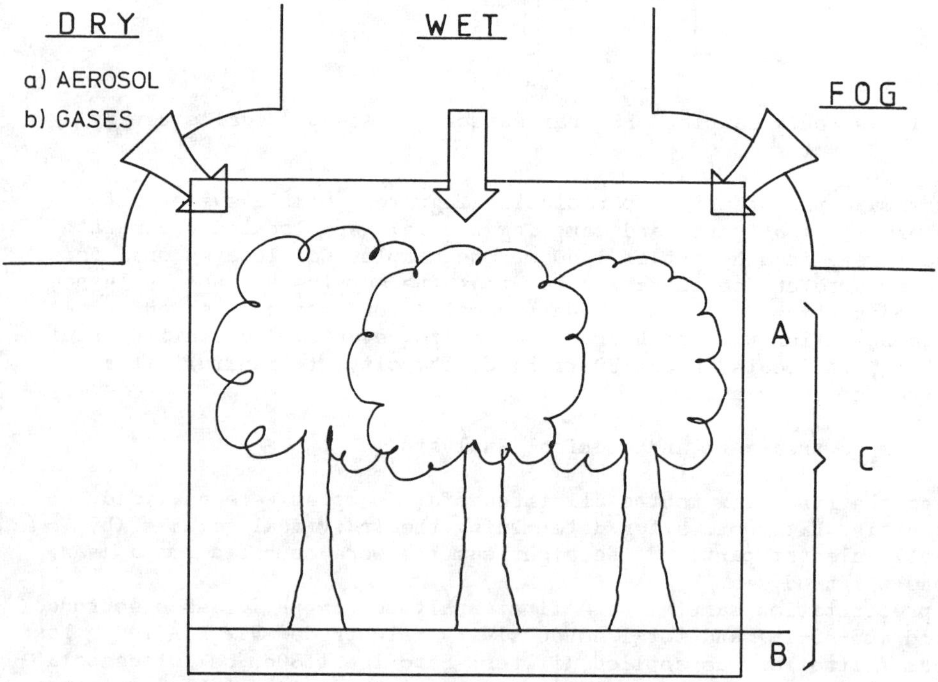

Fig. 2: Total atmospheric input into a forest area (C) differentiated due to different compartments (A: canopy, B: forest floor) and deposition pathways

Especially because natural surface conditions cannot be simulated sufficiently by artificial surfaces, direct measurements of dry deposition rates are always limited.
As mentioned above, considerable amounts of elements can also be deposited by fog interception.

This model (fig. 2) does not take into account possible element fluxes given by:

a) direct dry deposition on the forest floor
b) stemflow.

The first flux (a) is assumed to be negligible; stem flow (b) can be neglected for spruce but not for beech due to a very smooth bark and the different structure of the canopy.
Aerosol particles and gases are deposited within the canopy because of sedimentation, turbulent transport, molecular transport and impaction. This deposition will be washed down (totally or partly) with the next precipitation event leading to higher concentrations in the throughfall. This consideration is complicated because of the following processes:

a) absorption and adsorption on the leaves
b) release of material out of the leaves during precipitation (leaching effect)

For some elements (e.g. Mn, K, Ca, Mg) the leaching effect is well known and documented (Ulrich et al., Schmidt, 1987); whereas only little is known about ab- and adsorption effects. Anyway, for many elements it seems to be possible to give an estimation of the dry deposition.

Considering a certain forest area (fig. 2) as a closed system (C), the total atmospheric input is given by the deposition in an upper compartment (A) and a lower compartment (B) with a shift in time and space from A to B caused by the permanent change between periods of precipitation and dryness. Calculating the difference between wet element fluxes below the canopy and on the clearing, basicly leads to an approximation of the dry deposition within the canopy.
Fig. 3 illustrates the total mass-balance including all fluxes to be considered. Besides the ab- and adsorption (A) and leaching (L) effects, the additional fluxes Dc and Dc' appear in equation (1). Dc is the mass of elements wet deposited in the canopy with the current precipitation event. Dc' is the same kind of deposition from the previous precipitation event. The amout of rainwater kept in the canopy called "capacity" (C), can be calculated as the difference of precipitation amount on the clearing and below the canopy. A mean value was found to be about 3 mm for the spruce stand. This figure can be variable by at least a factor of 2.

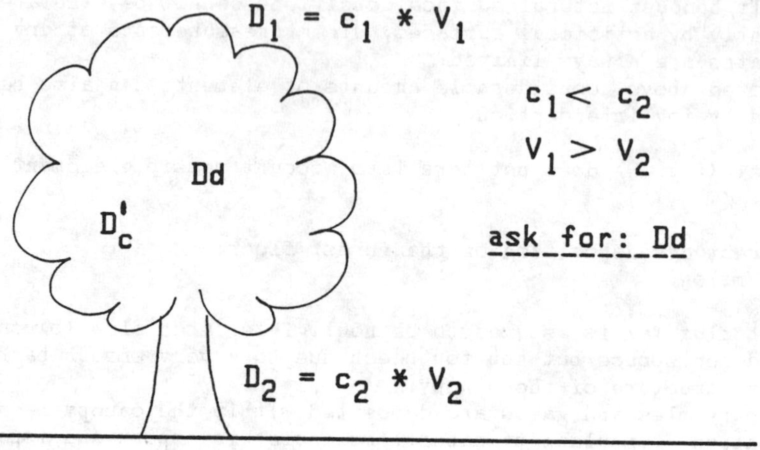

c = concentration — Index 1: clearing

V = precip. amount — Index 2: throughfall

D = wet deposition — Index c: canopy

Dd = dry depo. within the canopy

$C = V_1 - V_2 \quad \text{--->} \quad D_c = c_1 * C$

ESTIMATION OF THE DRY DEPOSITION:

$$Dd = D_2 - D_1 + D_c - D_c' + A - L \qquad (1)$$

Fig. 3: Flux balance to estimate the amount of dry deposition within a canopy for single dry periods.

4. RESULTS AND DISCUSSION

The following results are mainly restricted to data received from the spruce stand. At the same time the data base from the beech/oak-stand was not sufficient large enough for similar evaluations.

4.1. Sampling inhomogenity below the canopy

From analysis of the individual eight throughfall samples (spruce-stand) mean concentration values and standard deviations were calculated. Using the standard deviations of single precipitation events itself again as data base, it is possible to derive "mean standard deviations" with respective standard deviations for all analysed components. The result shows the mean standard deviation to be about 20-30% for all investigated quantities. The variability of these values again is about 10-15%.
In account of the naturally high inhomogeneous sampling conditions below the canopy, these deviations are still acceptable leading to reliable mean concentration values. Höfken and Gravenhorst (1982) showed for similar conditions (dense spruce-stand, eight samplers) that the mean concentration calculated by using not less than five samplers, already was not more than 5% different from the mean value derived from the complete number of samplers.

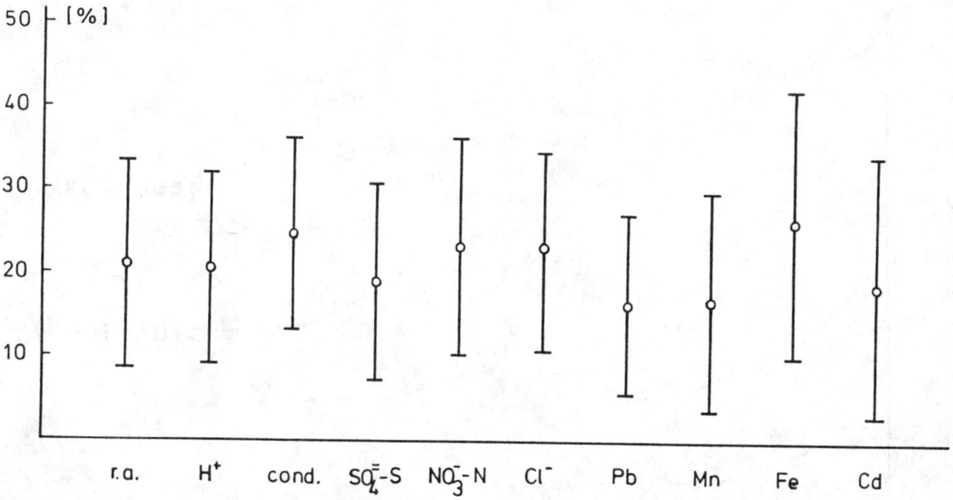

Fig. 4: Mean standard deviations of element concentrations in throughfall, characterizing the inhomogeneous sampling conditions below a canopy. 8 samplers per precipitation event.

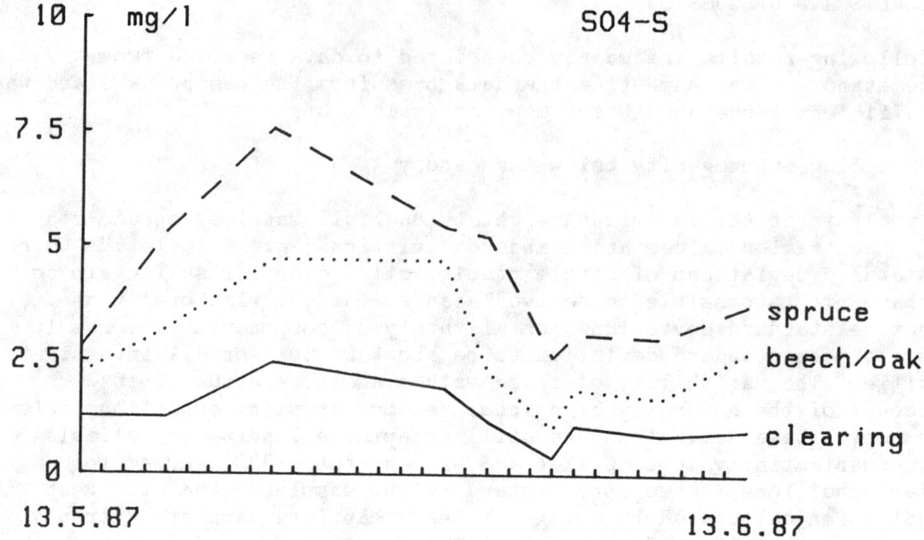

Fig. 5: $SO_4^{=}$- concentrations of throughfall below different tree species and of unaffected precipitation on the clearing

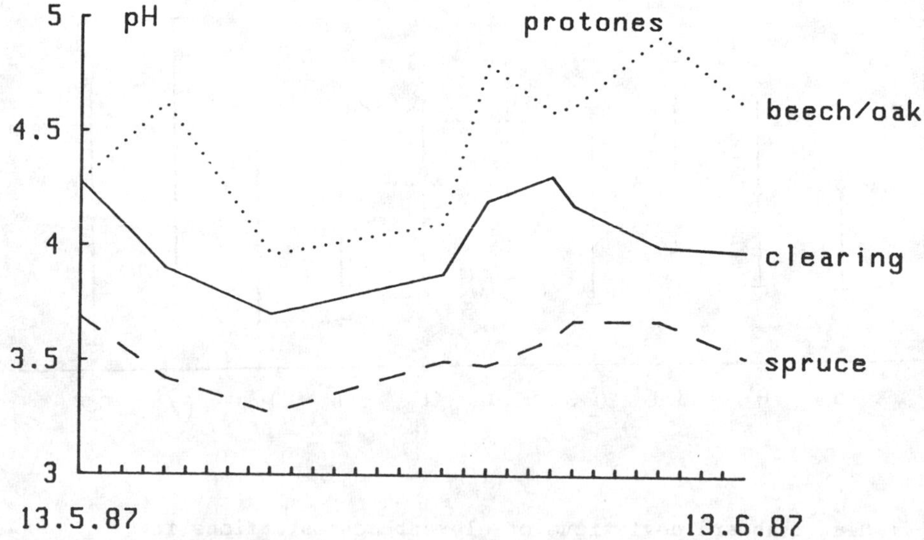

Fig. 6: pH values of throughfall below different tree species and of unaffected precipitation on the clearing

4.3 Comparative studies of throughfall below coniferous- and deciduous trees

Since May 1987 it was possible to compare throughfall samples from both forest stands (spruce and beech/oak) with the precipitation sampled on the clearing. For one month (13.5. - 13.6.1987; 9 events) the concentrations observed at all three sites are shown in fig. 5 and fig. 6 exemplary for $SO_4^{=}$-S and pH. The $SO_4^{=}$-S concentrations are small on the clearing followed by the beech/oak-stand with highest values for the spruce stand. A parallel trend in time is clearly to realize. The same behaviour can be found for all other elements. The pH-values on the clearing are higher than below the coniferous- but lower than to be found below the deciduous trees. This has to be interpreted as an effect of the so called "leaf-buffering"; known to be more effective for deciduous- than for coniferous trees (Ulrich 1983). This effect causes an exchange of protones to Mg^{++}- and Ca^{+}-ions on the leaf surfaces leading to higher Mg^{++}- and Ca^{+}- but lower H^{+}-concentrations in the throughfall. According to the pH-values shown in fig.6, proton concentrations in throughfall below these species can be different by a factor of about 10 !
The same general pattern found for other components is thought to be caused also by different dry deposition efficiencies.

4.4 Mean dry deposition and deposition velocities

Different gases and aerosol particles will be affected by dry deposition to a different extent. As a quantity describing this efficiency, the deposition velocity (v_d) can be used (Chamberlain 1955). This quantity correlates the deposition (D) to the mean, time integrated concentration (C) of the compound to be considered.

$$D = v_d \cdot c \qquad (2)$$

Using equation (1) dry deposition rates for single dry periods (times between precipitation events) were calculated finally leading to mean values for 1986 (spruce stand).

For the calculation of the mean value, only dry period > 10 hours were taken into account. Also the precipitation amount had to be > 5 mm to ensure that > 90 % of the dry deposition will be washed down (Höfken and Gravenhorst ,1981). These deposition rates are a result of aerosol as well as gaseous deposition. For the same period of time mean values of the aerosol concentration (of $SO_4^{=}$, NO_3^{-}, Cl^{-}, Pb, Mn, Fe and Cd) as well as the gaseous concentration (of SO_2 and NO_2) were derived from measurements on the clearing. For the elements with neglicable gasphase (Fe, Mn, Pb, Cd) deposition velocities for corresponding aerosol particles can be calculated by equation (2). Mn was excluded from further considerations because the leaching effect was recognized to be far dominating. From leaching experiments different authors found Mn in throughfall below spruce to be caused by leaching with 20-90% (Schmidt,1987). The deposition velocity for Cl^- was first also estimated by equation (2) considering the possible gaseous deposition (Cl, HCL) to be small. The deposition rate of Cl^- was reduced for 30 % according to leaching (Höfken and Gravenhorst, 1981).

		Pb	Fe	Cd	Cl^-
c	(µg/m³)	0.077	0.137	0.001	0.364
D_d	(mg/m²·d)	0.084	0.458	0.018	4.19
v_d	(cm/s)	1.26	1.67	2.08	14.0

Tab. 1: Deposition velocities (v_d) of aerosol particles for a spruce canopy calculated by equation (2). Dry deposition rates (D_d) estimated by equation (1). Aerosol particle concentrations measured on the clearing. Mean values for 1986.

The results (Tab. 1) show rather high deposition velocities of approximately 1 - 2 cm/s characterizing this spruce canopy. The extraordinary high v_d for Cl^- must be interpreted as gaseous deposition not to be neglicable and a possibly higher leaching. Similar results have been reported by Höfken and Gravenhorst (1981).

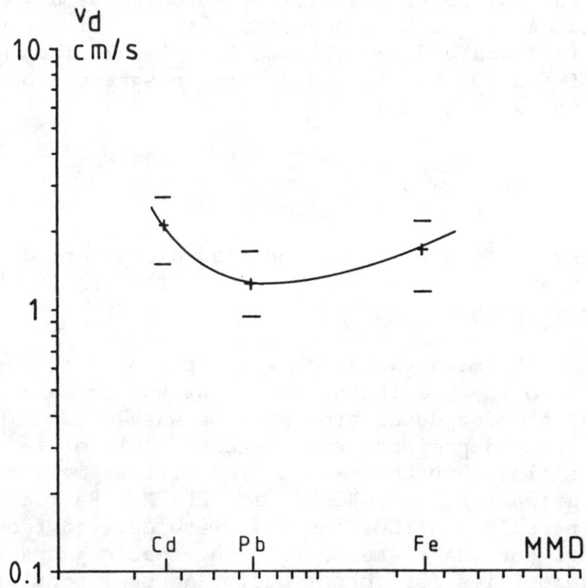

Fig. 7: Deposition velocity as function of the mass median diameter of the aerosol particles.

Knowing the mass median diameter (MMD) as characteristic quantity for the aerosol particle distribution from former investigations mainly applied at the same place (Ecker, 1985; Wolz, 1987; Höfken and Gravenhorst, 1981) the deposition velocity was plotted versus the MMD (Fig. 7).
Deposition processes of aerosol particles are predominantly depending on the particles size. Following this, it is possible to derive deposition velocities for aerosol particles of $SO_4^{=}$, NO_3^- and Cl^- from fig. 7.

Tab. 2 shows the MMD and the corresponding v_d. The deposition velocity derived for Cl^- on this theoretical way appears to be more reliable.
A certain part of the total $SO_4^{=}$-S dry deposition will be caused by SO_2 - deposition whereas the gases NO, NO_2 and HNO_3 might contribute to the total NO_3^-- deposition. Assuming the contribution from NO and HNO_3 as small, it is possible to derive deposition velocities for SO_2 and NO_2 from the data as well. Neglecting HNO_3 in principle might call for some error because of high v_d reported for this component (Höfken et al.,1986) although the concentrations usually are very small.
Reducing the dry deposition rates by the aerosol deposition ($SO_4^{=}$-S, NO_3^--N) remains figures related just to SO_2 and NO_2 respectively.

	$SO_4^{=}$	NO_3^-	Cl^-
MMD (μm)	0.61	1.6	2.2
v_d (cm/s)	1.26	1.45	1.6

Tab. 2: Deposition velocities (v_d) of aerosol particles for a spruce canopy according to the dependance v_d = f(MMD) shown in fig. 7. Mean values for 1986.

With mean SO_2- and NO_2-concentrations derived from measurements on the clearing, mean deposition velocities of 0.85 cm/s for SO_2 and 0.48 cm/s for NO_2 were estimated (Tab. 3).

	SO_2-S	NO_2-N
c (μg/m³)	12.5	8.4
D_d (mg/m² d)	9.16	3.49
v_d (cm/s)	0.85	0.48

Tab. 3: Deposition velocities (v_d) of gaseous compounds for a spruce canopy. Dry deposition rates for gases (D_d) estimated as the difference: (total D_d - aerosol particle D_d). Concentrations of gases measured on the clearing. Mean values for 1986.

According to fig. 2 the total atmospheric input to a "closed forest system" differentiated into wet and dry deposition can now be quantified.

Contributions from fog only have to be considered seperately if the fog event leads to a measureable amount of throughfall on the forest floor. Otherwise fog droplets impacted on leaves and branches will evaporate again and thus leading to an input, interpreted as part of the "dry deposition" after the following precipitation event.
The actual results, presented in tab. 4, demonstrate the important contribution of dry deposition of aerosol particles and gases for a complex surface like a spruce canopy. The percentage of dry deposition ranges from 73.6 % (Cl^-) to 83.3 % ($SO_4^{=}$-S). Even after reducing the dry deposition rate for Mn due to the leaching by 60 % (Höfken and Gravenhorst, 1981) the percentage fraction from the total input remains at 99 %, indicating an underestimation of the leaching effect in this case.
In comparison the portion of dry deposition to be found on the clearing since 1983, using the Bergerhoff sampling system is significantly lower for all elements.

	Fe	Mn	Pb	Cd	$SO_4^{=}$	NO_3^-	Cl^-
D_w (mg/m²·d)	0.141	0.008	0.023	0.0006	2.70	1.53	1.50
D_d (mg/m²·d)	0.459	0.80	0.084	0.0098	13.5	4.48	4.19
%- D_d	78.5	99.0	78.5	75.0	83.3	74.5	73.6
%- D_d (clearing)	60-70	50-60	10-20	<10	10-20	10-20	20-30

Tab. 4.: Mean wet (D_w) and dry (D_d) deposition rates occuring within a spruce stand. D_w equals the element fluxes with precipitation on the clearing. D_d equals the total dry deposition within the spruce canopy. Also listed are the percentages of D_d for this spruce stand and the range of mean percentages of D_d on the clearing, derived by using the sampling method after Bergerhoff

4.5. Additional input with fog events of high density

According to fig.2 the interception of fog has to be considered as a source contributing to the total atmospheric input. Fog droplets drifting through a canopy will be impacted at obstacles like leafs or branches. As long as this interception is smaller than the capacity defined in chapter 3, the deposition is restricted to the canopy. If it is exceeded, the fog interception furthermore leads to direct wet depositions on the forest floor. This phenomena was not observed at the sampling site were the main deposition measurements were carried out. Therefore it has not be taken into account for this certain area. At the mountain station, located about 300 m higher, the frequency of fog events as well as the duration and intensity is high enough to observe this special element flux. High wind speeds increasing the deposition efficiency by impaction also are characteristic for mountain stations like this. A rather simple model was used to describe the interception deposition of fog (equation 3). The total amount of intercepted elements by fogwater transported into a closed system like compartment C in fig. 2 is given by the amount of throughfall caused by fog, the capacity of the canopy and the concentrations in the fog.

$$FID_e = c_e \cdot (T + C) \qquad (3)$$

FID_e : Fog interception deposition of the element e

c_e : Concentration of the elememnt e in fog

T : Amount of throughfall caused by fog

C : Capacity of the canopy

To provide information about the element concentrations in fog an active fog sampler, permanently installed at this place was used. Detailed informations about the sampling system and the measurements itself are to be found in Schmitt (1986).
Preliminary investigations during the fall and winter season 1986 showed that the interception deposition during single fog events of high density and duration (> 10 h) will be in the same range as the amount of wet deposition from single precipitation events. This is demonstrated by fig. 8 showing the cumulative interception deposition of different elements with fog.

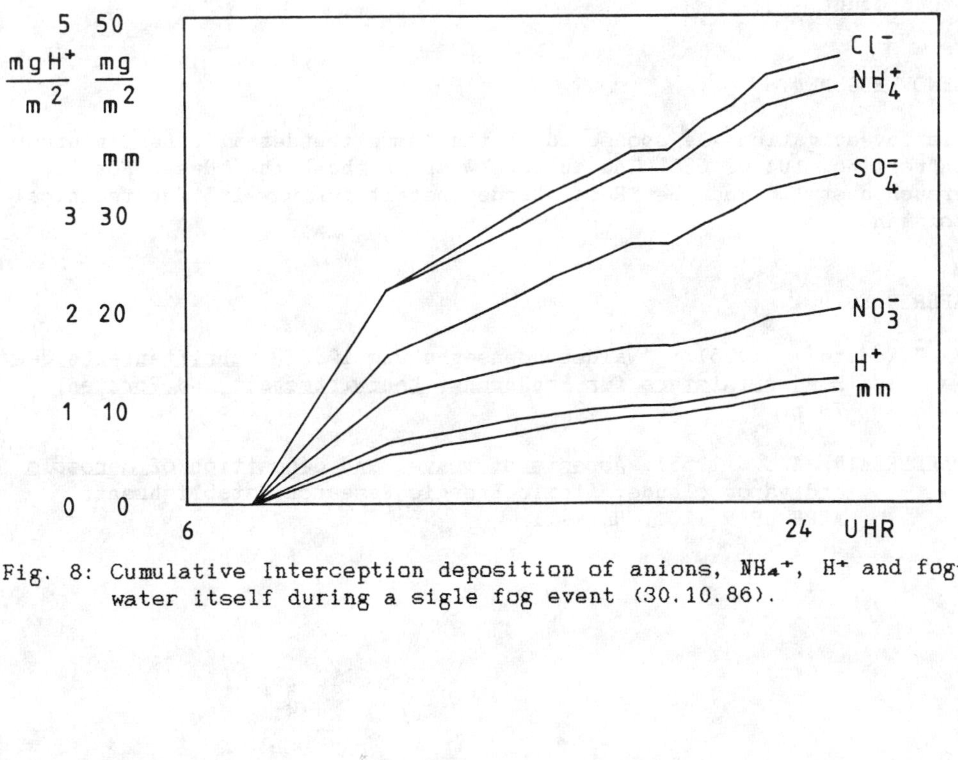

Fig. 8: Cumulative Interception deposition of anions, NH_4^+, H^+ and fog-water itself during a sigle fog event (30.10.86).

During this event 10 samples of throughfall were taken. The total amount of water was 11.7 mm. This includes an approximate mean value for the capacity of 3 mm. The element deposition, calculated by using equation (3) resulted in highest deposition rates for Cl^- (46 mg/m²) and NH_4^+ (43 mg/m²). The deposition of H^+ in this case (1.3 mg/m²) is interpreted as small because of relatively high pH values concerning fog of around pH 4.0.

5. CONCLUSIONS

A differentiated estimation of the atmospheric input into forest areas must consider the canopy as an effective sink for aerosol particles as well as for gases. Using this natural surface itself as a "sampling device" should lead to results of higher scientific and ecological evidence. By establishing a total mass balance for forest areas it is possible to approximate dry deposition rates. This estimation does not directly offer the possibility to differentiate between aerosol and gaseous contributions. Theoretical considerations can be used to seperate these different contributions for certain compounds. Uncertainties still arise from the lack of good knowledge about the relevance of ad- and absorption effects as well as the leaching effect. High deposition velocities, especially for aerosol particles (> 1 cm/s) and high dry deposition percentages (70-80%) demonstrate the efficiency of a spruce canopy to act as a sink for trace elements. For a complete description of the atmospheric input, the fog interception must be considered. If necessary the mass balance must be extended, taking the throughfall caused by fog into account.

ACKNOWLEDGEMENTS

This investigation was sponsered by the "Umweltbundesamt", Berlin under contract no. 104 02 635. The authors wish to thank the "Hess. Forstl. Versuchsanstalt" and the "Hess. Landesanstalt für Umwelt" for technical assistance.

REFERENCES

BMELF (editor) (1986) : "Waldschadenserhebung 1986", Schriftenreihe des Bundesministers für Ernährung, Landwirtschaft und Forsten, 73 p.

CHAMBERLAIN, A.C. (1955): Aspects of travel and deposition of aerosols and vapor clouds, Atomic Energie Research Establishment, AERE HP-R 1261, Harwell

ECKER, F.J. (1985): Experimentelle Untersuchungen zur Impaktion einiger Schwermetall-Aerosole in den Waldgebieten des Taunus, Diplom-Arbeit am Institut für Meteorologie und Geophysik, Universität Frankfurt/Main, 86 p.

GEORGII, H.W., S. GROSCH, G. SCHMITT, (1986): Feststellung der Schadstoffbelastung von Waldgebieten in der Bundesrepublik Deutschland durch trockene und nasse Deposition, Abschlußbericht Teil A Forschungsprojekt 104 02 715 im Auftrag des Umweltbundesamtes, Eigenverl. des Universitätsinstituts für Meteorologie und Geophysik, Frankfurt, 247 p.

GEORGII, H. W., C. PERSEKE, E. ROHBOCK (1984): 'Deposition of acidic components and heavy metals inthe federal republic of Germany for the period 1979 -1981', Atmosph. Environment, 18, no.3, p.581 - 589

GROSCH, S., G. SCHMITT (1985): 'Messungen des atmosphärischen Eintrags von Spurenelementen in Waldgebieten', VdI-Berichte 560, Waldschäden - Einflußfaktoren und ihre Bewertung, Kolloquium Goslar 18.6. - 20.6.1985, p.313 - 332

GROSCH, S. (1986): 'Wet and dry deposition of trace elements in forest areas', in: Atmospheric pollutants in forest areas, ed. by H.W. Georgii, Reidel Publ. Company, Dordrecht, p. 35 - 45

HESS. LANDESANSTALT FÜR UMWELT (1984): Waldbelastungen durch Immissionen, 1. Zwischenbericht; Schriftenreihe der Hess. Landesanstalt für Umwelt Nr. 8, 76 p.

HÖFKEN, K.D., G. GRAVENHORST (1981): Untersuchung über die Deposition atmosphärischen Aerosols an Buchen- und Fichtenwald; Bericht im Auftrag des Umweltbundesamtes, Forschungsprojekt 104 02 600, Eigenverlag des Universitätsinstituts für Meteorologie und Geophysik, Frankfurt/Main, 208

HÖFKEN, K.D., F. MEIXNER, P. MÜLLER, D.H. EHALT (1986): Untersuchungen zur trockenen Deposition und Emission von atmosphärischem NO, NO_x und HNO_3 an natürlichen Oberflächen, Berichte der Kernforschungsanlage Jülich, Nr. 2054, 207 p.

ROHBOCK, E., H.W. GEORGII (1982): Ein Depositionssammelgerät zur getrennten Erfassung der trockenen und feuchten Deposition atmosphärischer Schadstoffe; Bericht im Auftrag des Umweltbundesamtes, Forschungsprojekt 104 02 600 Eigenverlag des Universitätsinstituts für Meteorologie und Geophysik, 25 p.

SCHMITT, G. (1986): 'The temporal distribution of trace elements in fogwater during individual events', in: Atmospheric pollutants in forest areas, ed. by H.W. Georgii, Reidel Publ. Company, Dortrecht, 129 - 141

SCHMITT, G. (1987): Methoden und Ergebnisse der Nebelanalyse Berichte des Universitätsinstituts für Meteorologie und Geophysik, Frankfurt/M, 200 p.

SEHMEL,G.A. (1980): 'Particle and gas deposition: A Review', Atmosph. Environment, 14, p 983 - 1011.

ULRICH, B. (1983): A concept of forest ecosystem stability and of acid deposition as diving force for destabilization; in: Effects of accumulation of air pollutants in forest ecosystems, Hrsg. Ulrich, B., J. Pankrath, Reidel Publishing Company, Dordrecht, p. 1 - 29.

WOLZ, G. (1987): Untersuchung der Filtereffizienz eines Fichtenbestandes für Chlorid-, Nitrat- und Sulfataerosole in Abhängigkeit von der Partikelgröße, Diplom-Arbeit am Institut für Meteorologie und Geophysik, Universität Frankfurt/Main, 101 p.

EMISSION AND TRANSPORT OF FUGITIVE DUST FROM SLUDGE DEPOSITS

D. Etling
Institute of Meteorology and Climatology
University Hannover
Herrenhäuser Str. 2, 3000 Hannover 21, F.R.G.

ABSTRACT. River sediments are often contaminated with heavy metals due to waste of upstream industrial areas. If those sediments are removed and stored on sludge deposits they may cause an environmental problem due to dust emission. Difficulties in estimating source strength and transport of fugitive dust are pointed out and results from a case study are presented.

1. INTRODUCTION

Harbours located at rivers are often facing the problem to keep the fairway clear up to a minimum depth for ship traffic. For this reason the sludge, settled down at the bottom of rivers and harbour basins has to be removed regularly. After separation of silt and sand fraction and a subsequent drying out process the sludge, consisting mainly of silt with particle size between 2 - 60 µm (see figure 1), is often deposited in open compartments covering areas of 100 ha or more (Hensen, 1987).

Recent soil analysis revealed (Tent, 1982) that sludge from the river Elbe near the port of Hamburg located in Northern Germany is highly contaminated with heavy metals (Lead, Cadmium, etc.) mainly due to waste from industrial sources. Typical concentrations of heavy metals in sludge of the Elbe are given in Table 1. Additionally typical tolerable values of contamination for agricultural soils are also shown (Kloke, 1980). It can be concluded from Table 1 that in this case the river sludge should not be used for agricultural purposes, as was done at earlier stages of the sludge removal problem, but has to be treated as a waste storage problem.

Additional analysis has also shown that nearly all heavy metals are concentrated in the silt fraction (particle size < 6o µm) which accounts for about 70% of the total (dry) mass of the harbour sludge as can be seen in Figure 1. This is also nearly identical to the range of particle size (< 100 µm) where wind erosion takes place in the form of suspension (wind-blown dust) which can affect areas further downstream from a possible source of contaminated dust.

K. Grefen and J. Löbel (eds.), Environmental Meteorology, 217–226.

Table 1: Content of heavy metals in harbour sludge		
	Concentration in mg/kg dry substance	
heavy metal	I harbour sludge	II cultivated soil
Arsenik (As)	50 - 100	20
Cadmium (Cd)	7 - 25	3
Chromium (Cr)	150 - 300	100
Copper (Cu)	250 - 600	100
Lead (Pb)	150 - 250	100
Mercury (Hg)	10 - 20	2
Zinc (Zn)	1.500 - 2.500	300

I: measurements of harbour sludge (after Tent, 1982)
II: tolerable values for cultivated soils (after Kloke, 1980)

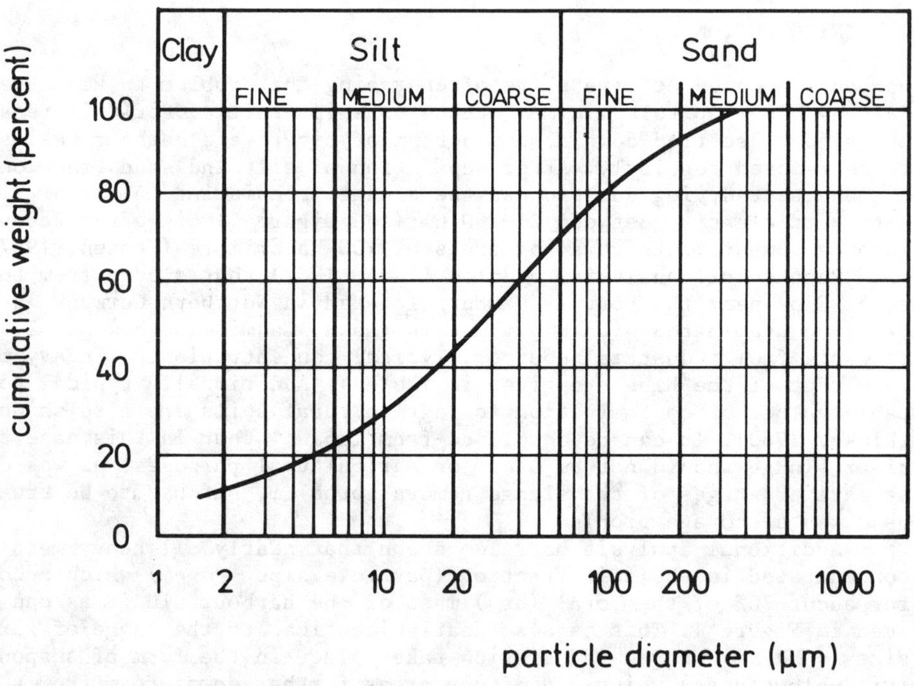

Figure 1. Typical particle size distribution of harbour sludge.

Measurements of particle deposition at stations located nearby an operating sludge deposit have revealed deposition rates of about 50 µg/m²d for Lead (Pb) and about 1 µg/m²d for Cadmium (Cd) on a yearly average. But those rates can be a factor of ten higher for short periods of one week or so, reaching critical levels as given by regulatory standards. It is of course not easy to distinguish between background pollution due to sources from industrial areas or from vehicle exhaust and fine particulates set free from sludge deposits. Therefore some estimates for source strength and deposition rates of fugitive dust from sludge deposits are given in the following chapters.

2. THE EMISSION PROBLEM

If the erosion of contaminated dust from sludge deposits is considered as a problem of air pollution one has to know the source strength (emission rate) of fugitive dust in order to give an estimate of deposition of heavy metals within the vicinity of a sludge deposit. But unless air pollution emissions from industrial sources the emission of contaminated dust from deposits is rather inhomogeneous in space and time and also dependent on weather conditions. Dust can be set free from deposits by either man made activities like dredge or haul operations of construction vehicles or by wind erosion. Hence time and duration of possible dust emission is depending on various factors like dry weather periods, strong winds, operation hours and season (usually spring to fall) of the sludge deposit. Additionally dust emission due to vehicle operations might be treated as point or line sources whereas emission due to wind forces may be qualified as an area source.

As no data on dust emission rate from sludge deposits are available to date one has to rely on measurements from other sources like open cut coal mines, unpaved roads, sand dunes or wind erosion from agricultural soils. With regard to dust emission from vehicles operating on a sludge deposit estimates from similar activities in coal mines (Axetell, 1978; Dean, 1986) or unpaved roads (Becker and Takle, 1979) give emission rates of 1 - 30 g/m vehicle track. Dust emission due to reloading operations on mines or storage places for coal and ore can be estimated as 50 - 1000 g/to of material handled (Dean, 1986; Davids and Lange, 1986).

With regard to dust emission due to the action of wind forces one might rely on the vast amount of literature on the more general problem of soil erosion. Hence one has to take into account the particle size distribution (see Figure 1), the aggregate state (dry or moist) and the surface state (crust or loose) of the deposited harbour sludge. As wind erosion takes place only above a certain threshold wind velocity also meteorological factors have to be considered. An example is given in Figure 2 where the threshold friction velocity u_* and wind speed u at 10m height above ground for surface particle movement due to wind forces is given (after Bagnold, 1954). For the silt fraction of harbour sludge (particle size ≈ 60 µm) e.g. a minimum wind speed of about 5 m/s (10 knots) is needed before fugitive dust can be set free. But this critical windspeed also depends on the type of soils considered (clay, loam, sand) and on the surface state, as was observed on natural desert soils by Gillette et al (1980). For hard crusted surfaces the threshold friction velocity was about $u_* = 60$ cm/s (corresponding to a wind velocity u = 18 m/s) in contrast to about $u_* = 40$ cm/s (u = 12 m/s) if the same soil had a disturbed surface (e.g. by vehicle action).

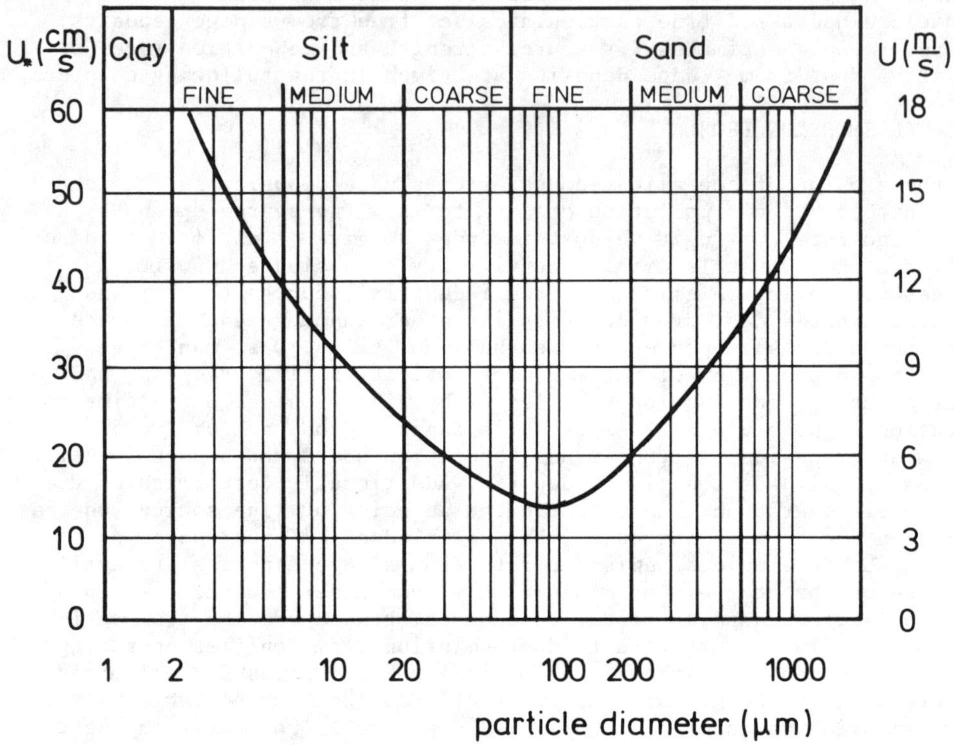

Figure 2. Threshold velocities for wind erosion as function of soil particle size. U_* is the friction velocity and U is the wind speed at 10 m height above the ground.

Even if particle size distribution, aggregate and surface state and wind velocity are known the evaluation of emission rate (source strength) of dust from sludge deposits is rather uncertain. Regarding the coarse particle fraction (> 100 µm), where particle motion is performed by creeping and saltation, numerous sand-transport formula have been proposed (see Zanke (1982) or Greeley and Iversen (1985) for a collection). They generally show that the particle emission increases with the third power of the friction velocity u_* and with the square root of particle size. Hence even for the same friction velocity (or equivalent windspeed) emission rates are higher for larger particles than for smaller ones. This might explain the large span of emission rates due to wind erosion as given in the literature. Values of 0.3 - 3 to/ha h (metric tons per hectares and hour) have been found for resuspension of particles containing Plutonium and Uranium (Anspaugh et al, 1975; Ciccone et al, 1987) which is of the same order as for loss of agricultural soils due to wind erosion, for which emission rates of 0.5 - 5 to/ha h can be found (Chepil and Woodruff, 1963; Zachar, 1982). But even for particle fractions of limited diameter range wind induced mass flux also depends on the soil texture. For wind erosion of fine particulates (< 20 µm) e.g. emission rates of 3 - 300 kg/ha h have been found depending on the relative portion of clay, silt and sand (Gillette, 1977). Similar conclusions regarding the soil texture have been derived from studies of soil erodibility of cultivated soils (Lyles, 1977), where e.g. the yearly soil loss due to wind erosion is estimated to be about 700 to/ha yr for fine sand and about 200 to/ha yr for clay or silty loam.

Summarizing the emission problem for dust sources, emission rates (source strength) due to vehicle action or wind erosion show a large scatter (about a factor of 10), as revealed from the references quoted above. With respect to emission of contaminated dust from sludge deposits one can, at present, only take a reasonable value for the source strength as derived for other wind-blown material. The emission rates, as summarized in Table 2, may be considered as a guidance for possible air quality and impact studies concerning the problem of fugitive dust.

Table 2. Dust emission rates due to vehicle operations and wind erosion

source	various materials(2)	harbour sludge(3)
vehicle track	1 - 30 g/m	10 g/m
shoveling, reloading	50 - 1000 g/to	500 g/to
wind erosion(1)	0.5 - 5 to/ha h	1 to/ha h

(1) wind speed $U > 5$ m/s

(2) desert sand, cultivated soil, coal, ore and other particulates

(3) assumed values due to lack of measurements

3. TRANSPORT AND DEPOSITION OF FUGITIVE DUST

Once contaminated dust has been set free from sludge deposits by either vehicle action or wind erosion it can be distributed over large areas by atmospheric transport and diffusion, because particle size is usually in the range of less than 60 µm where wind induced mass transport is due to suspension. The dust will be deposited downwind the source again and may affect housing or agricultural areas. In some cases regulatory standards demand that concentration and deposition rate of dust, containing heavy metals, should not exceed a certain limit. Hence for planning purposes the evaluation of deposition of fugitive dust from sludge deposits will be requested by public authorities in some cases. This evaluation could be done either by methods as prescribed through regulatory rules, e.g. by the German TA-Luft (Davids and Lange, 1986) or by applying results from the more general research on atmospheric diffusion, as can be found e.g. in Pasquill and Smith (1983).

Regardless the type of diffusion model used in calculating dust depositions, the correct evaluation of the source strength remains a major problem. Even if the emission rates from Table 2 are considered to give a proper estimate, dust emission due to vehicle action has to be treated as on instantaneous source and even wind erosion leads to a continuous emission only over a limited time. Hence the emission of contaminated dust from sludge deposits is rather episodic, depending on various factors like dry weather periods, strong winds or vehicle operation hours. On the other hand, regulatory standards usually demand pollution limits based on yearly averaged concentrations or deposition rates. This procedure is quite suitable for industrial sources, like SO_2 as emitted from power stations, since in those cases a more or less continuous source strength is maintained throughout the year. In order to apply this concept, including pollution concentration calculations, to the problem of dust emission from sludge deposits, one has to convert the instantaneous source strength to a continuous source strength with respect to a yearly averaged emission rate.

This is demonstrated for the dust emission due to vehicle track as an example. If one assumes a vehicle speed of 10 km/h on the sludge deposit one obtains a source strength of 10 kg/h for one vehicle using the emission rate of 10 g/m (vehicle path) from Table 2. But since construction vehicles are causing dust emission only during certain periods this source strength is reduced considerably if averaged over a year. Taking, as a first estimate, 6 months as total operation period on a sludge deposit, and assuming to have dry weather during half of these days with a working period of 8 hours daily, one would get about 640 hours of possibly dust emission due to vehicle action per year. Hence the yearly averaged source strength would be about 8 kg/h for one vehicle.

A similar estimate can be also given for dust emission due to wind erosion, although this problem is more complicated as one would have to know the joint statistics of dry days with strong winds and the related surface state. As those data are hardly available for a longer time period one has to make some more simple estimates based on the number of dry days and the total hours of winds exceeding a certain limit. For the area of Hamburg, e.g., such an estimate gives a yearly averaged source strenght of about 12 kg/ha h, based on the emission rate of 1 to/ha h as proposed in Table 2.

In order to carry out diffusion calculations still some information on the source type considered has to be given. Whereas wind erosion might be treated as an area source the dust emission due to vehicle track could be regarded as a line source. But for simple estimates of dust concentration in the vicinity of sludge deposits one could, as a first step, treat both emission mechanism as a continuous point source located in the centre of the deposit. In this case one has to know the total number of vehicles and the total area of the deposit in order to give a final estimate of the source strength. Such a source inventory is presented in Table 3 for a sludge deposit located near the city of Hamburg/Germany where the possible dust source area was about 10 ha with typically 10 vehicles operating. As the content of heavy metals in harbour sludge (see Table 1) is also known, the source strength for heavy metal particulate emission can also be estimated. Since regulatory limits exist only for dust (350 mg/m² d), Lead (250 µg/m² d) and Cadmium (5 µg/m² d) in the German standard TA-Luft, only emission rates for these substances are given in Table 3.

Table 3. Source strength Q for a sludge deposit as used in diffusion calculations after eq. 2.

	Dust	Lead	Cadmium
wind erosion	120 kg/h	48 g/h	1.8 g/h
vehicle action	80 kg/h	32 g/h	1.2 g/h
total	200 kg/h	80 g/h	3.0 g/h

Calculations of long term averaged airpollution concentrations are usually performed with a Gaussian-type diffusion model. Such a model is also prescribed for regulatory purposes by the German TA-Luft (Davids and Lange, 1986) and has been applied to the problem of dust emission from sludge deposits. The deposition rate of dust and heavy metals is calculated by:

$$D = CV_d \tag{1}$$

where D is the deposition rate, C the pollution concentration and V_d the deposition velocity. The surface concentration C is obtained by the Gaussian model for particulate dispersion:

$$C(x,y) = \frac{Q}{\pi u \sigma_y \sigma_z} \exp\left(-\frac{y^2}{2\sigma_y^2}\right) \exp\left(-\frac{h^2}{2\sigma_z^2}\right) \Phi(x) \tag{2}$$

where $\phi(x)$ is the correction for falling particles calculated from:

$$\Phi(x) = \exp\left[-\sqrt{\frac{2}{\pi}} \frac{V_d}{u} \int_0^x \frac{1}{\sigma_z(\xi)} \exp\left(-\frac{h^2}{2\sigma_z^2(\xi)}\right) d\xi\right] \tag{3}$$

Q is the source strength (taken from Table 3), h is the source height assumed to be h = 10 m, x and y are the downwind and crosswind distances from the source (located at x = y = 0). The values for the wind velocity u and the dispersion parameters σ_y and σ_z have to be taken from climatological statistics, as is described in more detail in Pasquill and Smith (1983) or Davids and Lange (1986). Although the deposition velocity V_d depends on the particle size, an average value of $V_d = 7$ cm/s after the TA-Luft has been adopted in these calcualtions.

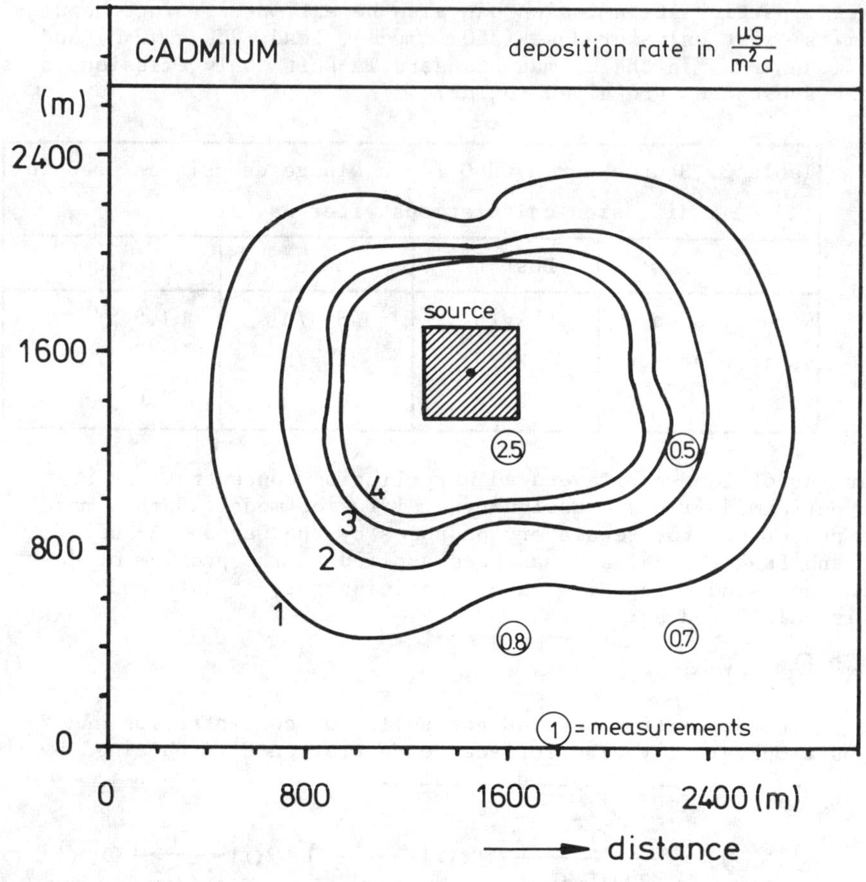

Figure 3. Calculated long term average of daily deposition of Cadmium caused by emission of contaminated dust from a sludge deposit.
○ = measurements by Bruckmann and Schwarz (1985).

Deposition rates for dust, Lead and Cadmium in the vicinity of a sludge deposit have been calculated for all wind directions using Equations (1) - (3) according to the proceedures as given in the TA-Luft. Results are shown for Cadmium as an example in Figure 3, where also the typical area size of the sludge deposit is shown, although the calculations have been performed assuming a point source in the centre of the deposit for convenience. Measurements of Cadmium deposition rates have been obtained by Bruckmann and Schwarz (1985) for several locations nearby the chosen sludge deposit and are also given in Figure 3. A comparison of observed and calculated deposition rates may be not quite suitable because observed values also contain depositions due to other sources of Cadmium not included in the calculations. In addition the dust emission rates (Table 3) are quite uncertain due to many assumptions adopted in their derivation. Hence one should regard Figure 3 as an order of magnitude estimate for both, observed and calculated values for Cadmium deposition rates for a specific location near the city of Hamburg. These values may be compared to the regulatory limit of 5 μg/m² d as given by the TA Luft.

4. CONCLUSIONS

The storage of dryed up contaminated sludge from rivers and harbours in open deposits may cause a problem in air pollution in case of emission of fugitive dust due to vehicle operation or wind erosion. The calculation of dust concentration and deposition rates in the vicinity of a sludge deposit by means of a diffusion model contains a great deal of uncertainty mainly due to the poor knowledge about the dust emission rate. As there are no measurements on the actual dust emissions from sludge deposits available at present one may, as a first step, rely on data on wind erosion of other materials. In contrast to standard situations like continuous SO_2 emission from elevated point sources, dust emission from deposits is rather episodic in space and time, depending on weather conditions and operation procedures on the deposit. Hence pollution level estimates will be also more uncertain than for standard industrial emission situations. On the other hand, this specific waste problem brings together different interesting scientific and practical aspects of soil mechanics, wind erosion and turbulent diffusion showing the need for interdisciplinary cooperation in solving complex environmental problems.

5. REFERENCES

Anspaugh , L. R., Shinn, J. H., Phelps, P. L. and N. C. Kennedy (1975): 'Resuspension and redistribution of Plutonium in soils'. Health Physics, 29, 571 - 582.

Axetell, K. (1978): 'Survey of fugitive dust from coal mines'. US Environmental Protection Agency, Report EPA - PB 283 162.

Bagnold, R. A. (1954): 'The physics of blown sand and desert dunes'. Methuen Co, London.

Becker, D. L. and E. S. Takle (1979): 'Particulate deposition from dry unpaved roadways'. Atmospheric Environment, 11, 661 - 668.

Bruckmann, P. and H. Schwarz (1985): 'Ergebnisse der Staub- und Schwermetallniederschlagsmessungen in Hamburg im Jahr 1984'. Hamburger Umweltberichte, 5/85, Umweltbehörde, Hamburg.

Chepil, W. S. and N. Woodruff (1963): 'The physics of wind erosion and its control'. Advan. Agron., 15, 211 - 301.

Ciccone, A. D., Kawall, J. G. and J. F. Keffer (1987): 'An experimental investigation of a wind-generated suspension of particulate matter from a tailings disposal area'. Boundary Layer Meteorol., 38, 339 - 368.

Davids, P. and M. Lange (1986): 'Die TA-Luft 86'. VDI Verlag, Düsseldorf

Dean, M. (1986): 'Appraisal of dust emission from open cut coal mines in New South Wales'. In: Hartmann, H. F. (Ed.): 'Proceedings of the Seventh World Clean Air Congress', Clean Air Society of Australia and New Zealand, Vol V, 231.

Gillette, D. A. (1977): 'Fine particulate emissions due to wind erosion'. Trans. ASAE, 20, 890 - 897.

Gillette, D. A., Adams, J., Endo, A. and D. Smith (1980): 'Threshold velocities for input of soil particles into the air by desert soils'. J. Geoph. Res. (C), 85, 5621 - 5630.

Greeley, R. and J. D. Iversen (1985): 'Wind as a geological process'. Cambridge University Press, Cambridge.

Hensen, U. (1987): 'Hamburger Baggergut auf Spülfelder aufhalden'. Umwelt, 6/87, 329 - 331.

Kloke, A. (1980): ' Richtwerte 80 - Orientierungsdaten für tolerierbare Gesamtgehalte einiger Elemente in Kulturböden'. Mittlgn. VDLUFA, 1-3, 9-11.

Lyles, L. (1977): 'Wind erosion: processes and effect on soil productivity'. Trans. ASAE, 20, 880 -884.

Pasquill, F. and F. B. Smith (1983): 'Atmospheric Diffusion'. Ellis Horwood, Chichester.

Tent, L. (1982): 'Auswirkungen der Schwermetallbelastungen von Tidegewässern am Beispiel der Elbe'. Wasserwirtschaft, 72, 2.

Zachar, D. (1982): 'Soil erosion'. Elsevier, Amsterdam.

Zanke, U. (1982): 'Grundlagen der Sedimentbewegung'. Springer, Berlin.

ESTIMATION OF WET DEPOSITION VIA FOG

G. Kroll, P. Winkler
Deutscher Wetterdienst
Meteorologisches Observatorium Hamburg
Frahmredder 95
D-2000 Hamburg 65

ABSTRACT. Measurements of the chemical composition of fog/cloud water at two mountain sites in West Germany are presented. The amount of deposited water in a fir forest by fog/cloud droplet interception is assessed using the cloud droplet deposition model of Lovett (1984). The model input data are taken from regular meteorological observations. The yearly amounts of cloud water intercepted by a forest vary between 25 and 150 % of the precipitation depth for heights between 840 and 1440 m a.s.l. Combination of intercepted amount with measured concentrations of sulfate, nitrate, and chloride show the trace substance deposition to be 1 - 4 times higher than the amounts deposited by rain.

1. INTRODUCTION

The importance of water that has been deposited to forests by interception of cloud or fog droplets to the hydrological circulation is well known since the last three decades. In forests in mountain top areas the total amount of intercepted cloud/fog water can be two or more times larger than the bulk precipitation at the same site (Grunow, 1957; Baumgartner, 1958). At the same time, fog water always shows a higher pollutant concentration than is measured in rainwater (Schemenauer, 1986; Waldman et al., 1985; Georgii et al, 1986). This indicates that forests in mountain top areas which are often enveloped in clouds, will obtain significantly higher pollutant deposition rates than the surrounding lowland.

It has been demonstrated from model calculations by Lovett (1984, 1986) that deposition rates by fog interception can vary appreciably with wind speed, fog density (LWC), and geometrical shape of the intercepting trees. In the real world, fog density, fog frequency and windspeed vary continuously. It was the purpose of this study to use

K. Grefen and J. Löbel (eds.), Environmental Meteorology, 227–236.

observed meteorological parameters from regularly meteorological observation stations to trigger the model output in order to come to a realistic picture of the cloud water amount intercepted by a needle forest. At various places (see below) fog water was collected by means of an active cloud water collector (Winkler, 1986). The samples were sent to Hamburg where they were analysed for pH, el. conductivity, $SO_4^{=}$, NO_3^{-}, Cl^{-}, and NH_4^{+}. The combination of measured concentrations with the calculated water interception provides a realistic assessment of trace substance deposition.

The seasonal variation of the fog frequency at four weather stations in West Germany can be taken from fig. 1, which is based on the hourly weather observations of the years 1982-1985. Especially at the mountain stations Grosser Arber (Bavarian Forest), Kahler Asten (Rothaargebirge) and Hohenpeissenberg (Alpenvorland) a marked maximum is observed during the winter season. The stations Grosser Arber and Kahler Asten are within clouds in 35 % of the total time during December.

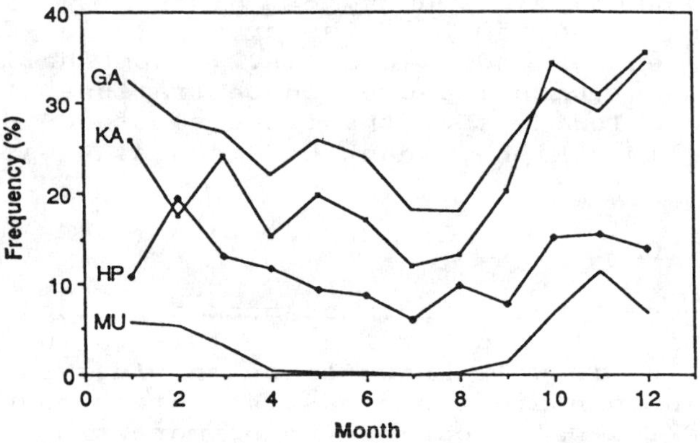

Figure 1. Seasonal variations of the frequency of foggy hours at four weather stations in the years 1982-1985. GA = Grosser Arber (1440m), KA = Kahler Asten (840m), HP = Hohenpeissenberg(980m), MU = Munich (530m).

2. FOG MEASUREMENTS

At the mountain stations Kahler Asten, Grosser Arber, Hohenpeissenberg and three others we installed active fog samplers operating on the impactor principle. One sample per fog event is taken. Table 1 gives the mean of the measured values from October 1986 through July 1987 at the mountain sites Kahler Asten and Grosser Arber, from where we got the most part of our samples. The data are normalized to a constant LWC (liquid water content) of 0.1 g/m^3 in order to eliminate variations which originate from different LWCs. It can be seen that the cloud water at Kahler Asten contains higher amounts of trace substances and hydrogen ions than at Grosser Arber. The anions and cations at the station Kahler Asten seems to be in balance, while at the station Grosser Arber more anions than cations are found in the fog water. A comparison of the mean measured conductivities with those conductivities, which were calculated from the measured concentrations of the ions, points out a 20 % difference at the station Grosser Arber and a 30 % difference at the station Kahler Asten. From literature data it is obvious that about 20 % difference can be attributed to the anions K^+, NA^+, Mg^{++} and Ca^{++} which we have not analyzed. The higher average concentrations at the Kahler Asten-site as compared to Grosser Arber is caused most probably by the larger emissions in the nearby Ruhr area.

All measured values show great variations from fog event to fog event. At the station Grosser Arber e.g. the pH value varies from 3.06 to 6.31 around a mean value of 3.80. The normalized sulfate concentrations vary from 1.2 - 118.0 mg/l with a mean value of 13.9 mg/l. About 10% of the fog events show sulfate concentrations above 35 mg/l.

	No	pH	LF (μS/cm)	$SO_4^{=}$ (mg/l)	NO_3^- (mg/l)	Cl^- (mg/l)	NH_4^+ (mg/l)
KA	56	3.51	265	12.2	17.0	3.1	5.3
		3.0-6.4	15-1212	0.1-53	0.1-76	0.1-28	0.4-45
GA	91	3.80	160	13.9	13.1	3.0	6.0
		3.1-6.3	14-1014	1.2-118	1.1-254	0.3-35	0.2-64

TABLE I
Average values and range of variation of trace substances in fog water at two mountain sites from October 1986 to July 1987. All data were normalized to a constant LWC of 0.1 g/m^3. The first rows of each site give average values and the second rows present the scattering ranges of the normalized values.

For effect related purposes it may be of interest to mention the really observed values: The lowest pH was 2.55 at

Kahler Asten, the highest pH was 6.4 at Grosser Arber. At Kahler Asten in a sample, collected in a fog rising from the valley, a pH of as low as 1.7 was even measured which, however, is not typical for clouds. The extremes for conductivity are 580 (Grosser Arber) and 3600 (Kahler Asten) µS/cm.

From trajectory analysis in the cases of very high sulfate concentrations (> 80 mg/l) it can be seen that these air masses had passed the highly industrialized areas either near Nürnberg, Munich, the Eger valley (CSSR), Saxonia (DDR), or upper Silesia (Pl). In all cases of high sulfate loadings the concentrations of the other substances are also very high, hence it follows that during single fog events the amount of deposited pollution will be considerably higher than estimated by assumption of mean concentrations. Up to now we got data from October to July, but neither the pH value nor the ion concentrations nor the conductivity show seasonal variations.

3. DEPOSITION BY FOG/CLOUD DROPLETS

3.1. Cloud droplet deposition model

For the estimation of the deposited amount of fog water in a given forest we are using the cloud droplet deposition model of Lovett (1984). It is a resistance type model, which simulates the transport of droplets from above the forest downwards and the impaction as well as sedimentation of droplets to the vegetation components (needles, twigs, branches). The canopy is divided into 1 m height strata and the vegetation surfaces in each stratum are composed of seven different canopy components: three age classes of needle-bearing twigs, three diameter classes of bare twigs, and boles. While Lovett partitioned the LWC only into three size classes, we are working with 20 size classes by which method a higher precision is obtained. The turbulent diffusion and the windspeed within the canopy are exponentially decreasing functions of the cumulative surface area index.

The LWC being necessary as input parameter for the model is not measured regularly. In order to work with actual data, we are parameterizing it out of the visibility. In that we go beyond Lovett, who feeds the model with mean values of LWC and windspeed for a given site. The correlation between the liquid water content and the visibility is not 100 %, but the results of our parameterization do harmonize quite well with the measurements of Chylek (1978), Pinnick (1979) and Garland (1971). Working with the parameterization and using hourly values of visibility and windspeed as input parameters, the model reproduces the mean LWCs which were measured at our stations.

At present the model is applied to calculate the fog water deposition to a 10m high fir canopy. The given forest has a total SAI of 7, about 80 % of the wettable surface are needles, the remaining 20 % are attributed to bare twigs and boles.

3.2. Sensitivity tests

The deposition flux of water depends mainly on the droplet size distribution and the wind speed in the canopy. The drop size distribution is calculated by applying Bests formula. Since several investigations (Deirmendjian, 1969; Blyth et al., 1980; Choularton, 1986) show drop size distributions which can better be described by Deirmendjian distributions than by Best distributions, tests were made in order to see how the model reacts to different forms of drop size distributions.

The four drop size distributions depicted in fig. 2 produce all the same LWC of 0.4 g/m^3 and the same visibility of 54 m. Since our method of parameterization does not allow to decide which drop size distribution occured in reality, this sensitivity test should show how large the errors could be arising from this ambiguity. Using these distinct distributions deposition fluxes differing from each other by only less than 10 % were calculated. The reason for this surprisingly little influence of the drop size distribution on the deposition flux is mainly due to the fact, that the farmost largest portion of the total amount

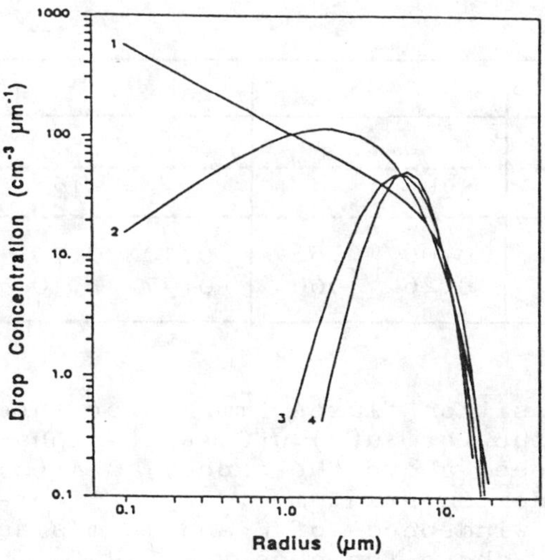

Figure 2.
Four drop size distributions, which produce a LWC of 0.4 g/m^3 and a visibility of 54 m: 1. Best (1952), 2.-4.Deirmendjian (1969).

of water is deposited by large droplets. In the range of the larger droplets the four curves do hardly differ from each other. In very dense fogs, droplets with a radius of 3 µm or smaller contribute less than 5 % to the deposited amount of fogwater.

Also the wind speed in a given stratum influences the droplet interception: the higher the wind speed, the higher the impaction probability for droplets of all sizes. The given windprofile was measured by Lovett, however, it assumes horizontal homogeneity.At the edge of a forest the wind profile changes gradually from the profile outside the forest to the equilibrium reached after a fetch sufficiently long. On the other hand, in inhomogenious terrain the windprofile could be locally reduced more than average e.g. in a dell. Since the windprofile can be described by an exponential formula a sensitivity test is most easily cunducted by varying the exponent ("extinction coefficient" alpha in table 2). In table 2 the calculated deposition fluxes for three windprofiles are compared for four different meteorological conditions. Case 3 represents the forest edge, case 2 describes the equilibrium and case 1 the "dell"-case. We see that the form of the windprofile has a large influence on the deposition flux: profile 1 reduces it, profile 3 increases it. The reduction of the deposition flux by using profile 1 instead of profile 2 is independent of the windspeed and the LWC above the canopy and amounts to almost 30 %. By using profile 3 the increase of the normal deposition flux is stronly dependent of LWC and windspeed above the canopy and is relatively higher for low windspeeds and LWCs. For example: switching over from windprofile 2 to 3 at the low LWC-case (0.15 g/m^3) increases the deposition flux at the low windspeed by 0.04 mm/h in

Profile no.	1		2		3	
alpha	0.47		0.27		0.07	
u(m/s)	2	12	2	12	2	12
0.15 (g/m^3)	0.04	0.37	0.06	0.56	0.10	0.75
0.45 (g/m^3)	0.18	1.15	0.26	1.66	0.37	2.10

TABLE II
Examples of calculated deposition fluxes (mm/h) for three different windprofiles of the form $u(z)=u(0)\cdot\exp(-\text{alpha}\cdot \text{CSAI}(z))$ (with u(0)= windspeed above the canopy, CSAI(z)= cumulative surface area index in stratum z, alpha = "extinction coefficient") for windspeeds of 2 and 12 m/s and LWC's of 0.15 and 0.45 g/m^3 above the canopy.

absolute terms, while at the higher windspeed the increase amounts to 0.19 mm/h.

3.3. Deposition rates of cloud water

Feeding the model with observed hourly values of windspeed and visibility yields the seasonal variation of the amount of fogwater, that is deposited **to** a given forest. In fig.3 monthly deposition rates of fogwater for the mountain stations Kahler Asten and Grosser Arber are compared with the monthly precipitation depths. At the higher situated station Grosser Arber the fogwater deposition in the given forest is markedly higher than at Kahler Asten because thick fogs occur more frequently (80 % of the fog hours show visibilities of less than 100 m) and windspeeds are usually higher (up to 30 m/s). The hourly deposition rates vary between 0.01 and 3.5 mm/h. Since from the weather bulletins we only know visibility classes instead of the observed exact values and since there are resulting uncertainties in the LWC, upper and lower limits of the deposition rates of fog water are presented in fig. 3. In order to reduce this uncertainties we are presently conducting frequency distributions of the observed visibilities in the single visibility classes. In fig.3 the upper line of the range of fog water deposition was calculated assuming a

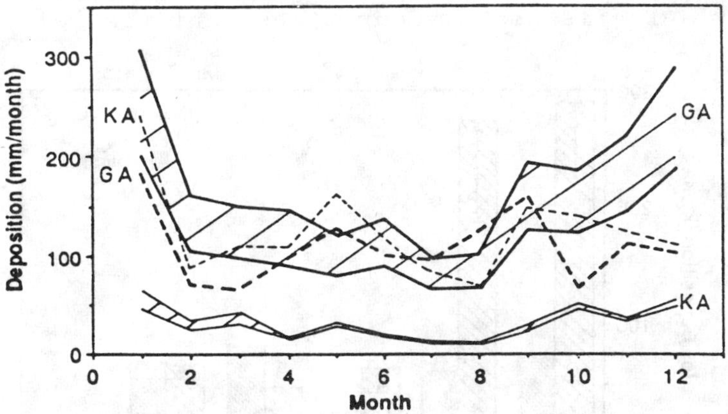

Figure 3. Range of calculated fog water deposition (solid lines) in a fir canopy at two mountain sites in comparison with precipitation deposition (dashed lines).

Deposition by	fog	precipitation
Station: GA	1400 - 2100	1300 (mm/a)
KA	330 - 400	1500 (mm/a)

LWC of 0.45 g/m^3 resp. a mean visibility of 50 in the first visibility class (below 100 m). The bottom line was calculated assuming a LWC of 0.2 g/m^3 corresponding to an average visibility of 80 m in the first class.

At both stations the fog water deposition exhibits a pronounced seasonal variation with minimum rates during summer and maximum rates during winter. It can be seen that at the higher mountain site Grosser Arber the deposition of fog water exceeds the annual rain deposition by 100 - 150 %, while at the mountain Kahler Asten it runs to 25 % of the total rain deposition.

4. ION DEPOSITION RATES

An evaluation of the trace substance deposition to the canopy can be obtained by multiplying the calculated water deposition rates by the average measured concentrations in the fog water at the site. In fig. 4 the deposition rates of the three major anions sulfate, nitrate and chloride at the sites Grosser Arber and Kahler Asten by fog and by rain are compared. Although the amount of deposited fog water at the station Kahler Asten made up only 25 % of the rain deposition the pollutant deposition by fog is almost as high as by rain because of the higher concentrations of the ions in fog water. Less surprising is the result for the station Grosser Arber, where even a conservative estimation yields that the sulfate, nitrate and chloride deposition via fog is nearly four times that via rain.

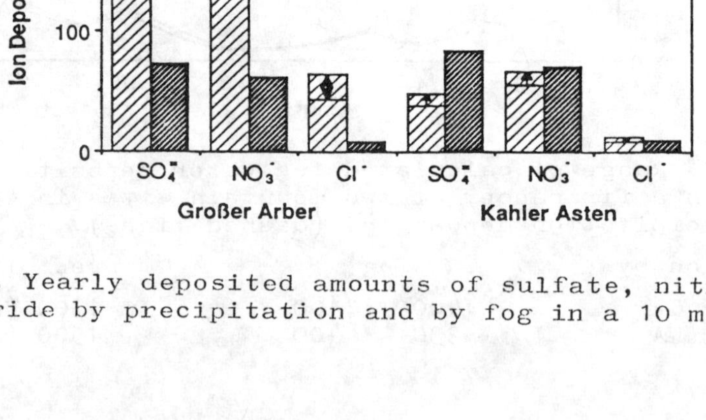

Figure 4. Yearly deposited amounts of sulfate, nitrate and chloride by precipitation and by fog in a 10 m fir canopy.

5. CONCLUSIONS

Our measurements on composition of cloud resp. fogwater indicate that concentrations of acidic substances can be so extreme (lowest pH in cloud water 2.5, in rising valley fog 1.7 !) that direct damages to plants are to be expected. At the same time fog interception represents another important pathway by which acidic substances can be deposited to soils at high elevation sites covered by forests. Our assessment of the pollutant deposition to a mountainous needle forest by combination of model calculations and fog measurements shows that the wet deposition by interception of fog/cloud droplets can be several times larger than the deposition via precipitation. The amount of deposition by cloud interception depends on height a.s. l., on the meteorological conditions and source strength and source distribution around the site. It is planned to extend these calculations to other mountain sites and other forest types in order to obtain a first assessment of the regional distribution of pollutant deposition by cloud water incerception.

ACKNOWLEDGEMENT

This work was supported by the Ministery of Research and Technology under grant 07431018.

6. LITERATURE

Baumgartner,A: Nebel und Nebelniederschlag als Standortsfaktoren am Großen Falkenstein (Bayerischer Wald). Forstw. Cbl.77 (1958), 257-320.

Blyth,A.M; A.Chittenden, J.Gatham: An optical device for the measurement of liquid water content in clouds. Quart. J. Roy. Meteor. Soc. 110 (1984), 53-63.

Choularton,T.W. et al: Field studies of the optical and microphysical characteristics of clouds enveloping Great Dun Fell. Quart. J. Roy. Meteor. Soc. 112, (1986), 131-148.

Chylek,P.: Extinction and LWC in Fogs and Clouds J. Atm. Sciences 35 (1978), 296-300.

Deirmendjian,D.: Electromagnetic Scattering on Spherical Polydispersions. New York, 1969.

Garland,J.A.: Some fog droplet size distributions obtained by an impaction method. Quart. J. Roy. Met. Soc. 97 (1971), 483-494.

Georgii,H.W. et S. Grosch, G. Schmitt: Feststellung der Schadstoffbelastung von Waldgebieten in der Bundesrepublik Deutschland durch trockene und nasse Deposition. Abschlußbericht des Forschungsprojektes 10402715, 1986.

Grunow,J.: Vergleichende Messungen des Nebelniederschlages. UGGI. Ass. Int. Hydr. Sci, Assemblée gen. Toronto, 1957.

Lovett,G.: Rates and mechanisms of cloud water deposition to a subalpine balsam fir forest. Atmos. Environ. 18 (1984), 361-71.

Lovett, Gary M. et W.A. Reiners: Canopy structure and cloud water deposition in subalpine coniferous forests. Tellus 38B (1986), 319-27.

Pinnick,G.R. et S.G. Jennings, P.Chylek, H.J. Auvermann: Verification of a Linear Relation between IR Extinction, Absorption and LWC of Fogs. J. Atm. Sciences 36 (1979), 1577-86.

Schemenauer,R.: Acidic deposition to forests: The 1985 chemistry of high elevation fog (CHEF) project. Atmosphere-Ocean 24 (1986), 303-28.

Waldmann,J. et al: Chemical characterisation of stratus cloud water and its role as a vector for pollutant deposition in a Los Angeles pine forest. Tellus 37B (1985), 91-108.

Winkler,P.: Observations on fog water composition in Hamburg. In: Georgii,H.W. (Ed.): Atmospheric Pollutants in Forest Areas. Reidel (1986), 143-151.

DEVIATION RANGE OF ANNUAL DRY AND WET DEPOSITION RATES

W. G. Huebschmann, S. Honcu
Institut für Meteorologie und Klimaforschung
Kernforschungszentrum Karlsruhe/Universität Karlsruhe
Postfach 3640, Weberstr. 5
D-7500 Karlsruhe 1

ABSTRACT. The rate of deposition of airborne pollutants on the ground downwind of a continuously emitting source is evaluated. The annual dry and wet deposition rates deviate from their long-term averages, due to variations of wind direction and velocity, diffusion category, precipitation and of others. Using 14 years of meteorological measurements at the Kernforschungszentrum Karlsruhe (KfK), the computer code ISOLA-IV was applied to calculate the annual deposition rates at the location of the maximum immission downwind of a 100 m high single source, following the German guideline "Allgemeine Berechnungsgrundlage ...". The deviation range is expressed by the variability, which is defined as the standard deviation being normalized to the average and given in %. The variability of the annual dry and wet deposition rates is up to ± 11 % and ± 27 %, respectively. The related long-term averages (time period of 14 years) are correct with in ± 3 and ± 8 % respectively. For the purpose of extrapolation the wet deposition rate is compared with the total annual precipitation, given for a period of 50 years.

1. PROBLEM

The assessment of the long-term atmospheric pollutant immission in the environment of a continuous source requires meteorological measurements over a time span which is representative for the lifetime of the source. In many cases such measurements are available for not more than one year. Due to variations of wind direction and velocity, diffusion category, precipitation and of other parameters the annual dry and wet deposition rate deviates from its longterm average. This report gives informations on the range of deviation and on the measuring period required for its evaluation. It refers to a continuous point source of aerosols at an effective height of 100 m.

K. Grefen and J. Löbel (eds.), Environmental Meteorology, 237–242.

2. THE CALCULATIONAL MODEL

The calculation is based on "Allgemeine Berechnungs-Grundlage .." (ABG) [1], a guideline recommended by the Ministry of the Interior, FRG, for nuclear installations. According to [1] the wind rose is divided into 10°- or 30°-sectors. The computer code ISOLA-IV [2] calculates the diffusion and deposition of continuously emitted pollutants, see Fig. 1 for an aerosol source of 100 m effective height.

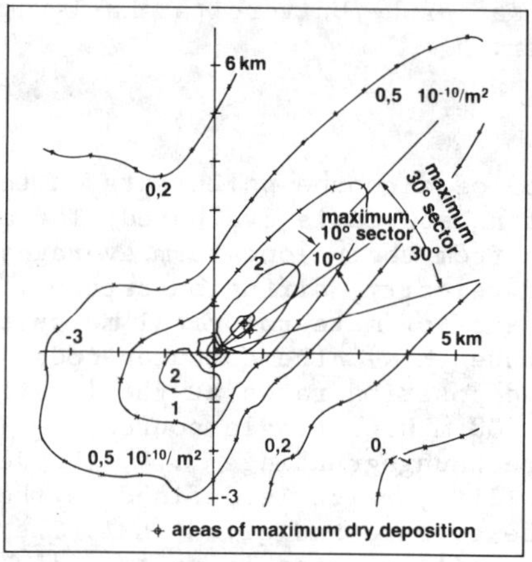

Figure 1. Isolines of dry deposition around a point source of 100 m effective height.

The meteorological input data are measured in the KfK [3]. The period of thoroughly checked measurements covers 14 years (1973-86) of the 3-parametric diffusion statistics (wind direction and velocity, diffusion category) and 13 years (1974-86) of the 4-parametric statistics (add the precipitation rate).

3. DRY DEPOSITION RATE

The dry deposition rate is given here, according to [1], as the fallout factor F, normalised to the source rate:

$$F = v\overline{\chi} \quad \text{in } 1/m^2$$

v = deposition velocity [m/s]

$v = 10^{-3}$ m/s for aerosol, see [1]

$\overline{\chi}$ = dispersion factor [s/m^3], annual average

F is evaluated for the long-term maximum, which is located on the centerline of the 50°-sector (10°-statistics) or the 60° sector (30°-statistics) at a distance of 1000 m, see the marked points in Fig. 1. Fig. 2 shows the annual F at the maximum point in the 14-years-period, the average F and its variability, which is the standard deviation normalized by the average value and given in %. The average can be expected to be correct within $\pm$ 3 %. It is concluded therefore, that the period of 14-years is sufficient to evaluate the long-term average and the variabilty.

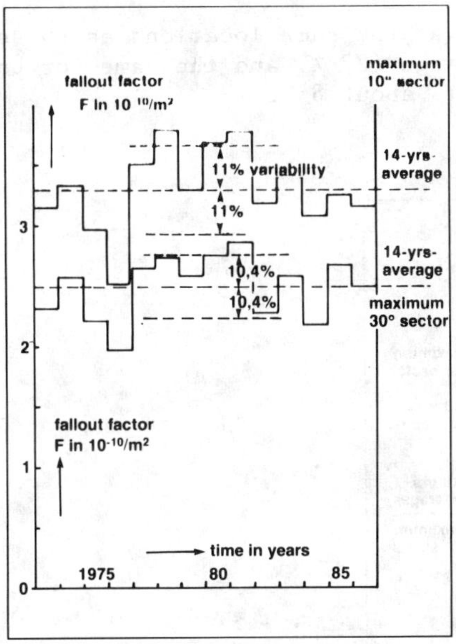

Figure 2. Annual fallout factor F 1973-1986.

4. WET DEPOSITION RATE

The rate of wet deposition from a plume is expressed by the washout factor W, normalised to the source rate:

$$W = \frac{\Lambda}{u\,\sigma_y \sqrt{2\pi}} \quad \text{in } 1/m^2$$

Λ = washout coefficient in 1/s
u = average wind velocity in m/s
σ_y = horizontal diffusion-parameter in m

Λ depends on the precipitation intensity I (in mm/h).

4.1. Model of "ABG" [1]

W_{ABG} is proportional to the amount of precipitation. This is equivalent to the formula

$$\Lambda = 0.3\ 10^{-4}\ (\frac{I}{mm/h}) \quad \text{in 1/s for aerosol}$$

Fig. 3 shows the anual W_{ABG} at the same locations as chosen for F, see Fig. 1. The variability is large (27 %) and the same for both sector sizes. The error of the average is about 8 %.

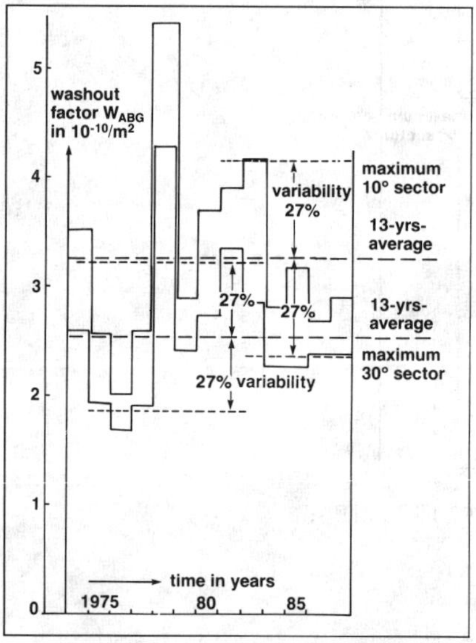

Figure 3. Annual washout factor W_{ABG} 1974-1986.

4.2. Model of Individually Determined Λ

A literature survey [3] yielded the formula

$$\Lambda = 0.8\ 10^{-4}\ (\frac{I}{mm/h})^{0.8} \quad \text{in 1/s,}$$

which determined the Λ of each range of rain intensity.

W_L is larger than W_{ABG}, see Fig. 4, but the variability and the error of the average are of the same order of magnitude.

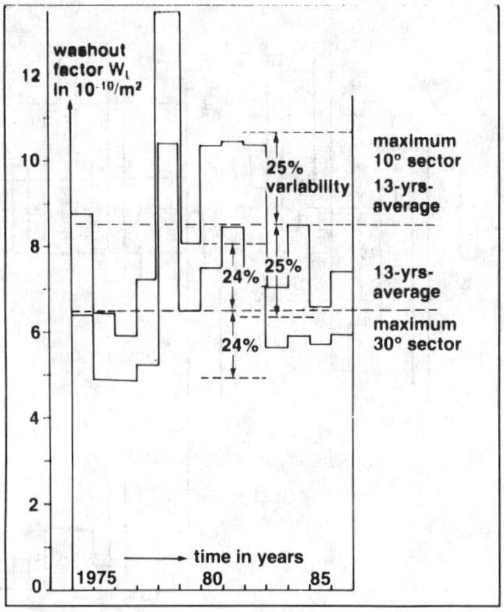

Figure 4. Annual washout factor W_L 1974-1986.

4.3. Expected long-term variability

In case of W the 13-years-period is too short as to yield sufficiently correct long-term values of average and variability. The total annual precipitation N is available for a 50-years-period [4], see Fig. 5. The average of N is correct within 3 %. For the period 1974-86 W_L and N show a qualitatively similar behaviour, see Fig. 5. Therefore the approach is made to compare both parameters. If also the average and variability of W_L and N behave in a similar way, then the average of W will decrease by up to 6 % with longer periods of time, but the variability will increase by about the same amount.

5. CONCLUSION

Dry deposition variability: about 11 %.
14 years are sufficient for long term assessment.

Annual precipitation variability: long-term 19 %
1974-1986 14 %

<u>Wet deposition</u> variability 1974-1986, ABG-model: 27 %
Λ -model: 24 %
long-term: larger by several percent

14 years not sufficient for long term assessment.

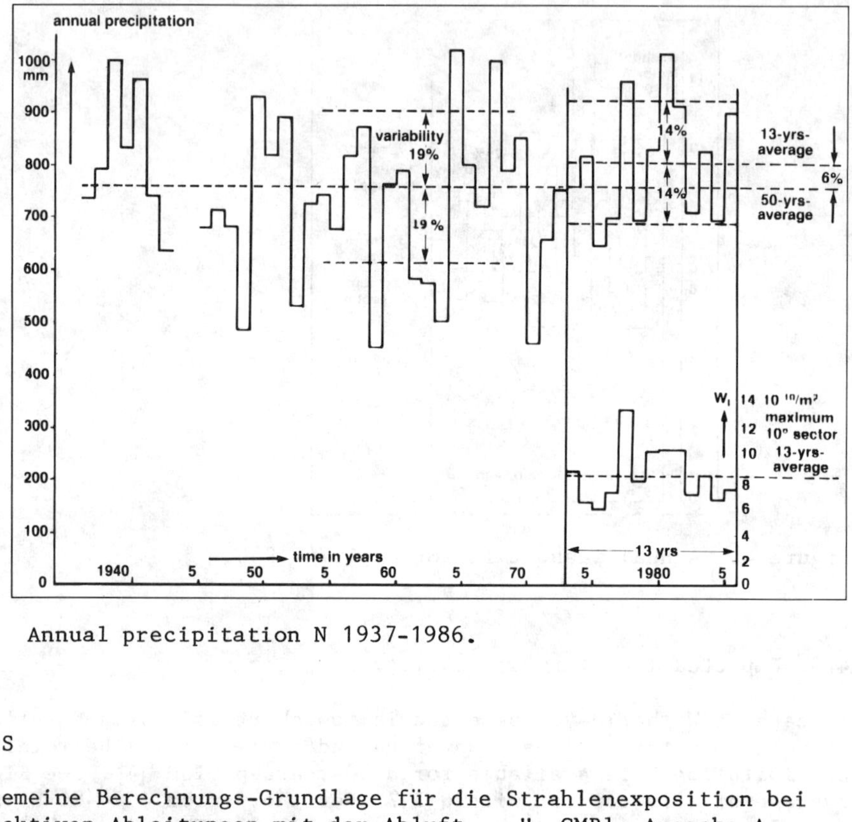

Figure 5. Annual precipitation N 1937-1986.

REFERENCES

1 "Allgemeine Berechnungs-Grundlage für die Strahlenexposition bei radioaktiven Ableitungen mit der Abluft ...", GMBl, Ausgabe A, **30**, 369 (1979); GMBl, Ausgabe A, **33**, 735 (1982)

2 W. Hübschmann, S. Honcu:
ISOLA IV - Ein Fortran 77-Programm zur Berechnung der langfristigen Dosisverteilung in der Umgebung kerntechnischer Anlagen, KfK 4146 (1986)

3 S. Vogt, W. Hübschmann, P. Wittek:
Niederschlag und Washout im Unfallfolgenmodell der Deutschen Risikostudie-Kernkraftwerke, KfK 3548 (1983)

4 K. Höschele, M. Kalb:
Das Klima ausgewählter Orte der Bundesrepublik Deutschland/Karlsruhe, in: Berichte des DWD Offenbach (in Vorbereitung)

WIND TUNNEL MODELING OF ACCIDENTALLY-RELEASED GASES HEAVIER THAN AIR

G. König-Langlo, M. Schatzmann
Meteorological Institute, University of Hamburg
Bundesstrasse 55
D-2000 Hamburg 13
Federal Republic of Germany

ABSTRACT Large quantities of gases usually in liquified or pressurised form are being transported and stored to a steadily increasing extent. Many of these gases are either flammable or toxic or both. The handling of these gases is therefore associated with safety hazzards which have to be minimized. Most of these gases are heavier than air because of their molecular weight or due to the low temperature which they would take on in the event of an accidentally caused release. In consequence, the dispersion of them cannot be calculated by conventional techniques which were developed for neutral and buoyant gases.

The objective of this paper is to present the concept and some results of a wind-tunnel study investigating the initial dispersion of heavy gases. Comparisons with full-scale data will confirm that data obtained in wind-tunnel experiments can be transferred to scales of practical interest. The wind-tunnel results can be applied to many real-world problems and give a contribution towards the establishment of safety regulations for inflammable and toxic gases heavier than air.

1. INTRODUCTION

Accidents involving heavy gases have already occurred many times in the past. Summarizing accounts of such accidents as they were published by, e.g., the German Environmental Protection Agency, Umweltbundesamt (1983), or the Association for Technical Monitoring, TÜV Norddeutschland e.V. (1982), indicate a large diversity among the cases reported on. They differ with respect to the following aspects:

- type of gas release (instantaneous or continuous),
- amount of the released gas,
- physical properties of the gas,
- atmospheric conditions,
- characteristics of the canopy around the source.

The diversity and complexity of accidents make them difficult to model. In recent years, several mathematical models for the prediction of heavy gas dispersion have been developed. For a survey see Wheatley et al. (1985). Although the models have been reasonably suc-

K. Grefen and J. Löbel (eds.), Environmental Meteorology, 243–254.

cessful in some cases, they are still limited in scope. The predominant insufficiency of mathematical models is their inability to handle problems of real practical interest properly in which the heavy gas cloud is released into an atmospheric boundary layer flow disturbed by complex bottom topography and obstacles.

An alternative approach is to investigate the heavy gas dispersion in a physical model, e.g., in a wind tunnel. Here, also complex structured arrays of obstacles can be simulated without any major difficulty. There is no need for fitting of constants or other calibration work known from mathematical modeling. As long as the predominant similarity parameters in the model flow can be matched to prototype values, the wind-tunnel experiments provide results directly applicable to the corresponding large-scale problem.

As will be shown in the following chapters, model/prototype similarity can easily be obtained for the so-called gravity spreading phase of the cloud (see Schatmann et al., 1985) which usually extends down to a few percent of the initial gas concentration. Such a dilution corresponds to the lower flammability limit of all major flammable gases. Consequently, the area potentially endangered by a fire hazard can be directly determined in the wind tunnel.

In case of toxic gases the area within which the cloud concentrations are hazardous is usually much larger and extends well into the far-field of the cloud dispersion. Here, buoyancy effects become insignificant and the cloud spreads like a neutrally-dense tracer. In many cases, the vertical extent of the far-field cloud is already large enough that obstacles can be included into the roughness parameterization. Thus, for accidentally released toxic gases a two-step procedure seems to be appropriate: A physical model should be used to cover the complex near-field dispersion of the cloud. The results of this model provide the input data for a simple mathematical dispersion model utilized to determine the gas concentrations in the far-field of the cloud.

The present paper concentrates on the near-field dispersion of heavy gas clouds. It starts with a derivation of similarity parameters governing the problem and describes a strategy to reduce the large number of parameters to a tractable quantity. It continues with a model/full-scale comparison and the presentation of further wind-tunnel results. Finally, an example is given which demonstrates the applicability of the results by determining lower flammability distances for butane.

2. SIMILARITY ANALYSIS

Certain similarity requirements must be fulfilled in order to transfer results from small-scale wind tunnel experiments to prototype scale. These similarity laws are usually obtained by dimensional analysis, a method which makes use of the fact that physical equations must be dimensionally homogeneous and hence the parameters occurring therein can only appear in certain combinations. This method will be outlined briefly as it applies to the present application.

The present problem is sketched in Fig. 1 at the example of an instantaneous release. It consists of a volume V_o of a heavy gas with density ρ_o and viscosity μ_o emitted into an atmospheric boundary layer flow characterized by a power-law wind profile n and boundary layer thickness δ. The initial shape of the cloud is unknown in reality, so we assume for simplicity a cylinder with height h_o and radius r_o. Since only the gravity-dominated near-field dispersion phase is of interest here, it is reasonable to assume neutral ambient stratification. The properties of the surrounding air are described by ρ_a and μ_a, z_R characterizes the surface roughness. The spread of the cloud is affected by several obstacles. There geometry is fully defined by the length scales $l_1, \ldots, l_n$ and the porosity Θ. The local dilution $\chi = \Delta c/\Delta c_o$ (with Δc and Δc_o the local and initial concentration excess above ambient, respectively) depends on the following variables

$$\chi = f_1(x, y, z, g, \rho_o, \mu_o, V_o, h_o, \rho_a, \mu_a, \bar{u}_{aR}, \delta, l_1, \ldots, l_n, \Theta) \tag{1}$$

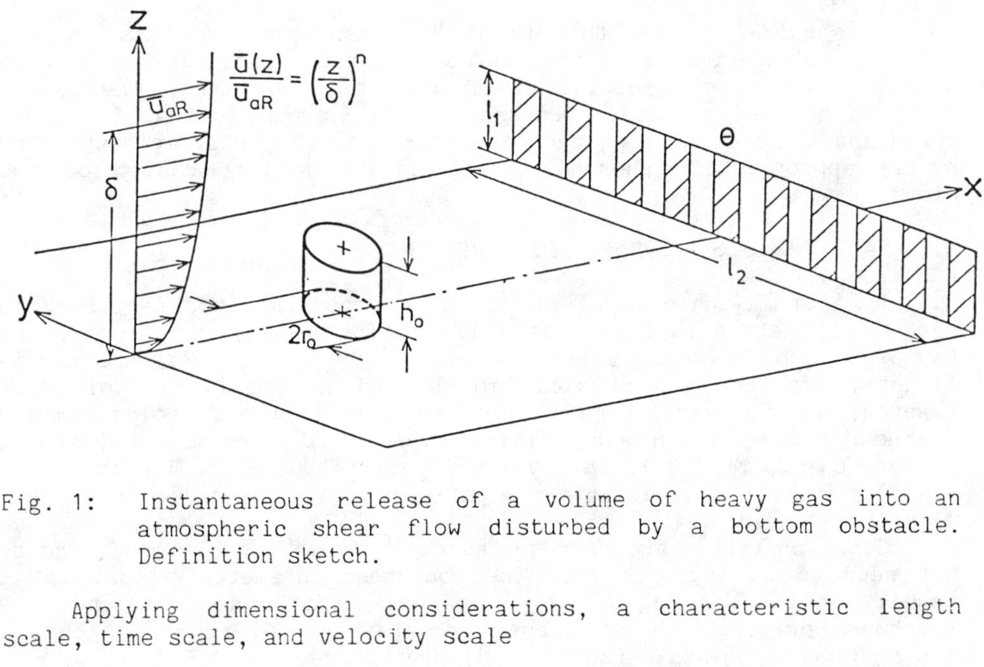

Fig. 1: Instantaneous release of a volume of heavy gas into an atmospheric shear flow disturbed by a bottom obstacle. Definition sketch.

Applying dimensional considerations, a characteristic length scale, time scale, and velocity scale

$$L_{ci} = V_o^{1/3} \qquad T_{ci} = \left(L_{ci}/g'\right)^{1/2} \qquad U_{ci} = \left(L_{ci}\,g'\right)^{1/2} \tag{2a-c}$$

can be defined with effective gravity $g' = g\,\Delta\rho_o/\rho_a$ and $\Delta\rho_o = \rho_o - \rho_a$. The subscripts c and i indicate that these scales are characteristic variables for instantaneous gas releases. For continuous releases, a somewhat different set of scales has to be chosen (see Schatzmann et al., 1985). Nondimensionalization of Eq. (1) results in

$$\chi = f_2\left(\frac{x}{L_{ci}}, \frac{y}{L_{ci}}, \frac{z}{L_{ci}}, \frac{\Delta\rho_o}{\rho_a}, \frac{L_{ci}\, U_{ci}}{\mu_o/\rho_o}, \frac{L_{ci}\, \bar{u}_{aR}}{\mu_a/\rho_a}, \frac{h_o}{L_{ci}}, \frac{z_R}{L_{ci}}, \frac{\bar{u}_{aR}}{U_{ci}}, \frac{\delta}{L_{ci}}, \frac{l_1}{L_{ci}}, \ldots, \frac{l_n}{L_{ci}}, \Theta \right) \tag{3}$$

Eq. (3) indicates that the local dilution χ in both model and full-scale takes on identical values of locations determined by the coordinates x/L_{ci}, y/L_{ci} and z/L_{ci} provided that all the remaining dimensionless parameters on the right of this equation can also be matched in the wind-tunnel experiment. These parameters are: the density excess ratio of the initial cloud, a cloud Reynolds number and an ambient Reynolds number, the aspect ratio of the initial cloud h_o/L_{ci}, the roughness parameter, the velocity ratio $\bar{u}_{aR}/U_{ci}$ (which has the quality of a Froude number, as can be seen when U_{ci} is replaced by Eq. (2c), the nondimensionalized mixing height, the dimensionless length scales, l_1/L_{ci} to l_n/L_{ci}, and finally the porosity Θ.

It should be noted that both the dimensional analysis and the wind-tunnel experiments do not include all those processes leading to the formation of the initial cloud (bursting of a tank, evaporation of the liquified gas etc.). Furthermore it is assumed that the molecular and thermodynamic properties of the gas are fully accounted for by an appropriately chosen initial density and viscosity of the cloud.

3. MATCHING OF SIMILARITY REQUIREMENTS

In a small-scale model it is neither possible nor necessary to match all the similarity numbers listed in Eq. (3). In general the Reynolds numbers (terms 5 and 6 on the right of Eq. (3)) are significantly smaller in a physical model than in the field. This has, however, only marginal implications as long as the Reynolds numbers are kept above a certain critical value. For the cloud Reynolds number, $L_{ci}\, U_{ci}/(\mu_o\, \rho_o)$, this value is about 400, according to Jansen (1981). In all our experiments the critical Reynolds number was clearly exceeded.

Since obstacles are represented directly in a scale model and do not need to be included into the roughness parameterization, z_R is usually small. With $z_R/L_{ci} \ll 1$ in both, model and prototype, the roughness criterion is at least approximately fulfilled as long as the profile exponents n (see Fig. 1) are matched.

Similar arguments hold for the boundary layer thickness δ/L_{ci} which is much larger than unity in both cases.

With the exception of extrem values which we do not consider here, the aspect ratio h_o/L_{ci} should effect the cloud development only in the direct vicinity of the source. For the determination of lower flammability distances it is sufficient to characterize the cloud by its initial volume alone.

With the provisions concerning the Reynolds number in mind, approximate similarity is to be expected by using an undistorted wind tunnel model and matching the density excess ratio and the velocity ratio. To prove this hypothesis model/full-scale comparisons have been carried out. The results are presented in the next chapter.

4. MODEL/FULL-SCALE COMPARISON

Wind tunnel validation experiments have been carried out using field data obtained in the HSE heavy gas dispersion field trials at Thorney Island, U.K., during 1982 and 1983 (see McQuaid, 1984). Using a fold-away tent, 2000 m^3 of denser-than-air gas was instantaneously released in each experiment. The experimental set-up was based on the assumption that a tank filled with liquid gas bursts and that its content evaporates without delay.

As an example, Fig. 2 contains a comparison of continuous monitor concentration measurements with model results for Thorney Island Trial No. 15. The operation conditions for this trial were: $\Delta\rho_o/\rho_a$ = 0.41, profile exponent n = 0.16 (see Fig. 1) and $\bar{u}_{aR}$ = 5.4 m/s in reference height 10 m. In the wind tunnel this trial was reproduced by an instantaneously released volume of 450 cm^3 SF_6 air-mixture. With the exception of internal and ambient Reynolds number which cannot be met in a small-scale experiment, all dimensionless parameters given in Eq. (3) were the same in model and prototype.

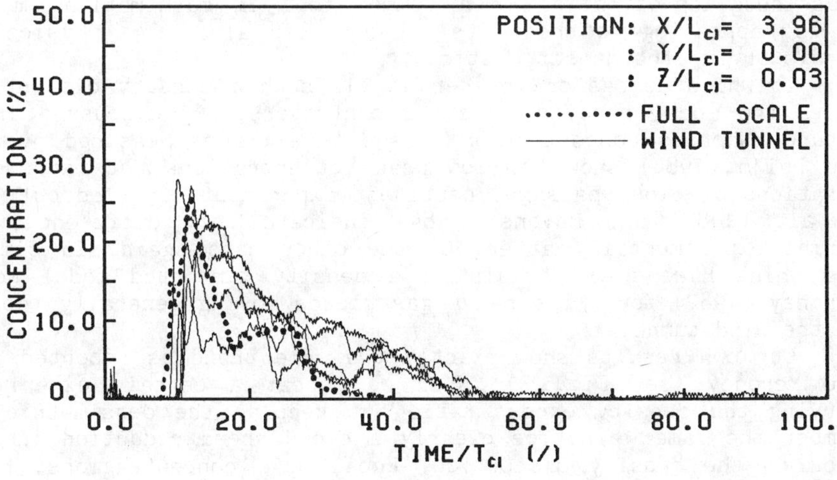

Fig. 2: Comparison of continuous monitor measurements with model results for Thorney Island Trial No. 15 at a fixed downwind position.

Due to the turbulence within and outside of the cloud there can be a high degree of naturally occurring repeat variability in the measurements. Although only "one-shot" data was available for the full-scale trials, model experiments were repeated five times to get some idea of the degree of variability. Judging from a practical point of view the agreement between model and full scale results shown in Fig. 2 is fully sufficient for the peak concentration as well as for the concentration variability with time.

Similar agreement was found also for all other locations and other Thorney Island Trials.

Further model/full-scale comparisons have been carried out with the Maplin Sands (Puttock, 1982) and TÜV Norddeutschland (Heinrich, 1986) continuous release field trials. The results are generally acceptable (see König, 1987).

5. EFFECTS OF DENSITY DISTORTION

The similarity requirement to keep both, the excess density ratio and the densimetric Froude number (identical with the velocity ratio, see Chap. 3) the same in model and prototype, has the consequence of inconvenient and sometimes even impractically low wind velocities in the tunnel. To match both parameters in a physical model corresponds to the application of a mathematical model utilizing the full, non-Boussinesq-approximated equations. On the other hand, relaxing the excess density ratio requirement and to using the Froude number as the sole scaling parameter would correspond to a Boussinesq-approximated mathematical model. Since the introduction of the Boussinesq-approximation is a common tool in mathematical modeling, it was appealing to check its consequences also in a physical model by distorting the density ratio.

To which degree or whether at all such a density distortion effects the model results is still controversially discussed in literature. Whereas concentration versus time traces measured by Heugten and Duijm (1984) show too low peak concentrations but extended time durations of cloud passage, pertinent experiments carried out by Hall et al. (1982) and Havens (1984) indicate only differences within normal experimental scatter. On the other hand again, in simulating the China Lake field trials in a density distorted model Neff and Meroney (1982) found the heavy gas clouds moving generally too slowly in the wind tunnel.

Our own results show exactly the same trend as reported by Neff and Meroney (see Fig. 3). When we repeat a certain experiment by varying the density excess ratio but keeping the densimetric Froude number the same we notice clearly a cloud speed reduction in proportion to the density distortion. Local peak concentrations, however, seem to be only marginally effected. Since a shift in cloud arrival and departure times does not change the lower flammability distances, and thus the area within which the gas/air mixture is hazardous, the subsequent experiments could be carried out with the Froude number as the dominant scaling parameter and keeping the density excess ratio fixed.

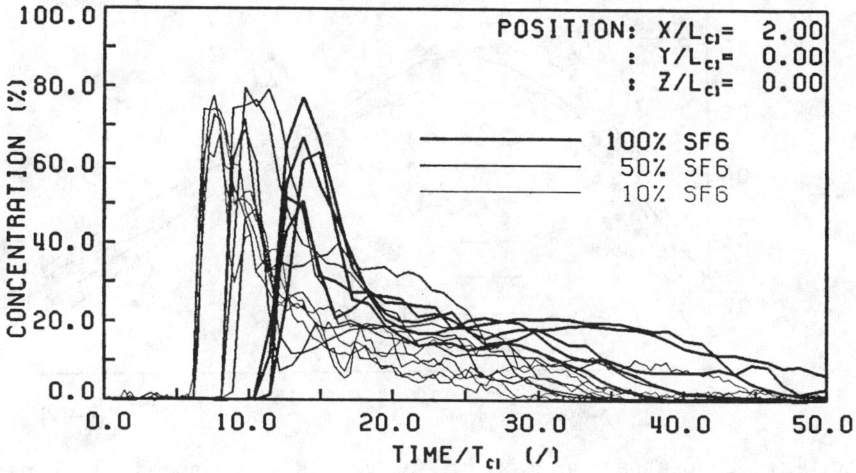

Fig. 3: Concentration of an instantaneously released heavy gas cloud at a fixed position with $\bar{u}_{aR} = U_{ci}$ for three different initial densities.

6. EFFECTS OF WIND VELOCITY

Risk prevention strategies are usually based on worst case assessments. To discover unfavourable conditions, the wind velocity ratio $\bar{u}_{aR}/U_{ci}$ was systematically varied and the concentration fields at ground level monitored. Fig. 4 shows the functional dependence of local maximum ground level concentrations C_M normalized with the maximum of all concentration maxima at this particular location for varying distances from the source x/L_{ci} and varying velocity ratios. As to be expected the concentrations are comparatively small for low winds, since the cloud spreads horizontally into all directions with the consequence of a small lower flammability distance. The concentration ratio is small again for large velocities due to the intense dilution of the cloud by ambient turbulence. Between these two extreme situations a maximum must appear. It occurs at about $\bar{u}_{aR}/U_{ci} = 1$, which means that the lower flammability distances are largest just then, when the front velocity of the heavy gas cloud (proportional to U_{ci} under stagnant ambient conditions) is of about the same magnitude as the advection velocity u_{aR} (taken at reference height L_{ci}).

On the basis of these results the experimental program for the investigation of heavy gas dispersion could be significantly reduced (see Chap. 7).

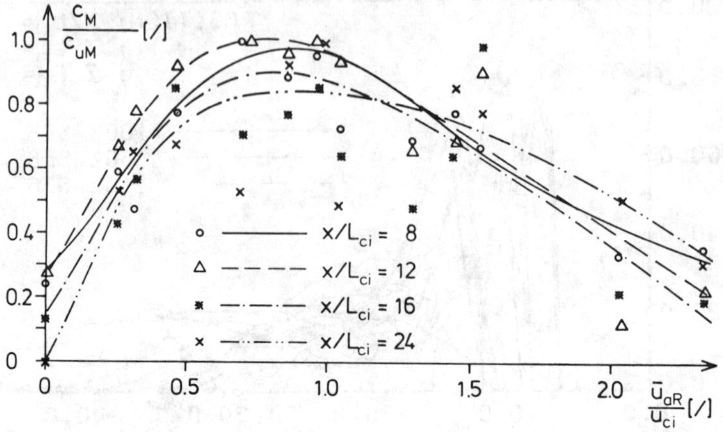

Fig. 4: Concentration maximum C_M versus wind speed at 4 different positions and the lee of the release point of a instantaneously released heavy gas cloud.

7. EFFECTS OF OBSTACLES

Surface obstructions can significantly modify the dispersion of a heavy gas cloud. Depending on the shape of the obstacles and their orientation towards the wind direction, the area within which hazardous gas concentration occur can be increased or reduced. Subsequently examples of these two cases will be given.

The first example demonstrates the sheltering effect of a 5 m high solid fence built at a distance of 50 m apart from an initially 2000 m^3 gas cloud with a density excess ratio $\Delta\rho_o/\rho_a$ = 0.92. The gas was instantaneously released. The experimental set-up realized in the tunnel in a scale of 1:165 corresponds to the Thorney Island field trial no. 20. In order to determine the natural variability of locally measured concentration-versus-time traces, the small-scale experiment was repeated five times. Fig. 5 contains a comparison of model/full-scale results for a location just behind the fence. The agreement is fair. There are also included into the figure the results from a wind-tunnel experiment without fence but otherwise identical conditions. As the comparison shows, the sheltering effect of an impermeable fence or wall is remarkable and leeds to a decrease in the concentration maximum by nearly an order of magnitude.

In the second example, instead to a reduction the obstacle effect leads to a significant increase in the lower flammability distances. Here, an infinite long street canyon parallel to the wind direction was investigated. The width of the canyon was 2 L_{ci}, the height of the bordering buildings was 1 L_{ci}. The experiments showed that the height of the buildings was already sufficient to trap the heavy gas cloud within the canyon completely. The release was instantaneous again, the wind velocity was chosen to be zero.

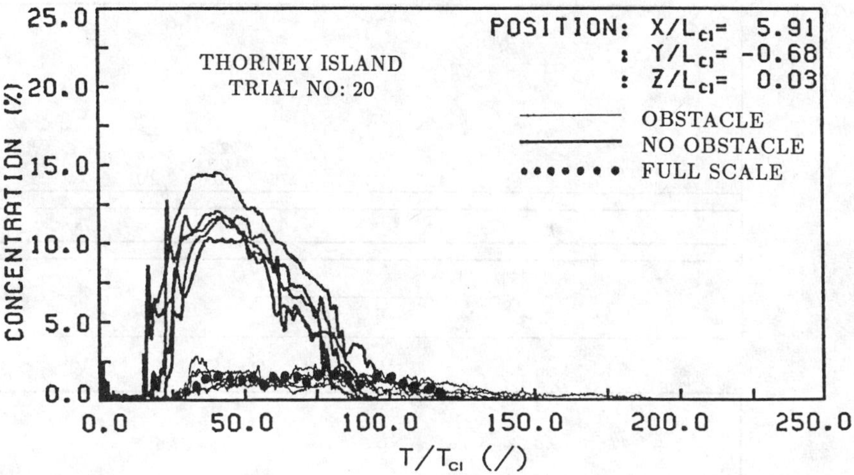

Fig. 5: Effect of a shelter wall on the dispersion of an instantaneously released heavy gas cloud. Model/full-scale comparison for Thorney Island trial no. 20.

The obstacle effect becomes evident when the results from the street canyon are compared with those from a corresponding experiment in flat terrain, also included into Fig. 6. Assuming, e.g., the instantaneous release of 2000 kg butane the actual increase of the mean lower flammability distance due to the street canyon can easily be calculated. The lower flammability level of butane is 1.5%. After evaporation the assumed 2000 kg form an initial gas volume of about 1000 m^3. Then, acc. to Eq. (2a), the characteristic length is L_{ci} = 10m. From Fig. 6 we obtain a mean lower flammability distance of 25 L_{ci} = 250 m for the street canyon and 10 L_{ci} = 100 m for flat, unobstructed terrain.

8. CONCLUSIONS

So far more than 300 single experiments have been carried out in the boundary layer wind tunnel of Hamburg University. The detailed results have been published in König (1987). The report contains further wind-tunnel/full-scale comparisons also for continuous sources and results from tests with a variety of additional obstacle configurations. Special emphasis was given to worst case conditions as they are needed for the development of risk prevention strategies.

The results are applicable to directly assess the lower flammability distances for accidentally released flammable gases heavier than air. In case of toxic gases the results are suitable to determine initial conditions for numerical models applied to the far-field of the dispersing cloud. Since here gravity effects are of

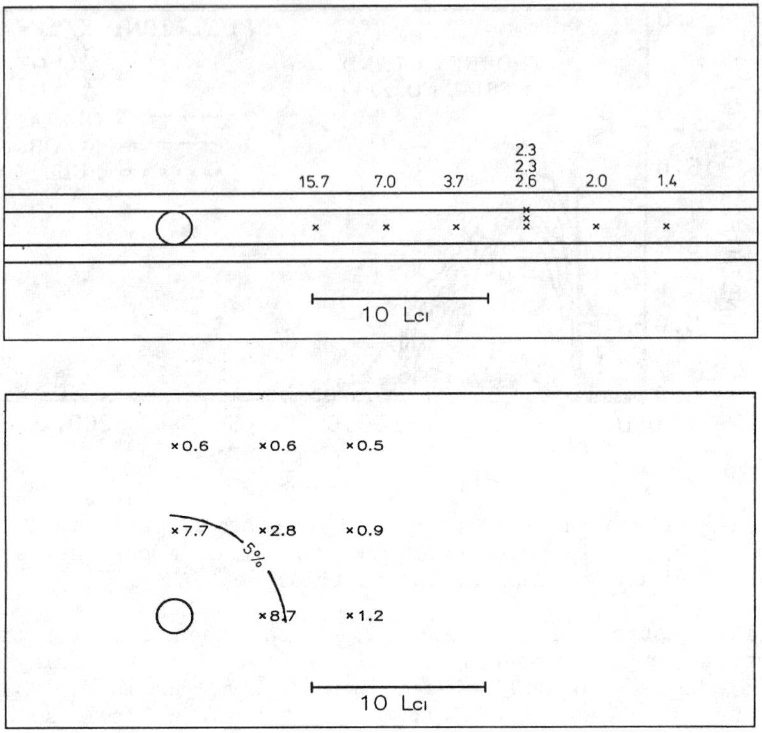

Fig. 6: Mean peak concentration at the ground for an instantaneous release of heavy gas with (top) and without (bottom) obstacle, $\overline{u}_{aR} = 0$. The crosses indicate probe positions, the circle the position of the gas source.

negligible importance, the structure of the far-field models can be rather simple.

The results are always being presented in form of non-dimensionalized graphs (see Fig. 6). As demonstrated in the last chapter, these graphs are easy to handle in practical applications. They apply to both release types (continuous and instantaneous) and any source size. The effect of most obstacle configurations of practical concern on lower flammability limits can already be quantified at least approximately. The work is presently being continued to cover further source conditions in combination with and without obstructions.

9. ACKNOWLEDGEMENTS

We are grateful to the German Ministry of Research and Technology (BMFT) for financial support.

10. REFERENCES

Hall, D.H., Hollis, E.J., and Ishaq, H. (1982):
"A wind tunnel model of the porton down dense gas spill field trials". Report LR 394 (AP), Warren Spring Laboratory, Stevenage, U.K.

Hartwig, S. (Ed.) (1982):
"Heavy gas and risk assessment II". D. Reidel, Dordrecht, Holland.

Havens, J.A., and Spicer, T.O. (1984):
"Gravity spreading and air entrainment by heavy gas instantaneously released in a calm atmosphere". In: Ooms, G., and Tenekes, H. (1984).

Heinrich, M., Gerhold, E. (1986):
Praxisgerechte Bestimmung der Zündentfernung bei der Freisetzung schwerer Gase (LPG). 3. Zwischenbericht, TÜV Norddeutschland.

van Heugten, H.H., and Duijm, N.J. (1984):
"Some findings based on wind tunnel simulation and model calculations of Thorney Island Trial No. 008". Journ. Hazardous Materials, Volume 11, pp. 409-416.

Janssen, L.A.M. (1981):
"Wind tunnel modelling of the dispersion of LNG vapour clouds in the atmospheric boundary layer". MT-TNO Rep. 81-07020.

König, G. (1987):
"Windkanalmodellierung der Ausbreitung störfallartig freigesetzter Gase schwerer als Luft". Hamburger Geophysikalische Einzelschriften, Verlag Wittenborn Söhne, Hamburg.

McQuaid, J. (1984):
"Large scale experiments on the dispersion of heavy gas clouds". IN. Ooms, G., and Tenekes, H. (1984).

Neff, D.E., and Meroney, R.N. (1982):
"The behavior of LNG vapor cloud". Report, Contr. No. 5014-352-0203, Gas Research Institute, Chicago, USA.

Ooms, G., and Tenekes, H. (1984):
"Atmospheric dispersion of heavy gases and small particles". Springer-Verlag, Berlin.

Puttock, J.S., Colenbrander, G.W., and Blackmore, D.R. (1982):
"Maplin Sands Experiments 1980: Dispersion results from continuous releases of refrigerated liquid propane". In: Hartwig, S. (Ed.) (1982): Heavy gas and risk assessment II.

Schatzmann, M., König, G., and Lohmeyer, A. (1985):
"Dispersion modeling of hazardous material in the wind tunnel". Proc. 15th Int. Techn. Meet. on Air Poll. Model. and its Appl., St. Louis, USA.

TÜV Norddeutschland (Northern Germany Technical Control Board) (1982):
Forschungsbericht zur Anfertigung einer Störfallstatistik im Rahmen des FE-Vorhabens "Untersuchung zur Ausbreitung von Luftverunreinigungen im Störfall" Produced on behalf of the Meteorological Institute of the University of Hamburg by TÜV Norddeutschland e.V.

Umweltbundesamt (Federal Environmental Agency) (1983):
"Handbuch Störfälle. Dokumentation über Störfälle in Chemieanlagen". Research Report 104 09 303 UBA-FB-83-023.

Wheatley, C.J., and Webber, D.M. (1985):
"Aspects of the dispersion of denser-than-air vapours relevant to gas cloud explosions". Report SR/007/80/UK/H, Safety and Reliability Directorate, Culcheth, Warringont, U.K.

TRACER EXPERIMENTS TO ESTIMATE DIFFUSIVE LEAKAGES AND TO VERIFY DISPERSION MODELS

Bjarne Sivertsen
Norwegian Institute for Air Research
P.O. Box 64, N-2001 Lillestrøm
Norway

ABSTRACT. Tracer experiments have been carried out for different purposes, using a tracer technique developed by the Norwegian Institute for Air Research (NILU). Several experiments were aimed at quantifying diffuse leakages of hydrocarbons at different petrochemical complexes in Norway and Sweden. Dispersion experiments in residential areas were undertaken especially to study cold, stable winter conditions. Tracer gas was also released from a tall stack at a coal fired power plant to investigate dispersion out to distances of 100 km.

Modelling was performed for different scales with different modelling techniques, also including simple proportionality models to estimate diffusive leakage emission rates. Various methods for the description of turbulence and dispersion was applied and verified with tracer gas concentration distribution on different scales.

Vertical and lateral spread was treated separately, and the choice of parameters for the dispersion models depended upon the actual state of the atmospheric boundary layer. The concentration distribution in the lateral direction was usually taken to be Gaussian, while for the vertical, several different approaches were suggested.

1. INTRODUCTION

The relationship between pollutant sources and concentrations of nondepositing inert gases at various receptor locations have been determined by multiple scales of atmospheric motions controlling the transport and dispersion of the gases. Usually meteorological dispersion models links the source emission rates to pollutant concentration distributions. During complex conditions within building complexes, in street canyons or in uneven rough terrain the use of tracers have been valuable in understanding the processes and linking emissions to concentrations. Several trace

K. Grefen and J. Löbel (eds.), Environmental Meteorology, 255–268.

experiments have been undertaken by the Norwegian Institute for Air Research (NILU) since 1978, to study and verify:

- emissions from complex diffusive sources
- transport and diffusion within and around buildings
- ventilation in streets and dispersion in build up areas
- dispersion in complex terrain

This paper briefly describes some of the results from such experiments and outlines models for establishing the relationship between emissions and downwind concentration distributions of inert gases.

2. THE TRACER EXPERIMENT SYSTEM

The tracer system; release equipment, samplers, calibration unit and portable gas chromatographs, was developed at NILU for extensive use in practical applications (Lamb and Sivertsen, 1978; Heggen and Sivertsen, 1983). The system was developed to meet the following requirements:

- easy to modify
- useful in different applications
- easy to handle (reload, prepare)
- light weights (portable) and sturdy
- preset timing for samplers
- rapid analysis in field.

These requirements have lead to less firm requirements on accuracy and low detection limits. For most practical purposes, however, the release rates can be adjusted to give concentrations within an optimum range ($\approx 10^{-6}$ p/p).

Two tracer gases have been applied in different experiments; sulphur hexafluoride (SF_6) and bromtrifluoromethane ($CBrF_3$). In most applications, when only one tracer is needed, SF_6 is always used. Only when there is a need for dual tracers (for identifying different sources) also $CBrF_3$ is applied.

2.1. The release systems

The tracer gases are usually released from 10 liters cylinders containing the liquefied gas. Through a reduction valve and a flow-meter the gas release is kept constant at the desired rate. At high release rates it is often necessary to supply heat to the cylinder to compensate for the heat loss due to evaporation. Syringe-pumps may be used for small gas releases (0-10 cm^3/min).

2.2. Sampling systems

Air samples are collected in inexpensive plastic syringes on either instantaneous or time-averaged basis. Time averages are usually taken over 15 or 30 minutes with automatic battery powered samplers. A number of such samplers collecting sequentially 15 min. average samples are available. The samplers can be programmed to start sampling at a given time before placement at the desired sampling locations.

Other automatic samplers has been designed for collecting several "instantaneous" air samples. Up to 90 samples can be taken at a predetermined interval. The shortest interval is 2 seconds and the sampling time is about 0.5 sec. These samplers were also used in air craft sampling.

2.3. Tracer sample analysis

The samples are usually analyzed immediate after each experiment using the portable electron capture gas chomatographs. If the analysis are not carried out within the first few hours, hypodermic needles on all syringes are exchanged with caps. The simple sampling and analysis methods permit the collection of a large number of samples during a tracer experiment. In a tracer field study, these techniques allow the results of one experiment to be used in the design of succeeding experiments.

3. TRACER EXPERIMENTS

Different tracer experiments have been carried out on different scales in the atmosphere. In the following section a brief outline of some tracer experiments is presented:

- diffuse leakages of air pollutants around building complexes (0.1-2 km)
- diffusion in residential areas during winter time inversion (0.1-5 km)
- pollutant transport in complex terrain (1-10 km)
- transport and dispersion modelling of tall stack emissions (1-100 km)

3.1. Diffuse leakages

A number of tracer experiments were undertaken in Norway and Sweden to study diffusive leakages of gaseous pollutants from various factories. Most of these studies dealt with hydrocarbon emissions from petrochemical complexes (Sivertsen, 1983a; Sivertsen, 1986; Andrén, 1987). Also emissions form a coke plant (Sivertsen, 1987) and from an aluminium smelter (Sivertsen and Vitols, 1980) were studied using this tracer technique.

To enable the diffusive leakage rates of the air pollutants to be estimated, the following conditions should be fulfilled:

- wind speed should be > 3 m/s to create mechanically induced turbulence for tracer and air pollutants to be sufficiently well mixed at the sampling points;
- periods with precipitation should be avoided;
- release time must adjust to wind speed and sampling point distances;
- sampling time must be long enough to avoid short period fluctuations due to atmospheric turbulence;
- sampling point location should assure at least one horizontal cross-section of the concentration distribution;
- tracer release points and release rates relative to each other should be located and proportioned to match actual leakages
- meteorological variables should be continuously recorded.

As an example of the experiments carried out, Figure 1 shows the 15-min average SF_6 plume due to releases from an ethylene cracker area. SF_6 was released at different heights above the ground in the cracker areas and at different release rates proportional to the expected leakage rates.

When the air pollution leakages are adequately simulated, the concentrations measured along one traverse (AB in Figure 1) should create similar distributions.

3.1.1. Emission estimates. The inert tracer gases usually applied in these studies are not depositing on the surface or washed out by precipitation downwind from the release. Applying simple proportionality models for estimating leakage rates requires that the air pollutants of interest are dispersed in the same way. Dry weather conditions, strong winds and short transport times should assure that no chemical or physical changes take place.

If the tracers (t) are released from the same point (or area) as the pollutants (p), the ratio of total fluxes downwind (F) will be the same as the ratio between the emissions rates (Q):

$$Q_p/Q_t = F_p/F_t \qquad (1)$$

Assuming that the experiments are carried out during high winds with mechanically induced turbulence and good mixing of tracers and pollutants and that there is no significant difference between tracer and pollutant release heights, the emission rate of pollutants (Q_p) can be estimated from the ratio of crosswind

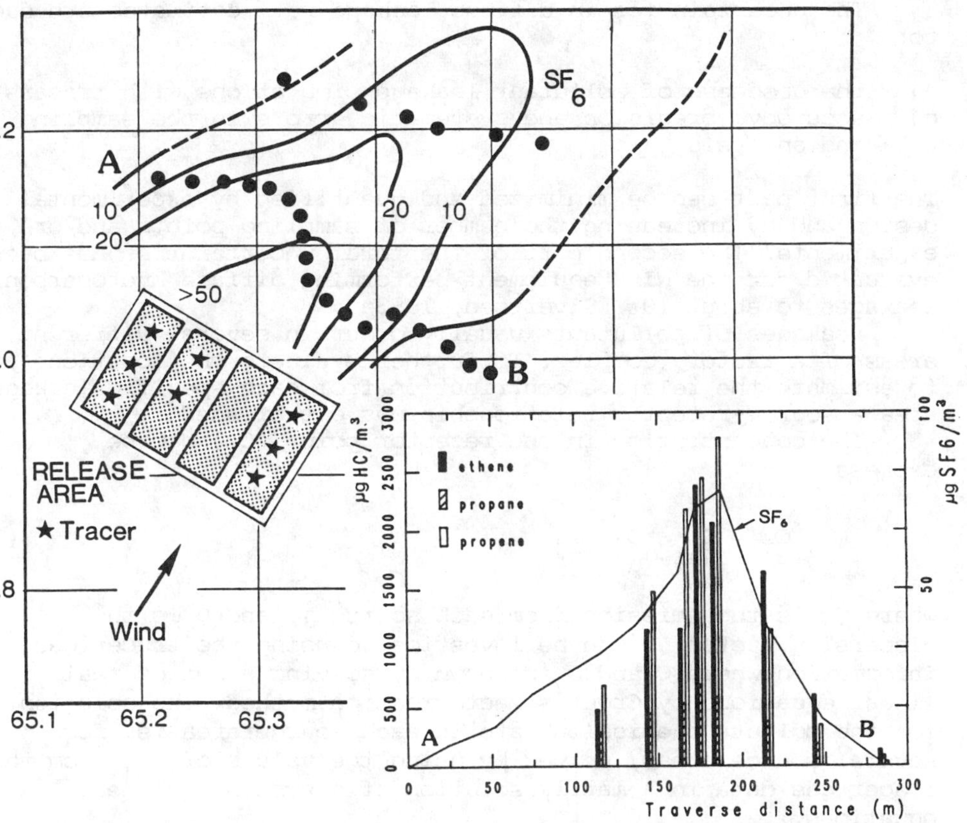

Figure 1. 15 min average concentrations of SF_6 and hydrocarbons ($\mu g/m^3$) measured downwind from a petrochemical complex.

integrated concentrations downwind from the source area:

$$Q_p = Q_t \cdot \left(\int_Y C_p dy / \int_Y C_t dy\right)$$

It is assumed that the total release rate of tracers is controlled and known (Q_t) and that concentrations of pollutants (C_p) are measured at selected points downwind. These measurements are taken along arcs perpendicular to the wint direction. Tracer concentrations (C_t) are measured at a great number of sampling location with the small NILU-samplers available. Tracers can thus be applied to establish a detailed mapping of the plume.

To verify the emission estimates, concentration measurements should be taken at more than one arc, and preferably during different wind direction. In this way a measure of the overall uncertainty of the emission estimate can be established.

The uncertainties in diffuse leakage rate estimates are due to:

a) the goodness of pollutant leakage simulations with tracers
b) accuracy, precision and systematic errors in the sampling and analysis

The first part can be minimized and quantified by experimental design and by increasing the number of sampling points and/or experiments. The second part of the total uncertainties has been evaluated for the NILU equipment performing diffuse hydrocarbon leakages to about 10% (Sivertsen, 1983a).

Leakages of pollutants usually occur in several different areas of a factory complex. To further evaluate the locations and to estimate the relative contribution from each specific leakage area a source-receptor relationship was established.

The concentration in one receptor point (C_i) can be expressed by

$$C_i = \sum_{j=1}^{n} Q_j \, D_{ij} \tag{2}$$

where Q_j is the emission from each source j, and D is the dispersion factor. D can be investigated using the tracer gas information where C and Q is known by solving a set of real linear equations by Crout´s factorisation method. The solution for the pollutant emission rate in each source area is, for several source areas, solved by using the values of D_{ij} from the tracer gas data in a matrix solution of a system of linear equations.

3.2. Diffusion in stable winter conditions

Dispersion experiments were undertaken during cold, stable winter conditions over flat homogeneous terrain (surface roughness $z_0 \sim 0.1$ m) and over a residential area ($z_0 \approx 0.3$ m) (Sivertsen, Bøhler, 1986; Tønnesen, unpublished).

Tracers were released both in the stable surface layer and in the local and z-elss scaling layers. Turbulence scaling in the stable boundary layer and parameterization of dispersion has been described by Gryning et al. (1987).

In the stable surface layer the Obhukov similarity theory was used to estimate the profiles of wind, temperature and turbulence where the diffusivity of matter is approximated by the diffusivity of heat:

$$K = k \cdot u_* z/\Phi_h \,(z/L) \tag{3}$$

Φ_h is the non-dimensional temperature gradient, k = 0.4, u_* is the friction velocity, z is the height and L is the Obukhov length. An analytical solution of the two-dimensional diffusion equation was used to estimate the crosswind-integrated concentration, while the distribution in the lateral is assumed to be Gaussian, in accordance with the results from the tracer experiments.

For releases into the upper stable atmosphere (the intermittancy layer or the z-less and local scaling layers) the dispersion has not been as well understood as in the stable surface layer. A combination of weak (suppressed) turbulence and waves seem to govern the transport and diffusion of pollutants. Local fluxes as presented by Nieuwstadt (1984) can be used to describe the turbulence, but experiments indicate that the tracers are dispersed vertically at a greater rate than the scaling parameters indicate.

Figure 2 shows observed and estimated crosswind-integrated concentrations (C_y/Q) as a function of the distance downwind, for one release in the z-less layer (SF_6) and another release ($CBrF_3$) in the stable surface layer (z_o = 0.1 m).

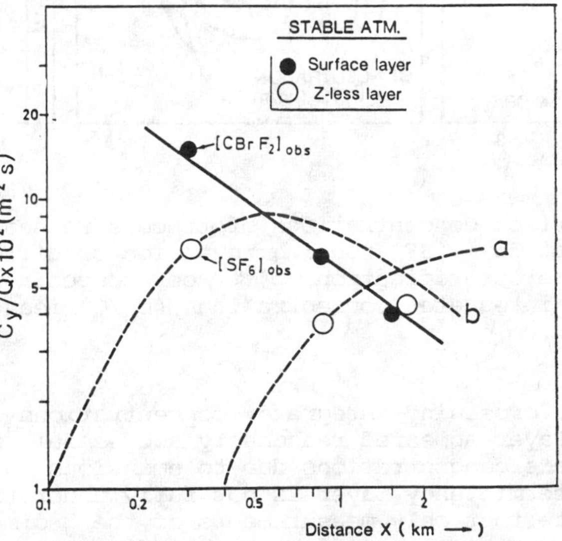

Figure 2. Crosswind integrated concentrations (C_y/Q) estimated (curves) and observed (circles) as a function of downwind distance.
$CBrF_3$ was released in the surface layer (1 m above ground)
SF_6 was released in the z-less layer (10 m above ground)
(C_y/Q) was estimated from σ_w estimates

a) $\sigma_w = 1.2 \cdot u_* (1-z/h)^{0.75}$
b) $\sigma_w = 1.3 \cdot u_*$

The SF_6 tracer gas emitted in the stable upper layer is reaching the surface closer to the source than could be expected from z-less layer turbulence estimation. This might be due to intermittent downwash over the built up area, or gravity waves in the valley bringing SF_6 to the surface more rapidly than by small scale turbulence.

Figure 3 shows a similar experiment carried out over the residential area ($z_o \simeq 0.3$ m).

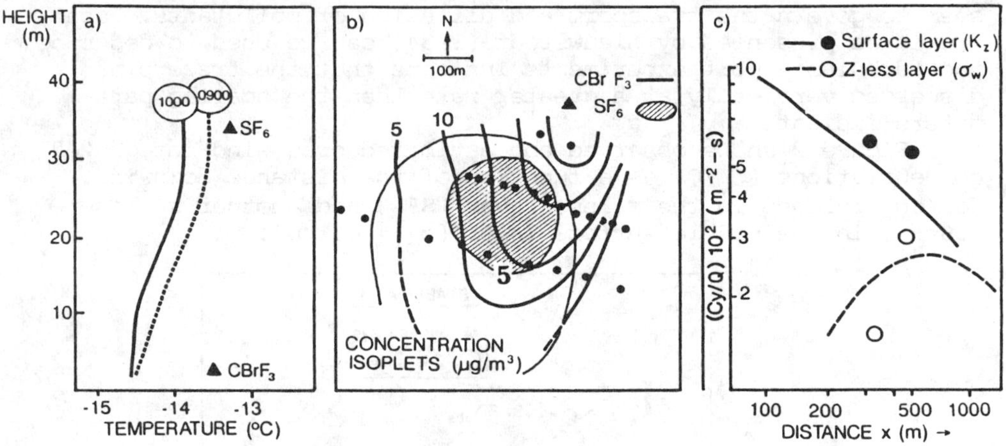

Figure 3. Surface concentrations of tracers released at 2 m ($CBrF_3$) and at 35 m (SF_6) during inversion conditions: a) temperature stratification, b) tracer concentration isoplets, c) crosswind integrated concentrations (C_Y/Q) measured and estimated.

Modelling the crosswind integrated concentrations for releases in the surface layer appeared reasonably successful, while estimation of surface concentrations due to emissions in the upper z-less or intermittency layer is not fully understood. Experiments indicate that only measurements of the 3-dimensional variation of wind in this layer might enable dispersion estimates.

A summary of dispersion experiments carried out during stable atmospheric conditions shown in Table 1, also support this statement.

TABLE I. Tracer experiments carried out during inversion conditions over a residential area (z_o=0.3 m)

	Experiment	Z_s (m)	U (m/s)	L (m)	σ_w (m/s)	Distance (m)	C_y/Q (m^{-2} s) observed	C_y/Q (m^{-2} s) estimated
Surface layer	070285	1	0.5	2.2	-	300	0.16	0.15
						700	0.065	0.073
						1500	0.044	0.047
	060187	2	0.6	14.5	-	260	0.062	0.055
						460	0.059	0.041
	070187	2	0.6	1.8	-	170	0.22	0.37
						550	0.19	0.20
						890	0.01	0.12
Z-less layer	060187	35	0.8		0.3	260	0.016	0.019
						460	0.033	0.024
Intermittency	070187	35	0.8		0.04	170	0.00065	-
						550	0.00074	0.0001
						890	0.00079	0.0015

3.3. Pollutant transport in complex terrain

Tracer experiments were carried out in different fjords and valleys in Norway (Sivertsen et al., 1983; Sivertsen, 1983b). The effectiveness of tracer techniques for the investigation of pollutant dispersion in complex situations has been demonstrated for:

- channelling of the valley wind
- land sea breeze air pollution impact
- thermal upslope winds giving rise to local exposure on the valley sides
- impact on hills and mountains

Simple statistical roll-back models have been applied for analysing the impact of planned emissions. The spatial concentration distribution of pollutants were estimated from:

$$C_p(x,y,o) = C_t(c,y,o) \cdot (Q_p/Q_t) \qquad (4)$$

where tracer concentrations (C_t) were measured at several locations downwind from a future source area. The emission rate of tracers released (Q_t) is known. The emission rate (Q_p) of pollutants is given.

Figure 4 indicates such a complex situation where the impact on the valley side during moring upslope winds, were to be estimated.

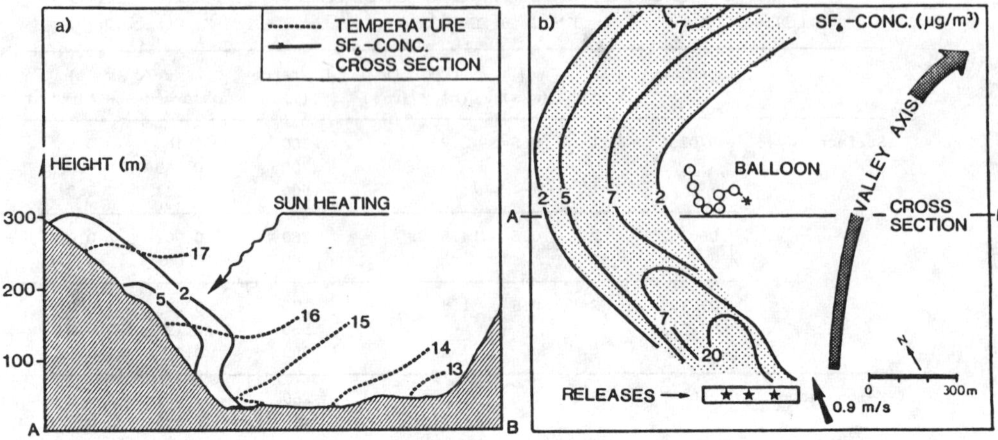

Figure 4. Morning hour upslope wind in a narrow valley during summer conditions.
a) Valley cross sections of temperature (^{0}C) and SF_6 concentrations ($\mu g/m^3$)
b) ground level concentrations of SF_6 ($\mu g/m^3$).

Several experiments were repeated during this typical situation, to establish the statistical basis for impact estimates.

3.4. Transport and dispersion modelling

The use of tracers to understand the physical processes in the atmosphere and to improve the modelling of air quality have been the main objectives of many tracer experiments on various scales.

Dispersion parameters were determined from measurements of wind direction fluctuations (σ_θ), temperature and wind profiles (Sivertsen, 1978). The horizontal parameter σ_y was estimated from

$$\sigma_y = \sigma_\theta \cdot x \cdot f\ (t/T_y) \tag{5}$$

where t is the transport time and T_y is the Lagrangian time scale for lateral dispersion. The function f was investigated from tracer data. Dispersion estimates were performed in stable atmospheric conditions using a socalled "split-sigma" approach, where σ_y was estimated from σ_θ-measurements and σ_z from similarity theory profiles using equation (3) for the diffusivity of matter (Sivertsen, 1980).

Later tracer experiments were applied to estimate the dispersion of nonbuoyant gaseous pollutants using characteristic scaling parameters in the planetary boundary layer (Sivertsen et al., 1986; Gryning et al., 1987).

Dispersion experiments with near surface releases within industrial complexes were also used to estimate the dispersion of gases to a distance of 10 km from the source area (Sivertsen, 1983a; Andrén, 1987).

Tracer gas was released from a 343 m high stack at a coal fired power plant to study the transport and dispersion of pollutants out to distances of 100 km. All experiments were undertaken during daytime convective conditions when the likelihood for the elevated plume to impact the surface was greatest. A commprehensive field experiment including mobile units on the ground and in the air, radiosonde launches, meteorological towers, SODAR-data and COSPEC measurements was performed (Sivertsen and Irwin, 1985).

The highest 30-min average SF_6 concentrations occurred at ground level when the wind speed at 60 m was 10-15 m/s.

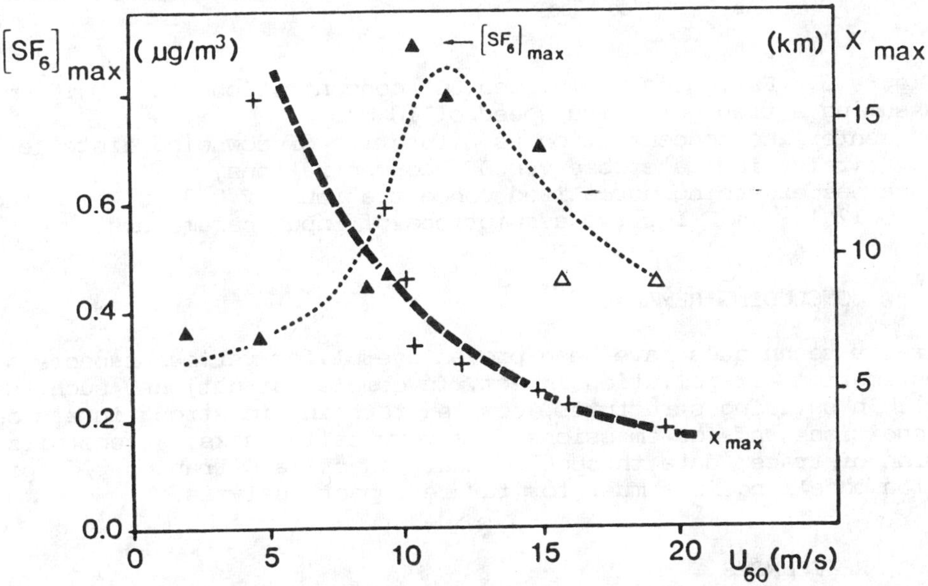

Figure 5. Maximum 30-min average SF_6 concentrations and the distance (x_{max}) to these maxima as a function of wind speed at 60 m above the ground.

Figure 5 also shows that the highest SF_6 concentrations at these winds occurred at distances of 5 to 10 km from the stack, indicating strong convective conditions. The results of the tracer experiments were generalized through modelling of the atmospheric boundary layer using estimated turbulence profiles for dispersion and measured SO_2/SF_6 ratios for depletion estimates.

In this way normalized SO_2 concentrations could be generated as shown in Figure 6.

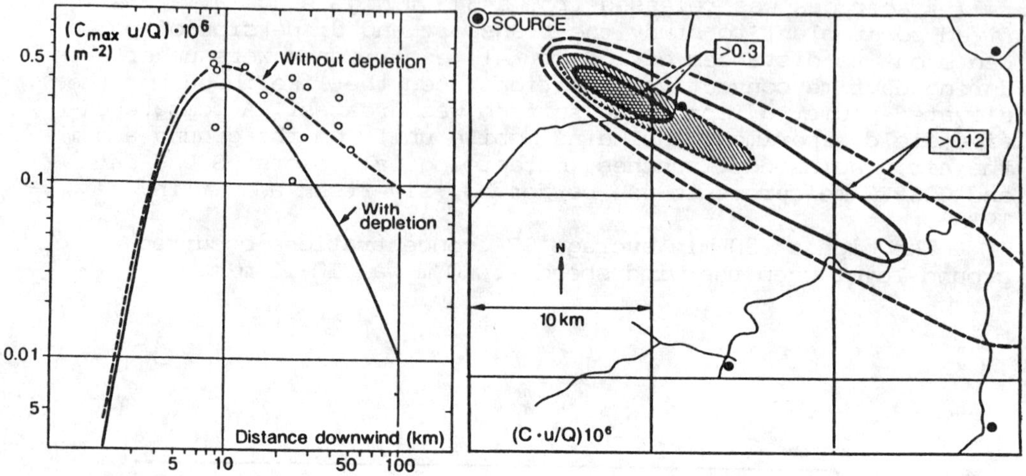

Figure 6. Estimated normalized SO_2-concentrations $(C \cdot u/Q)10^6\ m^{-2}$ assuming a transport wind speed of 10 m/s.

a) centerline concentration as a function of downwind distance (circles indicate observed SF_6 concentrations)

b) areas exceeding normalized concentrations of $0.3 \cdot 10^{-6}\ m^{-2}$ and $0.12 \cdot 10^{-6}\ m^{-2}$ for two assumptions of input parameters.

4. CONCLUDING REMARKS

Tracer techniques have been proven useful for various aspects of estimating air pollution impact. In complex situations, such as within building structures, complex terrain, in strong inversion conditions and for emissions from very tall stacks, a generalization of tracer data through boundary layer- and transport/dispersion modelling is a must for future impact analysis.

5. REFERENCES

Andrén, A (1987) A combined first-order closure/gaussian dispersion model. Atmos. Environ., 21, 1045-1058.

Gryning, S.E., Holtslag, A.A.M., Irwin, J.S. and Sivertsen, B. (1987) Applied dispersion modelling based on meteorological scaling parameters. Atmos. Environ., 21, 79-89.

Lamb, B.K., Sivertsen, B. (1978) Atmospheric disperison experiments using the NILU automatic weather station and SF_6 tracer techniques, Lillestrøm (NILU TN 12/78).

Nieuwstadt, F.T.M. (1984) Some aspects of the turbulent stable boundary layer. Boundary-Layer Meteorol., 33, 31-55.

Sivertsen, B. (1978) Dispersion parameters determined from measurments of wind fluctuations, temperature and wind profiles. In: Proceedings of the ninth international technical meeting on air pollution modelling and its application Toronto 1978 (NATO/CCMS no. 103) pp. 251-261.

Sivertsen, B. (1980) The split-sigma method applied to dispersion estimates. In Seminar on radioactiv releases and their dispersion in the atmosphere following a hypothetical reactor accident. Risø 1980. CEC, Luxemburg. pp. 141-153.

Sivertsen, B. and Vitols, V. (1980) A tracer study of dispersion patterns from Gränges Aluminium Smelter. Lillestrøm (NILU OR 36/80).

Sivertsen, B. (1983a) Estimation of Diffuse Hydrocarbon Leakages from Petrochemical Factories. J. Air Poll. Contr. Ass., 33, 323-327.

Sivertsen, B. (1984) Atmospheric transport in complex terrains. In: Course on atmospheric dispersion of hazardous/toxic materials from transport accidents. Amalfi 1983. Ed. by W.F. Dabberdt. Elsevier, Amsterdam. pp. 75-111.

Sivertsen, B., Lamb, B. and Grønskei, K.E. (1983) A tracer study of pollutant transport in a deep, fjord valley. Atmos. Environ., 17, 1915-1922.

Sivertsen, B. (1986) Tracer experiments to quantify the diffuse emissions of hydrocarbons from petrochemical factories in Stenungsund (in Norwegian). Lillestrøm (NILU OR 22/86).

Sivertsen, B. and Bøhler, T. (1986) Verification of dispersion estimates based upon atmospheric boundary layer parameterization. In: Proceedings of the Seventh Clean Air Congress. Sydney 1986. Eastwood, Clean Air Soc. of Australia and New Zealand. pp. 483-491.

Sivertsen, B., Gryning, S.E., Holtslag, A.A.M., Irwin, J.S. (1986) Atmospheric dispersion modelling based upon boundary layer parameterization. In: Air pollution modelling and its application, Vol. V. Ed. by C. De Wispelaere, F.A. Schiermeier and N.V. Gillani. N.Y., Plenum Publishing Corporation, 1986.

Sivertsen, B. (1987) Air pollution study in Mo 1983-85. Part D, Model estimates (in Norwegian). Lillestrøm (NILU OR 77/85).

Sivertsen, B. and Irwin, J.S. (1987) Data summary of 1985. SF_6 tracer experiments at Andorra (Teruel) power plant. Final report. Lillestrøm (NILU OR 49/85).

SMART - AN ADVANCED EMERGENCY RESPONSE TOOL

H. de Witt, H. D. Brenk, A.G. Knaup
Brenk Systems Planning
Heinrichsallee 38
D - 5100 Aachen
Fed. Rep. of Germany

D. Hesel, H. Schnadt
TÜV Rheinland e.V.
P.O. Box 10 17 50
D - 5000 Cologne
Fed. Rep. of Germany

ABSTRACT In case of accidental atmospheric releases of toxic substances it is an urgent need for decision makers to be provided with data as actual and precise as possible. In the past the assessment of pollutant impact based on measured meteorological parameters and hypothetical emission rates was associated with relatively great uncertainties. To improve this situation the new system SMART has been developed, which reduces the uncertainties substantially. This is in the first step achieved by feedback of actual measurements of the environmental impact into diagnostic calculations in order to adapt the calculations to the current measurements. The result is a more realistic set of model parameters, including the source term. Subsequently these parameters are used as input data for the second step of SMART which is real-time projection of the dispersion situation.

The performance of SMART is illustrated by means of two examples.

1. INTRODUCTION

An accidental atmospheric release of chemically toxic substances can take place in chemical factories, e.g. caused by a fire. It can also follow a transport accident with hazardous materials. In case of nuclear facilities radionuclides can be emitted to the atmosphere due to e.g. loss of coolant or criticality accidents. In all these cases emergency response forces as well as decision making authorities have to act properly. Whatever the reason for the contingency may be, a quick and most accurate estimate of the toxic air concentrations of the radiation exposures, in consequence of the accident must be made. This is a necessary prerequisite for decisions on protective measures and offsite emergency management.

K. Grefen and J. Löbel (eds.), Environmental Meteorology, 269–278.

In case of toxic chemical substances the decision as to which protection measures must be taken for the general public is generally left to the first on scene, the fire brigades. Once they have an idea of which material has been released, they require information on emission rates and meteorological data to estimate the consequences of the release to people in the vicinity of the accident site. To date the instrumentation and the tools to achieve this aim are very simple and unreliable. Identification of the material released is achieved by UN-number, or by questioning the manufacturer. Measuring is carried with absorption tubes, if it is carried out at all.

A more favourable situation is given for releases of radionuclides and their measurement in terms of absorbed doses. This is at least true for the Federal Republic of Germany where fully automated and computerized monitoring systems are installed around most of the commercial nuclear power plants. These systems have been designed to cover both routine operation and emergency situations. In the latter case it provides the licensing and surveillance authority, with certain in-plant key information about the actual operational status of the nuclear installation. It also provides more off-site related information, such as meteorological data, effluent doses from two stack monitors and dose rate measurements by GM-counters at about 30 locations in the vicinity of the plant. All monitors are connected to, and operated on-line by a centralized remote monitoring network. The data mentioned above, is generally, i.e. during routine operations, gathered in time periods of 1 hour. Depending on a certain dose threshold which may be exceeded in emergency situations, this collection cycle is automatically switched to time intervals of 10 minutes.

Despite of the multitude of available data during nuclear accidents, in both cases (releases of nuclear and chemical toxicants) the prediction of the propagation of the released material is effected with the aid of rather simple Gaussian plume models, usually in a precalculated form. If this is compared to the actual course of a released cloud one can see that the prediction made can lead to a severe misjudgement of the real conditions and thus to relatively great uncertainties for decision makers and emergency response forces.

To improve this situation, the German Ministry for the Environment, Nature Conservation and Nuclear Safety has sponsored the development of the system SMART (System for Measurement-based Assessments of Released Toxicants), which reduces the involved uncertainties substantially.

2. CONCEPT OF THE SYSTEM SMART

The system SMART permits optimal reduction of uncertainties connected with each of the two factors, dispersion calculation and measurement, by means of two key elements, cf. Figure 1. The first element is the continuous feed-back of current measurements of environmental impacts into diagnostic calculations in order to adapt the calculations to the measurements. This results in a more realistic set of model parameters, including the source term. Subsequently these parameters are used as input data for the second key element of SMART which is projection of the dispersion situation by means of a real-time calculation.

The dispersion calculations are based on an analytical quasi-instationary puff-trajectory model allowing for changing weather conditions [1].

The feed-back of current air concentration or dose rate measurements into diagnostic calculations is realized by the following adaption sequence:
After the release of air pollutants a first approximation of the dispersion situation is made as usual, without knowledge of measurements. In the second step the system SMART incorporates the current dose or concentration values measured at different locations and compares them to the modelled values. The fundamental calculation procedure, which is then initiated, can be described as quasi-continuous adjustment of the calculated to the measured impacts. This is repeated for each time interval, e.g. in 5 oder 10 minutes for which measurements are provided.

Depending on the specific question to be answered, the adjustment may be accomplished by evaluation of a number of model parameters such that the procedure results in a best fit within the scope of prediction accuracy of the dispersion model used. In mathematical terms this is realized with the aid of the following minimization procedure which is a modified least square fit

$$\left\{ \sum_{i=1}^{N} \left(\frac{C_{i,n}^{C} - C_{i,n}^{M}}{C_{i,n}^{C} + C_{i,n}^{M}} \right)^{2} \right\}_{k} \longrightarrow \text{MIN}$$

where C^C and C^M are the calculated and measured off-site impacts in terms of air concentrations or absorbed doses of each measurement lacation i at the end of the time interval n. N is the total number of the measurement stations and k is the number of trials to find the best fit. Investigations for the puff-trajectory model, used here, have shown that the following parameters are most sensitive:

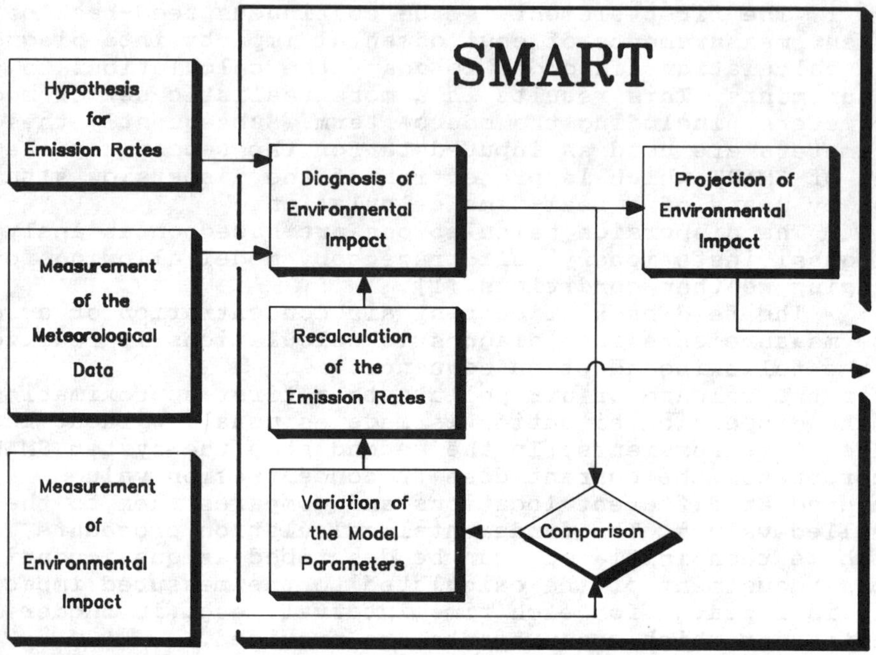

Fig. 1 Logical scheme of the corrective / predictive procedure of the System for Measurement - based Assessment of Released Toxicants **(SMART)**

BS

- wind direction
- wind velocity,
- standard deviation of horizontal and vertical wind direction fluctuations,
- the source term
- and a deposition parameter for radionuclide releases.

These parameters are evaluated according to a particular "trial and error" strategy, based on Monte Carlo techniques and guided by their sensitivity with respect to the dispersion calculations.

The adjustment provides a most actual and accurate set of effective model parameters to describe the current dispersion situation. This set is used to project pollutant concentrations or doses for subsequent time periods.

The system SMART is not restricted to a specific dispersion model such as the volume source model used here. Depending on the questions to be answered and the boundary conditions of the specific case, it is also possible to use numerical or Lagranian type models of various scales. The requirement, however, to keep the calculations within real-time on micro computers severely restricts the number of suitable models. This is especially the case with respect to the evaluation of external radiation exposure.

3. TEST APPLICATIONS

In order to illustrate the operational performance of SMART the system has been applied to both, a numerically simulated incident in a nuclear power plant near Brunsbüttel, and a tracer dispersion experiment with SF_6 in a large city.

In the case of the nuclear incident, dispersion took place during unstable atmospheric stratification over flat and relatively uncomplicated terrain with low roughness length in the range of 0.05 to 0.5 m. The following physico-chemical forms of radionuclides were released during 1.5 hours in an emission height of 100 m:

- noble gases
- elemental iodine
- aerosols

The resulting off-site doses were detected by a locally fixed monitoring network of 25 stations around the nuclear plant, cf. Fig. 2.

In this respect Fig. 2 shows absorbed isopleths due to external irradiation during 1.5 hours for two cases. The dashed curve shows a certain dose isopleth evaluated with the simple Gaussian plume model. The solid curve of the same isopleth illustrates the application of SMART. The difference between the two isopleths can be explained by two principal reasons: Firstly, SMART uses a quasi-insta-

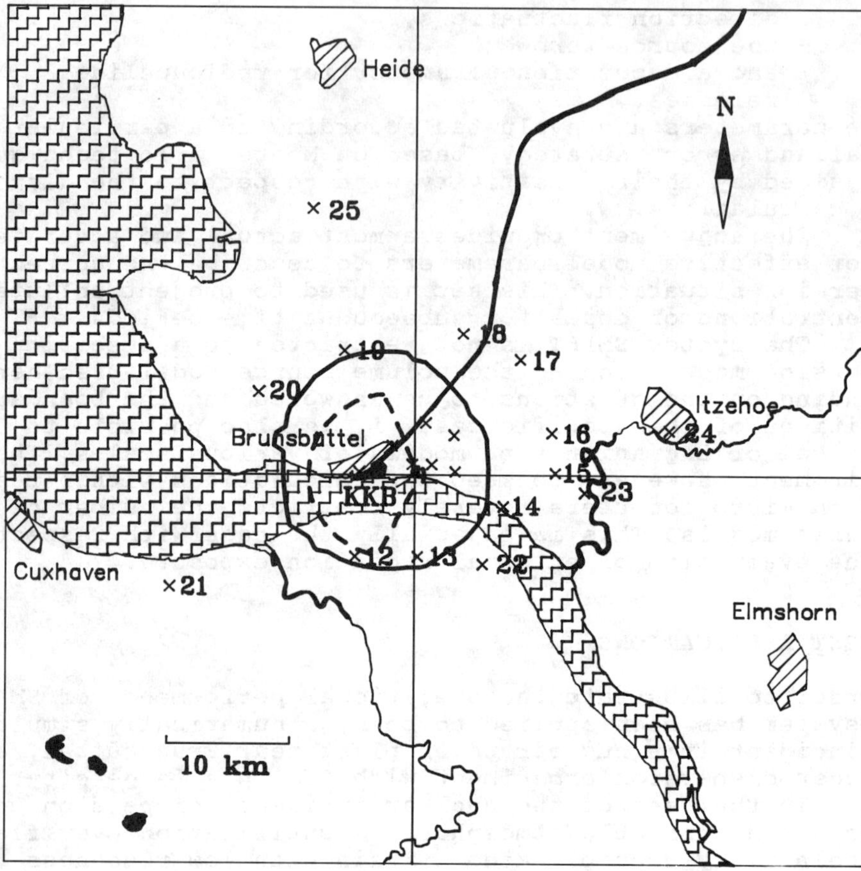

Fig. 2

Comparison of two certain dose isopleths calculated for an incidental radionuclide release of 1.5 hours. The dashed curve represents the procedure used to date. The solid curve shows the same isopleth after application of SMART.

tionary puff-trajectory model which allows for changing weather conditions. The second reason is the quasi-continuous feed-back of current measurement data from the monitoring network into the dispersion calculations. In this way the deviation between measured exposure and calculated exposure can be reduced to a minimum within the scope of the prediction accuracy of the model used.

For more detailed discussion of the problems involved, in the nuclear incident the reader is referred to publications [2], [3] and [4].

The tracer experiment with SF_6 was carried out in a metropolitan area under neutral atmospheric conditions and a rather high surface roughness in the range of 1.5 to 3 m. 33 kg of SF_6 were released from a point source in a height of 24 m. The resulting SF_6-concentrations in air were measured by 5 mobile units in a detection height of 1 and 18 m within a source distance of 0.5 to 1 km, cf. Fig. 3. The diagnostic adjustments of the calculated to the measured SF_6-concentrations at each measurement location were carried out in time steps of 10 minutes. For each time step the diagnostic calculations were started with a set of measured representative meteorological data (wind velocity, wind direction and stability class) and the related measurements of SF_6-concentration in the vicinity of the source. Knowledge about the amount of SF_6 released, was not used as input information for SMART. As already explained the system SMART then tries to find a new set of meaningful model parameters, including the source term, allowing the best fit of the concentration measurements. Fig. 3 shows the result of such an adaption procedure after a release time of 30 minutes (solid curves). Despite of the fact that the conditions for gas dispersion in metropolitan areas are extremely complex, the relative mean error of the calculated compared to the measured concentration data remained in a range between 20 % and 60 % for the sample used. The ratios of the adapted to the measured SF_6-releases for each time step varied between 0.5 and 2. For the time step depicted in Fig. 3 this ratio was smaller than 1.5.

In order to illustrate the conventional procedure for the assessment of toxic air concentrations, Fig. 3 also contains the results of a dispersion calculation based on a Gaussian plume model (dashed curves). These calculations were carried out with the same meteorological data and the same release time. Contrary to the SMART-calculations, however, the measured source term data have been used which usually are not available. Comparison of the calculated to the measured air concentrations, again, shows the serious deficiencies of this procedure: the dispersion model is inadequate and the absence of feed-back of off-site measurements into the dispersion calculations results in considerable and misleading uncertainties.

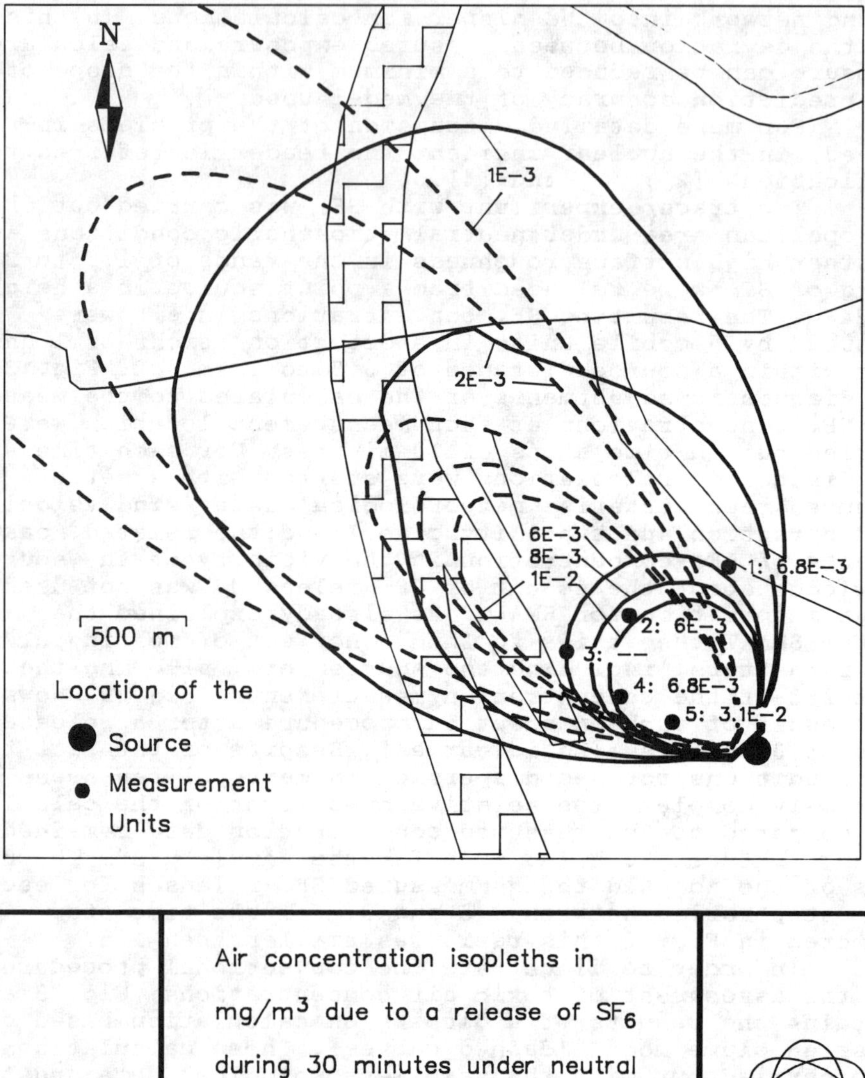

Fig. 3

Air concentration isopleths in mg/m^3 due to a release of SF_6 during 30 minutes under neutral atmospheric stratification calculated with (solid line) and without (dashed line) **SMART**

4. CONCLUDING REMARKS

Regarding the various deficiencies of present emergency response systems explained in chapter 1 and 3, it is evident that an improved emergency response system like SMART principally provides a higher diagnostic and projection accuracy than off-line calculations. This is mainly due to the fact that the projections of the system SMART are quasi-continuously corrected on the basis of measurements. This implies that hypothetical source terms, as being necessary for calculations without feed-back of off-site measurements, are not needed. It therefore forms a more reliable information base for decision makers and emergency response forces.

It would, however, be desirable to quantify the achieved improvement in prediction accuracy in a statistically more significant manner by application to suitable field experiments. But the number of rigorous validation studies with a relatively high resolution of temporal measurement data is rather limited. This is especially true for cases of radiation exposure after short-term releases of radionuclides.

The presently achievable projection accuracy may widely vary from case to case. This is caused by complex interaction for the following main influence factors:

- the accuracy of the dispersion and deposition model used,
- the accuracy and the number of measurements,
- the accuracy of the adjustment procedure,
- the meteorological boundary conditions,
- and the duration and the temporal course of the accidental releases.

But single test applications, two of which have been explained in chapter 3 as example, have revealed relatively good operational performance with sufficient adaption accuracy. Therefore, it can be stated that the application of SMART provides most actual and accurate information on dispersion situations if good and reliable impact measurements are incorporated into the system.

In the case of nuclear facilities this can be achieved, either by a monitoring network of GM-counters comparable to that explained here or a smaller network and additional mobile measurement units.

For accidents in chemical facilities or transport accidents, we found out, that all modern, currently available devices, such as Infrared-spectrometers, dosimeters for individual substances, portable GC/MS devices and so on, are not yet suitable for the use by fire brigades. The main reasons for this are the expertise required to operate and to maintain the systems, the lacking versatility of the devices and the costs. This disappointing result of our investigation suggests to remain with the widely used absorption tubes and incorporate them into the system, at least as far as fire brigades are concerned.

Additionally it may be noted that the results of our investigations reflect the current situation. If future developments show that absorption tubes can be substituted by other devices this can self-evidently be taken into account.

For chemical toxicants the system SMART is presently in its test phase in a joint effort with the Cologne Fire Department.

For radio toxicants it will be implemented next year into the monitoring network around the nuclear power plants in the federal state of Schleswig-Holstein.

REFERENCES:

[1] STRAKA, J., GEISS, H., VOGT, K.J.;
Diffusion of Waste Air Puffs and Plumes under Changing Weather Conditions
Contributions to Atmospheric Physics, Vol. 54, No. 2, May 1981

[2] BRENK, H.D., DE WITT, H., KRUSCHEL, K.-P., KNAUP, A.G., On-Line Accident Dose Projection Using Remote Monitoring Systems Around Nuclear Power Plants,
Paper presented at the Workshop on Real-time Computing of the Environmental Consequences of an Accidental Release to Atmosphere from a Nuclear Installation, Luxembourg, 17th-20th September, 1985

[3] DE WITT, H., BRENK, H.D., KRUSCHEL, K.-P., KNAUP, A.G., OLDES - An On-Line Dose Evaluation System for Measurement-based Assessments of Accidentally Released Radionuclides,
Paper presented at the Workshop on the Radiological Consequences of Chernobyl, Brussels, 3rd-5th February, 1987

[4] DE WITT, H., BRENK, H.D., KRUSCHEL, K.-P., KNAUP, A.G., OLDES - Ein KFÜ-integrierbares System zur Diagnose und Prognose der Strahlenexposition bei erhöhten Emissionen aus kerntechnischen Anlagen,
Schriftenreihe Reaktorsicherheit und Strahlenschutz BMU-1986-135, Hrsg.: Der Bundesminister für Umwelt, Naturschutz und Reaktorsicherheit, Bonn 1986

THE SANDOZ ACCIDENT - NUMERICAL SIMULATIONS OF THE TRANSPORT AND THE DISPERSION OF THE AIRBORNE RELEASES

B.Vogel, G.Adrian and F.Fiedler
Institut für Meteorologie und Klimaforschung,
Kernforschungszentrum Karlsruhe/Universität Karlsruhe
Postfach 3640, 7500 Karlsruhe 1, Germany

ABSTRACT. The wind conditions and the behaviour of the accidental releases of the Sandoz fire near Basel are described as they are available from observations. Since the airflow is strongly influenced by the orography it is obvious that a simple model like a Gaussian can not be used for the theoretical treatment of the problem.

Therefore the mesoscale model KAMM and the Lagrangian dispersion model TRAVELING are used to study the transport and dilution of the combustion products from the conflagration. Numerical simulations are carried out with different source heights. Brief descriptions of both models which take into account complex topography are given.

The results are compared with the observations. Qualitative good agreement is found. This shows that both models are helpfull tools to study the behaviour of accidental releases above inhomogeneous terrain. Possible applications of KAMM and TRAVELING are pointed out.

1. INTRODUCTION

In the early morning of November 1, 1986 a storehouse of the chemical plant Sandoz near Basel was destroyed by a conflagration. During the fire fighting a large part of the stored material was washed into the Rhine river. Since the compounds contained a lot of mercury, a large fish decease followed.

K. Grefen and J. Löbel (eds.), Environmental Meteorology, 279–290.

Afterwards it was estimated, that in addition about 1.5 tons of mercury were released into the atmosphere together with other combustion products during a span of three hours (AHRENS 1987).

The numerical mesoscale model KAMM and the dispersion model TRAVELING are used to calculate the transport and dilution of the airborne material released at Sandoz site.

The results of these simulations are compared with the available observations.

2. THE OBSERVATIONS

2.1. The observed flow field

During the fire west-southwesterly winds prevailed in the free atmosphere in the area of Basel as it can be seen from the synoptic charts.

Figure 1 shows the observed surface winds close to the source of the fire, which is marked by a triangle (AHRENS 1987). One can see, that there were easterly to south-easterly winds close to the ground. That means that the surface winds are blowing almost into the opposite direction of the flow in the free atmosphere and the cross isobar angle is nearly 180°. The local windfield is strongly influenced by the topography.

The change of the wind direction can also be noticed in the observed winds at the stations south of Basel (Figure 1), which recorded south-westerly winds. These stations have a higher elevation and therefore they are more influenced by the flow situation in the upper levels of the atmosphere.

The synoptic station at the Feldberg in the Black Forest registered also a surface wind around south-west.

So one can summarize, that a strong wind shift with height was observed in the Basel area.

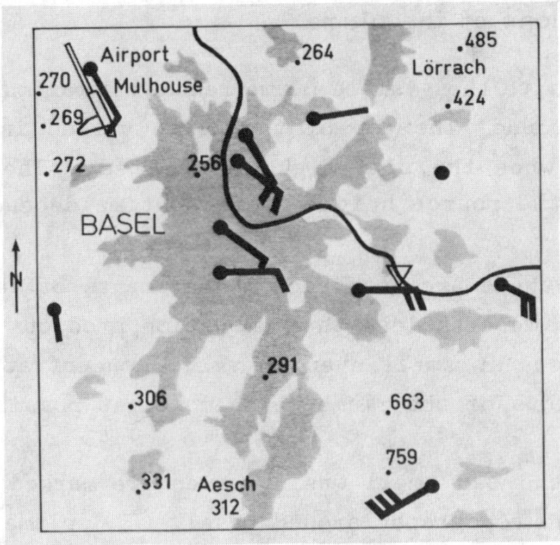

Figure 1: Observed surface winds in the area of Basel.
Numbers are indicating the height above sea level.
The Sandoz site is marked by a triangle.

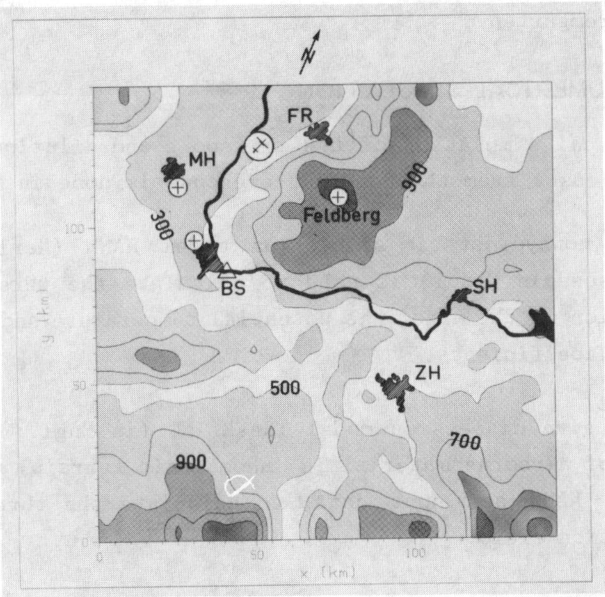

Figure 2: The model area. Places where the plume was observed are marked by a ⊕.
The Sandoz site is marked by a triangle.
⊕ = airport Bremgarten.

2.2. Observations of the plume

At the Sandoz site the visible plume reached a maximum height of about 400 m above ground. The top of the plume varied in time especially it decreased when the fire fighting started. Therefore it can be assumed that the source height was also time dependent.

Unfortunately there are no direct measurements of the concentrations in the plume. Nevertheless the combustion products were perceptible by their unpleasent smell even in small concentrations. So we have some observations of the plume which makes it possible to track it.

Places where the bad smell was observed are marked by ⊕ in Figure 2 which shows the topography around Basel. The plume was observed at Basel airport, in the city of Mulhouse and at the Feldberg in the Black Forest.

It should be mentioned here, that the plume wasn't observed at the airport Bremgarten.

3. THE NUMERICAL SIMULATION

The numerical simulation of the transport and dilution of the accidental releases from the Sandoz storehouse is done in two steps.

First the nonhydrostatic simulation model KAMM (Karlsruher Atmospheric Mesosacle Model) is used to calculate the three dimensional distribution of the wind, the potential temperature and the turbulent diffusion coefficients.

The Lagrangian dispersion model TRAVELING (in Engl.: Transport and dilution of airborne material in inhomogenic terrain) uses the model output of KAMM as input data, to calculate the three dimensional concentration field.

A short description of the models is given in the following.

3.1. The mesoscale model KAMM

The nonhydrostatic mesoscale-model KAMM was developed at the University of Karlsruhe (DORWARTH 1986).

KAMM is based on the equations of motion, the continuity equation and the first law of thermodynamics.

The turbulent fluxes are parameterized by a first order closure, using flux-gradient relationships and BLACKADARs (1962) mixing length hypothesis.

The equations are transformed into a coordinate system, which follows the topography.

The model KAMM is forced by a basic state, which is assumed to be hydrostatic and geostrophic. For the simulation, which is presented in the following, the basic state is determined from the operational numerical weather analysis of the German Weather Service with a method developed by ADRIAN (1987).

A radiation scheme is used for the outflow condition at the lateral boundarys. At the top of the model area a damping layer is introduced (for details see DORWARTH 1986).

3.2. The dispersion model TRAVELING

The Lagrangian dispersion model TRAVELING is used to calculate the three-dimensional distribution of the accidental releases. For this application the stationary version of the model is used, and 20000 particle trajectories are tracked.

The location of an individual particle at the time step t+Δt is given by:

$$\vec{x}_i(t+\Delta t) = \vec{x}_i(t)+\vec{v}_i(t)\Delta t \qquad (1)$$

where $\vec{x}_i$ is the location of the i-th particle, $\vec{v}_i$ is the particle velocity and Δt is the timestep.

The particle velocity is split up into the mean velocity $\overline{\vec{v_i}}$ and the velocity fluctuation $\vec{v_i'}$ and reads as:

$$\vec{v_i} = \overline{\vec{v_i}} + \vec{v_i'} \tag{2}$$

The mean velocity $\overline{\vec{v_i}}$ is calculated by linear interpolation from the flow field that is obtained by the nonhydrostatic mesoscale model KAMM.

The components u_j' of the turbulent velocity $\vec{v_i'}$ are calculated by a Markov-chain simulation for each timestep by:

$$u_j'(t+\Delta t)=a_j u_j'(t)+\sqrt{(1-a_j^2)}\ \sigma_{uj}\xi+T_{Lj}(1-a_j)\ \partial\sigma_{uj}^2/\ \partial x_j \quad , \ j=1,2,3 \tag{3}$$

a_j are the Lagrangian autocorrelation functions, T_{Lj} are the Lagrangian timescales and ξ is a random number taken from a Gaussian distribution with zero mean and unit variance.

The Lagrangian autocorrelation functions a_j are taken as:

$$a_j(\Delta t)=\exp(-\Delta t/T_{Lj}) \tag{4}$$

where T_{Lj} are the Lagrangian timesacles for the velocity components u_j.

According to VOGEL H. (1986) the velocity variances σ_{uj} are obtained from $\sigma_{uj}=\sqrt{2m_j E}$ where E is the turbulent kinetic energy. The factors m_j which depend on the thermal stratification were derived from measurements.

The local thermal stratification, the turbulent exchange coefficient of heat K_{hj} and the turbulent kinetic energy E are also available from KAMM.

The Lagrangian timescale T_{Lj} is given by :

$$T_{Lj} = K_{hj}/\sigma_{uj}^2 \tag{5}$$

Total reflection is used as the lower boundary condition.

3.3. The simulated flow field

Figure 3 shows the simulated windfield at an height of 3000 m above sea level. One can see a uniform airflow from WNW to WSW with an average windspeed of about 8 m/s. At this height no influence of the topography is to be recognized.

The simulated surface winds are plotted in Figure 4. Close to the ground a strong modification of the windfield caused by the topography can be seen. Easterly winds are simulated in the area around Basel which is in good agreement with the observations. There is also a good agreement between simulated and observed surface wind at the Feldberg where the winddirection is west-southwest.

Figure 5 shows two hodographs, where hodograph A is characteristic for the area of Basel and B for a mountaineous station where the flow conditions of the free atmosphere are dominating. The hodograph A shows the large wind shift with height, which is almost 180° in good agreement with the observations near Basel.

3.4. Simulated concentrations

Simulations with different source heights (50, 150, 250 and 450 m) were carried out, to take into account the timedependent source height during the fire. Since there is no information about the source strength, it is set equal 0.5 t/h.

Figure 6 shows the simulated surface concentration for a source height of 50 m. The dotted area shows concentrations larger than zero. Locations where the plume was observed are again marked by ⊕. The plume first reaches Basel, then it moves into the Upper-Rhine valley towards Mulhouse. Due to the confluence of the horizontal wind, the airborne material gets into heigher levels , where it follows the airflow in easterly direction. Then it moves across the Feldberg, where the plume was also observed. It should be mentioned here, that the simulated plume does not reach the airbase Bremgarten, where it was not observed.

A comparison shows a good agreement between the available observations and the simulation.

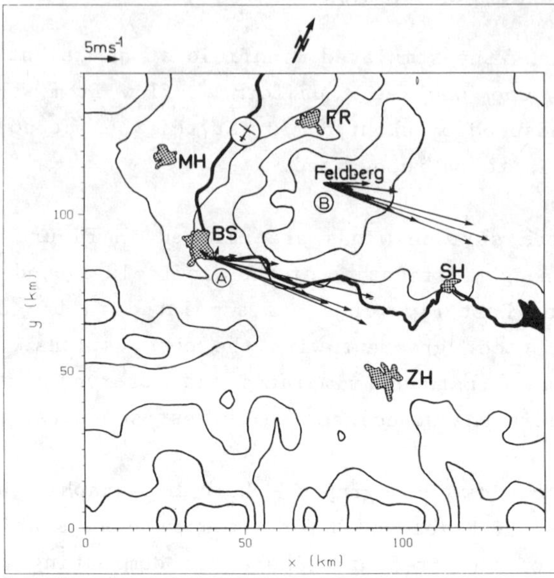

Figure 5: Simulated hodographs. (A at Sandoz site, B at the 'Feldberg')

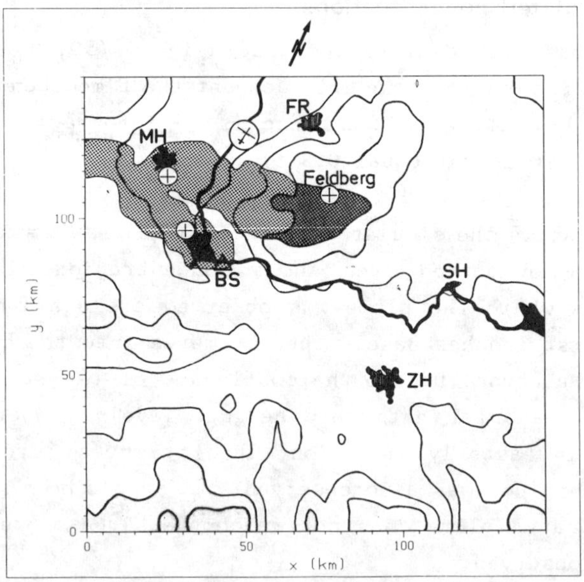

Figure 6: Simulated surface concentration. Source height: 50 m above ground. (Dottet area: concentration larger than 0.15 μg/m³)

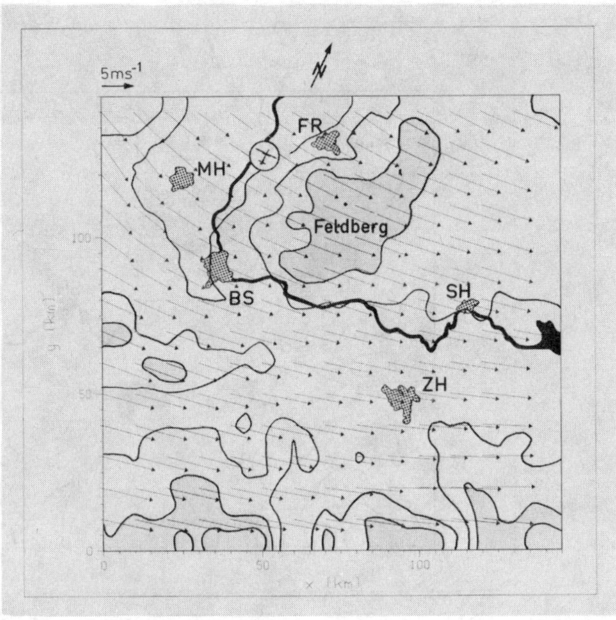

Figure 3: Simulated winds 3000 m above ground.

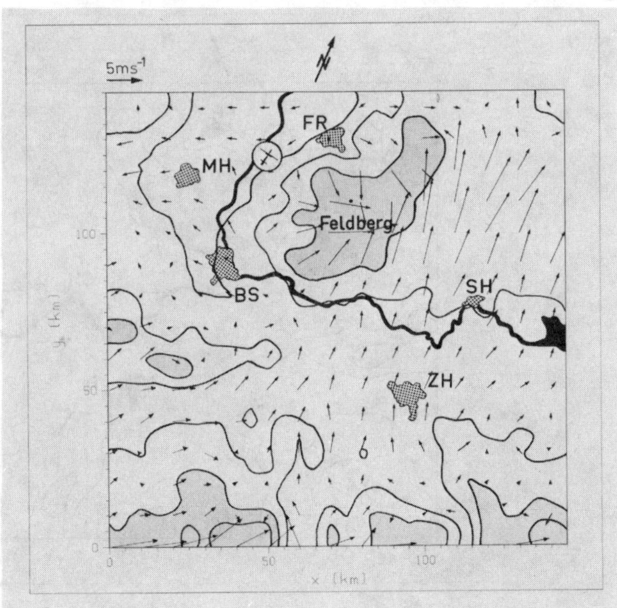

Figure 4: Simulated surface winds.

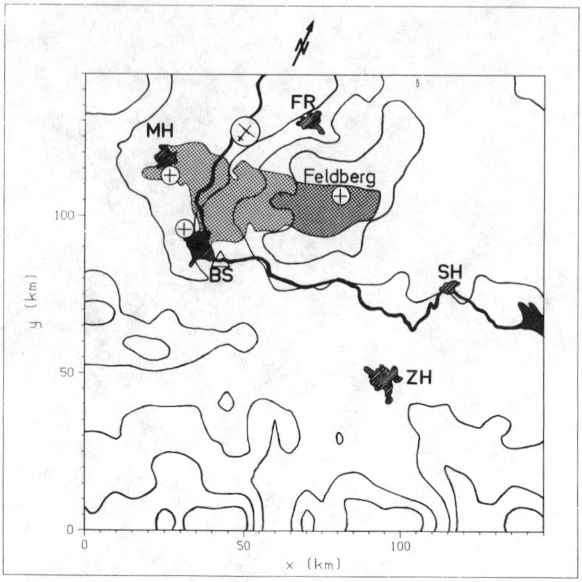

Figure 7: Simulated surface concentration. Source height: 450 m above ground.(Dottet area: concentration larger than 0.15 µg/m³)

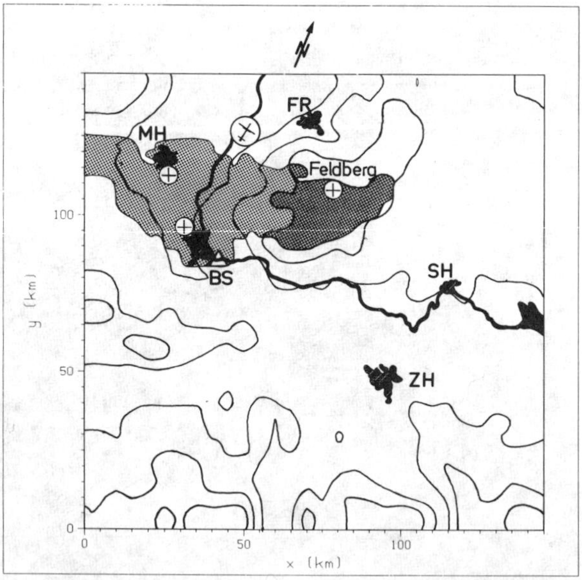

Figure 8: Simulated averaged surface concentration for the source heights of 50, 150, 250 and 450 m above ground. (Dottet area: concentration larger than 0.15µg/m³)

The surface concentration fields for the source heights of 150 and 250 m above ground look quite similar to the one shown in Figure 6 with decreasing values with increasing source heights.

The plume produced by a source height of 450 m above ground has a quite different behaviour. In this height, the wind turns into southwesterly directions (see Figure 5). Only the material that is mixed into the lower layers follows the airflow towards north or northwest. Figure 7 shows the surface concentration for that source height. Again a large part of the material moves across the Feldberg.

Finaly the concentration fields which were calculated with different source heights (50, 150, 250 and 450m) were averaged in order to determine the whole area, which is covered by the plume. Figure 8 shows again the obtained surface concentration. One can see, that a large area is touched by the plume, which is caused by the strong wind shift. There is a good agreement of the calculated plume with the observation.

4. CONCLUSION

A comparsion of the simulated and observed flow field during the accident shows a good agreement. Especially the complexity of the airflow in the area of Basel was reproduced by the numerical simulation.

Also the simulated plume is in good agreement with the available observations. From the calculated concentration distributions it can be seen, that it is impossible to get the right answers by using a simple model above inhomogeneous terrain. Especially if the simple model uses a uniform flow direction.

The data which are defining the basic state of KAMM were taken from the operational analysis of the prognostic synoptic model of the German Weather Service. So we have also the possibility to forecast the three-dimensional distribution of the wind, the temperature, the turbulence and the concentration above complex topography. This has to be shown in future works.

Beside this there is another helpfull application of KAMM and TRAVELING. After an accidental release measurements are to be carried out in order to determine, which areas have to be excluded from agricultural utilization. These measurements are in some cases rather expensive. So the results of the numerical models can be used to decide, at which places such measurements should be carried out.

REFERENCES

Adrian,G. (1987) 'Determination of the Basic State of a Numerical Mesoscale Model from Operational Numerical Weather Forecasts' Beitr.Phys.Atmosph. 60, 361-370

Ahrens,D. (1987) private comunication

Blackadar,A.K (1962) 'The Vertical Distribution of Wind and Turbulent Exchange in a Neutral Atmosphere' J.Geophys.Res. 67, 3095-3102

Dorwarth,G. (1986) 'Numerische Berechnung des Druckwiderstandes typischer Geländeformen' Wissensch. Ber. Inst. f. Meteorol. und Klimaforsch. der Univ. Karlsruhe, Nr. 6

Janicke, L. et al. (1984) 'Entwicklung eines Modells zur Berechnung des Staubniederschlages' Umweltbundesamt, Forschungsbericht 104 02 562

Vogel, H. (1986) 'Berechnung von Konzentrationsverteilungen mit einem Lagrange-Modell für mesoskalige Strömungsfelder' Diploma Thesis, Dept. of Meteorol., Techn. Univ. Darmstadt, 79 pp

WIND TUNNEL SIMULATION OF ACCIDENTAL RELEASES IN CHEMICAL PLANTS

W. Bächlin and E.J. Plate
Wind Engineering Division,
Institute for Hydrology and Water Resources
University of Karlsruhe
Kaiserstr. 12, D-7500 Karlsruhe 1, FRG

ABSTRACT. A concept is presented in which wind tunnel tests are used for determining the transmission function for the transmission of accidentally released gases from chemical plants. The flow field downstream from the release is divided into three regions. The first is the region of release, in which the source characteristics in conjunction with the characteristics of the flow field surrounding the source determine the dispersion of the spill, the second is the "near field" region, in which the local characteristics of the plant and its surroundings, but not the mode of release determine the spreading, and the third is the "far field" region in which the ground is characterized by an uniform roughness and the release by a simple (for example Gaussian) diffusion model. Some new experimental data are presented to illustrate the effect of the "near field" on the dispersion process.

1. INTRODUCTION

Some of the large scale accidents which occurred in recent times in chemical plants have sharpened the awareness for the dangers which can occur if a toxic or harmful gas is accidentally released from a chemical plant. In response to these accidents in many countries chemical plants which handle dangerous substances are required to provide warning systems which consist of fixed installed gas monitors in sensitive areas inside and outside of the plant, and of plans for meeting emergencies if an accident should happen. For both of these tasks, the wind tunnel can be a very useful design tool. Once the spill has occurred (hypothetically, as is assumed for the setting up of monitoring systems, or actually, as in the case of an accident), its further spreading is determined by the interaction of the wind field with the buildings in the vicinity of the release area.

For accidental releases, the quantity which has to be known for safety measures is the immission, which in the usual cases can be either the concentration of the gas at the ground level, or the dosage, i.e. the integrated amount of gas which passes by a certain point at

K. Grefen and J. Löbel (eds.), Environmental Meteorology, 291–303.

the ground level over a specified time. These quantities depend on two factors: first on emission, which is the released quantity in terms of a released mass or a release concentration in a release volume. It depends on the release mechanism which sets the initial conditions. The second factor is the transmission, which is the combined effect of the wind field which exists at the moment of release and thereafter, and of the transmission properties of the wind field, as is determined by the interaction of the atmospheric wind profile with the buildings and obstacles which form the roughness of the ground. In this chain of processes, the transmission is an important link which can be studied and quantified by means of wind tunnel tests. It is the purpose of this paper to report on a procedure by means of which a limited number of such wind tunnel tests can be used to obtain generalized transmission functions for the use in safety plans, or for the use in simplified mathematical models.

Regions of transmission and their use in safety models

We consider the case of a continuous release from a momentumless point source inside the building complex of a chemical plant, such as could occur for example from a leak in a pipeline, or from the defective safety valve of a containment vessel (This is the case which can be studied most readily in a wind tunnel, although extensions to other source configurations are, in principle, quite straightforward). If the release would occur over homogeneous terrain, the ground level concentration distributions downwind from the source are approximately Gaussian, as has been described in many publications (see for example Hanna (1982)), and as is assumed in many models, in which the spread of chimney plumes is calculated. In actuality, however, a very contorted concentration pattern might be expected near the source where the concentrations within the streets follow a different spreading law than the concentrations above street level. This is the "near field" of the diffusion process, shown schematically in Fig. 1, which at this time cannot be described with reasonable accuracy by any analytical model. There exists, for any wind direction β, a distance $R(\beta)$, which we may call the "radius of homogenisation", in which the assumption of Gaussian plumes leads to completely erroneous results, and for which no simple rules for the spread rates or the profile shapes can be given. It is our opinion that within the radius of homogenisation the only feasible way to obtain information on the diffusion of the gas is by means of wind tunnel tests - tests which are to be done with the actual arrangement of buildings, and at orientations corresponding to the actual wind directions. These tests, although elaborate, are simplified by the fact that in the near field for even the slightest winds the flow field is so strongly affected by locally generated turbulence caused by the separation of the air flow on the buildings, that thermal effects become negligible. Therefore, the tests can be done with ordinary boundary layer wind tunnels and according to the scaling laws which are valid for neutrally stratified flow (see Plate, 1982,1, 1982,2). On the other hand thermal effects can affect the diffusion in the so called "far field". Investigations for describing the influence are proposed in a wind tunnel like shown in Figure 3.

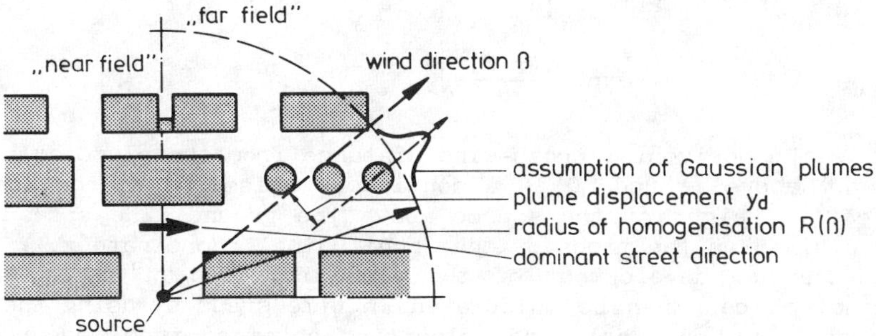

Figure 1. Definition sketch for the regions of diffusion from accidental releases.

It is evident from this description that one cannot expect that any two chemical plants will yield identical results for the near field. There exists however the possibility that small differences in building arrangements will have only small consequences in the resulting values of radius of homogenisation and spread rates. And since there exist many chemical plants which have similar functions it is possible that typical building arrangements for typical plants can be defined.

Working from this assumption we have built a model of one of the largest chemical plants in the world, the Badische Anilin und Soda Fabrik (BASF) in Ludwigshafen, FRG. On this model, wind tunnel tests are being performed for different wind directions and at different release points. It is our intention to perform wind tunnel tests on this model and to determine ground level concentration fields for typical plant situations and building arrangements. Also, for each configuration and wind direction we want to find the radius R(β), and the virtual origines of the spread rates in all directions. We hope that the different groups of buildings which exist within the BASF are typical enough to represent a large variety of other chemical plants, and we shall develop a catalog of ground level concentration fields for typical plant situations. Since usually the release rate of harmful gas is known only approximately, the uncertainty derived from the application of a similar, but not identical model may usually be quite acceptable for the purposes of a warning system.

At distances beyond the radius of homogenisation the cloud has spread out over such a large vertical area above the roughness elements formed by the buildings and the topographic micro structure that the concentration is practically constant over the height of the roughness elements, so that the main change in shape of the concentration cloud takes place for above the roughness. In this region, which is called the "far field" the plume behaves like a cloud released from a point source at ground level. This point source, however, is not located at the actual release point, but it has a virtual origine which is likely to be found far behind the actual source, and it has initial spread parameters which may appear to have different origines for the vertical and lateral spread rates, i.e. the plume obeys (approximately) a law:

$$c(x,y,z) = \frac{Q}{2\pi \cdot u \cdot \sigma_y \cdot \sigma_z} \exp[-\frac{y^2}{2\sigma_y^2}] \cdot [\exp[-\frac{(z-h)^2}{2\sigma_z^2}] + \exp[-\frac{(z+h)^2}{2\sigma_z^2}]] \quad (1)$$

where y is the horizontal cross-wind distance from the plume axis, z is the height above ground, Q is a continuous emissions source and h is the effective height of the source above the ground. The spread rates σ_y, σ_z are given as functions of the longitudinal coordinate x.

The further development of the plume downwind of the radius of homogenisation can then be calculated in time steps by using the local wind directions and local wind velocities obtained from measurements, or from a suitable model.

The transmission functions are to be used in a siting model for a warning system in conjunction with wind statistics, and perhaps in conjunction with wind tunnel studies in which the modification of the wind field by topography or roughness changes is evaluated. They are particularly useful for real time warning systems after the occurrence of an accidental release. In identifying beforehand typical transmission regions within a plant for given wind directions, the safety manager has a set of transfer functions from the catalog which we plan to prepare. One can also assume that he has available instantaneous values for the important components of the wind field, and a suitably accurate estimate of the emission. Warning systems in which the development of the cloud of hazardous gas as function of space and time is traced in real time on a computer map exist, and such systems can be improved in the accuracy of their prediction capability if the catalog of concentration profile parameters described above is available.

In the remainder of the paper we give some experimental results from the first wind tunnel tests which were performed on the model of the BASF.

2. EXPERIMENTAL CONFIGURATION

The experiments were performed in a boundary layer wind tunnel. Its test section is about 12.0 m long, 2.0 m wide and 1.0 m high. A thick boundary layer was generated along the floor of the tunnel by vortex-generators at the entrance of the test section and sustained by a roughness consisting of a random arrangement of "LEGO" blocks on the floor of the wind tunnel. The meteorological parameters wind velocity profile, turbulence intensity and turbulence spectra were chosen to be similar in full scale and in wind tunnel. Figure 2 shows a wind tunnel for neutral stratified flow as is required for the near field, and Figure 3 presents one for modelling stratified boundary layers. Both types of wind tunnel are available at the Institute for Hydrology and Water Resources at the University of Karlsruhe.

A geometrically similar model of the industrial plant was constructed for this study. The model scale was 1:500.

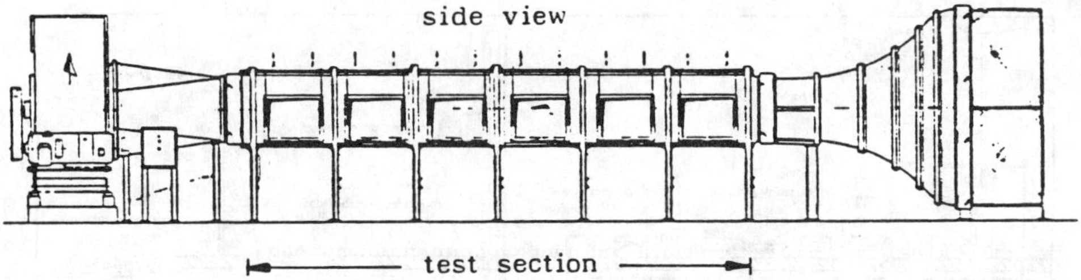

Figure 2. Boundary layer wind tunnel for neutral stratified flow.

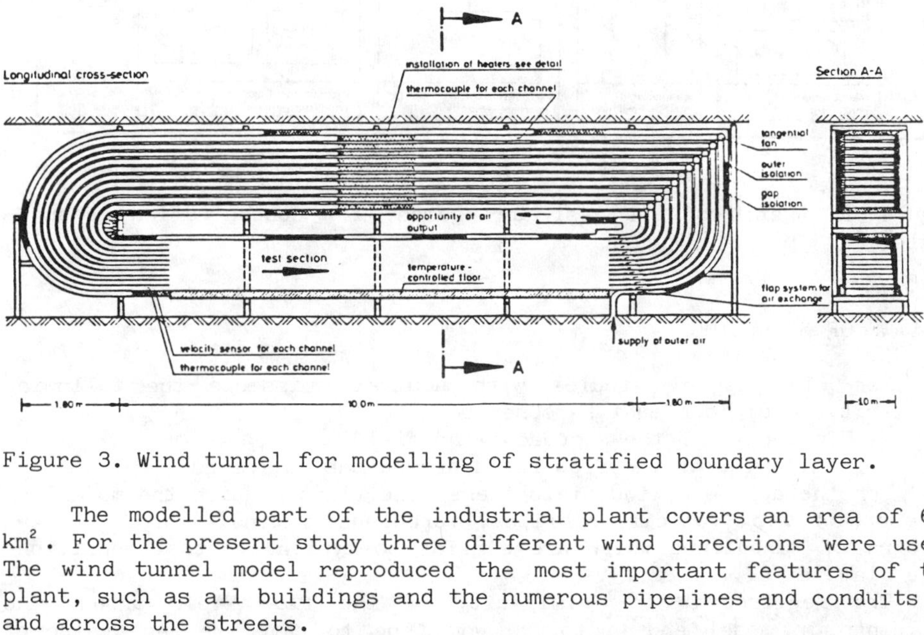

Figure 3. Wind tunnel for modelling of stratified boundary layer.

The modelled part of the industrial plant covers an area of 6.6 km^2. For the present study three different wind directions were used. The wind tunnel model reproduced the most important features of the plant, such as all buildings and the numerous pipelines and conduits in and across the streets.

The releases were located at typical positions identified in the actual plant. The continuous source consisted of a horizontal tube. The velocity of emission of gas was set equal to the wind velocity at the location of the tube. The gas neutral, i.e., ρ_{Gas} is equal ρ_{Air}. A tracer gas (SF_6) was admixed to the emission. At different distances from the source the gas transverse concentration profiles were measured by a leak detector. The experimental arrangement is shown in Figure 4.

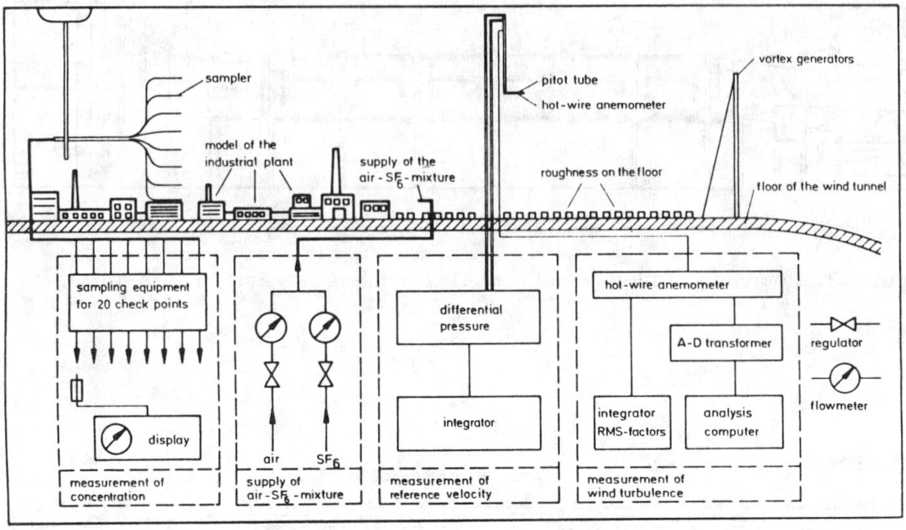

Figure 4. Sketch of the test arrangement for concentration measurements.

3. LAWS OF MODELLING

For wind tunnel studies with neutral emissions the following similarity conditions must be observed:

Similarity of the approach wind field

Similarity of the flow field around the buildings .

For the approach wind field these conditions require the modelling of approach flow velocity profiles, turbulence intensity profiles, and spectra of turbulence which are similar to typical neutral conditions (see Plate et al., 1982,1).

The requirement for similarity of the flow field around the buildings in model and prototype would be to have the same value of Reynoldsnumber in both cases. It is not possible to meet this requirement. But with sharp edged bluff bodies the pressure distributions around the buildings are similar if the Reynolds number $Re=u \cdot L/\nu$, with ν = kinematic viscosity of air, and L = a typical length, is above a critical value of abount $5 \cdot 10^3$ zu 10^4.

With these assumptions, results of concentrations obtained from wind tunnel experiments can be converted to relative concentrations in the field by means of the conversion

$$\frac{c(z/H,\ y/H)}{c_o} \cdot \left(\frac{u_{ref} \cdot H_{ref}^2}{u_o \cdot A_o}\right)\Big|_{model} =$$

$$\frac{c(z/H,y/H)}{c_o} \cdot \left(\frac{u_{ref} \cdot H_{ref}^2}{u_o \cdot A_o}\right)\Big|_{nature} \tag{2}$$

where H is a typical building dimension,
H_{ref} is the reference height
u_{ref} is the velocity in the height H_{ref}
z,y are the coordinates
A_o is the cross section at the outlet of the leakage
c is the gas concentration at the control point
c_o is the gas concentration at the release point.

4. RESULTS

The present study was divided into two parts. In the first part wind velocities were measured in a given street of the model. In the second part concentration profiles were measured at ground level (z = 0 m) and at a height of z = 30 m. Since the purpose of this study is to show only relative effects of built up areas vs. uniformly rough areas, the concentration data are plotted as relative concentrations c/c_o.

Measurements of gas concentration were conducted for both uniformly rough area and built up area. On the basis of the ratios c/c_o for identical release conditions the relative effect of different ground surface configurations can be demonstrated.

Essentially the streets in the investigated plant form a rectangular grid, whose axes dominate the wind field channeling near the ground. For the present study two interesting wind directions were selected according to the orientation with respect to this grid. In the first case the wind direction is the same as the dominant street direction (β = 161°), in the second case wind direction and dominant street direction are intersecting at an angle of 45 degrees (β = 26°).

4.1 Wind velocity measurements in one of the street canyons

Although there exist some studies about wind velocity measurements in idealized two dimensional street canyons there are only few investigations about measurements in really existing built up areas.In the present study such investigations were conducted for a given street canyon which is shown in Figure 5. There, the very complex flow conditions in this street region are recognizable.

4.2 Case studies concerning to dispersion of pollutants in built up areas

a) Gas concentration measurements in uniformly rough area
For reference purpose the diffusion conditions for uniformly rough area were considered. In this case the floor of the wind tunnel was

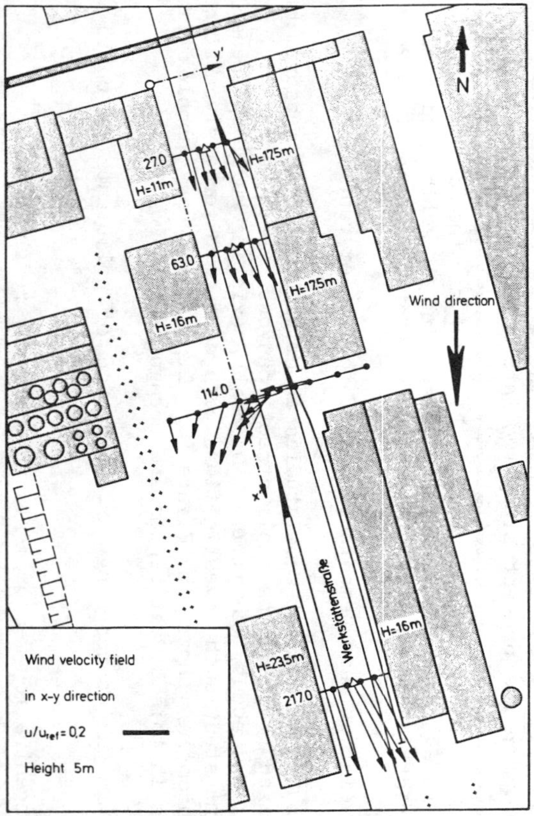

Wind velocity conditions in x-y-direction

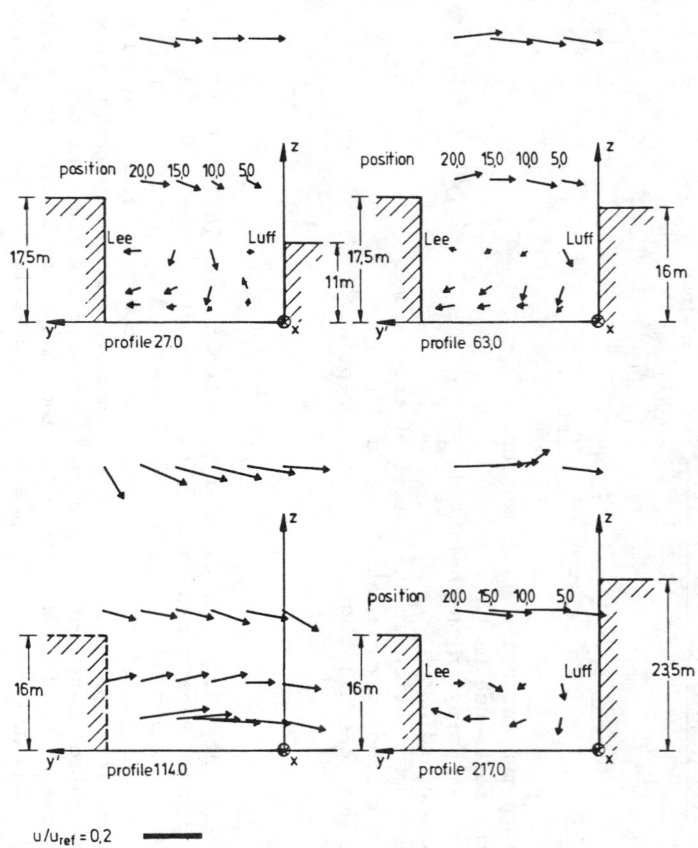

Wind velocity conditions in y-z-direction

Figure 5. The complex wind velocity conditions in a given street canyon.

covered with a random array of LEGO-roughness. Horizontal gas concentration profiles across the flow direction were measured at different distances from the source. Typical results of gas concentration measurements are shown in Figure 6. The peaks of the profiles and the diffusion parameters σ_y were determined.

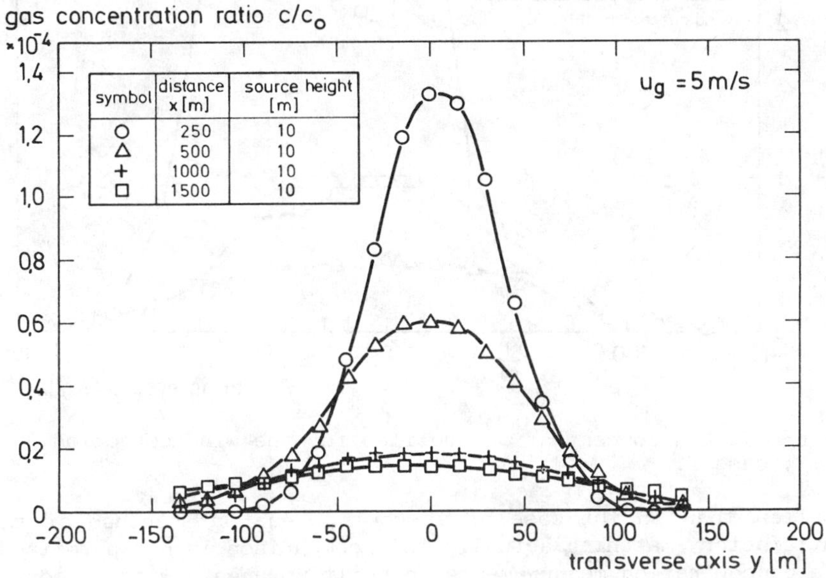

Figure 6. Gas concentration profiles across to the flow direction at different distances x from the source (uniformly rough area).

b) Gas concentration measurements in built up area – wind direction β = 161°

In this case the wind direction is equal to the dominant street direction. The source was located within a street canyon. Horizontal gas concentration profiles across the flow direction were measured in three cross roads downstream from the source position. Typical results for a source height of H_Q = 10 m are shown in Figure 7. The positions of the peaks of gas concentration in the vicinity of the leak are to be found within streets parallel to the wind. At large distances from the release point the profiles are more homogeneous. The radius of homogenisation is also dependent on the height of the source. The data for different cases are shown by Bächlin and Plate (1986,1). The comparison between the peak gas concentration values for built up area and for smooth area shows large increases (e.g. for the distance 300 m, source height 10 m the ratio is greater than 2), because the diffusion is obstructed by the buildings parallel to the street. In the investigated area the diffusion parameters σ_y are

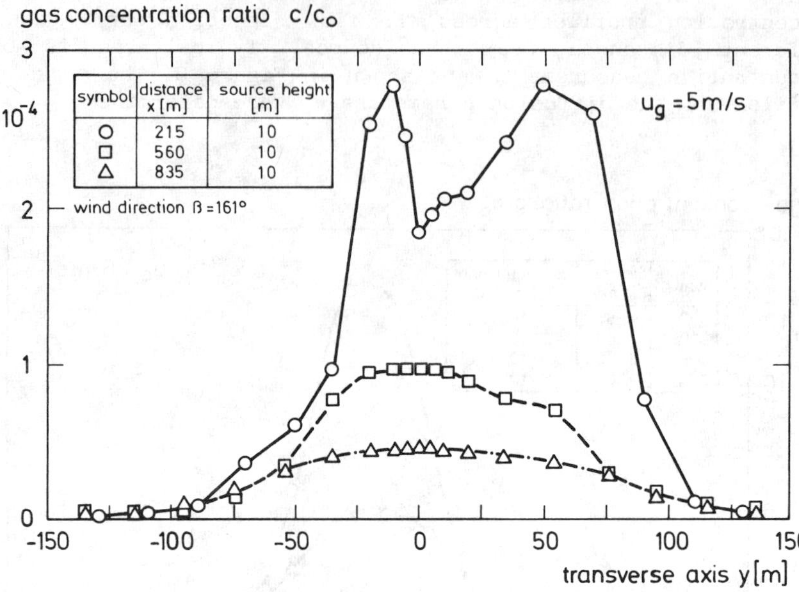

Figure 7. Gas concentration profiles for the wind direction β = 161°, case 1.

smaller than in the case of a smooth area. But because of the unsatisfactory matching actually this comparison is not permitted.

c) Gas concentration measurements in built up area - wind direction β = 26°

In the second case the wind direction and dominant street direction are intersecting at an angle of 45 degrees. Again horizontal gas concentration profiles were measured at different distances from the source. Whereas in the first case (β = 161°) the location of the peaks of gas concentration correspond to the wind direction, in this case the plumes are displaced. This displacement described by the shift parameter y_d as function of the distance from the source is shown for different source heights in Figure 8. It is worthy of note that e.g. for the source height 10 m the magnitude of displacement is increasing over the whole distance between 0 m and 1500 m from the leak whereas for the height 20 m the displacement is constant for distances greater than ~ 800 m. The maximum shift parameter measured in the experiments is approximately 230 m over the distance of 1500 m.

The flow which approaches the dominant street direction under an angle causes an increase of turbulence near the ground. Accordingly at greater distances from the leak the peaks of gas concentration profiles are decreasing more rapidly than for the uniformly rough case while the diffusion parameters σ_y are increasing. In some cases the parameter σ_y for built up area is twice the amount for uniformly rough area.

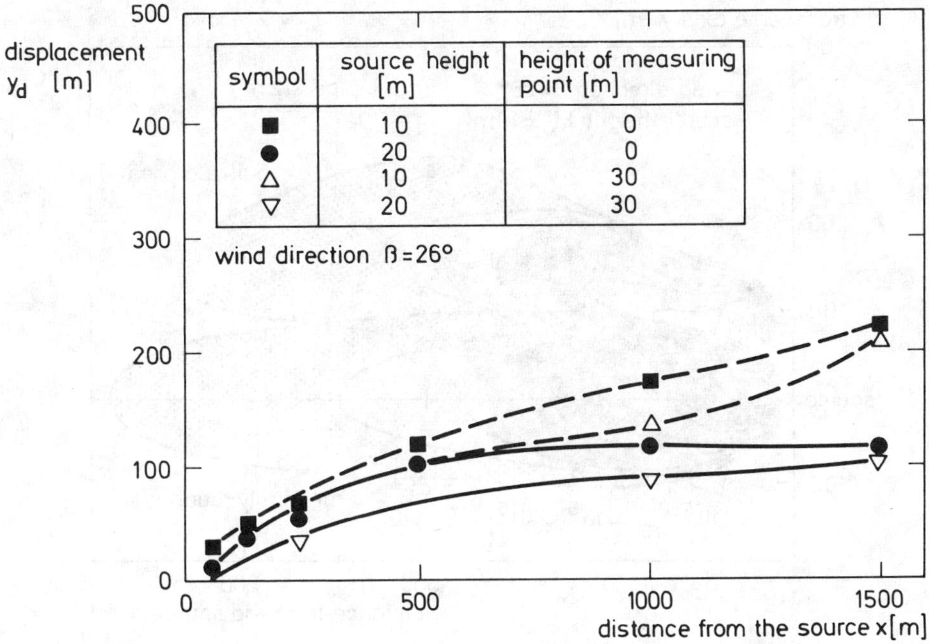

Figure 8. Plume displacement (shift parameter y_d) as function of the distance from the source.

CONCLUSIONS

In case of an accidental release it is important to know the exact area of danger zone created by the released gas. In Figure 9 are shown smoothed isolines of gas concentration for both uniformly rough area and built up area. The considerable displacement of the plume and the high gas concentrations in the vicinity of the leak for the case of built up area are recognizable. For the determination of the dangerous zones for different configurations of buildings and for different wind directions such isolines of gas concentration should be known. The ultimate aim of the present research project is to provide a catalog of typical cloud configurations as functions of typical building configurations to be used in safety assessment and planning of operations during an acute hazard occurrence.

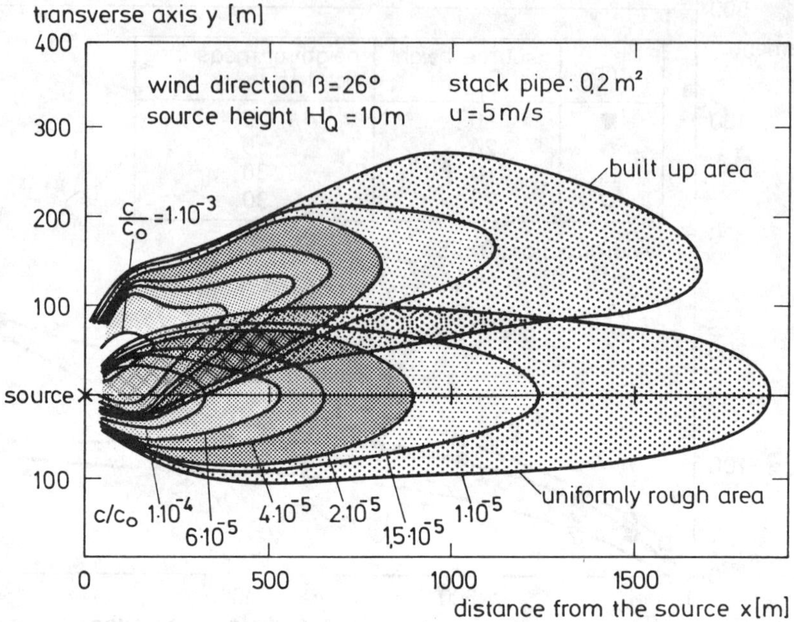

Figure 9. Isolinies of gas concentration for the case of uniformly rough area and the case of built up area.

REFERENCES

Bächlin, W., Plate, E.J., 1986,1; "Ausbreitung von Störfallemissionen in bebautem Gelände". Forschungsbericht Bundesministerium für Forschung und Technologie.

Bächlin, W., Plate, E.J., 1986,2: "The dispersion of accidentally released gases in a built up area". 5th International Symposium – Loss Prevention and Safety Promotion in the Process Industries –, Cannes, p. 27.1 - 27.15.

Cermak, J.E., Lombardi, D.J. and Thompson, R.S., 1974: "Applications of physical modelling to the investigations of air pollution problems in urban areas". Proc. of the 67th Annual Meeting of the APCA, Paper No. 74-160, Denver, Colorado.

Hanna, S.R., 1982: "Engineering Meteorology", Studies in wind engineering and industrial aerodynamics, Vol. 1, Chapter 10. Elsevier Scientific Publishing Company, Amsterdam - Oxford - New York.

Hoydysh, W.G., Griffiths, R.A. and Ogawa, Y., 1974: "A scale model study of the dispersion of pollution in street canyons". Proc. of the 67th Annual Meeting of the APCA, Paper No. 74-157, Denver, Colorado.

Plate, E.J., 1971: "Aerodynamic characteristics of atmospheric boundary layers". AEC Critical Review Series, Wash. DC.

Plate, E.J., 1982,1: "Engineering Meteorology". Studies in wind engineering and industrial aerodynamics, Vol. 1, Chapter 13. Elsevier Scientific Publishing Company, Amsterdam - Oxford - New York.

Plate, E.J., 1982,2: "Windkanalmodellierung von Ausbreitungsvorgängen in Stadtgebieten". Kolloquiumsbericht. Verlag TÜV Rheinland GmbH.

Wedding, J.B., Lombardi, D.J. and Cermak, J.E., 1977: "A wind tunnel study of gaseous pollutants in city street canyons". Journal Air Pollution Control Assoc., Bd. 27, p. 557- 566.

SPATIAL AND TEMPORAL STRUCTURES OF THE URBAN CLIMATE - A SURVEY

W. Kuttler
Universität - GHS - Essen
Fachbereich 9, Landschaftsökologie
Postfach 103 764
FRG - 4300 Essen 1

ABSTRACT. In this paper the elements of the urban climate are described and discussed in an overview. After having dealt with the bioclimatical consequences of the air-hygienic problems the "heat-island" effects are investigated in view of the formation of low-lewel breezes. Furthermore some examples show possibilities to convert climatological data into planning processes.

1. INTRODUCTION

The term "town" means a highly populated area which with its rows of adjoining buildings constitutes a built-up area. A division of labour has taken place among its citizens and the town itself exercises an attraction on its surrounding countryside.
Despite the apparent socio-economic advantages rarely being realized because of the what is often a drastic deterioration in the environment, even in the last decade the development of the urbanization process has njoyed unchecked growth from the pedestrian period (up to about 1880) to the telecommunication era (since 1975) (BURDACK 1985). This applies to both the continuously increasing use of land (from about 20 m^2 per inhabitant in the 50's to 40 m^2 in the 80's, BOESLER 1987) and to the increasing urban populations worldwide.
Whereas only about 20 % of the entire world population lived in cities or towns at the beginning of the 20th century we must expect a figure of 40 % at the beginning of the 21st century (OKE 1986; SCHÖLLER 1983).

In contrast to the developing countries, however, the industrialized nations traditionally have a higher percentage of urban population. This share is expected to reach 80 % in the year 2000 (fig.1).

Conflicting aims result from the intense spatial entanglement of a dense urban population and the immissions into the ground air layers of agglomerations. The only solution is the ensurance of a permanent supply of fresh air and favourable human-bioclimatic conditions for the population.
It is the special function of applied urban-climatology in the fields of

K. Grefen and J. Löbel (eds.), Environmental Meteorology, 305–333.

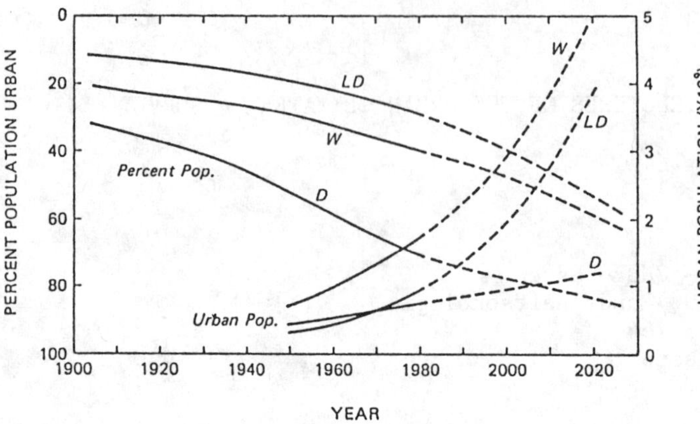

Fig.1:

LD = Africa,Asia, Latin America and Oceania

D = Europe, North America Australia, Japan and USSR

Data and projections using the United Nations sources according to SASSIN (1980) & PETERSON (1984)

Fig.1: Percentage of the population in urban areas in the Less Developed (**LD**), Developed (**D**) and total World (**W**) communities, and the number of urban inhabitants according to the same groupings.

inventory and interpretation of urban-ecological parameters, both to analyse the micro- and mesoclimatic conditions and to find answers to the questions still open in the area of town planning.

The factors which lead to the mesoclimate of our towns and cities are mainly, an increase in surface-roughness, a high amount of highly built-up areas and too few green areas. The result: In contrast to the surrounding rural areas, the ground levels of urban atmosphere show a different structure as can be shown schematically in fig.2.

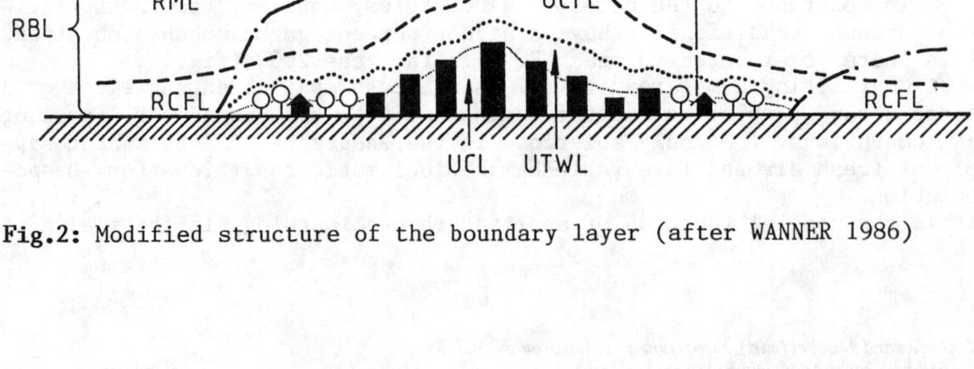

Fig.2: Modified structure of the boundary layer (after WANNER 1986)

2. THE STRUCTURE OF THE URBAN ATMOSPHERE

The Rural Boundary layer (RBL) above the flat surrounding area can be subdivided into the PRANDTL-layer, or Rural Constant Flux Layer (RCFL), and the EKMAN-layer or Rural Mixed Layer (RML).

The RCFL lies like a "skin" on the ground and reaches a maximum height of 100 metres. The RML lying above this, however, reaches a height of some several hundred metres, whereby the change in wind directions increases with height as the influence of shearing-stress decreases allowing gradient power and Coriolis-acceleration to dominate. When approaching the atmospheric structure of a town one can particularly observe an increased height in the CFL. While the RML becomes an UML, the UCFL "fans out" into the UCL above the ground and into the TWL. The UCL is the "skin" that covers the built-up area, whereby the dynamic of the air-skin is uncoupled from the remaining boundary layer up to about 2/3 of the height of ground objects (e.g. buildings, trees) (WANNER 1986, p.71) i. e. the exchange processes occur in micro-scale.

As mentioned, above UCL there is a turbulent transitional layer incorporated in UCFL which in general reaches twice or three times the width of buildings (OKE 1984).

The UBL - which is situated above UCL as a meso-scalic boundary layer - is larger in the daytime (0.6 to 1.5 km) than at night (0.1 to 1.3), as radiation causes convection and thus increases the thermic turbulence (OKE 1983). However these conditions occur only during weather conditions with neutral stratification of the atmosphere. During windless weather conditions, especially on days with high amounts of radiation, different behaviour in the layers can be observed.

The vertical structure and heights of the layers shown in fig.2 are influenced by widely differing factors determined by the balance of radiation and energy.

3. PARAMETERS INFLUENCING URBAN CLIMATE

Thus, the changed characteristics of urban surface (tab.1) lead to e.g. a decrease of short-wave albedo, long-wave emissivity, volume heat, soil permeability and pore volume as well as evaporation, water storage capability and the sky view factor. Higher values (with partial restrictions) are only determined for heat conductivity and the roughness length (z_o) of the built-up area. Typical values of the roughness length of a city amount to i.e. z_o = 0.02 - 3.0 m (MALBERG 1985; OKE 1973) quotes values for urban areas of z_o equals 1 % to 10 % of the height of the building. According to BAUMGÄRTNER et al. (1977) z_o can be determined with the approximation formula $\log z_o = 1.03 \log h - 0.86$, whereby h equals the height of rough elements in metres.

As an example of change in the biosphere one can easily see (group 2 in tab.2) that the towns and cities have fewer plants as well as a smaller share of plant coverage, which is e. g. for downtown West- Berlin about 32 % but reaches 95 % in the outskirts of the city. Even the range of species changes in the towns and cities: the share of hemerochores reaches 50 % whereas only about 30 % predominate in the countryside (SUKOPP 1983). Densely built-up areas combined with a lack of vegetation result in lower values of latent heat, specific humidity and evapotranspiration, but on the other hand also cause more sensitive heat-fluxes. The third group incorporates technical facilities that partially influence the UBL directly. In this group we mainly refer to the exhaust fumes from smoke stacks, but also to these form different types of transport and the high number of domestic heating systems.

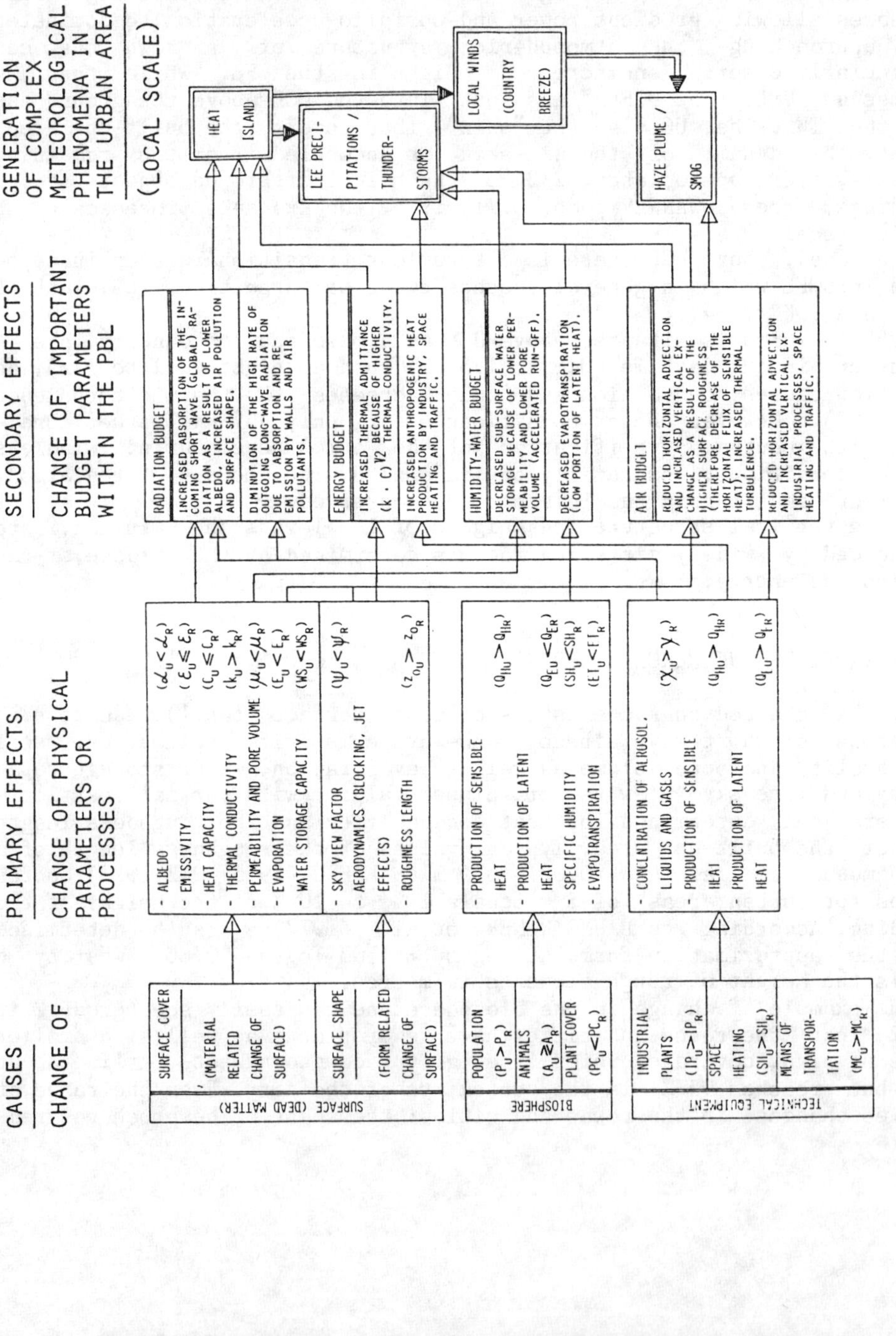

Tab. 1: Urban climate: pattern of interactions (after WANNER 1986)

All these factors are partially responsible for strongly increased flow densities of sensitive and latent heat as well as - from the point of view of air hygiene - increased concentrations of solid, fluid and gaseous particles.
The prominent mark of urban - industrial agglomerations visible from afar are the smog covers above the municipal areas consisting of a wide spectrum of gaseous and solid particles emitted from the agglomerations. Their existence is dependent on the type, height and strength of the source as well as the photochemical changes of the pollutants during their transmission.

3.1 Air Pollution

The numerous emissions in the urban atmosphere can be classified into distinct groups of pollutants e. g. sulphur, nitrogen and halogenous compounds as well as hydrocarbons, oxidants and dusts.
As far as immissions are concerned SO_2, NO_x, O_3, CO, C_m, H_n, dust and heavy metals have - often continuously - been a matter of investigation for several years in a number of agglomerations, sometimes with a large-scale network of measuring stations.

An analysis of the groups of polluters for the Rhine-Ruhr-District with regard to the whole spectrum of emissions in the five polluted areas, shows an average of 18.8 % for domestic heating systems, 22.8 % for traffic and 58.4 % for industry and power stations (KUTTLER 1987). With these immissions, it must be said that one can observe different trends depending on the measuring station and the measuring period. The strong dependency of e.g. the ground level SO_2 concentration on the heating habits of the population, can clearly be seen in the latest data from the municipality of Stuttgart - (BAUMÜLLER & HOFFMANN 1987) and here we must remember that in the FRG, 59 % of individual heating systems are fired by heating oil or coal (SOLFIAN et al. 1987). It shows a highly significant correlation to the mean monthly temperature per day with a given variance of $r^2 = 0.98$. In West-Berlin the share of emissions from domestic heating make up 24 % while its immissions account for a share of 45 % of total pollution (HASSEMER 1983). High concentrations of immissions can be observed for a short time during prevailing windless weather conditions as the occurrences of smog in many German agglomerations showed last winter .

Even a simple analysis of the concentrations e. g. for SO_2 and NO_x, divided into inversion and non- inversion days in a town (fig.3), shows that the values for inversion days are higher than those for days without inversions by a factor of 2.1 for SO_2 and 2.2 for NO_x. The figure also shows that these conditions mainly occur during the winter-half of the year. With the help of a flow-chart similar factors of increase have also been calculated for SO_2-concentrations during low-lying inversions in West-Berlin (ground-level). The main polluter during periods with very high concentrations of pollution given here as being a low lying spatial source, is above all domestic heating (KUTZNER 1983).

In contrast to SO_2, O_3 has the highest values in summer in particulars around noon because of the dependence on different "predecessor-gases" and the intensity of sunlight.

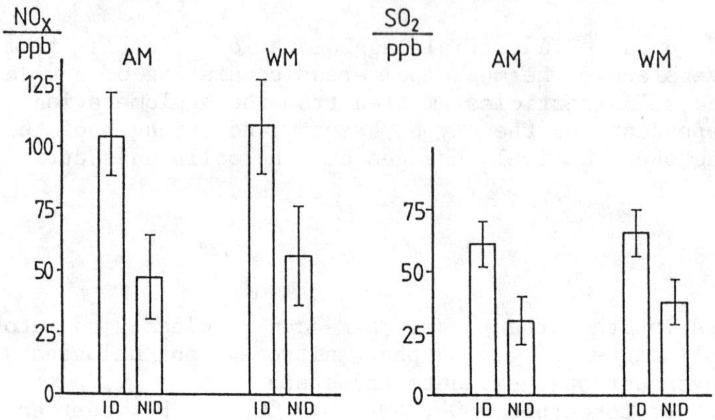

Fig.3:

Average concentration of $\mathbf{SO_2}$ and $\mathbf{NO_x}$ of days with (ID) and without (NID) temperature inversion

(height = 700 m above surface; high polluted city in the Rhine-Ruhr-Area, 1982-1985)

(**AM** = annual mean; **WM** = winter mean)

In addition, higher O_3 concentrations with less significant day-time trends can be observed in clean-air areas where pollution is mainly caused by long-distance transport (PFEFFER 1985).

Besides the contingencies influenced meteorologically, certain other influencing factors not caused by meteorological exchange conditions can also be proven for some highly polluted urban areas, such as, weekly trends in the concentration of immissions traceable purely to the consumption of particular types of fuel.

Fig.4 shows what is in part distinct differences between weekdays (Monday to Friday) and weekend values for immensely polluted measuring-stations in the Rhine-Ruhr-Area. The greatest differences have been found for NO which shows values as being 55 % higher on weekdays than at weekends, followed by dust at 14 % and SO_2 at 12 %. Increased values on weekdays measured at a station near the centre have also been proven for Munich. The findings were an additional 46 % for NO and 22 % for SO_2 (NOACK et al. 1986).
The trends for O_3-concentrations in the Rhine-Ruhr-Area, however, are quite the reverse. The weekend-values were 19 % higher than during the week.
If the differences seen between weekdays and weekends can be explained with the higher energy consumption and increased use of private vehicles typical for weekdays, the higher O_3-concentrations on weekends are at first sight, all the more surprising.
This could be linked to the fact that the ground-level atmosphere contains less NO_x on Saturdays and Sundays than on weekdays, so that less O_3 is reduced. However, this is only plausible as the process of synthesis and degradation is very complex (BRUCKMANN et al. 1980).

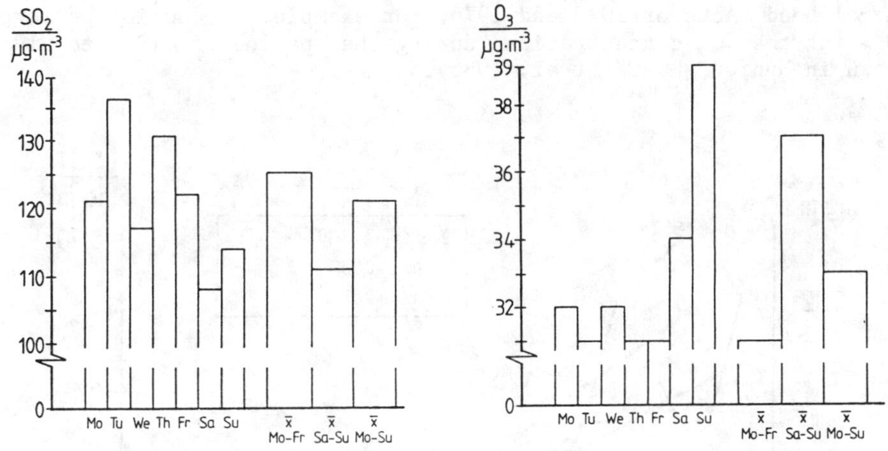

Fig.4: Average daily concentration of air pollutants at a high polluted station in the Rhine-Ruhr-Area (measuring period: Jan.-Dec. 1982) (data after PFEFFER et al. 1985)

Long-term measurements in polluted areas taken over a number of years show different trends in some pollutants: while the CO-values have decreased very slowly (DEIMEL 1982), the NO_x-values have been increasing considerably, particularly since the beginning of the 80's. Considerable decreases in concentrations can be observed in SO_2, SST, B(a)P in dust, (HEINRICH 1982) as well as in lead, this being a direct result of the

Petrol Lead Acts of 1972 and 1976. For example a reduction of more than 80 % in the SO_2-concentration during the period of 1966 to 1984 was shown in Munich (NOACK et al. 1986).

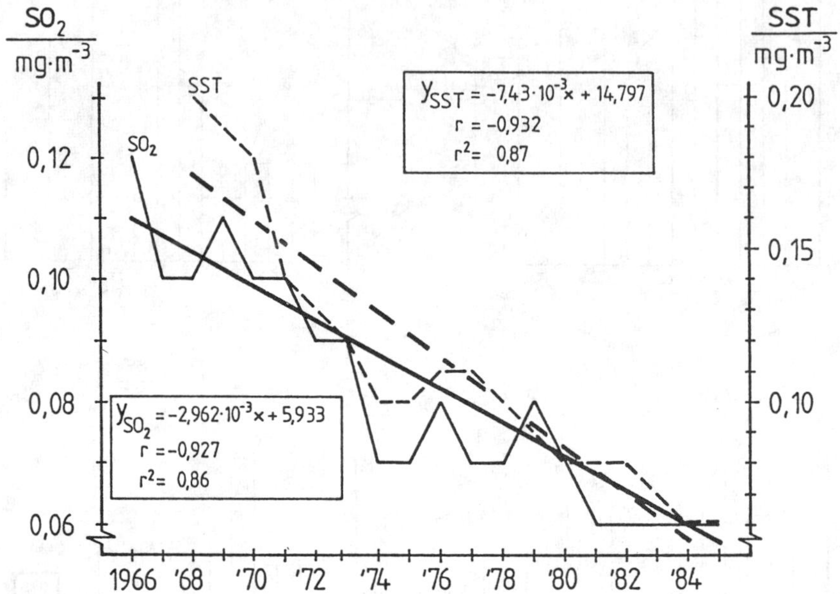

Fig.5: Yearly mean of **SO_2**- and **dust**- concentration (SST) in the Rhine-Ruhr-Area (data after BUCK et al. 1982 and LIS, Essen)

Fig.5 contains two examples from the Rhine-Ruhr-Area, showing a decrease of 50 % for SO_2 and 78 % for dust in the period between 1966 to 1984. For SO_2 - of which more than 90 % is produced by domestic heating, industry and power stations - it became obvious that the decreases occur both in summer and in winter, but with a considerably higher decrease during the cold season. As there is no demand for domestic heating in summer, industry and power stations are the only polluters of ground-level atmosphere in this period. In winter, however some 50 % of SO_2 emissions can be ascribed to domestic sources and the remainder to industrial emissions. It is therefore reasonable to assume that the decrease is due to a reduction in both industrial and domestic emissions. This marked decrease in winter is also a result of the policy of replacing short smoke-stacks, still widespread in the 60's, with higher ones. This enabled the pollutants to spread themselves above the relatively low upper boundary of the exchange layer in winter, about 400 to 600 metres above sea-level, thus removing them from the local ground-level atmosphere (BUCK et al. 1982; KUTTLER 1979).

4. BALANCE OF RADIATION AND ENERGY

The input and output of the urban fluxes of radiation and energy have been considerably changed by the aerosols and gases released into the urban atmosphere, as well as by the structure of the built-up area on the surface (fig.6). The smaller share of global radiation (S+H) supplied to the ecosystem city is absorbed or reflected by gases and particles in the smog-cover. In the Ruhr-area, using a sunshine-autographer, we calculated an average loss in the hours of sunshine of 0.3 h/d (KUTTLER 1985 a). A reduction of global and ultraviolet rays in a town are therefore caused by the urban smog cover which incidently reached 10 to 20 % in Munich (NOACK et al. 1986). Most of the radiation on the other hand penetrates to the urban surface resulting in absorption ($(S+H)_{abs}$) and - depending on the different materials on the surface - a lower or higher rate of reflection $(S+H)_t$. Using recent readings a value of urban albedo of 14 % can be determined (NOACK et al. 1986).

The long-wave radiation from the surface (E_g) is dependent on the surface temperature and the constants of material emissions. It is partly absorbed (E_{abs}) and also re-emitted by the smog layer. The long-wave counter-radiation is transmitted (G_t) and partly absorbed (G_{abs}) by the smog cover and again re-emitted onto the urban surface as well as into the hemisphere above the smog cover.

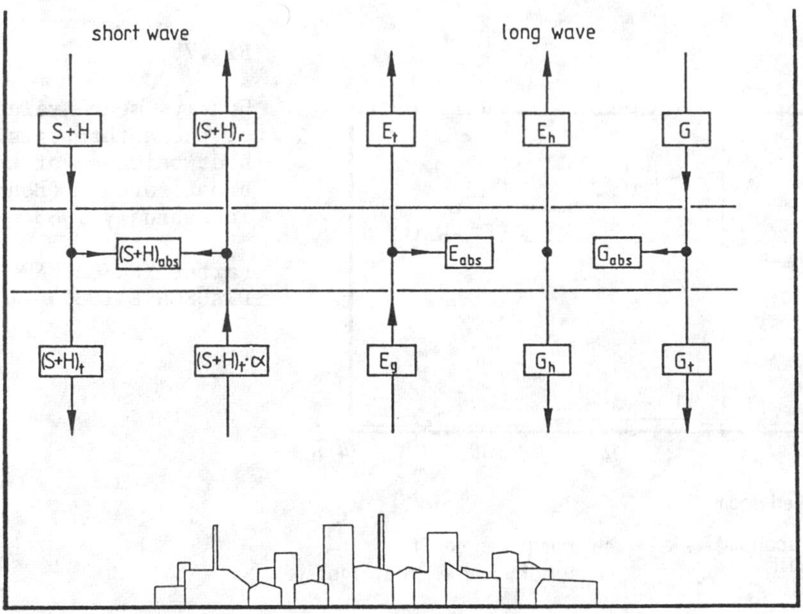

$$(S+H)_t \cdot (1-\alpha) - (E_g - G_h - G_t) + B + L + LE + N + A = 0$$

Fig.6: Radiation and energy balance of an urban atmosphere

The thermic processes engineered by global radiation cause the building structures to heat up - and here the buildings can increase - the area available for energy transfer by factor 6 (ERIKSEN 1975). They also cause the flux of sensible (L) and latent heat (LE) as long as sufficient water is available for evaporation.
The artificially produced "anthropogeneous" energy (A) is released near to the ground into the UCL or by smoke stacks directly into the UBL and remains in the urban smog cover.
The "anthropogeneous heat production" is to be seen as a form of energy that does not result from the natural energy surplus, but, for example from the flux of energy through badly insulated houses and industrial plants or heat emissions from vehicle-motors or exhaust fumes of all types as well as heat deriving from the metabolism of living organisms (HÖPPE 1984). This artificially produced heat becomes particularly apparent on cold winter days, as KERSCHGENS & DRAUSCHKE (1986) were able to show using Bonn as an example. Their analyses were based on the urban energy consumption of oil and electric energy, traffic censuses and estimated metabolic energy production based on the distribution of the population in the town. It was shown that with regard to the area of the town, (fig.7) the amount of energy released metabolically did not reach 5 Wm^{-2}. This value is supported by Japanese findings (FUKUOKA 1983). Slightly higher values resulted from traffic with maximum values in rush-hours of up to 10 Wm^{-2}. Estimated waste heat from vehicles in the

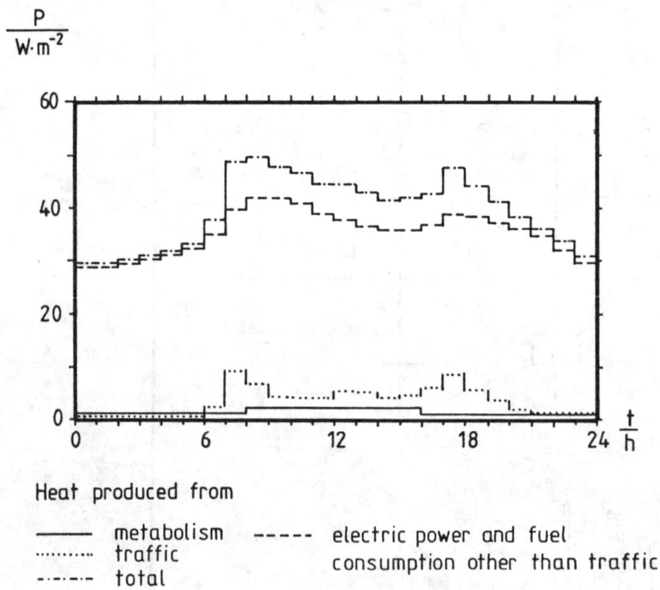

Fig.7:

Hourly mean values of the anthropogenic heat release of the urban area (Bonn) in February 1983

(after KERSCHGENS & DRAUSCHKE 1986)

town of Bochum gave an annual mean value of about 5 Wm^{-2}, corresponding roughly with the total global radiation of 4.4 % (KUTTLER 1985 b).
The decisive elements in "anthropogeneous heat release" are of course those flows of energy related to the consumption of oil, coal and

electricity which, in proportion to consumer demand - for example in Bonn, amount to 30 Wm^{-2} at night and 40 Wm^{-2} during the day. A rough estimate for the city of Essen carried out recently with the help of a heat requirements forecast, obtained a similar result of 40 Wm^{-2} as an annual mean value.

The amounts of anthropogeneous heat production calculated for some municipalities in different climatic zones partially showed differing values: a factor of between 0.2 and 0.4 times the balance of radiation for cities like Berlin or Hamburg; for Moscow, during the cold season however 3 times the total balance of radiation (compare classification in KUTTLER 1985 b).

Further factors which additionally increase the balance of energy of urban ecosystems, is the lack of vegetation and the high degree of built-up areas which lead to accelerated precipitation into sewerage and thus reduced evaporation. The energy that is not used for evapotranspiration is therefore available for heating the air. This can also be seen in the BOWEN-ratio calculated for towns which are, in general higher than the surrounding countryside.

In contrast to natural surfaces like grassland and forests (fig.8) which show negative values in the early morning and the evening (as (H) becomes positive) ß remains positive in the city (as (H) and (LE) are negative) and reach values of more than 1.0 around noon. This means that most of the energy reaches the atmosphere via the sensitive heat flux. The urban ß-values in the morning and the evening of less than 1.0 prove that "the city is not the desert for which it is sometimes held". (GARSTANG et al. 1975, p.155).

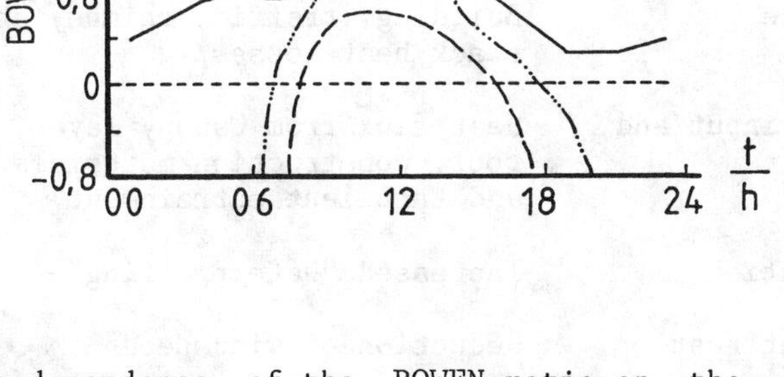

Fig.8:

Diurnal variation of the BOWEN-ratio over various surfaces.

(after OKE et al. 1972; from GARSTANG et al. 1975)

The dependence of the BOWEN-ratio on the share of green-areas (G) (fig.9) shows that the increase of (G) to 100 % ß at noon only amounts to about 0.8. On the other hand ß increases when (G) decreases. If the share of green-area is lower than 25 %, ß exceeds 1.0 at noon.

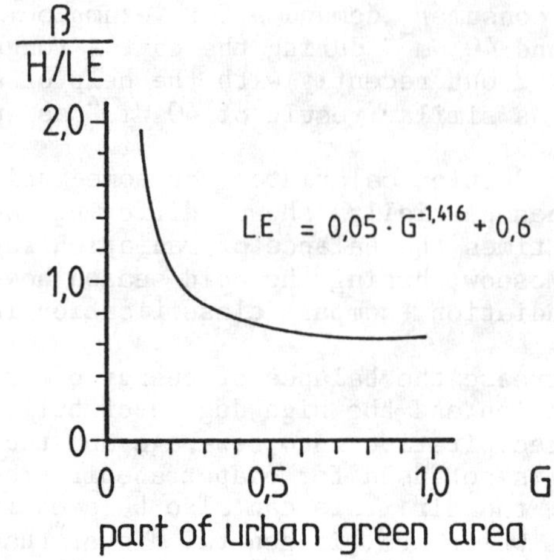

Fig.9: Dependency of the BOWEN-ratio (β) on the part of urban green area (G) (high pressure area, summer, at noon)

(after KERSCHGENS & HACKER 1985, from NOACK et al. 1986)

5. THE URBAN EXCESS HEATING

The following factors can be summarized as being the causes of excess heating in urban areas (tab.2):

Altered Energy Balance terms in the urban Canopy and the Boundary layer	Features of urbanization underlying Energy Balance changes
1. Increased absorption of short-wave radiation	Increased surface area multiple reflection, air pollution
2. Increased long-wave radiation from the sky	Air pollution
3. Decreased long-wave radiation loss	Reduction of sky view factor
4. Anthropogenic heat source	Building, traffic, chimney and stack heat losses
5. Increased sensible heat input and storage	Heat flux from Canopy layer, roof, construction materials and turbulent entrainment
6. Decreased evapotranspiration	Increased "waterproofing"
7. Decreased total turbulent heat transport	Reduction of windspeed

Tab.2: Suggested "causes" of the urban heat island (not rank ordered) (from OKE 1982, shortened)

These result in differences between heating and cooling rates of air temperature for town and countryside, as shown in the seasonal trend in fig.10. The heating and cooling rate at the rural station is much higher throughout the whole year than in the city. At sunrise there is a relatively rapid warming up in the rural environment (in summer: 3.5 K/h; in winter: 1.5 K/h) as the heat conductivities of the rural surfaces are smaller than those of urban surfaces. At sunset there is a relatively rapid cooling-down (in summer: -3.5 K/h; in winter: -1.2 K/h) as the surface experiences a net radiative energy drain which draws heat from the shallow layer of stable air directly above it. While the surface temperature drops and the rate of radiative emission decreases, the cooling rate also declines as the night progresses. As a result rural air temperatures at night exhibit a decay curve until just after sunrise when the pattern is abruptly interrupted by warming up.

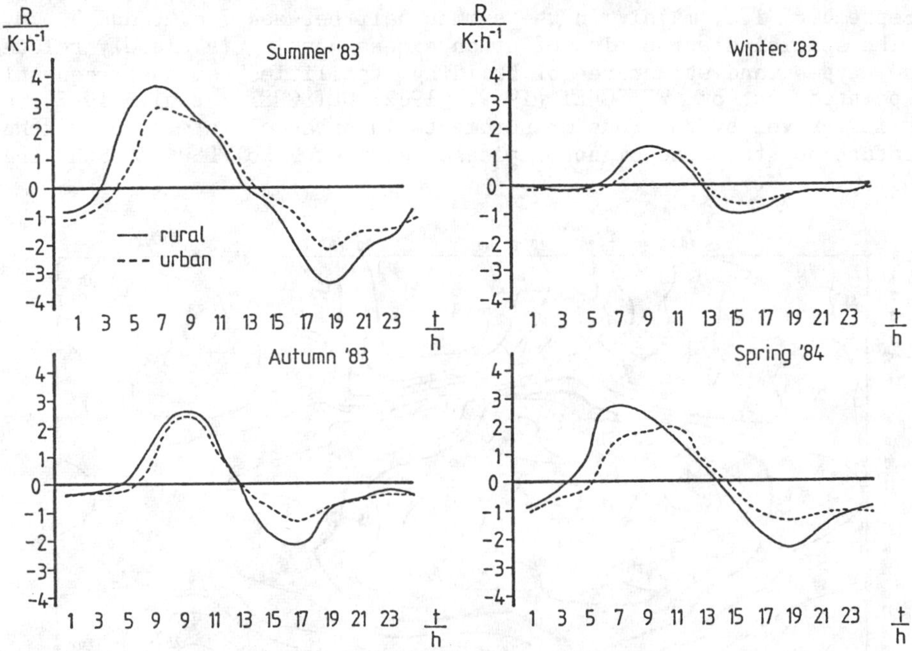

Fig.10: Heating and cooling rates (**R**) for different seasons in the city (Bochum) and rural surroundings

Fig.11 shows a varied image of the daily and annual trends of the intensity of the "heat island" in a mid-latitude town. It shows the mean hourly differences in air temperature between an urban and rural station. According to this figure, during the daytime, street canyons only experience a slight difference to the rural environment. The reason for this is that the increased radiation at noon causes a convective exchange. Further causes would be the shadows cast by buildings and the height of roof-top levels.

Differences in air temperature between urban and rural environments arise through a prevailing lack of cloud so that high radiation values from the surface can be registered as, for example for the months of July and August between 9 p.m. and 4 a.m..
Consequently, urban excess heating is mainly a night-time phenomenon. According to measurements taken by MIESS (1974) it is also dependent on the fact that the exchange values at night in the lowest layer above the surface decline considerably i.e. from 8 g cm^{-1} s^{-1} to 1 g cm^{-1} s^{-1} as compared to those during the day. Thus, only a reduced amount of the heat stored in buildings can be transported. The result: higher air temperatures than in the rural surroundings. On the basis of an analysis done in Bochum we were able to discover that the downtown area is, in about 80 % of all hours in the year up to 3.5 K, warmer. For the remaining 20 % it has the same temperature as the rural environment or is up to 1.2 K cooler, especially in the period between mid-February and mid-September, i.e. mainly in the summer half between 7 a.m. and 5 p.m.

The spatial distribution of urban excess heating is closely related to the type and structure of building facilities, as has frequently been pointed out by WEISCHET (1979, 1982; WEISCHET et al. 1977) and which is proven by numerous measurements in urban climate. For further literature on this phenomenon, please refer to KUTTLER & SCHREIBER (1984).

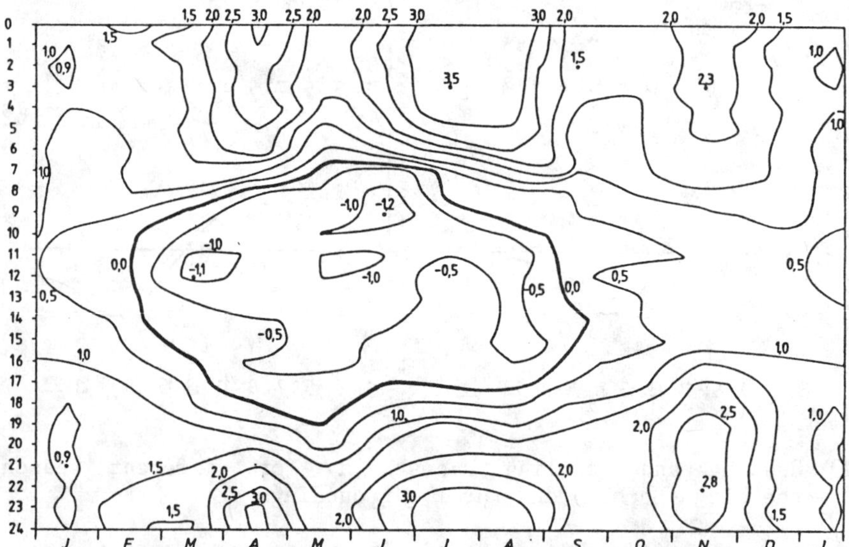

Fig.11: Diurnal and annual variation of the heat island intensity ($\Delta t = t_u - t_r$)(Bochum 1984) (KUTTLER 1985)

Windless or low wind velocities and a stable stratification of layers in urban ground-level atmosphere promote the generation of the urban "heat island". Measurements taken in Vienna (BÖHM & GABL 1978) show for example (fig.12), the close connection between the vertical distribution of wind velocities above a city and the generation of the urban "heat island".

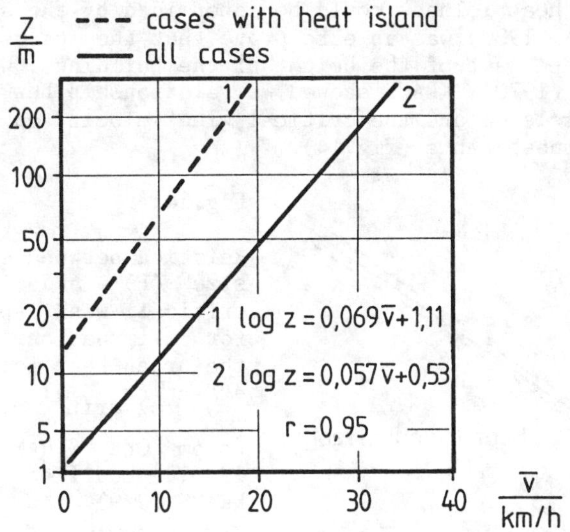

Fig.12:

Vertical profile of wind speed over the city of Vienna

(after BÖHM & GABL 1978)

As you can see in fig.13 under stable vertical gradients a "heat island" is formed in the UBL, (here I refer you to LUDWIG 1970 and KRAUS 1979), whereas with neutral or unstable strata, the town is cooler than the surrounding countryside.

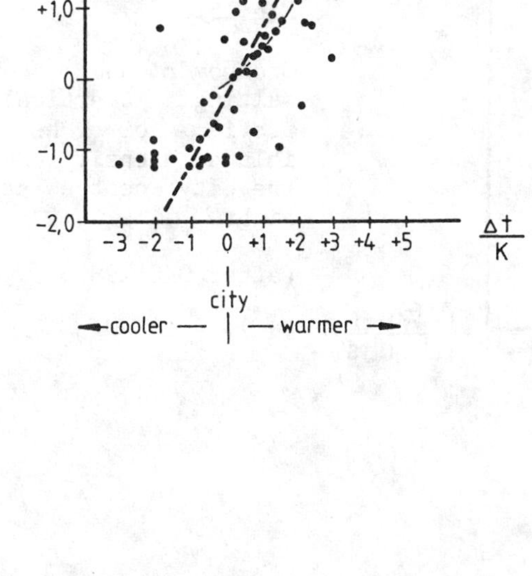

Fig.13:

Relations between temperature gradient and heat island intensity

(after AHRENS 1981)

- - = regression curve
— = theoretical curve

(after MUNN 1973)

There have been some attempts made to find whether the intensity of the "heat island" bears any relationship to the size of a city and its population.
FUKUOKA (1983), YOSHINO & KAI (1973) and OKE (1973) were able to show that the intensity of the "heat island" could be determined by the size of the population. YAMASHITA (1980) was able to prove that the intensity of the "heat island" is dependent of the height of the buildings. With reference to OKE & HANNELL (1970) KRAUS showed a relationship between the generation of a heat island and the critical wind velocity on the logarithm of the number of inhabitants (fig.14).

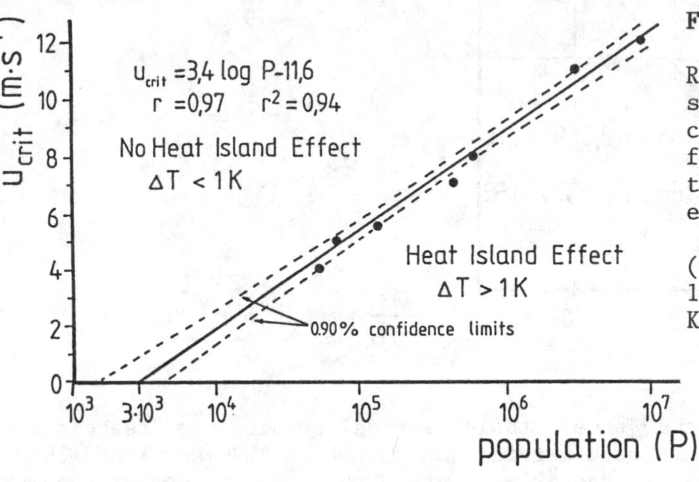

Fig.14:

Relation between city size (**P**) and the critical wind speed for elimination of the urban heat island effect (**u_{crit}**)

(from OKE & HANNELL 1970; modified by KRAUS 1979)

Apart from the formation of heat islands in the horizontal plane a three dimensional structure can also be observed, only typical, however, during weather conditions with high amounts of radiation. As you can see from fig.15 with a fall in the different temperatures in town and surrounding countryside, the urban "heat island" reaches a height of 200 m - in this example at night. Above this level which is called "crossover" it is cooler above the city than above the rural environment.

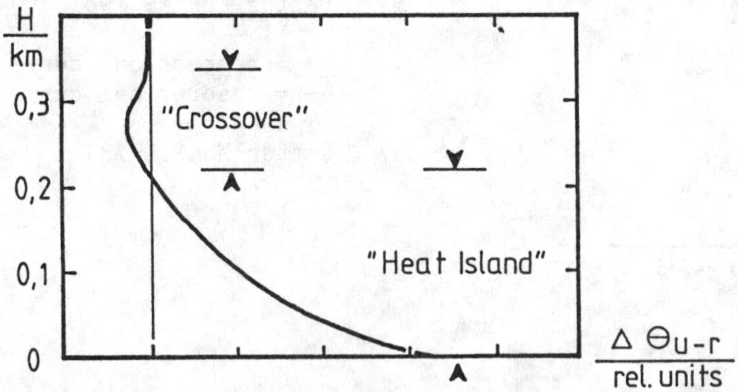

Fig.15:

Crossover and resulting vertical profil of heat island intensity in the city centre at night

(after OKE 1982)

According to investigations made by DUCKWORTH & SANDBERG (1954), and often proved on several later occasions (BRÜNDL et al. 1986; THOMMES 1986; DELAGE & TAYLOR 1970) the following reasons can be given for the formation of a "crossover":
"The intense surface inversion (in the rural area) will cause some subsidence and consequently some adiabatic warming in the layer just above the inversion. In the city the formation of a pollutant layer is somewhat lifted by the low-level instability. This layer can cause heat loss by outgoing radiation and hence cool below the temperature prevalent at the same height in the cleaner rural air column" (LANDSBERG 1981, p.109).
This effect, however, does not influence the formation of the "crossover" at night significantly. Another possibility under discussion is that due to the stronger convection above the town, a large scale vertical exchange takes place which could result in lower temperatures at higher altitudes.
As the infra-red pictures show, during the daytime the convection above urban structures is caused mainly by the outgoing radiation from hot roofs, which can easily be proved by SODAR-measurements.
Based on 10 minutes averages, and using the city of Essen as an example, fig.16 shows an increase in the vertical wind velocities of more than $u = 0.75\ ms^{-1}$ above a height of 100 metres.

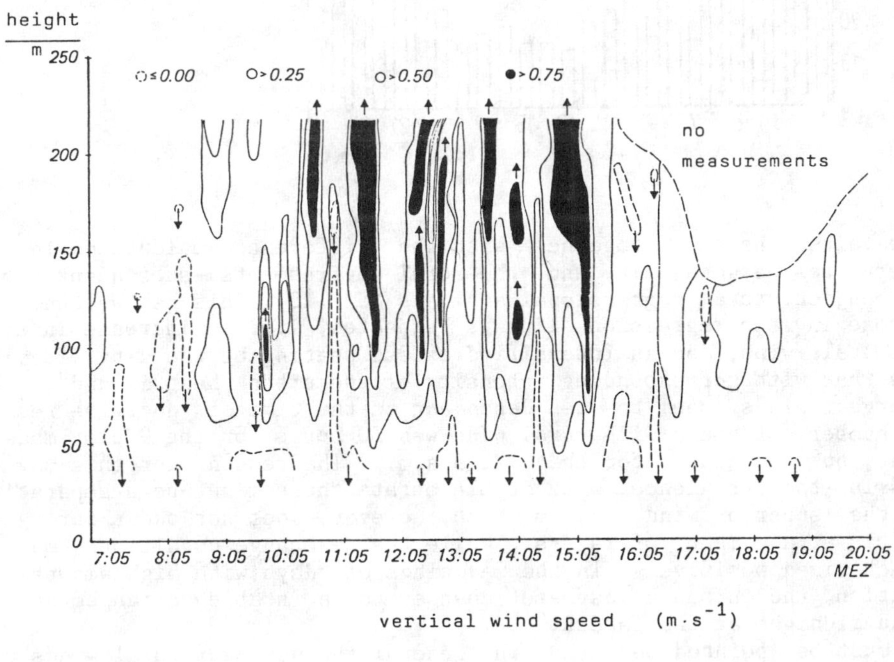

Fig.16: Vertical wind speed over the city of Essen, 27.09.1983 (after BECKRÖGE & FRANK 1986)

Using mesurements taken at the same time, the horizontal velocities above the heated city were, in part, shown to exceed those of the rural surroundings because of higher impulse exchanges caused by the higher temperatures. The exchange of air masses during low-gradient weather conditions with high amounts of radiation is only enforced by the thermic imbalance.
The resulting intermittent advection which is near to the surface and directed at a city is usually known as "low level breeze" or rural wind, and has mean velocities of only $u = 2\ ms^{-1}$ (KIESE & OTTO 1986).
On the basis of measurements taken in Dortmund, at set times during the day, the frequency of rural winds reached their maximum around 5 a.m. and minimum around noon (fig.17).

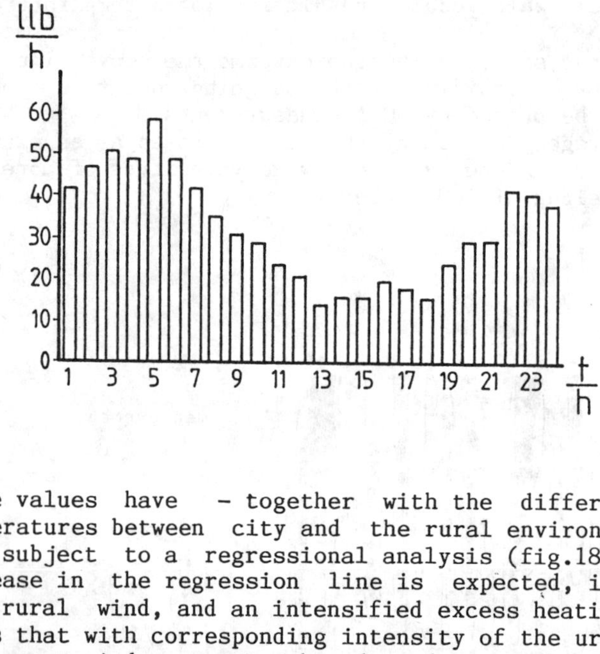

Fig.17:

Hourly distribution of low-level-breezes (Jan.-Dec. 1985)

(after KIESE & OTTO 1986)

These values have - together with the differences calculated in air temperatures between city and the rural environments -subsequently become subject to a regressional analysis (fig.18). This shows that an increase in the regression line is expected, i.e. an increase in time with rural wind, and an intensified excess heating of the city. It also seems that with corresponding intensity of the urban "heat island", the convergent winds seem to be dependent on the time of day. The resulting number of hours with rural wind was 30 hours for the 9 p.m. measurement, but 60 hours for that at 5 a.m.. The reason for this may be found in the turbulence. With stable strata there might be a separation from the superior wind system which, however, does not occur during an unstable arrangement of layers. While the exchange conditions may be characterized positively in the evenings of days with high amounts of radiation, the urban atmosphere changes to a stable arrangement at around midnight at the latest.
It must be pointed out that in general the hours of rural winds are above the regressional line in the early morning and before noon, while fewer hours of rural wind can be observed in the afternoon despite the same amount of excess heat. The occurrence and intensity of rural wind

is bound to a certain pattern of ventilation which supplies the city with fresh air, but only then, when the fresh air from the city's surroundings is not yet polluted.

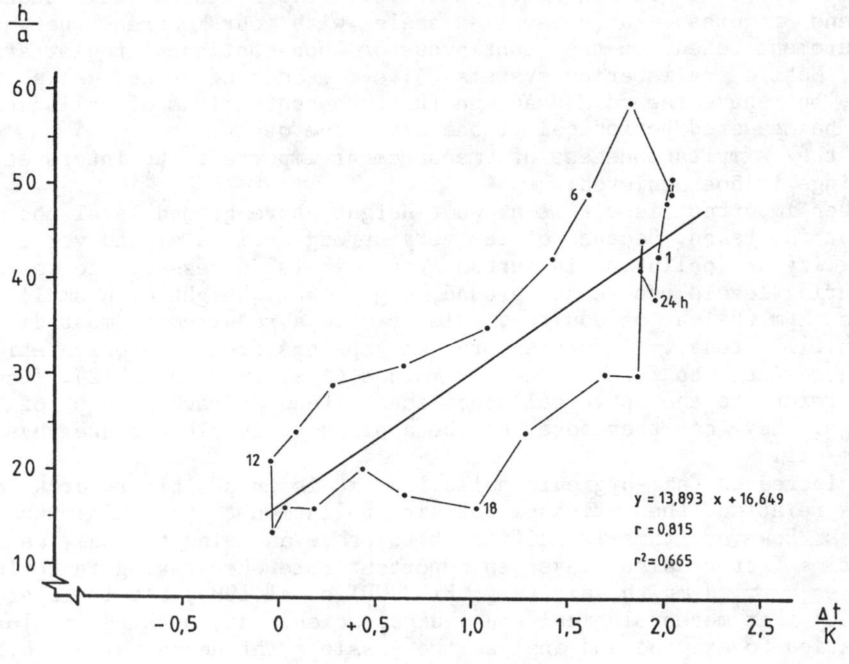

Fig.18: Dependency of low-level breezes hours on the heat island intensity; measuring period Jan.-Dec. 1985 (city of Dortmund, Rhine-Ruhr-Area)

VENT-SCHMIDT (1985) pointed out the difficulties of an evaluation analysis for cold-air flux.
Quantitative results of the potential cooling rate of an urban building structure which is cooled down by the cooler air from the surroundings, were presented by MIESS (1974).
Furthermore he was also able to differenciate between the dependence of these data on the intensity of the "heat island" and the velocity of the cool-airflux.

6. EVALUATION OF THE URBAN CLIMATE

Of the three elements in bio-climatology, actinic, air-hygienic and thermic complex of efficiency, the thermic and air-hygienic components, enjoy relatively high importance.
Registering and evaluating the air-hygienic situation involves several difficulties. It is, in fact, extremely difficult to talk about air hygiene figures on a small- scale with our current network of measurement based on the continuous or non-continuous registration of data. Mobile registering systems (i.e.: measuring vans) are a limited help, but have the disadvantage that concentrations of pollutants can only be measured periodically, one after the other.
Thus the simultaneousness of measurement important to interpret the findings is not achieved.
Another important aspect is at what height above ground level the measurement is taken. Because of the very strong horizontal and vertical variability of pollution in urban areas it is necessary to establish measuring levels above the ground, e.g. 0.5 m (height of a small child) up to 2 m (height of adults). This is also a requirement most difficult to fulfill today. A further problem springs from the orientation of measured data to fixed threshold values (i.e. IW 1 and IW 2). These values refer to the physical condition of an average group of human beings. They can therefore not be applied to people who are physically worse-off.
With increased air-hygienic pollution it is plausible to draw conclusions relating the influence of air pollution to potential human illnesses. However this is difficult to prove as being the only cause, as the time-factor also plays an important role when taking in toxic substances carried in the air (e.g. EINBRODT et al.1982). It is interesting to note that meteorological and medical science are working in close cooperation to examine and analyse the possible influences of air-hygienic components in urban areas.
For further information on this, I refer you to the literature available.
In the work carried out in the field of urban climatology hardly any characteristic values for the thermic influences based on the stationary effect of the climatic parameters have been used. In this area, the following standards are frequently used in measuring: the equivalent temperature, the effective temperature and the physiologically equivalent temperature based on GAGGE (1980). These characteristic values are widely used, as the essential meteorological basic values only have limited complexity and can thus be used by people outside the field of climatology more easily.
Since using these values, the bio-climatical patterns for human beings which have developed (FANGER 1972, PMV-value; JENDRITZKY et al. 1979; HÖPPE 1986, MEMI) have made it possible to make essentially differentiated statements on the basis of complex basic values. These model calculations are of course based on stationary parameters of the climate and the human body.
Because of the diversity of micro-climates (BURT et al. 1982) - which have e.g. certain influences on pedestrians in street canyons (i.e.

differing conditions of radiation because of differing heights of buildings and possible shadows from tree-tops), it is impossible even with these patterns, to draw any valid conclusions about the effects on pedestrians in permanently changing micro-climate conditions. The application of a pattern which also considers mobile basic values is thus an important, and necessary developement. The mobile Munich pattern of energy balance (IMEM) developed by HÖPPE (1986) has clear advantages in this context as was proved by comparative measurements and calculations. However - and this must be added with reservations - this model is only able to help drawing any valid conclusions about individuals. To solve practical problems, which are frequently concerned with making statements about the inhabitants of a municipal district, and consequently a collective statement, it is better to use the models given by FANGER (1972) and JENDRITZKY et al. (1979).

7. URBAN CLIMATOLOGY ORIENTATED TO PLANNING

If one reviews the urban-climatological history of research, three fields of investigation marking its development spring to mind. Starting with the works of SCHMIDT, HORWARD, RENOU, PEPPLER and KRATZER (all quoted in LANDSBERG 1981) the differences in temperatures between town or city and the surrounding areas always stood in the fore-front of their investigations. The results were determined on the basis of comparative stations in- and outside of a city. Mobile measurements had already been taken in early years (e.g. SCHMIDT in Vienna (1930). The purpose was to get away from the restricting results given by selective determination of single climatic parameters and to move towards spatial and representative findings.

In the period following, scientists turned to spatial, three-dimensional examinations in order to study the exchange conditions between city and rural surroundings and thereby obtained a clearer picture of the dynamics of ventilation systems for different weather conditions. In addition to the purely meteorological work carried out since about 1970 (BRÜNDL et al. 1986), air-hygienic components have been taken into consideration more and more. Since the beginning of the 80's this development has led to the utilization of the wind tunnel for the simulation of spreading processes in urban areas as well as to an intensified inclusion of the numerical simulation of mesoclimatically relevant parameters - or small spatial planning, in particular the well-known microscale pattern called MUKLIMO, which can simulate air flows above block-shaped structures (BECKRÖGE 1987, OKE 1982, SIEVERS et al. 1987).

The efforts made in research in the field of applied "reality" must finally result in a spatial description of the area examined (which must be as complete as possible). However, there are, and have always been problems in transposing the data one has ascertained from single measurements onto the total area. Here a useful method is to group all those areas with the same utilization and surface structure into so-called "climatic topes" (climatopes) (HORBERT et al. 1986). Climatopes, plots showing "uniform, long-term features of land and micro-climate even during different weather conditions" (LESER 1978, p.139), are mapped as

synthetic and functional maps of the area's climate (HABERL & STOCK 1983). In addition to the description of the spatial "distribution" of the climatic parameters relevant to the urban area and its causes, these maps also incorporate a differentiated list of the various climatic effects as was required by HORBERT et al.(1985).
The necessary data is based on the establishment of special measuring networks, the analysis of blocks of buildings with similar structures and the incorporation of infra-red thermal pictures (STOCK & LEHNER 1986).
With these aids, it must be remembered that only temperatures equivalent to radiation are given. Before they can be used for further utilization, these temperature values, however, have to be corrected - as had not been done before and which was quite rightly criticized by BAUMGARTNER et al.(1985).
A synthetic and functional map of the urban climate based on a detailed analysis of the micro- and mesoscale conditions (there are already a growing number of examples) should give detailed information on at least the following three factors (KUTTLER & SCHREIBER 1984):
1. the location and intensity of over-heated areas in the municipal area
2. records of areas with cold air sources
3. records of the courses of cold-air fluxes into areas with excess heat

These charts could then form the basis of maps illustrating the importance of the factors climate and air-hygiene and could serve as a source of information for town planners (STOCK et al. 1986).

8. FURTHER DEMANDS ON AN APPLIED URBAN CLIMATOLOGY

The demands made by urban climatologists and town planners can be summarized as follows (FRANKE 1977, ERIKSEN 1980, KUTTLER & SCHREIBER 1984, WANNER 1986):
1. The temporary excess-heating of downtown areas in summer must be prevented or - at least - be reduced considerably. This can be realized by an increased coverage of vegetation. The two main questions here are:
- "How large must the green-area in the municipality be to have the desired effect?" and
- "How can the green-area be designed to be climatically most efficient for its environment?" (POTTHOFF 1984).

In this field v. STÜLPNAGEL (1987) carried out a detailed examination using the City of Berlin as an example.
Examinations carried out by HORBERT et al.(1985) in Berlin, showed that on extremely wind-still days with green-areas of 30 ha, temperatures fell in the immediate vicinity up to distances of from 150 to 600 metres. Under similar conditions, the influence of the 212 ha -"Tiergarten" could be measured up to a distance of 900 metres on its leeside. SPERBER (1974), found, however, that although the measurable influence increases with the size of the green-area, the decrease in temperature even with large green-areas can only be determined up to a distance of about 250 metres. Nevertheless, the positive influence of such large green-areas on the cleanness of the air is much greater than their cooling effect (WILMERS 1985).

In this context we must not forget that the 'greening' of houses is of special importance. Intentional planting of creepers on the outer walls of houses and a 'greening' of roofs could be used as a not inconsiderable area for climatic melioration - without having to use additional land (HOYANO et al. 1985, YAMASHITA 1980).

2. The air-exchange inside the municipal area must be guaranteed as follows: Firstly by keeping aisles for fresh air free, through the optimum positioning of building structures in relation to important air flow systems (i.e. to the most frequent wind direction, the lowest air exchange and the periodical wind systems of a day)
and Secondly by restricting the height of building structures to only a few floors and also by restricting their width.

3. The share of particles and gaseous pollutants in the air must be reduced through furthering the use of heating systems in compound networks, filters in smoke stacks and catalytic converters.

References

AHRENS, D. (1981): 'Untersuchungen über die Wärmeinsel und die Mischungsschicht einer Großstadt'. - In: Arch. Met. Geoph. Biokl., Ser. B, **29**, pp. 29 - 36.

BAUMÜLLER, J. & U. HOFFMANN (1987): 'Verbesserung der Infrastruktur des Stadtklimas - Erfahrungen im Raum Stuttgart'. - In: Umweltschutz in großen Städten, VDI-Berichte **605**, pp. 191 - 221.

BAUMGARTNER, A.; MAYER, H. & W. METZ (1977): 'Rauhigkeitsparameter z_o mit Anwendung auf die Energiedissipation an der Erdoberfläche'. - In: Meteor. Rdsch., **30**, pp. 43 - 48.

BAUMGARTNER, A.; MAYER, H. & M. NOACK (1985): 'Thermalkartierungen in bayerischen Großstädten'. - Abschlußbericht zum Teilprogramm "Thermalkartierungen", Bay. Staatsminist. f. Landesentw. u. Umweltfragen

BECKRÖGE, W. (1987): 'Einsatz eines Mikroklimamodells in der Humanbiometeorologie und praktische Anwendung bei der Planung'. - In: Wiss. Mitt. Meteorol. Inst. Univ. München.

BECKRÖGE, W. & J. FRANK (1986): 'Einsatz von SODAR-Geräten bei Messungen in der städtischen Grenzschicht'. - In: Staub-Reinh. Luft, **46**, pp. 523 - 527.

BOESLER, K.-A. (1987): 'Die Stadt als Lebens- und Belastungsraum'. - In: Umweltschutz in großen Städten, VDI-Berichte **605**, pp. 23 - 41.

BÖHM, R. & K. GABL (1978): 'Die Wärmeinsel einer Großstadt in Abhängigkeit von verschiedenen meteorologischen Parametern'. - In: Arch. Met. Geoph. Biokl., Ser. B, **26**, pp. 219 - 237.

BRUCKMANN, P.; BUCK, M & P. EYNCK (1980): 'Modelluntersuchungen über den Zusammenhang zwischen Vorläufer und Photooxidantienkonzentrationen'. - In: Staub-Reinh. Luft, **40**, pp. 412 - 417.

BRÜNDL, W.; MAYER, H. & A. BAUMGARTNER (1986): 'Untersuchung des Einflusses von Bebauung und Bewuchs auf das Klima und die lufthygienischen Verhältnisse in bayerischen Großstädten (Stadtklima Bayern)'. Abschlußbericht zum Teilprogramm "Klimamessungen München".

BUCK, M.; IXFELD, H. & K. ELLERMANN (1982): 'Die Veränderung der Immissionsbelastung in den letzten 15 Jahren im Rhein-Ruhr-Gebiet'. - In: Staub-Reinh. Luft, **42**, pp. 51 - 58.

BURDACK, J. (1985): 'Entwicklungstendenzen der Raumstruktur in Metropolitan Areas der USA'. - In: Bamberger Geographische Schriften, 2

BURT, J. E.; O'ROURKE, P. A. & W. H. TERJUNG (1982): 'The relative influence of urban climates on outdoor human energy budgets and skin temperatures. II: Man in an Urban Environment'. - In: Int. J. Biometeor., **26**, pp. 25 - 47.

DELAGE, Y. & P. A. TAYLOR (1970): 'Numerical studies of heat island circulations'. - In: Boundary Layer Meteor., **1**, pp. 201 - 226.

DEIMEL, M. (1982): 'Kfz-Abgasbelastungen in Straßen und Unterführungen im Vergleich zur Allgemeinimmission'. - In: TÜV-Rheinland (Hrsg.): Abgasbelastungen durch den Kraftfahrzeugverkehr im Nahbereich verkehrsreicher Straßen, pp. 115 - 142.

DUCKWORTH, F. S. & J. S. SANDBERG (1954): 'The effect of cities upon horizontal and vertical temperature gradients'. - In: Bull. Amer. Met. Soc., **35**, pp. 198 - 207.

EINBRODT, H. J.; PRAJSNAR, D. & TH. EIKMANN (1982): 'Möglichkeiten der Wirkungsforschung zur Beurteilung Kfz-bedingter Schadstoffbelastungen'. - In: TÜV-Rheinland (Hrsg): Abgasbelastungen durch den Kraftfahrzeugverkehr im Nahbereich verkehrsreicher Straßen, pp.15-125.

ERIKSEN, W. (1975): 'Probleme der Stadt- und Geländeklimatologie'. - In: Erträge der Forschung, **35**.

ERIKSEN, W. (1980): 'Klimamodifikationen im Bereich von Städten. Grundlagen und städtebauliche Aspekte'. - Veröff. Joachim-Jungius-Ges. Wiss., Hamburg, **44**, pp. 161 - 175.

FANGER, P. O. (1972): 'Thermal Comfort'. McGraw Hill, New York.

FRANKE, E. (Hrsg.) (1977): 'Stadtklima. Ergebnisse und Aspekte für die Stadtplanung'. Seminar der Württ. Verwaltungs- und Wirtschafts-Akademie, Stuttgart, April 1976. - In: Beitr. z. Umweltplanung, **108**.

FUKUOKA, Y. (1983): 'Physical Climatological Discussion on Causal Factors of Urban Temperature'. - In: Mem. Fac. Integrated Arts and Sciences, Hiroshima Univ., Ser. IV, **8**, pp. 157 - 168.

GAGGE, A. P. (1980): 'The new Effective Temperature ET - an Index of Human Adaption to Warm Environments'. - In: Environmental Physiology: Aging, Heat and Altitude, pp. 59 - 77.

GARSTANG, P. D. ; TYSON, G. & D. EMMITT (1975): 'The Structure of Heat Islands'. - In: Reviews of Geophysics and Space Physics, **13**, pp. 139 - 165.

HABERL, S. & P. STOCK (1983): 'The Synthezised Climatic Function Map'. In: Proc. EARSeL/ESA Symp. on Remote Sensing Appl. Env. Stud., Brussels, **26 - 29**, pp. 273 - 281.

HASSEMER, V. (1983): 'Luftreinhaltung als langfristige Aufgabe der Stadtentwicklungsplanung - dargestellt am Beispiel Berlin'. - In: Reinhaltung der Luft in großen Städten, VDI-Berichte **477**, pp. 15 - 20.

HEINRICH, U. (1982): 'Tierexperimentelle Befunde zur Wirkung von polyzyklischen aromatischen Kohlenwasserstoffen (PAH) unter besonderer Berücksichtigung des Kfz-Abgases'. - In: TÜV-Rheinland (Hrsg): Abgasbelastungen durch den Kraftfahrzeugverkehr im Nahbereich verkehrsreicher Straßen; pp. 315 - 327.

HÖPPE, P. (1984): 'Die Energiebilanz des Menschen'. - Univ. München, Meteorolog. Inst., Wiss. Mitt., **49**.

HÖPPE, P. (1986): 'Die thermische Komponente des Stadtklimas'. - In: MAYER, H. (Hrsg.): Bioklima in der Stadt; Münchener Univ. Schriften, Wiss., Mitt., **53**, pp. 94 - 108.

HORBERT, M.; KIRCHGEORG, A. & A. v. STÜLPNAGEL (1985): 'Ergebnisse stadtklimatischer Untersuchungen in Berlin (West)'. Internationale Tagung für Human-Biometeorologie vom 2. bis 4. Oktober 1985 in Freiburg. - In: Ann. d. Meteorol., N.F., **22**, pp. 137 - 139.

HORBERT, M.; KIRCHGEORG, A. & A. v. STÜLPNAGEL (1986): 'Klimaforschung in Ballungsgebieten, dargestellt am Beispiel Berlin'. - In: Geogr. Rdsch., **38**, Heft 2, pp. 71 - 80.

HOYANO, A.; CHATANI, M. & K. YAGI (1985): 'Experimental study on solar control by an ivy covered wall'. - In: Journal of Architecture, Planning and Environmental Engineering, **351**, pp. 11 - 19.

HOYANO, A. & T. YAMASHITA (1986): 'Experimental Study on solar control effects of vine sunscreen at the Veranda'. - In: Heat & Environment, **14**, **6**, pp. 1 - 14.

JENDRITZKY, G.; SÖNNING, W. & H.-J. SWANTES (1979): 'Ein objektives Bewertungsverfahren zur Beschreibung des thermischen Milieus in der Stadt- und Landschaftsplanung'. - In: Beitr. ARL, **28**.

JENDRITZKY, G. & W. NÜBLER (1981): 'A model analysing the urban thermal environment in physiologically significant terms'. - In: Arch. Met. Geoph. Biokl., Ser. B, **29**, pp. 313 - 326.

KERSCHGENS, M. J. & R. L. DRAUSCHKE (1986): 'On the energy budget of a wintry-mid-latitude city atmosphere'. - In: Beiträge Phys. Atm., **59**, Nr. 1, pp. 115 - 225.

KIESE, O. & G. OTTO (1986): 'Windanalyse der Stadt Dortmund'. - Endbericht für den Kommunalverband Ruhrgebiet, Essen, unveröff. Manuskript.

KRAUS, H. (1979): 'Die Wärmeinsel'. - In: promet, **9**, Heft 4, pp. 7 - 11.

KUTTLER, W. (1979): 'Einflußgrößen gesundheitsgefährdender Wetterlagen und deren bioklimatische Auswirkungen auf potentielle Erholungsgebiete'. - In: Bochumer Geogr. Arbeiten, **36**.

KUTTLER, W. (1985 a): 'Stadtklima. Struktur und Möglichkeiten zu seiner Verbesserung'. - In: Geogr. Rdsch., **37**, H. 5, pp. 226 - 233.

KUTTLER, W. (1985 b): 'Neuere Arbeiten zur angewandten Klimageographie in der Bundesrepublik Deutschland'. - In: Ber. z. Dt. Landeskde. **59**, H. 1, pp. 59 - 84.

KUTTLER, W. (i. print): 'Immissionsklimatologische Probleme und Entwicklungen im Rhein-Ruhr-Gebiet'. - In: Der Raum Dortmund - Entwicklung, Strukturen und Planung im östlichen Ruhrgebiet, Spiecker-Landeskdl. Beiträge u. Berichte, **32**.

KUTTLER, W. (1987): 'Das Stadtklima und seine raum-zeitliche Struktur'. - In: Ökologische Probleme in Verdichtungsräumen, Hohenheimer Umwelttagung vom 23.1.1987.

KUTTLER, W. & D. SCHREIBER (Hrsg.) (1984): 'Stadt- und geländeklimatische Untersuchungen im südlichen Münsterland'. - In: Materialien zur Raumordnung, Geogr. Inst. Univ. Bochum, **25**.

KUTZNER, K. (1983): 'Der Hausbrand als bodennahe Emissionsquelle - Flächendeckende Emissionen und ihre Bedeutung für die Lufthygiene'. - In: Reinhaltung der Luft in großen Städten; VDI-Berichte **477**, pp. 61 - 72.

LANDSBERG, H. E. (1981): 'The urban climate'. - Academic Press, New York

LESER, H. (1978): 'Landschaftsökologie'. - Stuttgart.

LUDWIG, F. (1970): 'Urban temperature fields'. - In: WMO Techn. Note 108, WMO **254**, TP 141, pp. 80 - 112.

MALBERG, H. (1985): 'Meteorologie und Klimatologie'.- Berlin, Heidelberg

MIESS, M. (1974): 'Planungsrelevante und kausalanalytische Aspekte der Stadtklimatologie'. - In: Landschaft + Stadt, **6**, pp. 9 - 16.

NOACK, E.-M.; MAYER, H. & A. BAUMGARTNER (1986): 'Quantifizierung der Einflüsse von Bebauung und Bewuchs auf das Klima in der urbanen Biosphäre'. - (Abschlußbericht), Nationales Klimaforschungsprogramm der Bundesrepublik Deutschland.

OKE, T. R. (1973): 'City Size and the Urban Heat Island'. - In: Atm. Environm., **7**, pp. 769 - 779.

OKE, T. R. (1982): 'The energetic basis of the Urban Heat Island'. - In: Quart. J. Roy. Met. Soc., **108**, pp. 1 - 24.

OKE, T. R. (1983): 'Boundary Layer Climates'. - London, New York.

OKE, T. R. (1984): 'Methods in urban climatology. Applied Climatology'. - In: Zürcher Geographische Schriften, **14**, pp. 19 - 29.

OKE, T. R. (1986): 'Urban climatology and the Tropical City: An Introduc tion'- In: OKE, T. R.(Hrsg.): Urban climatology and its Applications with special Regard to tropical Areas; WMO **652**, pp. 1 - 25

OKE, T. R. & F. G. HANNELL (1970): 'The Form of The Urban Heat Island in Hamilton, Canada'. - In: WMO Techn. Note 108, WMO **254**, TP. 141, pp. 113 - 126.

PFEFFER, H.-U. (1985): 'Immissionserhebungen in quellfernen Gebieten Nordrhein-Westfalens'. - In: Staub-Reinh. Luft, **45**, pp. 287 - 293.

POTTHOFF, H. (1984): 'Ökologisch-kleinklimatische Messungen in Bonn unter besonderer Berücksichtigung der Wirkungen der Vegetation'. - Diss. Landwirtschaftl. Fak. Univ. Bonn.

SCHMIDT, W. (1930): 'Kleinklimatische Aufnahmen durch Temperaturfahrten'. - In: Meteorol. Z., **47**, pp. 92 - 106.

SCHÖLLER, P. (1983): 'Zur Urbanisierung der Erde'. - In: Tag.ber. u. Wiss. Abhandl. 43. Dtsch. Geographentag Mannheim, **43**, pp. 25 - 34.

SIEVERS, U.; MAYER, J. & W. G. ZDUNKOWSKI (1987): 'Numerische Simulation des urbanen Klimas mit einem zweidimensionalen Modell. Teil II'. - In: <u>Meteorolog. Rdsch.</u> **40**, pp. 65 - 83.

SOLFRIAN, W.; SCHADE, H. & E. R. SCHRAMEK (1987): 'Ermittlung der Emissionen des Sektors Haushalte und Kleinverbraucher im großstädtischen Lebensraum, dargestellt an den Luftreinhalteplänen im Ruhrgebiet'. - In: <u>Umweltschutz in großen Städten</u>, VDI-Berichte **605**, pp. 121 - 139.

SPERBER, H. (1974): 'Mikroklimatisch-ökologische Untersuchungen an Grünanlagen in Bonn'. - Diss. Landwirtschaftl. Fak. Univ. Bonn.

STOCK, P. & M. LEHNER (1986): 'Die städtische Wärmeinsel in Thermalbildern verschiedener Aufnahmemaßstäbe'. - In: ENDLICHER, W. & H. GOSSMANN (Hrsg.): <u>Fernerkundung und Raumanalyse</u>.

STOCK, P. ; BECKRÖGE, W. ; KIESE, O.; KUTTLER, W. & H. LÜFTNER (1986): 'Klimaanalyse Stadt Dortmund'. - In: <u>Planungshefte Ruhrgebiet</u>, PO 18, Kommunalverband Ruhrgebiet, Essen.

STÜLPNAGEL, A. v. (1987): 'Klimatische Veränderungen in Ballungsgebieten unter besonderer Berücksichtigung der Ausgleichswirkung von Grünflächen, dargestellt am Beispiel von Berlin (West)'. - Diss. Fb 14, TU Berlin.

SUKOPP, H. (1983): 'Städtebauliche Ordnung aus der Sicht der Ökologie'. - In: <u>Reinhaltung der Luft in großen Städten</u>, VDI-Berichte **477**, pp. 163 - 172.

THOMMES, W. (1986): 'Untersuchung der dreidimensionalen Wärmeinsel im Bereich der Städte Ludwigshafen und Mannheim während ausgewählter Wetterlagen'. - Deutscher Wetterdienst, Zentralamt, Offenbach/M., unveröff. Manuskript.

VENT-SCHMIDT, W. (1985): 'Die Darstellung lokalklimatischer Phänomene und daraus resultierende bioklimatische Aussagen in der Regionalplanung'. - In: <u>Ann. Meteor. (N. F.)</u>, **22**, pp. 106 - 108.

WANNER, H. (1983): 'Stadtklimatologie und Stadtklimastudien in der Schweiz'. - In: <u>Jahrbuch der Schweizerischen Naturforschenden Gesellschaft</u>, wiss. Teil 1983, pp. 96 - 111.

WANNER, H. (1986): 'Die Grundstrukturen der städtischen Klimamodifikation und deren Bedeutung für die Raumplanung'. - In: <u>Jahrb. d. Geogr. Gesellsch. von Bern.</u>, **55**, 1983 - 1985, pp. 67 - 84.

WEISCHET, W. (1979): 'Problematisches über die städtische Wärmeinsel und die Notwendigkeit einer Baukörperklimatologie'. - In: <u>Siedlungsgeogr. Studien</u>, pp. 407 - 423.

WEISCHET, W. (1982): 'Der Vorteil einer Baukörperklimatologie für Zwecke der Stadtplanung'. - In: Akademie für Raumforschung und Landesplanung (ARL) (Hrsg.): Thermalluftbilder für die Stadt- und Landesplanung, Beitr. d. Akad. f. Raumforsch. u. Landesplanung, **62**, pp. 23 - 38.

WEISCHET, W.; NÜBLER, W. & A. GEHRKE (1977): 'Der Einfluß von Baukörperstrukturen auf das Stadtklima am Beispiel von Freiburg/Br.'. - In: FRANKE, E. (Hrsg.): Stadtklima; Stuttgart, pp. 39 - 63.

WILMERS, F. (1985): 'Ansprüche des Menschen an seine Klima-Umwelt'. In: Landschaft + Stadt, **17**, H. 1, pp. 30 - 42.

YAMASHITA, S. (1980): 'On causal factors of heat islands. Applied Geographical Study on Causal Factors of urban Climates'. - In: MONBUSHOKAKEN Report, pp. 88 - 94.

YOSHINO, M. & K. KAI (1973): 'Change of air temperature in Japanese cities in the recent years and its relation to the synoptic patterns and population'. - In: Tenki, **20**, pp. 489 - 497.

ENERGETIC FEEDBACK BETWEEN URBAN CLIMATE AND DIFFUSION OF ATMOSPHERIC POLLUTANTS

M. J. Kerschgens, H. Suer
Meteorologisches Institut der Universität Bonn
Auf dem Hügel 20
D 5300 Bonn
FRG

ABSTRACT. The turbulent sensible heat flux at the top of an urban or rural canopy layer forms a key parameter for the thermal structure of the overlying Boundary Layer. The thermal structure of the atmosphere, i. e. the turbulence intensities and the height of the BL, on the other hand determine the spread of atmospheric particles of natural and anthropogenic origin.

From BL-experiments characteristic scaling parameters of the energetic differences between convective urban and rural boundary layers are evaluated. With the help of similarity theory it is shown how the influence of the urban structure can be incorporated into the basic turbulence parameterization of air pollution models. From this basic stage it makes no difference whether the model is of a simple Gaussian type or a more elaborate one. It is shown that model results with energetically based parameterizations differ considerably from conventional ones (e. g. TA - Luft).

1. INTRODUCTION

It is a common approach to declare the causes of the differences between the state of the urban and rural atmosphere with the help of the energy fluxes leaving or entering the respective system (cf. OKE, 1982). The systems under consideration for an energetic study of the urban and rural atmosphere are

- the Urban Canopy Layer (UCL),
- the Rural Canopy Layer (RCL),
- the Urban Boundary Layer (UBL), and
- the Rural Boundary Layer.

The canopy layers extend vertically between the level of zero net heat flux in the ground up to an arbitrary upper level. This upper level should be chosen in a way, that all structures of the surface contributing to the energy storage of the CL lie below. The top of the UCL is consequently situated atop roof level, while the top of the RCL is generally the surface of the ground, if there are no crops or trees which cause significant shading with all inherent consequences.

K. Grefen and J. Löbel (eds.), Environmental Meteorology, 335–344.

The vertical extent of the Convective Boundary Layer (CBL) reaches under fair weather conditions from the top of the canopy layer up to the base of the subsidence inversion. Figure 1 gives a rough impression of the scenario including some measuring equipment which is necessary for the evaluation of energy fluxes as described below.

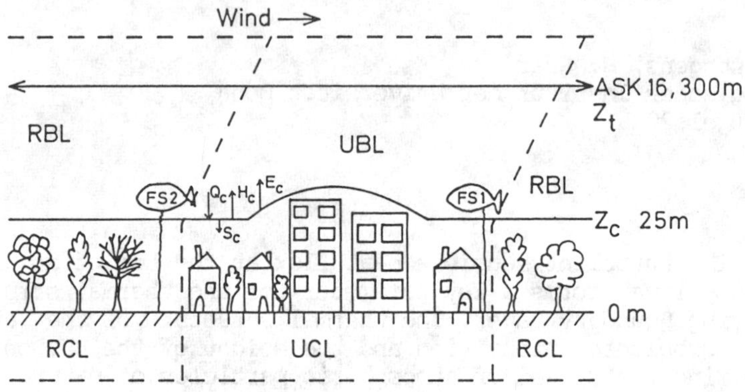

Figure 1: Schematic picture of the scenario. ASK 16, FS1, and FS2 indicate operated measuring systems as there are an instrumented motor glider and two thethered sonding systems. Also symbolized are the energy fluxes of the energy balance equation (1).

In the following chapters it will be shown how the diffusion of particles depends on the energetic behaviour of a convective boundary layer and the underlying surface. For this purpose some equations, though very simplified, may give a rough idea of the complexity of the problem.

2. THEORY

2.1 Energetic features

The energy balance at the top of the canopy layer (equation 1) determines the thermodynamic behavior of the CL and forms the lower boundary condition for the energy budget of the boundary layer.

$$Q_C = H_C + E_C + S_C \tag{1}$$

with Q_C = Net radiation flux,
H_C = Turbulent sensible heat flux,
E_C = Turbulent latent heat flux,
S_C = Net heat storage change.

The sensible (equation 2) and latent (equation 3) heat budget of a differential volume may be written (cf. KERSCHGENS and HACKER (1985)):

$$\bar{\rho}\, c_p \frac{\partial \bar{\Theta}}{\partial t} = -\bar{\rho}\, c_p \bar{\mathbb{V}}\cdot(\mathbb{V}\,\bar{\Theta}) - \mathbb{V}\,\mathbb{H} + Q_\Theta \qquad (2)$$

(I) (II) (III) (IV)

$$\bar{\rho}\, L \frac{\partial \bar{q}}{\partial t} = -\bar{\rho}\, L\, \bar{\mathbb{V}}\cdot(\mathbb{V}\,\bar{q}) - \mathbb{V}\,\mathbb{E} + Q_q \qquad (3)$$

(I) describes the lokal change of the respective variable of state (q or Θ) due to advection (II), the divergence of the turbulent energy flux (III), and due to sources in the volume (IV). These sources of energy are, e. g., heat releases due to phase changes of water vapour or to anthropogenic sources and to the divergence of the net radiation flux.

ρ, c_p, Θ, t, L, q, $\mathbb{V}$, and $\mathbb{V}$ denote density of air, heat capacity of air at constant pressure, potential temperature, time, heat of vaporization, specific humidity, del operator ($\partial/\partial x$, $\partial/\partial y$, $\partial/\partial z$) and threedimensional windvector respectively. $\mathbb{H}$ and $\mathbb{E}$ are symbols for the above mentioned threedimensional fluxes. Considering the heat-budget of a boundary layer equation (2) and (3) have to be integrated by volume, with equation (1) as lower boundary condition.

A conceptual frame for the evaluation and measurement of equations (1) to (3) with experimental case studies can be found in detail in KERSCHGENS (1987). The interaction between the energy fluxes of the above mentioned equations and the movement of inert particles becomes evident from an inspection of the simplified equation of Turbulent Kinetic Energy (TKE).

2.2 Convective Scaling

Starting point is the TKE equation for an horizontal homogeneous atmosphere:

$$\frac{1}{2}\frac{\partial e}{\partial t} = -\overline{u'w'}\frac{\partial u}{\partial z} + \frac{g}{\Theta}\overline{w'\Theta'} - \frac{\partial}{\partial z}\overline{\left(w'\frac{1}{2}u_i'^2 + \frac{p}{\rho}\right)} - \varepsilon \qquad (4)$$

(I) (II) (III) (IV)

with $e = \overline{u_i' u_i'}$. The index i in the argument of a function represents the whole vector and repeated indices indicate summation over all coordinate directions. Ensemble averages are indicated by an overbar and fluctuations by a prime, e. g., $u = \bar{u} + u'$. The change of TKE is composed of production by the shear of wind (I), the buoyancy (II), a combined transport and pressure term (III) and the dissipation of turbulent energy (IV).

With the restriction of a stationary convective boundary layer and the vertical integration from the surface to the top h of the BL the energy budget may thus be approximated by

$$\int_0^h \frac{g}{\Theta} \overline{w'\Theta'}\, dz = \int_0^h \varepsilon \, dz \tag{5}$$

Assuming the fundamental approach (cf. TENNEKES and LUMLEY, 1972)

$$\varepsilon = c_\varepsilon \sigma_w^3/h, \text{ with } \sigma_w^2 = \overline{w'^2} \tag{6}$$

and a linear vertical decrease of $\overline{w'\Theta'}$, σ_w may be estimated as

$$\sigma_w = c \left(h \frac{g}{\Theta} \overline{w'\Theta'_o}\right)^{1/3} \tag{7}$$

in which c is an undetermined constant and $\rho\, c_p\, \overline{w\Theta'_o}$ the turbulent sensible heat flux at the surface.

From (7) it seems convenient to introduce a convective velocity scale w^* (cf. TENNEKES, 1970):

$$w^* = \left(h \frac{g}{\Theta} \overline{w'\Theta'_o}\right)^{1/3} = \left(\frac{hg}{\rho c_p \Theta} H_o\right)^{1/3} \tag{8}$$

From similarity considerations and numerical experiments (cf. PASQUILL and SMITH, 1983) parameterizations follow for the use in diffusion models of Gaussian type:

$$\begin{aligned} \sigma_y &= c_1\, w^*\, t \\ \sigma_z &= c_2\, w^*\, t \end{aligned} \tag{9}$$

There exist different vallues for the constants c_1 and c_2. E. g. LAMB (1978) suggests $c_1 = 0.33$, $c_2 = 0.5$, while OMSTEDT (1984) promotes $c_1 = c_2 = 0.56$. IRWIN (1979) and KAIMAL et al. (1976) recommend parameterizations for Monte - Carlo calculations:

$$\begin{aligned} \sigma_w &= c_3\, w^* \\ T_L^w &= 0.15 \frac{h}{\sigma_w} \left(1 - \exp(-5z/h)\right) \end{aligned} \tag{10}$$

where T_L^w denotes the Lagrangian time scale (cf. PASQUILL and SMITH, 1983) and c_3 is either a constant or a function of z/h.

GODOWITCH (1986) shows that the characteristics of the spectrum of energy in the convective boundary layer over uneven terrain are similiar to those over homogeneous terrain. Therefore the same parameterizations (equations 9 and 10) should also be valid over an horizontal inhomogeneous surface. The above equations demonstrate that, concerning the diffusion of particles in the CBL, the basic scaling parameters are w^* and h. However, as w^* depends on the surface heat flux H_o it should differ between urban and rural surfaces.

3. RESULTS

3.1 Canopy layer fluxes

The basic features of the energy balance of the urban canopy layer are meanwhile well known, especially because of the pioneering studies of T. R. OKE and collaborators (e. g. OKE et al., 1981 or YAP and OKE, 1974). Figure 2 gives an example of typical energy fluxes during the summer season over an urban and rural canopy layer.

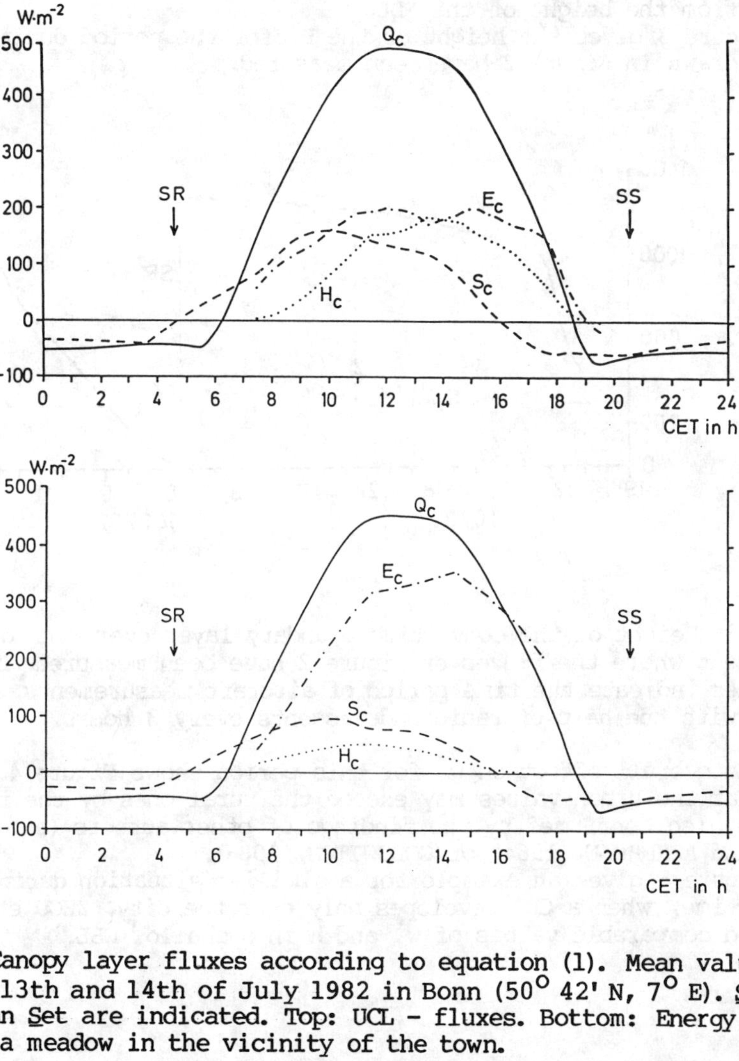

Figure 2: Canopy layer fluxes according to equation (1). Mean values during the 13th and 14th of July 1982 in Bonn (50° 42' N, 7° E). Sun Rise and Sun Set are indicated. Top: UCL - fluxes. Bottom: Energy balance over a meadow in the vicinity of the town.

The essential difference for the purpose of this study between RCL and UCL lies in the higher intensity of the sensible heat flux over the urban environment. w^* shows the immediate consequences.

3.2 Scaling parameters

As outlined above, the urban environment is under convective situations characterized by an, in comparison to the rural surroundings, higher surface heat flux H_o. The other variables of equation (8) do not differ considerably between an urban and a rural CBL. Measurements over St. Louis (SHEA and AUER, 1978) show that the height of the UBL differs only by 10% from the height of the RBL.

Figure 3 gives the height of the BL for the period during which the fluxes drawn in Figure 2 have been measured.

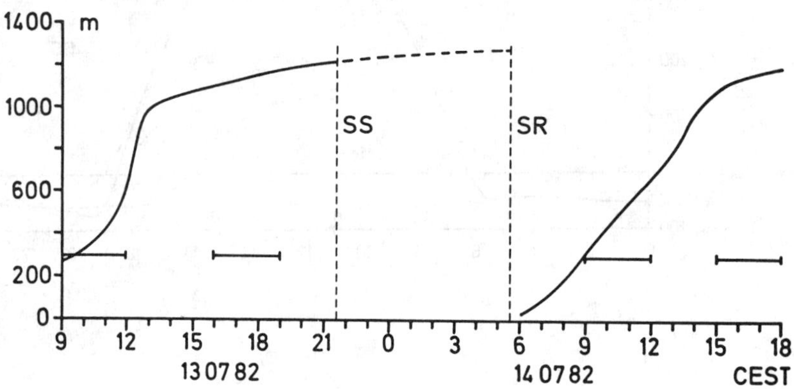

Figure 3: Height of the convective boundary layer over Bonn during the experiment where the fluxes of Figure 2 have been measured. The horizontal lines indicate the time period of aircraft measurements. h was evaluated with the help of radiosonde ascents every 3 hours.

The overall effect on w^* for this period shows Figure 4. It can be seen that the urban values may exceed the rural ones by the factor of 2. This is also confirmed by the findings of other authors (e. g., HILDEBRAND and ACKERMAN, 1984, or GODOWITCH, 1986).

Figure 5 gives an example for a similiar situation during the winter time, when a CBL developes only over the city. ZHOU et al. (1985) measured comparable values of w^* and h in a shallow CBL.

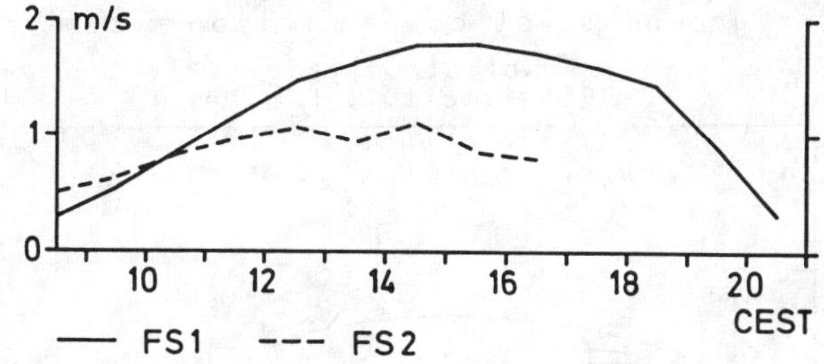

Figure 4: Mean w^* calculated from a composit of the values drawn in Figure 2 and 3. The solid line indicates the urban, the dashed line the rural values.

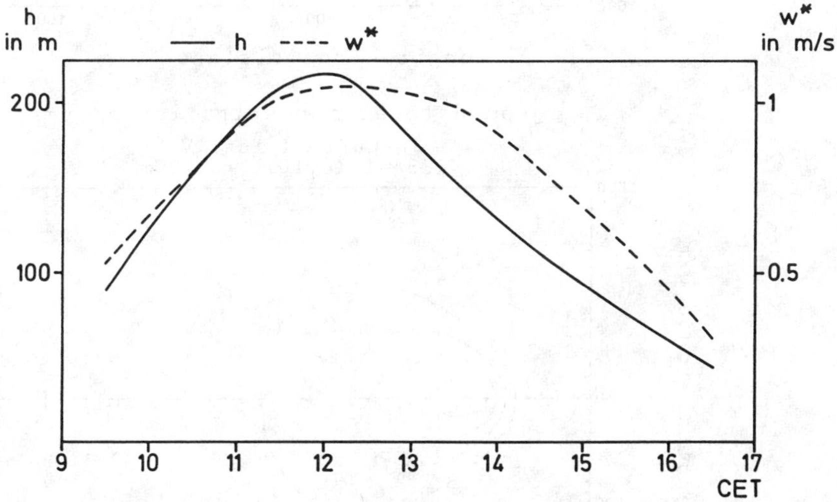

Figure 5: Height of the CBL and corresponding w^* for a typical boundary layer in the winter season, measured during an experiment in February 1983 over Bonn.

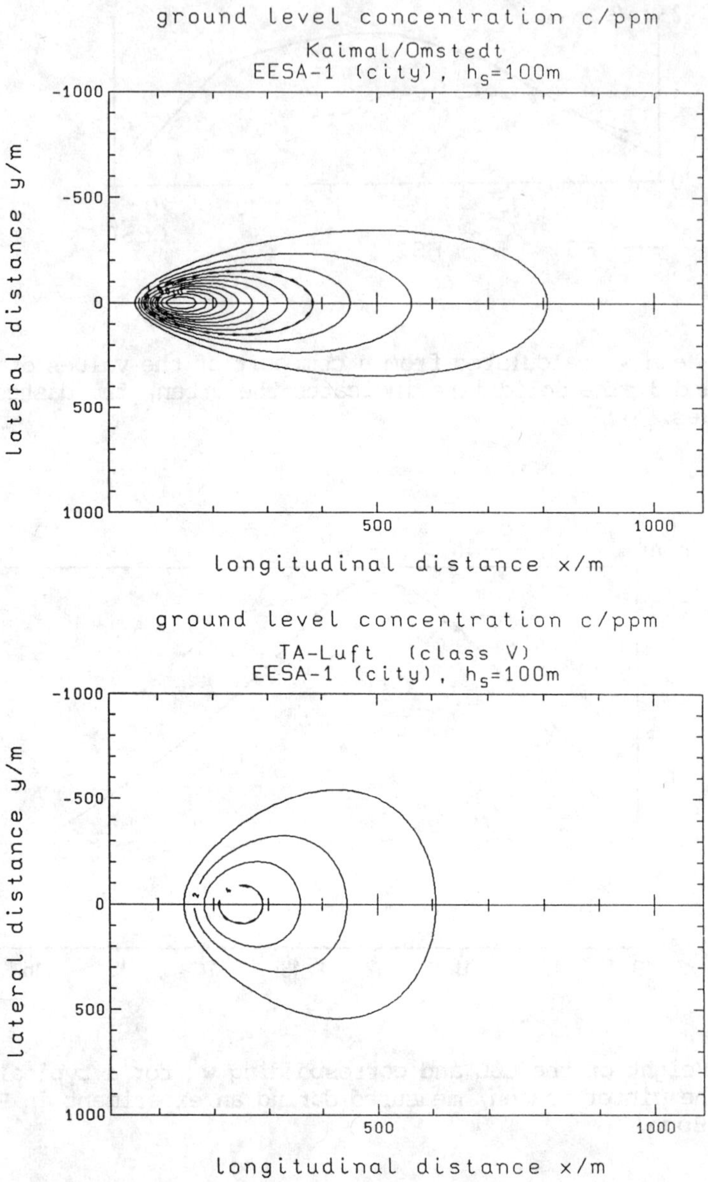

Figure 6: Surface concentrations of a comparison of σ - parameterizations according to equation (9) (top) with the empiric formulas of "TA - LUFT" (bottom). The source height h_s was set to 100m above the surface ,the height of the CBL to 1000 m, the mean wind speed to 2 m/s, w_* to 2 m/s. The concentration is drawn in steps of 1 ppm, the 4 ppm line is dashed.

3.3 Consequences for the diffusion of particles

LAMB (1982) has shown, that the parametrizations of equation (9) and (10) lead to similiar results when using either a Gaussian or a Monte - Carlo model for the diffusion of particles. It is therefore of more interest to compare the parameterizations of equation (9) to the empiric power law formulations of the "TA - Luft". The results of this intercomparison are shown in Figure 6.

It can be seen, that the empiric parameterizations in maximum surface concentrations which are too low and too far away from the source. The uncertainties due to experimentally determined constants of equation (9) are in the same order of magnitude as the differences of w^* between an urban and rural boundary layer. However, test calculations show no pronounced effect on the surface concentrations by these uncertainties.

4 CONCLUSIONS

It can be shown, that the energetic behaviour of the CBL immediately influences the spread of particles in the atmosphere if reasonable σ - parameterizations from energetic and similarity considerations are used.

These parameterizations result in strong deviations in comparison to conventional Gaussian dispersion calculations.

The differences in modelling the transfer of particles in an urban and rural boundary layer are beyond the accuracy of those formulations.

The interaction between the energetic state of the urban atmosphere and the diffusion of atmospheric pollutants in the convective boundary layer, however, is only one aspect of the wide field of the dispersion of particles over inhomogeneous terrain. Therefore other mechanisms as, e. g., effects due to orography or interaction with the surface by variable deposition velocities may be of a higher importance.

5. REFERRENCES

GODOWITCH, J. M., 1986: 'Characteristics of vertical turbulent velocities in the urban convective boundary layer. Boundary Layer Meteorology, **35**, 387 - 407.

HILDEBRAND, P. H. and B. ACKERMAN, 1984: 'Urban effects on the convective boundary layer'. J. Atmos. Sci, **41**, 76 - 91.

IRWIN, J. S.,1979: 'Estimating plume dispersion - a recommended generalized scheme'. Preprints Fourth Symposium on Turbulence, Diffusion and Air Pollution. Am. Meteorol. Soc., **45**, 62 - 69.

KAIMAL, J. C. and colloborators, 1976: 'Turbulence structure in the convective boundary layer'. J. Atmos. Sci., **33**, 2152 - 2169.

KERSCHGENS, M. J., 1987: 'Die Energetik der Stadt'. Bonner Meteorologische Abhandlungen, **35**, 105 pp. Dümmler Verlag Bonn.

KERSCHGENS, M. J. and HACKER, J. M., 1985: 'On the Energy Budget of the Convective Boundary Layer over an Urban and Rural Environment'. Beitr. Phys. Atm. **58**, 171-185.

LAMB, R. G., 1978: 'A numerical simulation of dispersion from an elevated point source in the convective planetary boundary layer'. Atmos. Environ., **12**, 1297 - 1304.

LAMB, R. G., 1982: 'Diffusion in the convective boundary layer'. Atmospheric Turbulence and Air Pollution Modelling, ed. F. T. M. Nieuwstadt and H. van Dop, D. Reidel Publishing Co., 159 - 230.

OKE, T. R., KALANDA, B. P. and D. G. STEYN, 1981: 'Parameterization of heat storage in urban areas'. Urban Ecol., **5**, 49 - 54.

OKE, T. R., 1982: 'The energetic basis of the urban heat island'. Quart. Journ. Roy. Met. Soc., **108**, 1 - 24.

OMSTEDT, G., 1984: 'An Operational Air Pollution Model using Routine Meteorological Data'. SMHI, 35 pp.

PASQUILL, F. and F. B. SMITH, 1983: Atmospheric Diffusion. Ellis Horwood Ltd., 437 pp.

SHEA, D. M. and A. H. AUER JR., 1978: 'Thermodynamic Properties and Aerosol Patterns in the Plume Downwind of St. Louis'. J. Appl. Meteor., **17**, 689 - 698.

TENNEKES, H., 1970: 'Free convection in the turbulent Ekman layer of the atmosphere'. J. Atmos. Sci., **27**, 1027 -1034.

TENNEKES, H. and J. L. LUMLEY, 1972: A first Course in Turbulence. M.I.T. Press Cambridge, Mass. 300 pp.

YAP, D. and T. R. OKE, 1974: 'Sensible heat fluxes over an urban area - Vancouver, B. C.'. J. Appl. Meteor., **13**, 880 -890.

ZHOU, M. J. and collaborators, 1985: 'Wave and turbulence structure in a shallow baroclinic convective boundary layer and overlying inversion'. J. Atmos. Sci., **42**, 47 - 57.

NUMERICAL MODELLING OF THE URBAN CLIMATE

J. Eichhorn, R. Schrodin *, W.G. Zdunkowski
Johannes-Gutenberg-Universität
Institut für Meteorologie
6500 Mainz
Bundesrepublik Deutschland

ABSTRACT. Three-dimensional numerical models of the urban climate have been developed in order to simulate climatic changes and air pollution in urban regions due to anthropogenic activities. Two versions are presented. Model A covers mesoscale γ and uses the roughness length to characterize surface structure. The microscale version B approximates the actual urban build-up by rectangular blocks. Model A results agree reasonably well with observations. For complex building structures, the microscale model must be applied. Model B results on neutral flow and pollutant transport within a particular building configuration are discussed.

1. MODEL DESCRIPTION

Since the urban population is steadily rising worldwide and is expected to exceed 80 % in developed areas by the end of the century, a comprehensive investigation of the climatic properties of urban regions is of increasing interest. An important aspect of such an investigation is the prediction of changes of the urban climate due to anthropogenic activities.

For this purpose, a three-dimensional urban climate model has been developed at the University of Mainz (Eichhorn et al., 1987). Figure 1 supplies an overwiev of the physical content of the model. The non-hydrostatic three-dimensional flow model uses the equation of motion in anelastic Boussinesq-approximation. The thermal model predicts temperature changes due to advective and turbulent heat transport, heat sources and radiative effects. At the present time, phase transitions are excluded. The hydrological model therefore reduces to a transport equation of the vapor phase. The air pollution model may be used if desired to predict the concentration of chemically inert pollutants due to advection and turbulent diffusion.

---

* Present adress: Deutscher Wetterdienst, Abt. F, 6050 Offenbach

K. Grefen and J. Löbel (eds.), Environmental Meteorology, 345–358.

The diagnostic parts of the atmospheric model determine the turbulent exchange coefficients for momentum, heat and matter and the radiative fluxes. Exchange coefficients are modelled by means of K-theory, using Blackadar's (1962) mixing length formula. The diffusive fluxes at the earth's surface are calculated by use of an external formulation of the Monin-Obukhov-laws proposed by Clarke et al. (1970). The radiation model (Zdunkowski et al., 1982) includes shortwave and infrared radiative transport in the presence of atmospheric gases and aerosols. Multiple scattering is accounted for also. To keep the computational effort small, semi-empirical expressions (Zdunkowski et al., 1975a,b) may be used instead of the complete radiation model.

At the present state of development, the soil model predicts the vertical temperature profile within the soil. Atmospheric and soil processes are coupled by the surface energy budget and an evaporation formula.

The input block provides all the information which cannot be recognized by the model itself such as the urban structure and the synoptic state.

The 'alternating direction implicit (ADI)' - procedure proposed by Douglas and Rachford (1956) is employed for the time integration of the prediction equations. The scheme makes use of a time splitting technique. The time derivatives are discretized by use of forward differences while centred differences are used for the diffusion terms. For the advective terms an upstream donor-cell scheme is employed in the flow model. In the transport equations and the thermal model, the advection scheme of Smolarkiewicz (1983) is used.

The flow model requires the solution of Poisson's equation for the dynamic pressure perturbations at each timestep. This is performed by an SOR-procedure.

The model has been developed in two versions. Model A refers to mesoscale γ which extends from 2.5 to 25 km. In this scale, surface structures may be properly described by use of the roughness length concept. Results of Model A may be interpreted as an image of the average conditions in the grid square. If detailed information is required, then the actual urban structure must be approximated in high resolution which is subject to the microscale model B.

2. RESULTS

2.1. Model A

To illustrate the different concepts, in the following figures a comparison of computed and observed diurnal variations of temperature and moisture for two stations in Munich is presented. The measurements were obtained from 'Stadtklima Bayern' data. This study (Mayer and Bründl, 1984) included continuous recordings of the meteorological conditions at several urban measuring sites.

In the first example (Figure 2) the agreement between theory and observation is excellent for both variables, temperature and humidity. One reason for this is that the station was placed in a suburban region and the measuring instruments were properly exposed to the flow. In contrast to this, the simulation for a station in the urban center (Figure 3) is not of the same quality. In this particular case, the instrument shelter was placed in a narrow courtyard characterized by very little ventilation and shadowing of the solar radiation by the surrounding buildings. These conditions are not representative of the model grid square which is always characterized by uniform roughness length and surface properties.

2.2. Model B

In order to obtain more accurate information, the high resolution model B must be used. This model approximates the actual urban build-up by rectangular blocks. In the next few figures, Model B results on neutral flow and pollutant dispersal for the configuration shown in Figure 4 will be discussed.

For the vertical cross-section which is marked by the dotted line in Figure 4, the projection of a simulated stationary flow field is displayed in Figure 5. In the luff of the building complex the vertical velocity is of the same order of magnitude as the corresponding horizontal component. Two vortices have formed, one within the cavity between the buildings, a less intense one behind the right building.

Similar results may, of course, be obtained from two-dimensional simulations. In the horizontal projection of the flow field pertaining to a height of 2 m (Figure 6), however, several effects occur which can be modelled only three-dimensionally (e.g. the channeling of the flow between the buildings and behind the right block as well as the appreciable accelerations near the front corners).

Next, the stationary distribution of a polluting gas as resulting from the arrangement of pollution sources indicated in Figure 4 will be discussed. The location and intensity of sources simulate carbon monoxide release by a stream of automobiles.

In the vertical cross section (Figure 7) the maximum concentration is found in the left half of the cavity. The reason for this is the vortex transporting the pollution upward near the wall. The concentration decreases rather quickly with height due to downwind transport and due to mixing with unpolluted air from the left. These results are in reasonable agreement with available observations.

The corresponding horizontal cross section for $z = 2$ m shows maximum concentrations near the right building caused by the additional pollutant transport through the gaps at the left and front side of the building complex (Figure 8). Again, this can be simulated only in three dimensions.

In order to illustrate the importance of using an accurate advection scheme in numerical experiments of this type, the simulation has been repeated using a simple upstream scheme which is uncorrected for numerical diffusion instead of Smolarkiewicz's scheme. The 2 mg/m^3 - isoline in Figure 8 (dashed line) indicates the 'smoothing' effect of the numerical diffusion. The location of the maximum concentration is not only shifted by one grid cell but its value is also reduced by roughly 10 %.

Finally, the effects of stability on the pollutant transport shall be discussed briefly. For comparison purposes, an additional simulation has been carried out for the case of stable stratification representing a weak inversion. In Figure 9, the difference between the concentrations for the stable case and for neutral conditions are displayed. The increase of pollution for stable stratification is not uniformly spread over the entire model region but has a maximum in the lee of the single building, where the concentration is nearly twice as high as in the neutral case.

3. CONCLUDING REMARKS

Two versions of a three-dimensional urban climate model have been presented together with a brief discussion of model results. The mesoscale version A yields satisfactory results as long as the surface structure is of suburban character. For strongly structured and densely built-up terrain, the high resolution microscale model B must be used. Results on neutral flow and pollutant dispersal are realistic and plausible. The microscale model is not yet complete. The next step in the development will be the inclusion of the thermal model and the surface energy budget to enable the model to handle the feedback between the temperature field and the air flow.

4. ACKNOWLEDGEMENT

This work was sponsered by the German Department of Research and Technology, Bonn. The authors wish to express their gratitude to Dr. H. Mayer, Munich, for supplying the 'Stadtklima Bayern' data.

References

Blackadar, A.K.; 1962: 'The Vertical Distribution of Wind and Turbulent Exchange in Neutral Atmosphere.' J.Geophys.Res., 67, 3095-3102.

Clarke, R.H.; 1970: 'Recommended Methods for the Treatment of the Boundary Layer in Numerical Models.' Austr.Met.Magazine, 18, 51-73.

Douglas, J., Rachford, H.; 1956: 'On the Numerical Solution of Heat Conduction Problems in Two and Three Variables.' Trans.Amer.Math.Soc., 82, 421-439.

Eichhorn, J., Schrodin, R., Zdunkowski, W.; 1987: 'Three-dimensional Numerical Modelling of the Urban Climate.' Beitr.Phys.Atmos., submitted for publication.

Mayer, H., Bründl, W.; 1984: 'Stadtklima Bayern - ein Forschungsvorhaben zum Klima in der Stadt.' DMG-Mitteilungen, 2/84.

Smolarkiewicz, P.K.; 1983: 'A Simple Positive Definte Advection Scheme with Small Implicit Diffusion.' Monthly Weather Review, 111, 479-486.

Zdunkowski, W., Paegle, J., Fye, F.; 1975a: 'The Short-Term Influence of Atmospheric Carbon Dioxide on the Temperature Profile in the Boundary Layer. Pageoph, Vol. 113, Birkhäuser Verlag, Basel.

Zdunkowski, W., Paegle, J., Reilly, J.; 1975b: 'The Effect of Soil Moisture upon the Atmospheric and Soil Temperature near the Air-Soil Interface. Arch.Met.Geoph.Biokl., Ser. A., Bd. 24, H. 3, 245-268, Springer-Verlag.

Zdunkowski, W., Panhans, W.-G., Welch, R., Korb, G.; 1982: 'A radiation scheme for circulation and climate models. Beitr.Phys.Atmos., 55, 215-238.

FIGURE 1

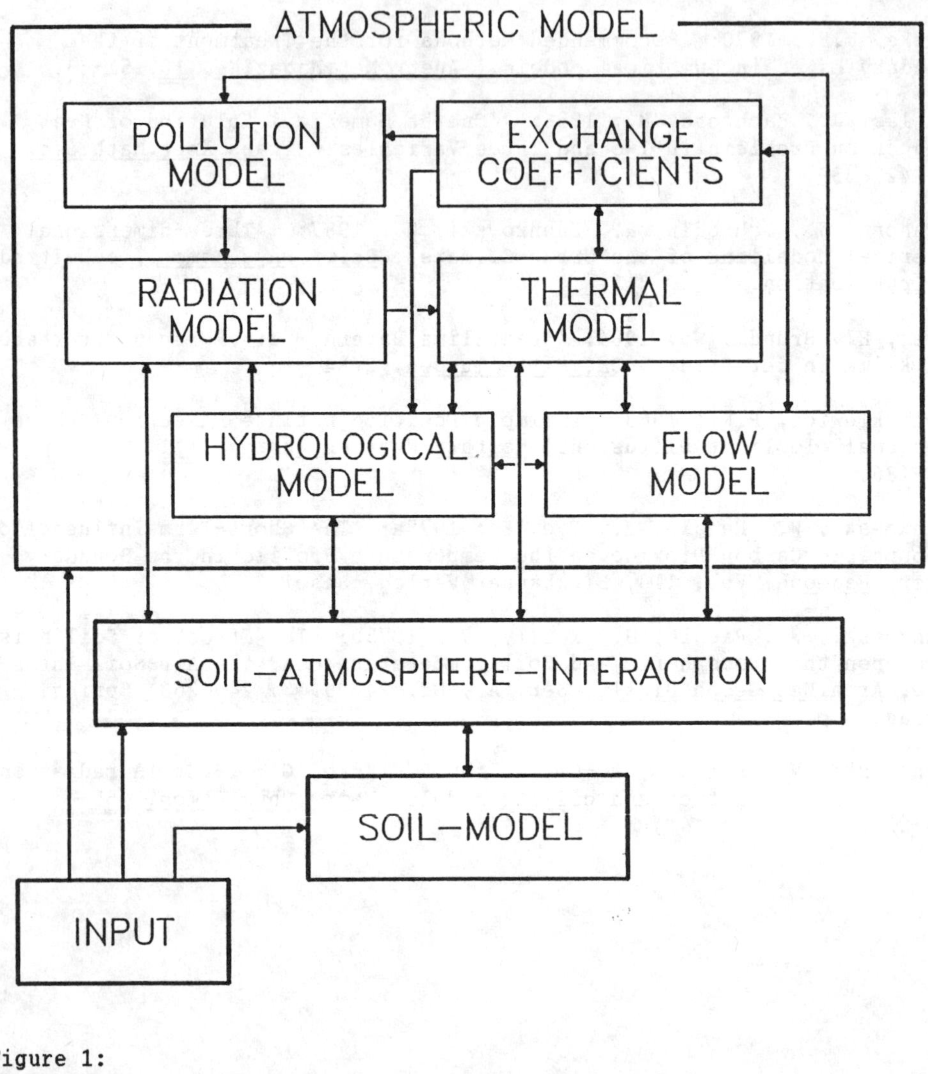

Figure 1:
Physical content of urban climate model. Arrows indicate transfer of information between model parts.

FIGURE 2

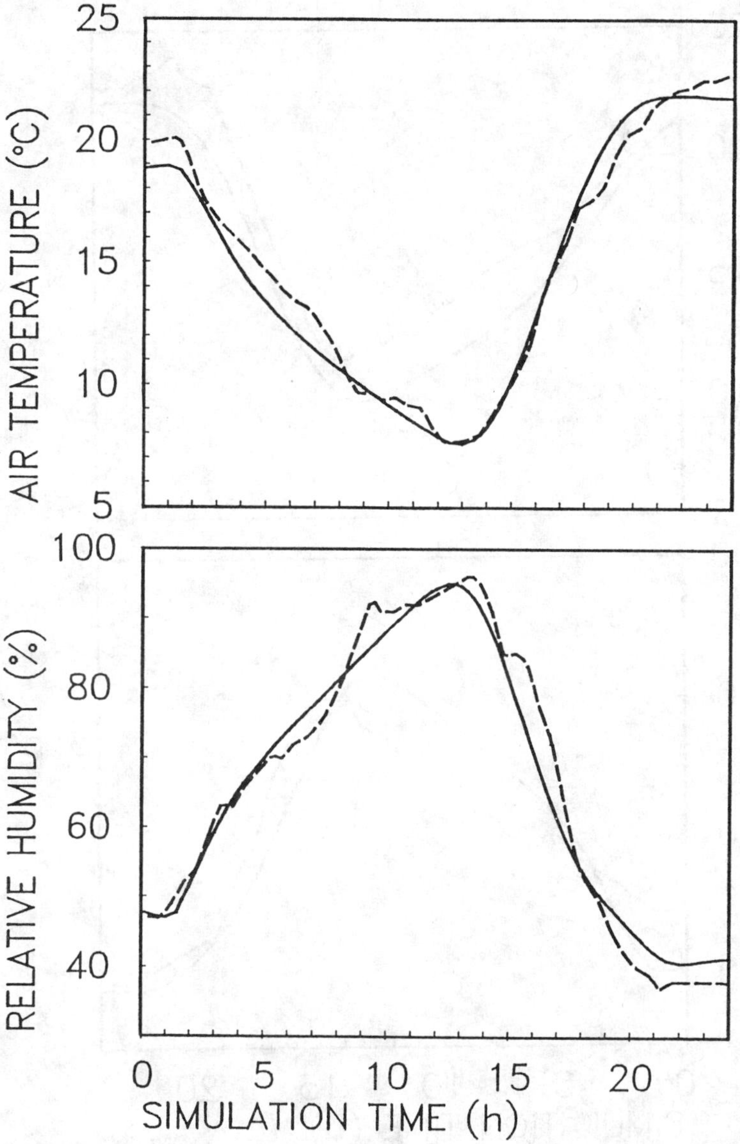

Figure 2:
Diurnal variations (Aug 20 4pm - Aug 21 1984 4pm) of air temperature and relative humidity at shelter height. Solid line: Model A results. Dashed Line: 'Stadtklima Bayern' observations. Suburban station.

FIGURE 3

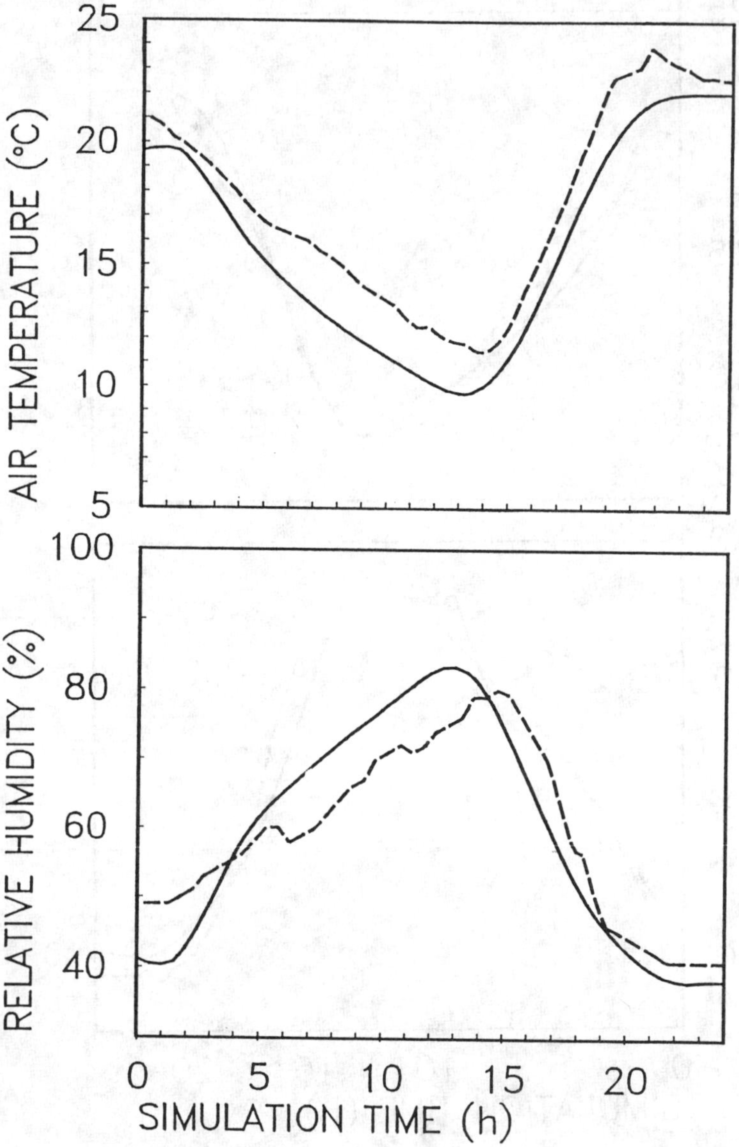

Figure 3:
As Figure 2, but for station in central city.

FIGURE 4

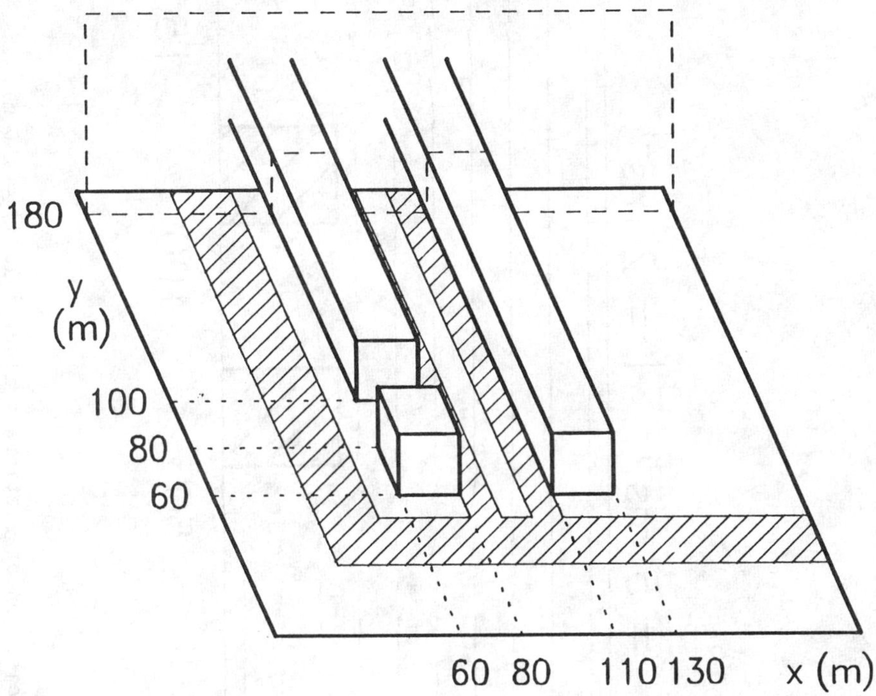

Figure 4:
Building configuration for Model B simulations. Building height: 12 m. Roughness length: 0.1 m (ground), 0.01 m (walls and roofs). Dashed line: x-z-cross section, see Figures 5 & 7. Hatched area: pollutant sources (intensity: 1 $mg/m^2/s$), see Figures 7 - 9.

FIGURE 5

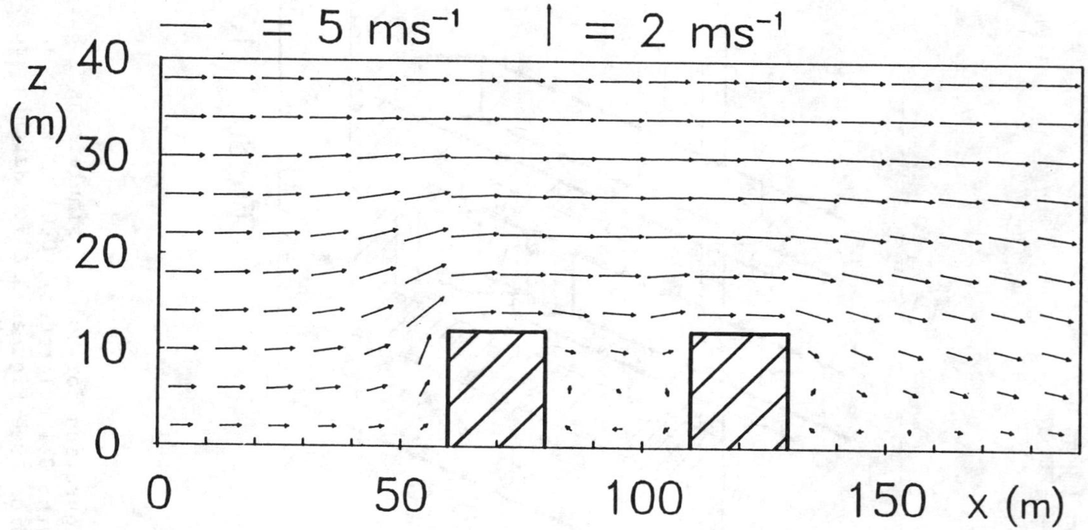

Figure 5:
Vertical projection of stationary flow field (y = 180 m).

FIGURE 6

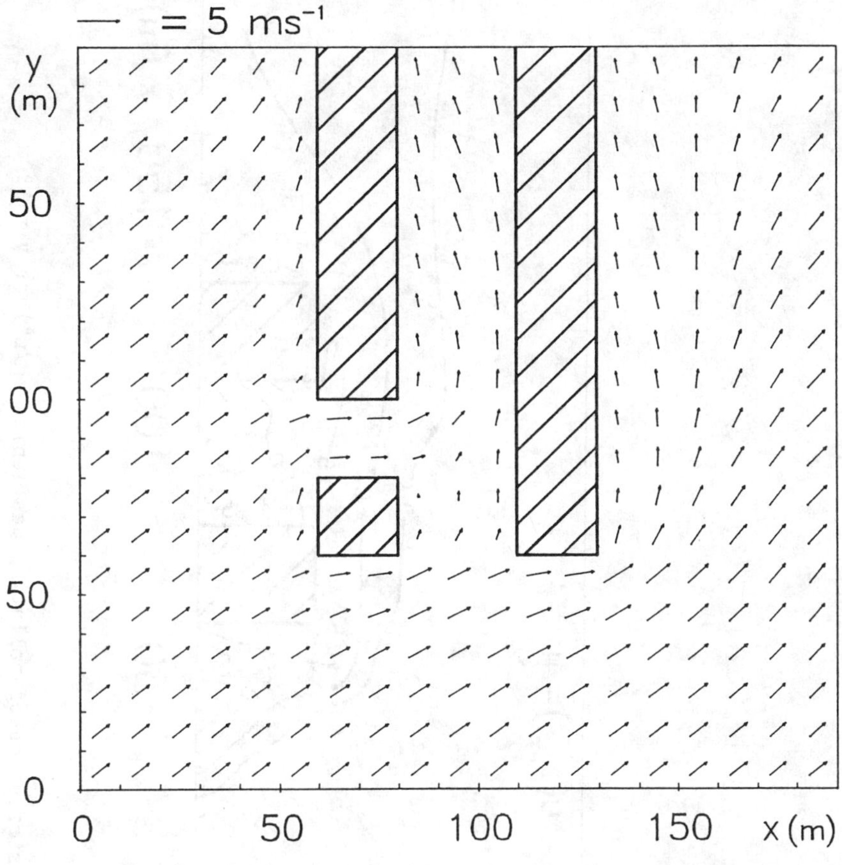

Figure 6:
Horizontal projection of stationary flow field (z = 2 m).

FIGURE 7

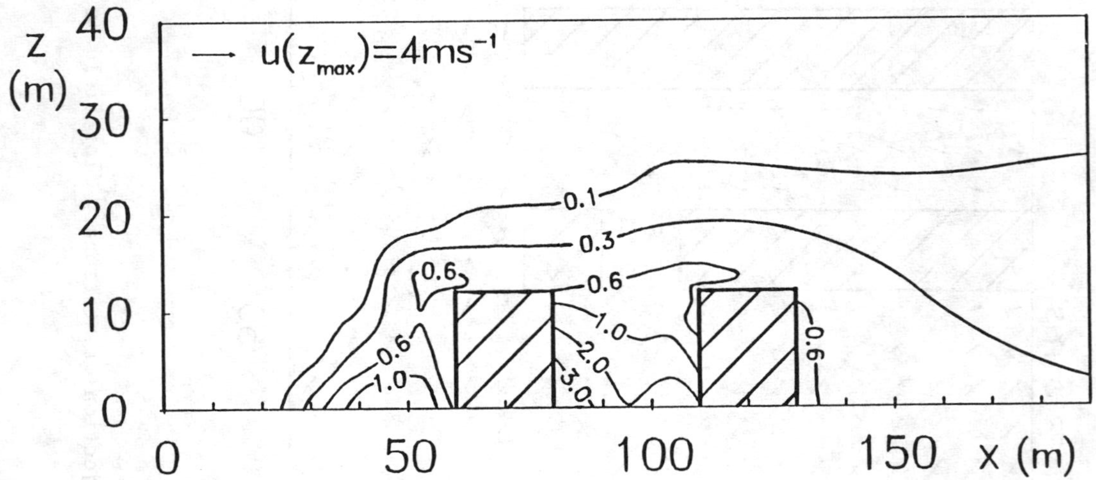

Figure 7:
Vertical distribution of polluting gas (units mg/m^3) at y = 180m.

FIGURE 8

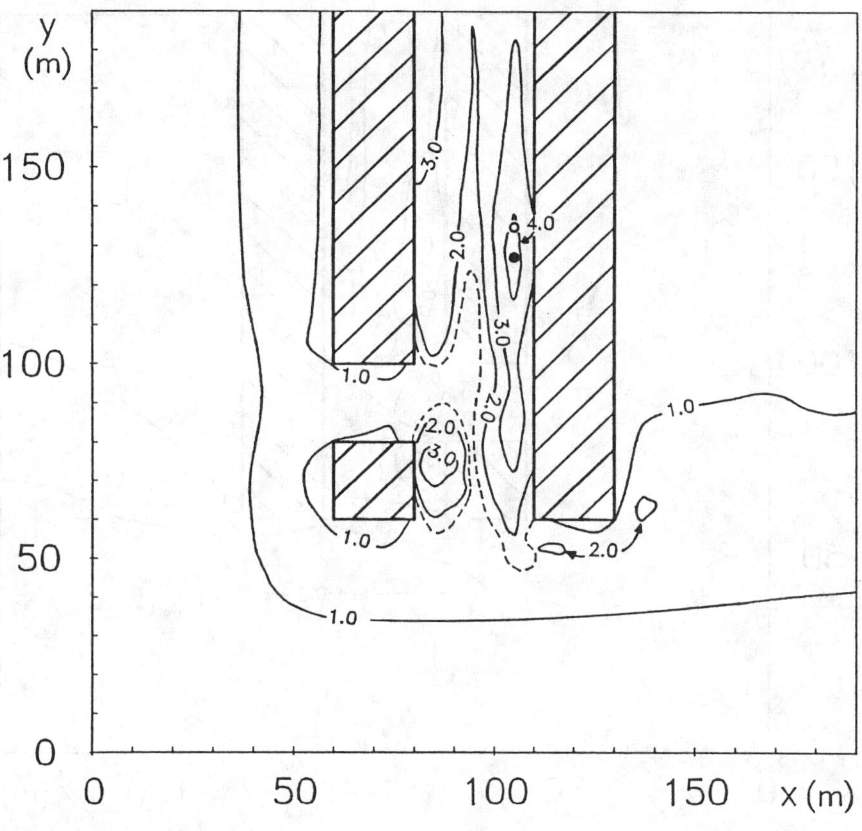

Figure 8:
Horizontal distribution of polluting gas (units mg/m^3) at z = 2m. Dashed line: 2 mg/m^3 - isoline obtained from uncorrected upstream advection scheme. ●, o: positions of maximum concentration obtained from Smolarkiewicz (4.3 mg/m^3) and upstream scheme (3.9 mg/m^3), respectively.

FIGURE 9

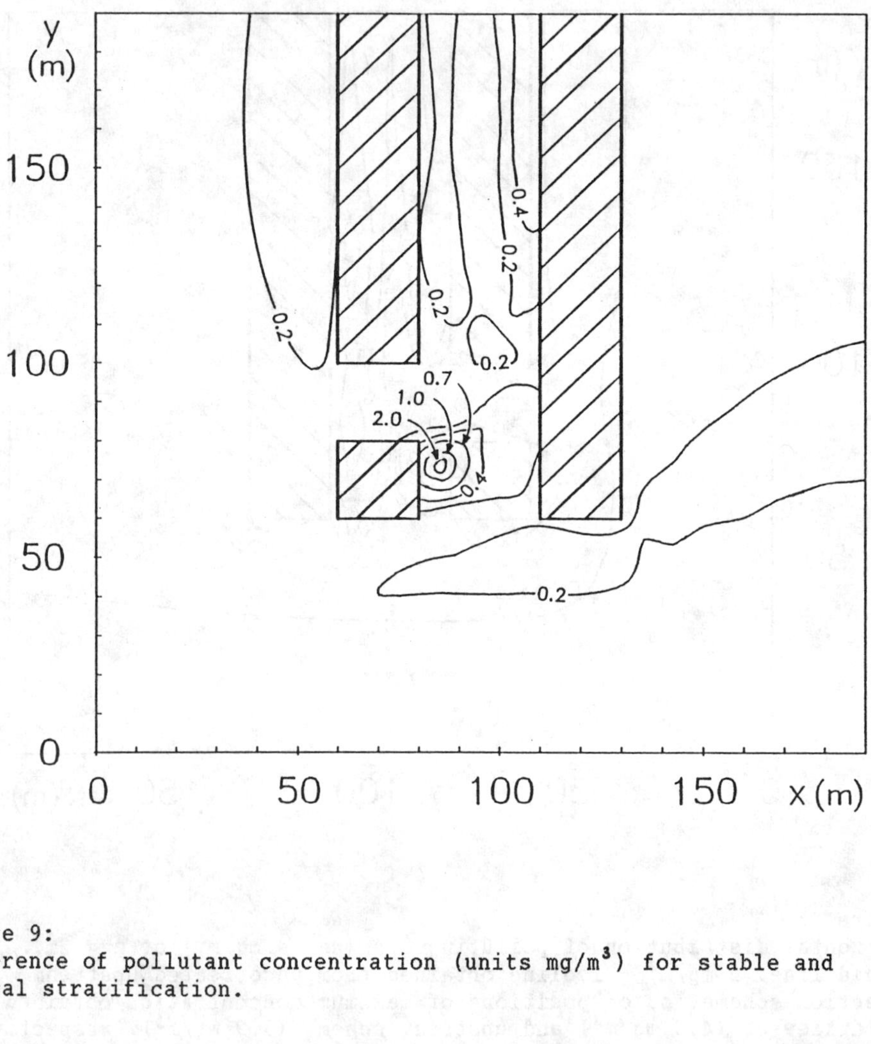

Figure 9:
Difference of pollutant concentration (units mg/m^3) for stable and neutral stratification.

GREEN IN URBAN CLIMATE

Dr. Fritz Wilmers
Dipl.-Met. Klaus-Dieter Scholz
Institute for Meteorology and Climatology
University of Hannover
Herrenhaeuser Strasse 2
D-3000 Hannover 21
F. R. G.

SUMMARY. Two negative aspects of urban climate are the urban heat island (= the overheating of the city) and air pollution, both of which can be diminished by areas of green vegetation. Only the former will be discussed here.

It is shown that the effect of diminishing the difference of temperature in its range and its reach increases with the size of a green area. But this effect on the wider environment should not be overrated. Even in the environment of very large green areas the effect can be noticed only within a relatively small distance. Convection reduces and suppresses wider effects.

The special climates of green areas, their effect on air temperature resp., were proved in measurements with different scales in Hannover <4,Beckröge,1980; 45,Wilmers,1972; 48,Wilmers,1978b; 49,Wilmers,1982> and as well in Bonn <37,Sperber,1974>. These show that even in small spaces between buildings green areas have favourable bioclimatic effects.

The ranges of temperature, that occured in the environment were examined by measuring the surface temperatures with handheld scanner. Some examples of measurements of infrared temperature in garden courts in Hannover make clear that vegetattion surfaces show considerably different temperatures than other surfaces, above all dry and inanimate ones <45,Wilmers,1972; 48,Wilmers,1978b; 51,Wilmers,1986>. In this treatise we mention some of the values of infrared radiation temperatures, which may not be absolutely precise, because the grades of emissivity of some surfaces are not exactly known. Those values are near 1 for all natural surfaces, and small deviations from the mean extreme temperature differences. However we have measured great dissimilarities in temperature. There were ranges of surface temperature, during periods of intense solar insolation, of more than 20 Kelvin, dependent on the types of surface.

At night the differences are rather small, but the parts, which

K. Grefen and J. Löbel (eds.), Environmental Meteorology, 359–379.

got less direct insolation, remain cooler, except for those parts which are protected against effective outgoing radiation.

The demonstrated measurements show important dissimilarities in the thermal behaviour of our surroundings. They were done in small areas but nevertheless they have a good indication of the different thermal zones of the urban environment.

1. INTRODUCTION

In principle, the effect of vegetation on the climate of agglomeration areas is well known, especially that on temperature <24,Kessler,1971>. But the importance of "green" matter and "green areas" for the climate of cities is more comprehensive. In a diagram, <33,Oke,1972> showed that the maximum and minimum temperatures sink with the increase of the percentage of green in a settlement, and, proceeding from a synopsis by <33,Oke,1972>, <12,Fezer,1976> proved that the effects of the urban heat island increase with the size of the settlement, but that, on the other hand, this effect can be reduced by building towns with green areas and with buildings in an optimal arrangement.

As easy as it might be to estimate the special climate of a green area on itself, so it is difficult to prove that a green area changes the climate of its wider environment <13,Finke,1976>. This subject was treated by <12,Fezer,1976; 37,Sperber,1974>, too, and completed by <39,v. Stülpnagel,1980>. Here, also, it is shown that the effect in its range and its reach increases according to the size of the green area. But this effect on the wider environment should not be overrated. Even in the environment of very large green areas the effect can be noticed only within a relatively small distance. Convection reduces and suppresses wider effects <6,Bründl,Mayer and Baumgartner,1986>.

For the climate of built-up areas the open spaces are most important whether covered with vegetation or not <13,Finke,1976>. Green areas are very effective for urban climates. The percentage of green coverage often varies, but open spaces have in any case some definitive effects on the urban climate.

Open spaces are characterized by the lack of built-up structures. According to the size of the area, the inventar and the usage, there are eight different grades in the organization of open spaces <13,Finke,1976; 16,Gesamthochschule,1974; 47,Wilmers,1978a>, TAB.1 .

The different use and the width of the open spaces are responsible for the kind of surface and the texture and usage of the open areas, especially the green part. Therefore the effects on the different aspects of urban climate vary widely. A glance at the feature of open spaces is to be found in <47,Wilmers,1978a>.

The surface is the boundary between the air above and the soil underneath. The properties of the surface, especially the roughness, are also responsible for the deviations and transductions of energetic and dynamic processes in the boundary layer of the atmosphere. Furthermore many of such open spaces include relief elements such as walls and dykes, hills and valleys, buildings and bridges etc. Open spaces are not only situated in planes but also in areas with more relief energy. then the exposition and inclination may have important effects on the topoclimate and also on the possibilities of the effectiveness on the urban climate.

The different categories of open spaces have very different properties in the water balance, what is responsible for dissimilarities in the energy balance. Also the effects on the wind profile and the turbulent convection form certain groups of surface types.

2. PHYSICAL PROPERTIES OF OPEN SPACES

2.1 ENERGY BALANCE

The interaction between surface and atmosphere in the different open spaces, as elsewhere, is dependent on the different surface roughness, structures and percentage of green vegetation. Therefore the manifold physical properties of those different parts of the surface should be characterized.

The most important variations occur with the terms of the energy balance beginning with the radiation fluxes <5,Berger-Landefeldt,1967; 9,Eils,1972; 40,Tajchman,1967>.

The equation of energy balance (1)

$$Q + S + H + E + P = 0 \qquad (1)$$

with Q := Radiation balance
S := Flux of soil energy
H := Thermal heat flux
E := Latent heat flux, evapotranspiration
P := Energy of photosynthesis

is well known <15,Geiger,1961v1965>.

The actual parts of the components in the energy balance are to be determined by measurements or by calculation. The first step is to decide the relationship of the magnitude of the terms to each other. Great differences occur between surfaces covered with vegetation and those without <18,Höschele and Schmidt,1974>. <46,Wilmers,1976> compared the combination of the terms for surfaces with different vegetation cover and different water saturation with maximum radiation balance.

With dense plant stands, whether with tall or with low growth, the term of soil heat flux is the least one. It is usually less than 10 %. The diversion of available energy into thermal heat flux and latent heat flux is determined by the conditions of the available humidity. With optimal water conditions the evapotranspiration can grow over 80 %. But if the actual evapotranspiration is less than the potential one the thermal heat flux increases and may be dominant. This effect can be seen on surfaces without any vegetation. In these examples without evaporation the soil heat flux and the thermal heat flux share the energy <29,Miller,1965; 31,Monteith and Szeicz,1961> according to the thermal conductivity of the soil. The thermal conductivity depends on the physical properties of the underground. As well as the dry substance, it is dependent on the parts of air and liquid water content in the soil. Values can be found at <15,Geiger,1961v1965; 28,Miess,1968>.

Dense plant stands insulate very strongly as their physical density is very small. <1,Baumgartner,1952; 2,Baumgartner,1956> computed the whole vegetation matter of a young spruce forest as a layer of about 2 cm opposed to the 6 m elevation of the trees (biomass 189 m**3/ha). According to <11,Ellenberg,Mayer and Schauermann,1986> the biomass of a 125 years old beech forest with a height of 27.2 m in the Solling mountains was summarized to 293.102 kg/ha. That means the volume of a dense forest is by biomass less than 1 %.

2.2 EFFECTS OF VEGETATION SURFACE

Surfaces covered with vegetation are most effective in the transformation of energy, as the surface of the foliage covers a greater area than the simple surface. The LEAF AREA INDICES (LAI) lay at about 5 times the surface. Beech trees in thr Solling had a LAI of about 5.74 <11,Ellenberg,Mayer and Schauermann,1986>. With single standing trees LAIs of 7 were measured <32,Möller,1945> and maximal values of 10 <42,Vareschi,1951>. Some authors showed the correlation between light, energy and CO2 consumption and an increasing LAI <7,Chang,1968; 36,Rosenberg,1978>.

Also Voeikov, in the last century (cit. a. <15,Geiger,1961v1965>), declared the upper third of the canopy as the active layer. Therefore the diurnal march of all components of radiation and energy balance is much higher in the upper part of the canopy than below the canopy layer of a dense plant stand. So <10,Van Eimern,1986> showed the increase of temperature in a beech forest going from the canopy into the stand. But on hot days with strong insolation into a wood without leaf the temperature increases from the surface of the ground <24,Kessler,1971; 26,Lorenz,1969>.

During the night the energy of the outgoing net radiation from a vegetation surface is fed from the thermal heat flux and the latent heat flux. Therefore above a vegetation surface the temperature becomes lower than elsewhere. Also the fall of dew is higher above the canopy, if there is any <30,Monteith,1976>.

2.3 EFFECTS OF GREEN COVER WITH WIND

The enlargement of the surface according to the LAI is also effective with ventilation. Each single leaf gets an internal surface layer which is especially effected by its internal boundary layer <30,Monteith,1976>. The roughness parameter zo and the zero-plain-displacement layer d increase according to the height of a plant stand and with the density of roughness elements above an open area <3,Baumgartner,Mayer and Metz,1977; 15,Geiger,1961v1965; 38,Stanhill,1969>. <3,Baumgartner,Mayer and Metz,1977> computed world maps of surface roughness according to the equation (2) log zo = 1.03 log h − 0.86 (2).

According to it zo lays at about 1/10 to 1/30 of the height of the roughness elements. Effect of green plants with wind depends on their structure as well as on their roughness <17,Grace,1977; 20,Houghton and Carruthers,1976>. As shown above the classification of surface roughness poses some problems. The following model may help.

According to <14,Frost and Nowak,1977> (cit. <22,Justus,1985> in <20,Houghton and Carruthers,1976>) flat terrain has the following characteristics (<20,Houghton and Carruthers,1976>, P. 924): "(1) elevation differences between the site and surrounding terrain within a radius of 12 km no larger than 60 m; (2) all hill height-to-width (length) ratios h/l < 0.02 within 4 km of the site."

The relation of zo to hc is related by (3).

zo = 0.056 hc ** 1.37 (** := exponent) (3)

with:

hc := average height of certain type of surface feature (e. g. trees, shrubs, buildings) densely spread over the terrain in a random fashion hc and zo in meters <14,Frost and Nowak,1977>.

With equation (3) the values of zo become greater than with equation (2) if the height of obstacle is more than 15 m. If it is smaller, the values of zo are less than those with equation (2).

<34,Rauner,1974> gave some physical characteristics of different plant stands especially the vertical distribution of leaf and non-leaf surfaces with height showing that there are three main types of phytoclimate, the inversion type , convective type , and the mixed type , as did <1,Baumgartner,1952> and <8,Dirmhirn,1964>. They are according to the vertical distribution of phytomass and LAI.

The inversion type means plant stands with a dense canopy but an open stem area such as a beech forest or a potato field. The main radiant and turbulent energy exchange takes place in the upper canopy.

The convective type has the maximum phytomass and LAI in the centre of its height, such as low dense stand of young spruce. Therefore the ratio of radiant energy exchange diminishes to about 50 % at the centre of the height, but the turbulent energy exchange increases near the ground.

The mixed type has the maximum phytomass and LAI in the part near the ground such as shrubs or a flowering meadow. The energy exchange lies here above the dense part of the vegetation stand.

In dimensionless numbers (better dimension 1) LAI/z for short plants have higher densities of phytomasses than forests <34,Rauner,1974>. That means that shrubs and grasslands can have the same or even greater evaporation than trees. Nevertheless their effects on wind are not the same.

2.3.2 GREEN OBJECTS AND THE AIR FLOW

Rather dense trees, shrubs and woods with dense outskirts are important elements of roughness in parks and other open areas. They are also able to lead the stream of air above the surface especially during stable temperature conditions. Therefore they are suitable for forming pathways for fresh air into towns. <35,Richter,1981> proposed systems of green areas for the continous circulation of fresh air within urban areas.

3. MEASUREMENTS OF SURFACE TEMPERATURES

The ranges of temperature really occuring can be examined by measuring the surface temperatures with an infrared thermometer <26,Lorenz,1969> operating from an airborne platform <43,Wienert,1980; 52,Wilmers,Wienert,1981> or with a handheld scanner <24,Kessler,1971; 51,Wilmers,1986>.

Some examples of measurements of infrared temperature in garden courts in Hannover point out that vegetattion surfaces show considerably different temperatures more than other surfaces, above all dry and inanimate ones. The above mentioned infrared radiation temperatures are not absolutely precise numbers, because the grades of emissivity of some surfaces are not exactly known <26,Lorenz,1969; 41,Tetzlaff,1974; 43,Wienert,1980; 52,Wilmers,Wienert,1981>. Those values are near 1 for all natural surfaces, but small deviations mean great temperature differences. So we have measured important diversities in temperature. There were maximum ranges, during periods of intense insolation, of more than 20 Kelvin.

At night the differences are rather small, but the parts which got less direct solar insolation, remain cooler, except those parts which are protected against effective outgoing radiation.

The demonstrated measurements show important dissimilarities in the thermal behaviour of our surroundings. They were carried out in small areas but nevertheless they give a good summary of the different thermal zones of the urban environment.

The measurements of the surface-temperature were made by an AGA shortwave Infrared-Scanner Type THV 782 combined with a GESOTEC computer BI 800.

Depending on the exposition of the walls and the sizes of the courtyards there were sites with sun or shade during nearly the whole day, and other places with direct insolation only in the morning or in the afternoon <45,Wilmers,1972; 48,Wilmers,1978b>.

In combination with the vegetation - which occured as climbing vines on walls, different kinds of vegetation above the ground, shrubs, trees, and vines on the pergolas respectively - plus water ponds of different sizes - distinct interdiurnal marches of surface temperature are evaluated. The curves vary widely in time and magnitude of peaks dependent on the energy balance of the surrounding of each site <51,Wilmers,1986>.

The measurements took place on September 28th and 29th 1984 during a synoptic situation with bright sunshine - radiation weather type. The air temperatures were measured 8 and 19 oC, surface temperatures between about 5 and more than 50 oC. The radiation temperature of the sky - partly to be seen in the figures - was not evaluated.

4. SOME EXAMPLES OF INFRARED MEASUREMENTS

4.1 MEASUREMENTS ON TWO WALLS WHICH ARE EXPOSED TO THE SOUTH AND WEST

Two walls made of concrete - one exposed to the south, the other one to the west - show the different diurnal march of temperature for the expositions. Both of them are partly covered with an evergreen vine - LONICERA HENRYI - and with fern in front, FIG. 1 + 2 .

In the morning at 6:35 h a.m. on the south wall and at 6:42 h a.m. on the west wall there are the same ranges of temperature. The walls are relatively warm, the vegetation temperature is below air temperature, plants on earth have minimum of temperature.

In the early afternoon at 1:22 h p.m. the south wall with bright incident sun radiation is rather hot, whereas the parts which are shaded by vegetation as well as by timber of the pergola are cooler, the surfaces of vegetation cover in the sun are warmer thann the air temperature, but 10 K below the temperature of the wall, FIG. 3 . The shaded westwall at 1:26 h p.m. is about 20 K cooler than the southwall, FIG. 4 . The plants on this wall are cooler, too, ferns in the sun are warmer.

In the evening at 9:41 h p.m. the southwall is a little bit cooler than the westwall at 9:48 h p.m., FIG. 5 + 6 . The plants are cooler than the walls; their temperature is about air temperature.

4.2 MEASUREMENTS IN A LAWNGARDEN

Some measurements in a smaller courtyard will demonstrate the importance of direct insolation and of shading, too. By infrared scans taken during night and day the diurnal march of surface temperatures on different parts of this small site is shown.

There is also a westward oriented wall without any vegetation with a passage into a sunlit courtyard, FIG. 7 . A pergola is situated on the left hand side, an appletree on the right hand side, the soil surface is covered with lawn. The site is strongly shaded by the tree and the surrounding walls.

FIG. 8 shows the diversion of the infrared-temperature during the night at 2:16 h a.m. The tree strongly influences the temperature in the yard. The angle at the base of the wall has the highest temperatures of all, as it is protected against outgoing net radiation. The temperature of the wall decreases with the distance to the base, as does the temperature of the lawn. In the parts of the foreground with the maximum of effective outgoing net radiation are the deepest surface temperatures.

Later on in the morning at 6:37 h a.m. the diversion of surface temperature has hardly changed, FIG. 9 . The influence of the coverage by the tree is easy to be seen.

At 10:35 h a.m. in the morning the maximum of surface temperature is found in the appletree and in the passage through the wall. At 2:36 h p.m. in the afternoon the tree and the partly sunned timber of the pergola on the left side have a higher temperature, the lawn has a lower temperature than the air.

Later in the afternoon at 4:14 h p.m. the tree, the lawn and the pergola are partly exposed to the sun and therefore remain warm, but the wall has a low temperature.

In the evening at 10:40 h p.m. the pillars of the pergola as well as the sheltered walls remain warm. The lawn and timber of the pergola have the lowest temperature, FIG. 10 .

4.3 RESULT OF INFRARED MEASUREMENTS

The temperatures of the surfaces fluctuate more strongly than the air temperature. It is dependent on the energy balance, especially on the thermal conductivity of the underground and on the surface matter.

On direct sun the highest values of surface temperature occur. The maximum values are measured on sunlit walls. There the most significant differences from sunned to shaded parts are to be found. At night the differences are rather small, but parts which are seldom or never in the sun remain cooler, except those parts which are protected against outgoing net radiation. In those places the nightly temperatures are higher than elsewhere.

5. DISCUSSION

According to <15,Geiger,1961v1965> autochthon microclimate is optimal with weak winds but with strong solar radiation <23,Kaiser,1958>. In such a case the main parameters of the microclimatic differences in urban climate are the exposition to direct solar radiation, its duration and the matter of the underground and the surrounding walls <15,Geiger,1961v1965>. Vegetation plays an important role by its special properties with energy balance and wind. <53,Wilmers,Wittler,1981> showed the relationship of foliage temperature on the status of transpiration. <21,v. Hoyningen-Huene> also measured the leaf temperature during the day in connection with the energy balance. Depended on their evapotranspiration, vegetation cover and water surfaces diminish the peaks of temperature during the day.

The bioclimatic effects on human beings are combined with some different bioclimatic complexes. One of them, of most importance to human comfort, is the thermal complex which can be demonstrated by the enrgy balance <50,Wilmers,1985>. Therefore the knowledge of climatic conditions in urban areas which rule the energy balance - such as the surface temperature - enables the possible to lay out some sites which are suitable in the morning or in the afternoon and other ones which enable one to stay there even during the hottest hours of a summer day.

The usage of green elements in town planning is a very important planning device to optimize urban climate, especially in the temperature climates but also in hot climate with greenary is given in <50,Wilmers,1985>.

Ackknowledgements. The authors are very much indebted to Prof. G. Nagel, director of the "Institute of Greenplanning and Garden Architecture" for his great interest and his permission to the measurements in the garden courts. Grateful thanks go to Mrs. Wilmers for their great help during the measurements and the latter evaluation; also to Mrs. R. Lorenz for preparing the figures.

6. LITERATURE

<1> Baumgartner, A. 1952: Untersuchungen zum Wärme- und Wasserhaushalt junger Fichtenbestände. III. Die Strahlungsbilanz einer Fichtendickung, Forstwiss. Cbl. 71. 337-349.

<2> Baumgartner, A. 1956: Untersuchungen über den Wärme- und Wasserhaushalt eines jungen Waldes. Ber. DWD 5. Nr. 28.

<3> Baumgartner, A., H. Mayer, W. Metz, 1977: Weltweite Verteilung des Rauhigkeitsparameters zo mit Anwendung auf die Energiedissipation an der Erdoberfläche. Meteor. Rdsch. 30. 43-48.

<4> Beckröge, W. 1980: Auswirkungen von Parkanlagen auf die städtische Wärmeinsel. Dipl-Arb. unpub. Hannover.

<5> Berger-Landefeldt, U. 1967: Der Energieumsatz am Standort. Botan. Jahrb. 86. 402-448.

<6> Bründl, W., H. Mayer, A. Baumgartner, 1986: Untersuchung des Einflusses von Bebauung und Bewuchs auf das Klima und die lufthygienischen Verhältnisse in bayerischen Großstädten. Stadtklima Bayern. Abschl.ber. "Klimamesssungen München". LBAM. München.

<7> Chang, J. H. 1968: Climate and agriculture - an ecological survey. Aldine. Chicago.

<8> Dirmhirn, I. 1964: Das Strahlungsfeld im Lebensraum. Akadem. Verlagsges. Frankfurt a. M.

<9> Eils, W. 1972: Der Wärmehaushalt einer Wiese in Abhängigkeit von unterschiedlicher Bewuchshöhe. Ber. Inst. Met. u. Klimat. TU Hannover. Nr. 7.

<10> Van Eimern, J. 1986: Oberflächen- und Lufttemperatur in einem Buchenwald zu verschiedenen Jahreszeiten. Meteor. Rdsch. 39. 13-24.

<11> Ellenberg, H., R. Mayer, J. Schauermann. 1986: Ökosystemforschung. Ergebnisse des Sollingprojekts 1966-1986. Ulmer. Stuttgart.

<12> Fezer, F. 1976: Wieweit verbessern Grünflächen das Siedlungsklima? Ruperto Carola. 57. 77-79.

<13> Finke, L. 1976: Zuordnung und Mischung von bebauten und begrünten

Flächen. Schrift.R. Städtebaul. Forsch. 03.044. BMRBS. Bonn.

<14> Frost, W., D. K. Nowak. 1977: Handbook for WTG siting relative to small scale terrain features. ERDA report on Contract E(45-1)-2443. Washington. D. C.: Energy Research and Development Administration.

<15> Geiger, R. 1961/1965: Das Klima der bodennahen Luftschicht. 4. Ed., Braunschweig. The climate near the ground. 611 pp. Harvard University Press. Cambridge. Mass.

<16> GHS Kassel. 1974: OEG. Projekt F" Freiraumplanung Bettenhausen. Kassel.

<17> Grace, J. 1977: Plant response to wind. Academic Press. London.

<18> Höschele, K., H. Schmidt. 1974: Klimatische Wirkungen einer Dachbegrünung. Garten und Landschaft. 84. 334-337.

<19> Houghton, D. D. (ed.) 1985: Handbook of applied meteorology. Wiley. New York.

<20> Houghton, D. D., N. B. Carruthers. 1976: Wind forces on buildings and structures. Arnold. London.

<21> Hoyningen-Huene, J. v.: Mikrometeorologische Untersuchungen zur Evapotranspiration von bewässerten Pflanzenbeständen. Ber. Inst. Meteor. u. Klimat. Univ. Hannover Nr. 19.

<22> Justus, C. G. 1985: Wind energy. Chapt. 33. In: Houghton, D. D. (Ed.) 1985: Handbook of applied meteorology. Wiley. New York.

<23> Kaiser, H. 1958: Über den "Strahlungstyp" und den "Windtyp" des Mikroklimas. Meteor. Rdsch. 11. 162-164.

<24> Kessler, A. 1971: Über den Tagesgang von Oberflächentemperaturen in der Bonner Innenstadt an einem sommerlichen Strahlungstage. Erdkunde 15. 13-20.

<25> Kiese, O. 1972: Bestandsmeteorologische Untersuchungen zur Bestimmung des Wärmehaushalts eines Buchenwaldes. Ber. Inst. Met. u. Klimat. TU Hannover Nr. 6.

<26> Lorenz, D. 1969: Temperature measurements of natural surfaces using infrared radiometers. Appl. Optics 7. 1705-1718.

<27> Mahringer, W. 1962: Das Temperaturregime einer Betonoberfläche. Arch. Met. Geophys. Bioklimat. B. 11. 533-559.

<28> Miess, M. 1968: Vergleichende Darstellung von meteorologischen Meßergebnissen und Wärmehaushaltsuntersuchungen an drei unterschiedlichen Standorten in Norddeutschland. Ber. Inst. Meteor. u. Klimat. TU Hannover. Nr.2.

<29> Miller, D. H. 1965: The heat and water budget of the earth's surface. Advances in Geophys. Vol. 11. 176-302. Academic Press. London & New York.

<30> Monteith, J. L. 1976: Vegetation and the atmosphere. Vol. 2. Academic Press. London & New York.

<31> Monteith, J. L., G. Szeicz. 1961: The radiation balance of bare soil and vegetation. Quart. J. Royal Meteor. Soc. 87. 159-170.

<32> Möller, C. M. 1945: Untersuchungen über Laubmenge, Laubverlust und Stoffproduktion des Waldes. Forstl. Forsogswes. Danmark. 17. 1-287. Kopenhagen.

<33> Oke, T. R. 1972: Evapotranspiration in urban areas and its implication for urban climate planning. Preprints. CIB Colloq. Stockholm 4.-6. 9. 1972.

<34> Rauner, J. L. 1974: Some aspects of physical phytoclimatology. In: Proc. Symposium on physical and dynamic climatology. Leningrad. August 1971. WMO No. 347. 28-41. Leningrad.

<35> Richter, G. 1981: Handbuch Stadtgrün - Landschaftsarchitektur im städtischen Freiraum. BLV. München.

<36> Rosenberg, N. J. 1978: Microclimate: The biological environment. Wiley. New York.

<37> Sperber, H. 1974: Mikroklimatisch-ökologische Untersuchungen an Grünanlagen in Bonn. Diss. Landw. Fak. Univ. Bonn.

<38> Stanhill, G. A. 1969: A simple instrument for the field measurement of turbulent diffusion flux. J. Appl. Meteor. 8. 509-513.

<39> v. Stülpnagel, A. 1980: Planungsrelevante Aspekte des Stadtklimas - Literaturanalyse. Diplomarb. unpub. Univ. Hannover.

<40> Tajchman, S. 1967: Energie- und Wasserhaushalt verschiedener Pflanzenbestände bei München. Wiss. Mitt. Nr. 12. Meteor. Institut Univ. München.

<41> Tetzlaff, G. 1974: Der Wärmehaushalt in der zentralen Sahara. Ber. Inst. Meteor. u. Klimat. Univ. Hannover. Nr. 13.

<42> Vareschi, V. 1951: Zur Frage der Oberflächenentwicklung von Pflanzengesellschaften der Alpen und Subtropen. Planta 40. 1.

<43> Wienert, U. 1980: Versuch einer quantitativen Oberflächentemperaturanalyse von Flugzeugscanneraufnahmen im thermischen Infrarot. Diplomarb. unpub. Univ. Hannover.

<44> Wiesner, C. J. 1970: Hydrometeorology. Chapman & Hall. London.

<45> Wilmers, F. 1972: Temperaturstudien in Gartenhöfen. Das Gartenamt. 21. 677-681.

<46> Wilmers, F. 1976: Klima und Vegetation im Gelände. Probleme der Geländeklimatologie. Naturwiss. Rdsch. 29. 118-123.

<47> Wilmers, F. 1978a: Grünflächen-Meteorologie. Akumet. Koll. Deutsch. Meteorol. Ges. Angewandte Stadtklimatologie am 19. 1. 1978. 7.1-7.6.

<48> Wilmers, F. 1978b: Temperaturen in und an einem künstlichen Teich. Verhandl. Gesellschaft f. Ökologie. 7. 413-426.

<49> Wilmers, F. 1982: Small scale thermal effects of "green" and "shade" in an urban environment and their bioclimatic significance. In: Symposium on building climatology, Sept. 20-24. 1982. Proc. P.2. Moscow. CIB 1982. 624-643.

<50> Wilmers, F. 1985: Ansprüche des Menschen an seine KLima-Umwelt. Landsch. u. Stadt. 17. 30-42.

<51> Wilmers, F. 1986: Vegetation in architectural courtyards affecting human bioclimate recognized by IR-scanning of surface temperature. Internat. symposium on urban and local climatology Freiburg. February 1986. Freiburger Geogr. H. 26. 155-164.

<52> Wilmers, F., U. Wienert, 1981: Calibration of the Heat Capacity Mapping Radiometer in the thermal infrared at Lake Steinhude in June 1979. Ber. unpub. Inst. f. Meteor. u. Klimat. Univ. Hannover. 1-60.

<53> Wilmers, F., P. Wittler, 1981: Leaf temperatures in a glasshouse. In: Abstr. Vol. 9th Internat. Congress of Biometeorology. Osnabrück. 187-188.

TABLE 1: ORGANISATION OF OPEN SPACES

1. near dwelling areas: home gardens, play grounds, balconies, roof gardens
2. open spaces between buildings: open and enclosed
3. traffic areas: road areas, traffic areas, foot paths, roads
4. open spaces between quarteers: street squares, green areas
5. areas of suburbs and rest spaces: waste land, fallow land
6. open spaces due to the suburbs or quarteers: play grounds and sporting areas, parks, cemetries, garden allotments
7. open spaces - recreation areas within easy reach: urban woods, river banks, meadows, farm land
8. open spaces of the countryside: agricultural land, fields, meadows, fallow land, marsh and moor land, heath land, woods, lakes, rivers.

FIG. 1. Southward exposed wall of concrete, on the right hand covered by an evergreen vine Lonicera Henryi, with fern in front of it, sometimes shaded by a pergola.

FIG. 2. IR-temperature of westward exposed wall of concrete, like Fig. 1. covered with vine and fern in front of it.

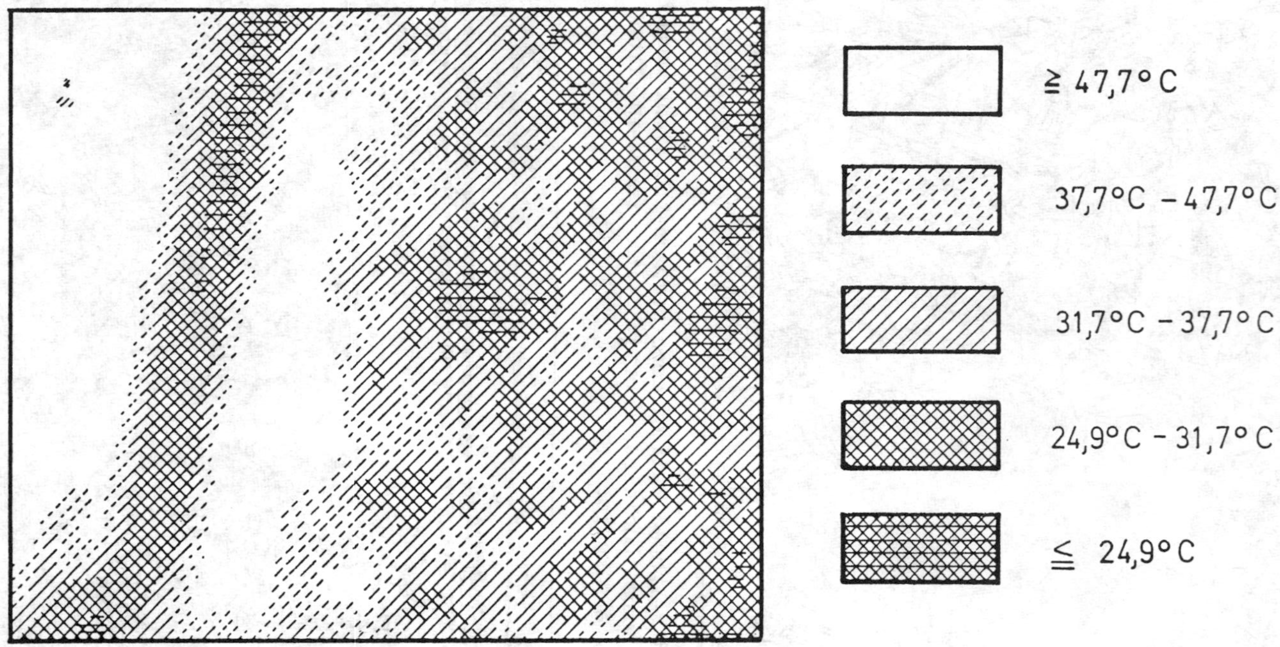

FIG. 3. IR-temperature of southwall at 13:22 h. Sunlit wall rather hot, shaded parts cooler also the vegetation cover.

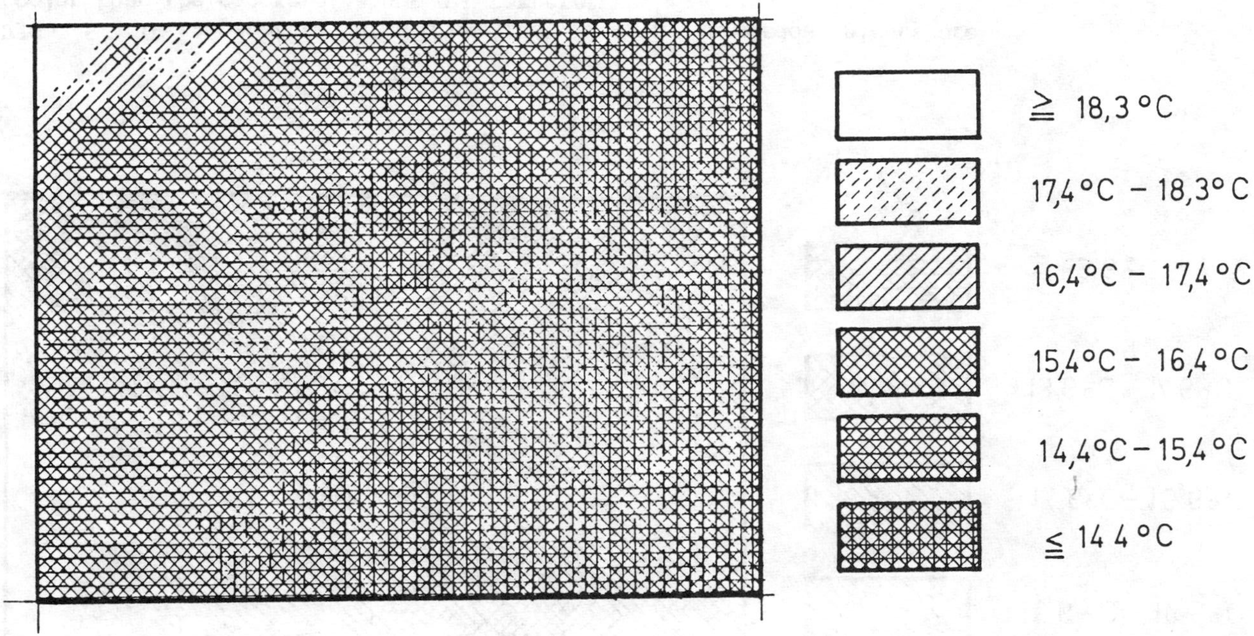

FIG. 4. IR-temperature of westwall at 13:26 h. The shaded wall is cooler than the sunlit wall.

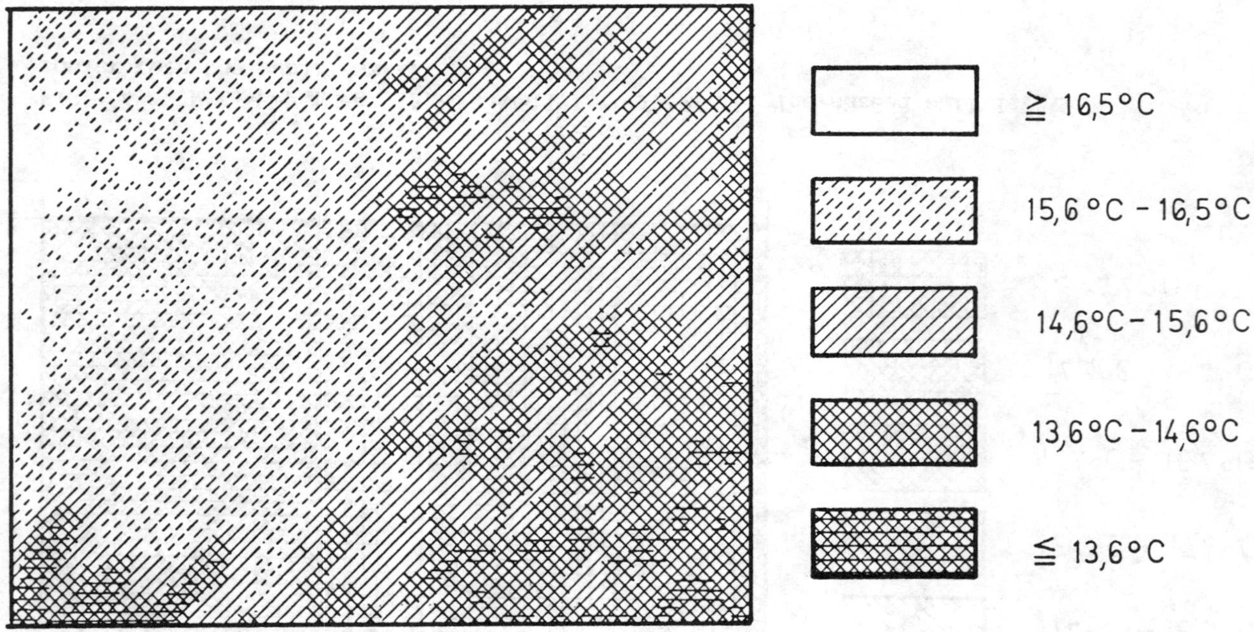

FIG. 5. IR-temperature of southwall at 21:41 h. In shadow, plants are cooler than the concrete, about air temperature.

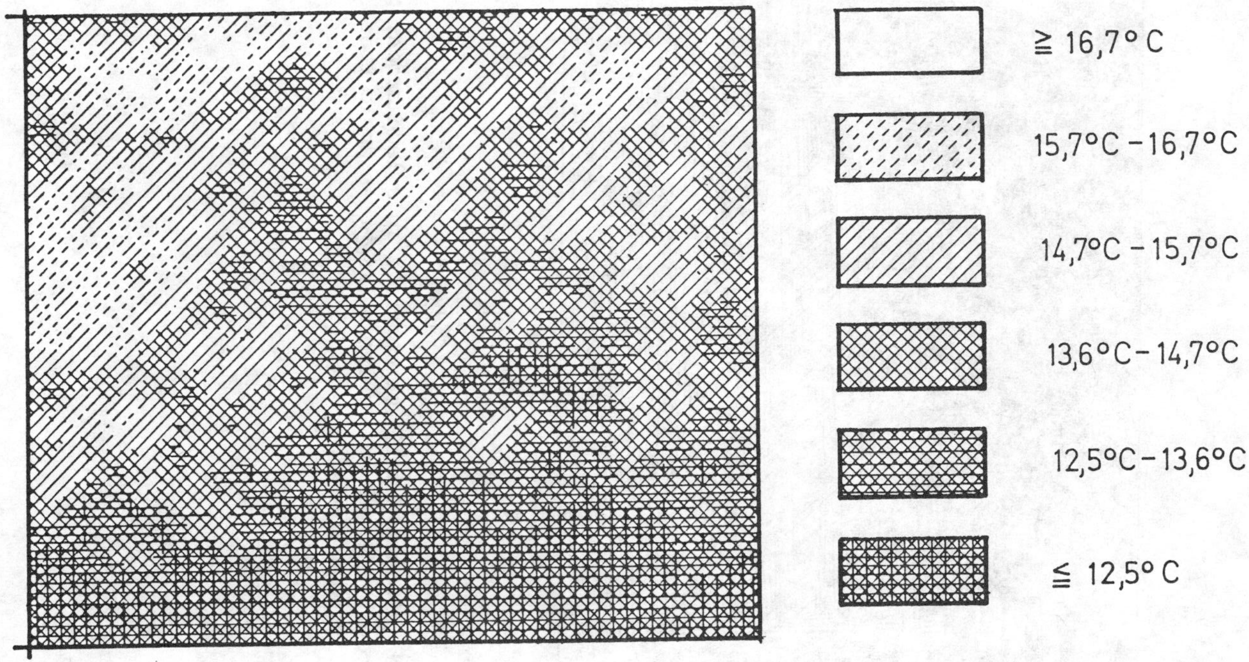

FIG. 6. IR-temperature of westwall at 21:48 h. After the last sunrays the wall is relatively warm, vegetation cooler about air temperature.

FIG. 7. Westward situated wall of concrete in a lawngarden, passage into a sunlit courtyard, left a pergola, right an apple tree, ground covered with lawn, strong shaded.

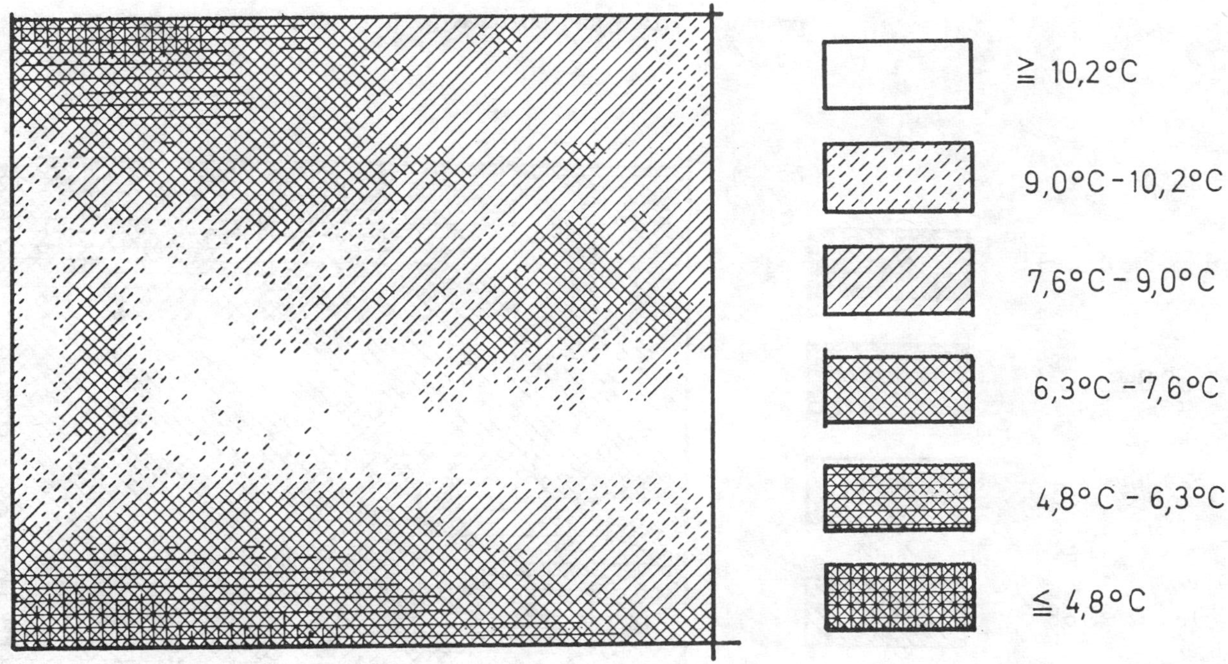

FIG. 8. Diversion of IR-temperature of site

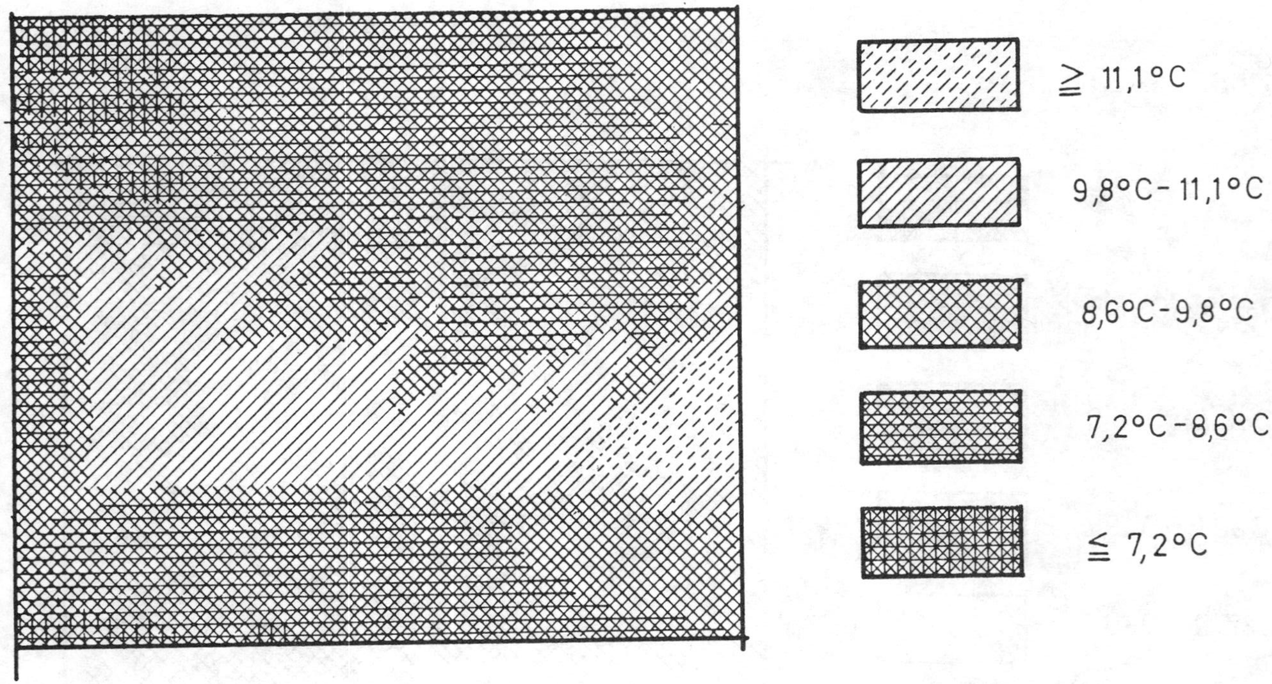

FIG. 9. IR-temperature at 6:37 h. Diversion has hardly changed.

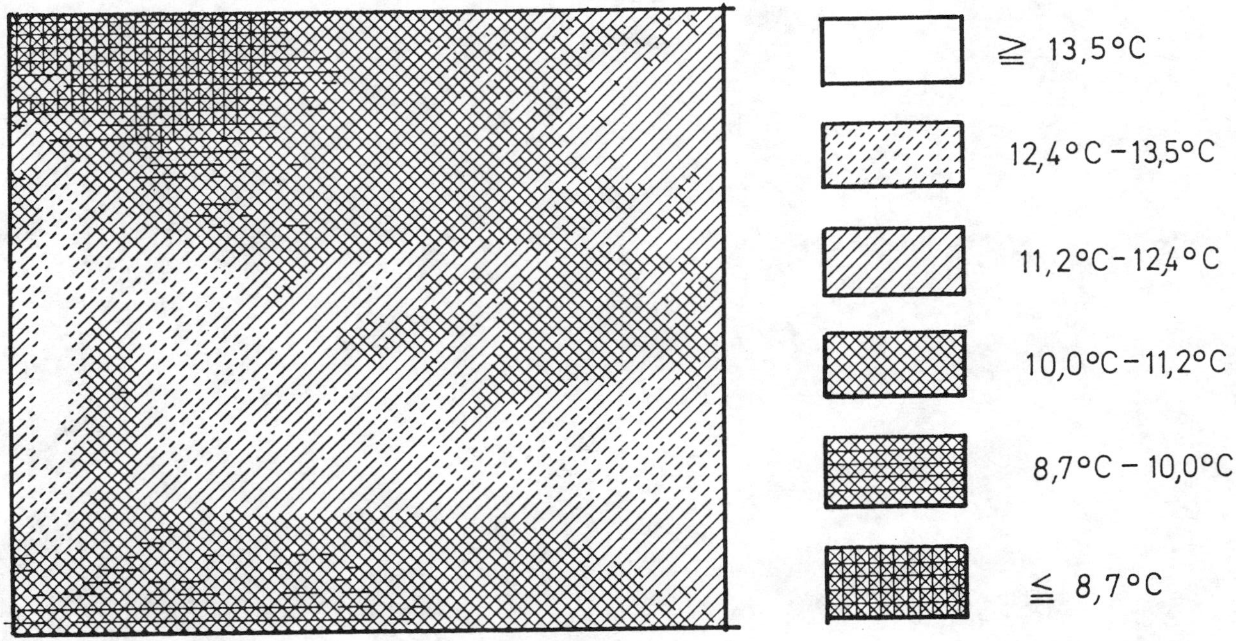

FIG. 10. IR-temperature at 22:40 h shows sunheated supports of the pergola and the sheltered wall warm, lawn and timber have lowest temperature.

ANALYSIS OF AN OROGRAPHICALLY INFLUENCED WIND SYSTEM

Claus-Jürgen Richter
Rheinisch-Westfälischer TÜV
Abteilung Vorprüfung Umweltschutz
Postfach 10 32 61
4300 Essen
FRG

Abstract

Thermally induced wind systems developing in a small valley are discussed. After presenting two case studies, the problem is handled from a more statistical point of view. It is shown that the main parameters governing the development of mountain wind systems are 1. cloud cover and 2. wind velocity in the free atmosphere.

The importance of such thermally induced wind systems for immission predictions is discussed and proposals are made to account for these when long-term diffusion statistics have to be generated.

1. Introduction

Immissions caused by near-ground sources can be rather well predicted by simple diffusion models. This is especially true if long-term concentration-distributions in the nearby surroundings of the emittents are required. The main parameters entering those predictions are the long-term distributions of wind velocity, wind direction and diffusion categories. As is well known, in valleys these parameters may be very different from those in the undisturbed atmosphere. Thus methods have to be found to estimate them.

There exist a great number of models which can describe the long-term distribution of the flow field and the coupled concentration distribution in valleys refer to for example to HEIMANN/5/, FIEDLER et. al. /2/.

K. Grefen and J. Löbel (eds.), Environmental Meteorology, 381–395.

Although these models are able to simulate channeling effects rather well, a deficit remains in predicting thermally induced wind systems. As there is a variety of parameters entering the simulation of these effects, model runs are normally restricted to case studies and rather time consuming three dimensional models are used e. g. GROSS /4/, MC NIDER and PIELKE /7/.

Nevertheless, the frequency of occurrence especially for mountain-wind systems has to be known, because, as was already pointed out by DAVIDSON /1/, these winds exhibit

- strong directional stability
- reduced turbulence
- rather low wind speeds, especially in small valleys

As is known by experience, the above mentioned characteristics often are the main factors which cause problems in the surroundings of e. g. odour emittents.

The scope of this contribution is to show the conditions under which such thermally induced wind systems can develop in a small valley. Finally, proposals are made on how to consider such thermally induced wind systems for generating the long-term frequency distribution of wind direction, wind velocity and diffusion categories in such a valley.

2. Monitoring sites

Figure 1 shows the topographic map of the area and the measuring stations. The aim was to investigate thermally induced wind systems in the valley which is situated at the right top of the map. Altitude differences between the valley bottom and the neighbouring ridges are of the order of 50 m only. The direction of the valley is from south-west to north-east, where the valley flows into a larger valley 2 km downstream. Most of the area was covered with grass and a few isolated trees. Some houses were located in the upper part of the valley.

Only results of the measuring stations MP1 and MP3 will be presented here. The following parameters were measured (as 10-min-averages):

- at MP1:

  wind direction and wind velocity in 10 m above ground

  temperature and relative humidity in 0,1 m and 2 m above ground

- at MP3:

  wind direction and wind velocity in 40 m above ground (tower)

  temperature and relative humidity in 40 m and 2 m above ground.

At the remaining points, temperature and relative humidity were measured.

MP3 was located on the highest point in the surroundings, so it should represent the undisturbed atmosphere rather well.

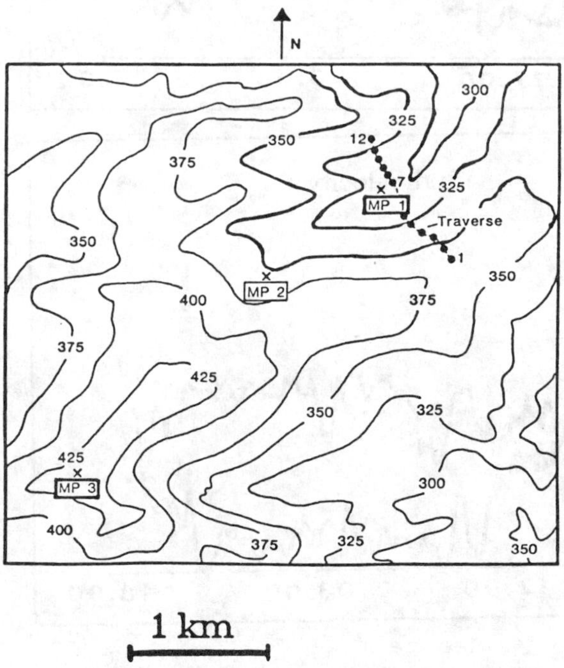

Figure 1. Measuring stations

3. Results

3.1 Case studies

Figure 2 shows the wind direction and velocity measured during the period from the 5th to the 6th March 1987. This period was characterized by cloudless sky and northeasterly to easterly winds, as can be seen from the record of MP3 (Tower Station). Wind velocities at MP3 ranged from about 3 m/s to 6 m/s.

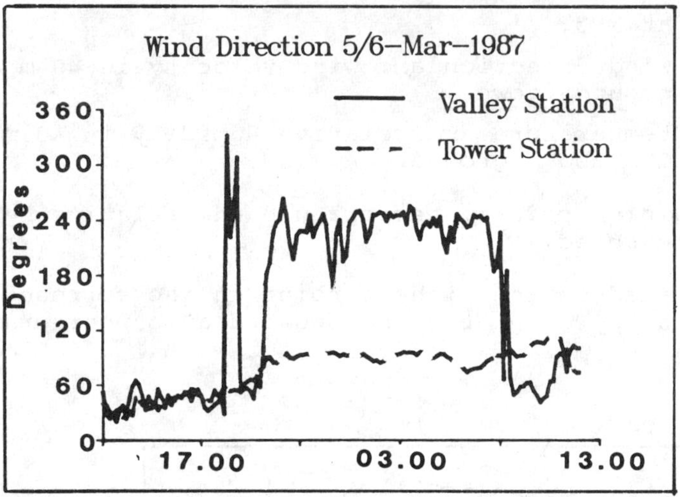

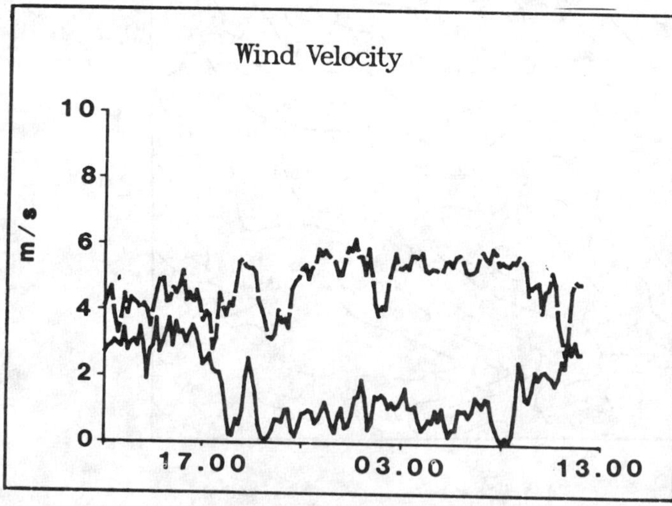

Figure 2. See text

At MP1 wind direction abruptly changes from northeast to south-west soon after sunset. Apart from a small interruption around 19.00 hours, the wind blows from the south-west during the whole night. The interruption at 19.00 hours can perhaps be explained by an increase of mechanically induced turbulence due to higher wind speeds at MP3 at that time. Low-level-jet phenomena may play a role here. After sunrise the wind direction at the valley bottom again changes from south-west to north-east.

While wind speeds at MP1 range from 2 m/s to 4 m/s during the day, they are of the order of 1 m/s throughout the night. Minima of wind speed occur when the wind direction is changing at sunset and sunrise.

The situation some days later is shown in Figure 3. The period is again characterized by cloudless sky but wind speeds were much higher. As a consequence, no mountain-wind systems could develop and the wind direction at the valley bottom was north-east throughout the observation period. Nevertheless, the wind speed decreased approximately exponentially during the night.

The Froude-Number

$$F = U/(N*H)$$

with U = windspeed at MP3

N = Brunt-Vaisala-frequency (calculated from the temperatures at MP3, 40 m above ground and MP1, 2 m above ground)

H = altitude difference (taken as 100 m)

for the two periods is shown in Figures 4 and 5. Its value is of the order of 1 for the first and 10 for the second period (only night hours are considered). Thus the Froude-Number could be taken as a parameter describing the development of mountain-wind systems. Unfortunately, the wind velocity U and the Brunt-Vaisala frequency N are not independent from each other, making things more complicated.

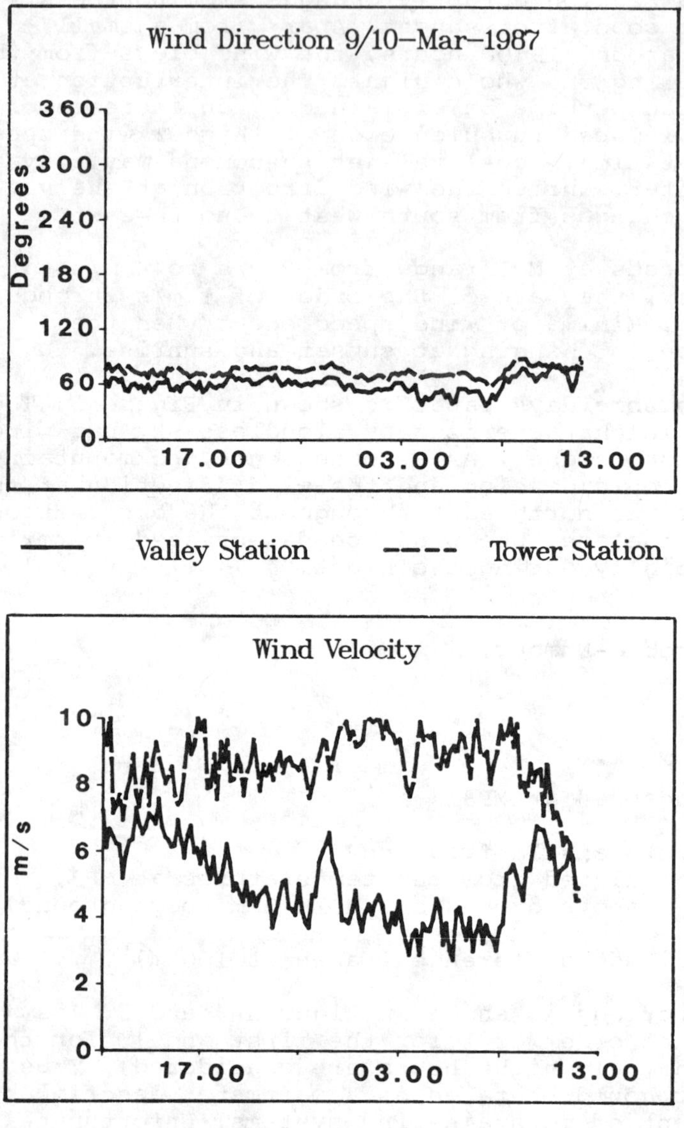

Figure 3. See text.

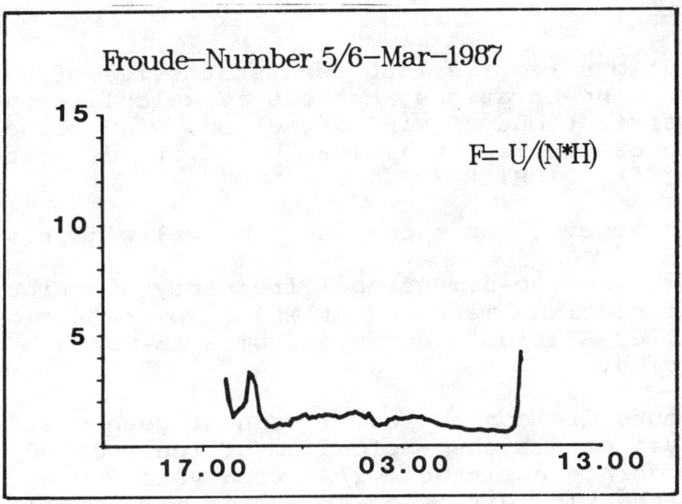

Figure 4. Froude-Number for the first period.

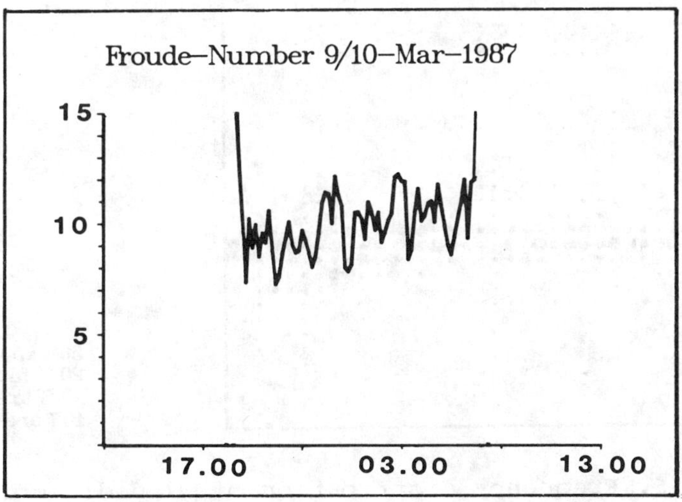

Figure 5. Froude Number for the second period.

3.2 Statistics

German regulations require long term statistics of the immission concentrations. Methods to calculate the long-term distributions of wind direction, wind velocity and diffusion categories at a given point are described in e.g. KOLB /6/, RICHTER /8/.

These methods however don't consider thermal wind systems.

Figure 6 shows the two-dimensional frequency distribution of the wind directions measured at MP3 ("Tower-Station") and MP1 ("Valley Station") derived from a 15-month measuring period.

Only night hours are considered. It can be seen that under the given conditions - cloud cover (only clouds of low and moderate altitude) ≤ 2/8, wind velocity at MP3 ≤ 5 m/s - the wind direction at MP1 is more or less independent from the wind direction at MP3.

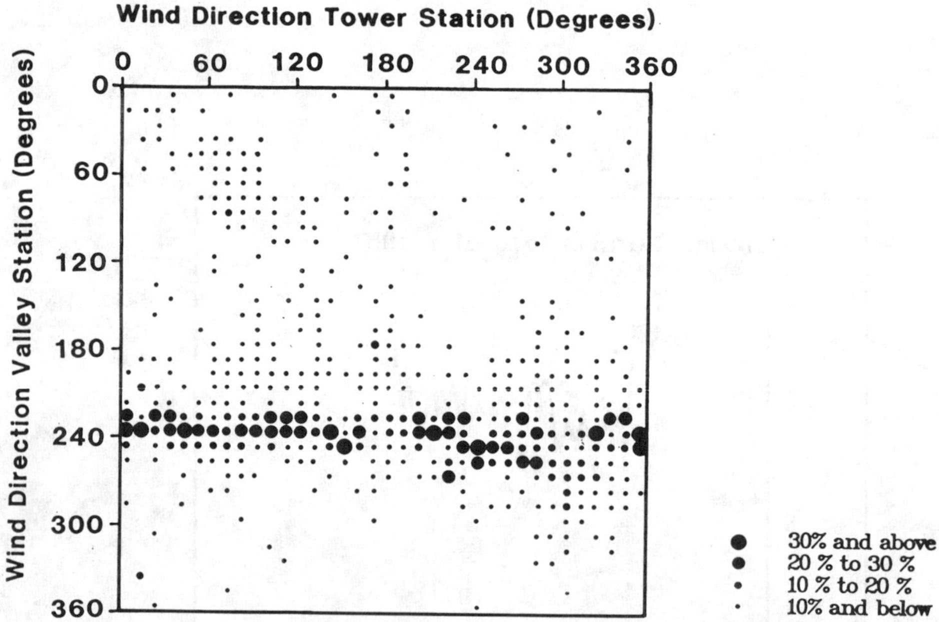

Figure 6. Frequency distribution of wind directions. The sum in each column is 100 %. Cloud Cover: ≤ 2/8. Wind velocity (tower): ≤ 5 m/s

That means, that no matter were the wind comes from at MP3, the wind direction at the valley bottom at the same time is nearly always south-west. This is an important fact to be recognized for immission predictions, as these situations occur rather often.

Figure 7 represents the situation, when the cloud-cover is still less than 2/8, but the wind velocity at the tower station ranges from 5 to 7 m/s. Here a "transition zone" ist present, were mountain winds still develop rather frequently.

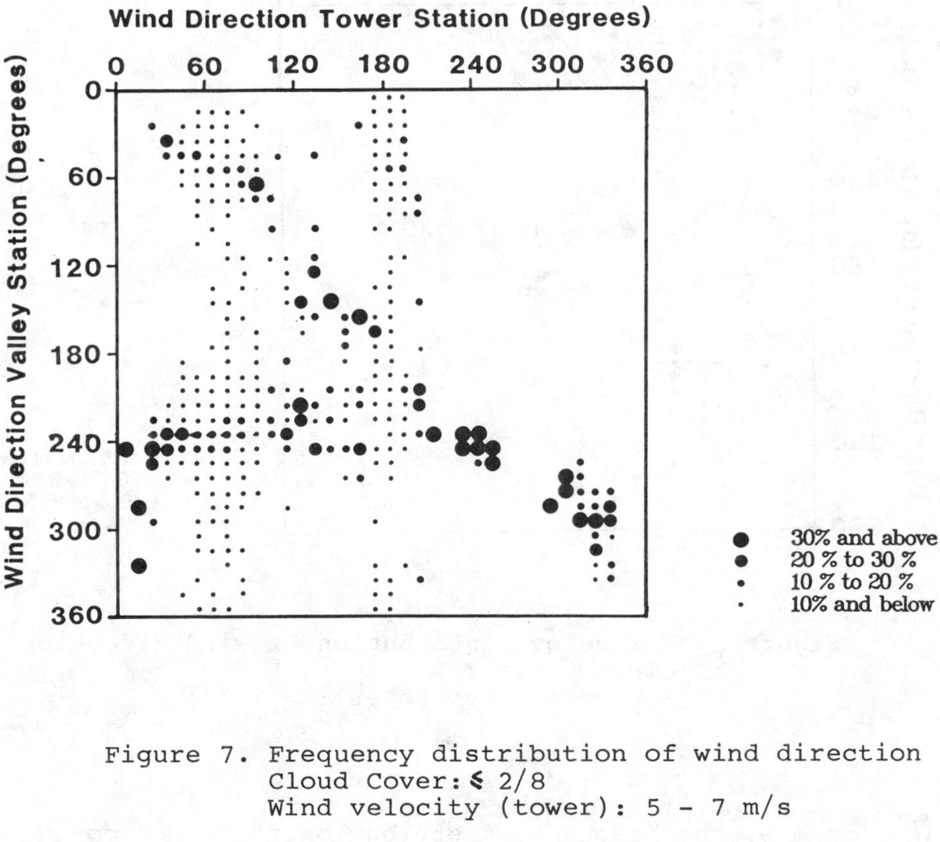

Figure 7. Frequency distribution of wind direction
Cloud Cover: ≤ 2/8
Wind velocity (tower): 5 - 7 m/s

Figure 8 shows the frequency distribution, when the wind velocity at MP3 is above 7 m/s. Despite a lack of data, it can be seen that mountain wind systems develop rather seldom.

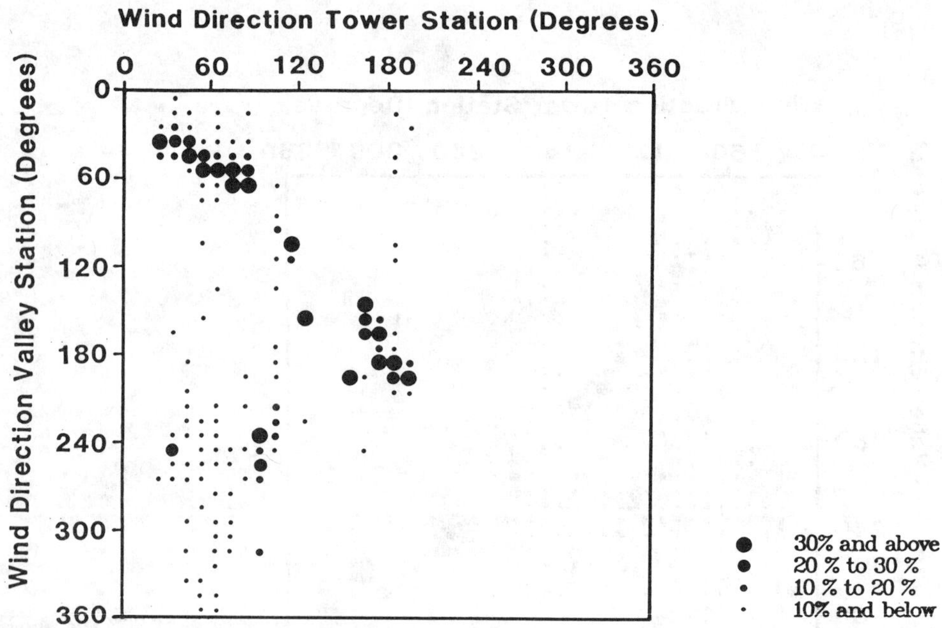

Figure 8. Frequency distribution of wind directions
Cloud Cover: ≤ 2/8
Wind velocity (tower): > 7 m/s

In Figure 9, the frequency distribution for cloud cover from 3/8 to 5/8 and wind velocities less than 5 m/s at MP3 is shown. Thermal wind systems still occur especially for wind directions from north to east.

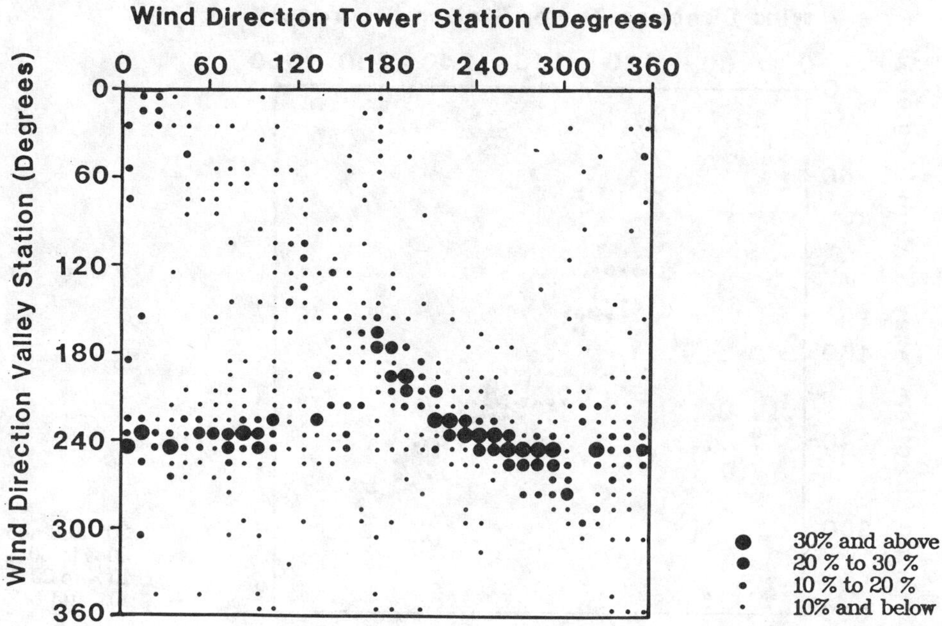

Figure 9. Frequency distribution of wind directions
Cloud Cover: 3/8 - 5/8
Wind velocity (tower): ≤ 5 m/s

The remaining figures (10- 12) show conditions, under which practically no mountain winds can develop. Note that most of the frequencies lie in the neighbourhood of the diagonal.

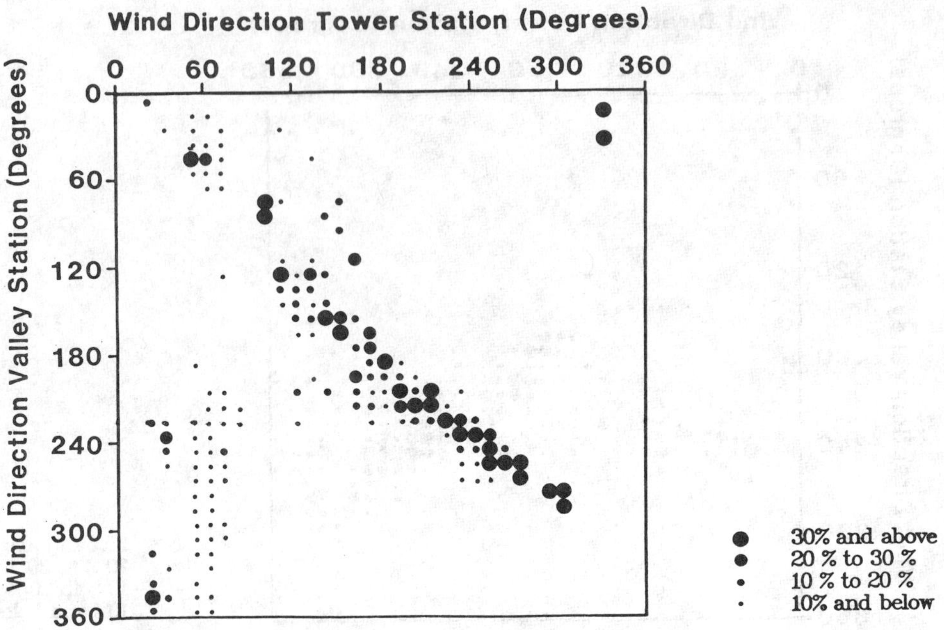

Figure 10 . Cloud Cover: 3/8 - 5/8
Wind velocity (tower): > 5 m/s

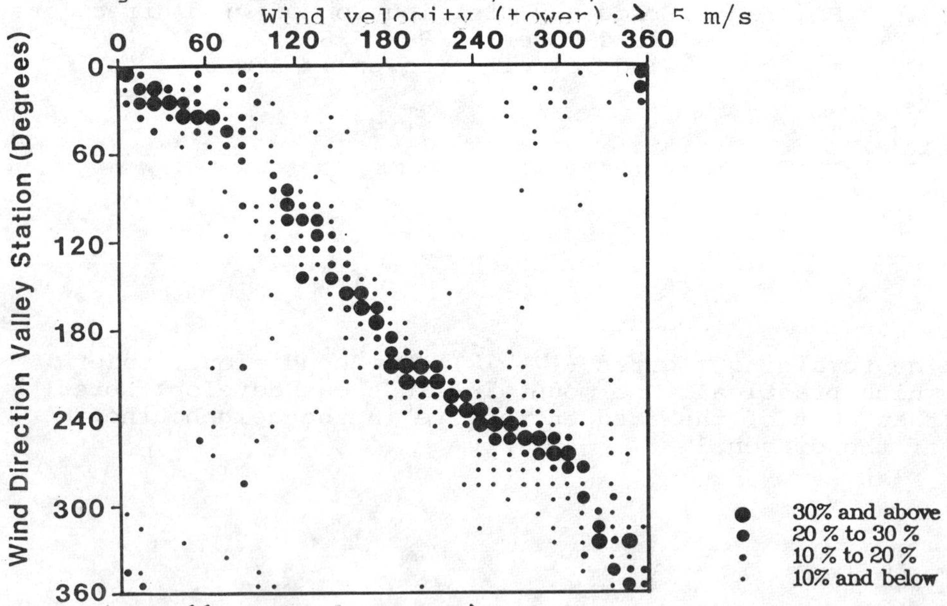

Figure 11 . Cloud Cover: ≥ 6/8
Wind velocity (tower): ≤ 5 m/s

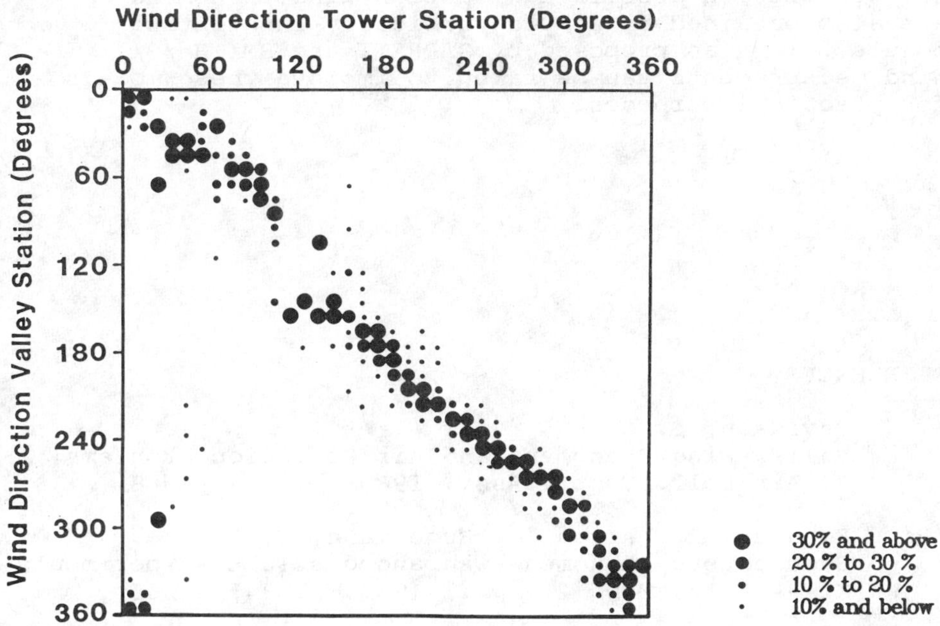

Figure 12. Cloud Cover: ≥ 6/8
Wind velocity (tower): > 5 m/s

4. Conclusions

These results are valid for the investigated valley only. Nevertheless, they illustrate rather well the parameters which influence the development of mountain wind systems. Different threshold values for cloud cover and wind velocity have to be taken for other valleys.

Once the long-term distributions of wind direction, wind velocity and cloud cover at some suitable station in the nearby surroundings of such a valley are known, the occurrence of mountain wind systems can at least be roughly estimated. The reference station should not be influenced by thermal wind systems and should be representative of the undisturbed atmosphere.

The diffusion categories related to mountain winds can always be estimated to be stable.

Good results in describing mountain wind phenomena are also obtained by simple vertically integrating models, e. g. as proposed by GARRETT and SMITH /3/. Wind measurements can be taken to improve these models for practical purposes.

REFERENCES

/1/ Davidson, B.
Valley Wind Phenomena and Air Pollution Problems.
J. Air Poll. Contr. Ass., 1960, p. 364 - 368

/2/ Fiedler, F.; Arian, G; Hugelmann, C. P.
Beobachtete Phänomene während des TULLA-Experiments.
KfK-PEF 12, Band 2, April 1987, p. 347 - 365

/3/ Garrett, A. J.; Smith, F. G.
A Two Dimensional Dynamical Drainage Flow Model with Monte Carlo Transport and Diffusion Calculation.
DP-1639, Savannah River Laboratory, Aiken, SC 29808, 1982, 67pp

/4/ Gross, G.
Numerische Simulation nächtlicher Kaltluftabflüsse und Tiefsttemperaturen in einem Moselseitental.
Met. Rdsch., 38, 1985, p. 161 - 171

/5/ Heimann, D.
Estimation of Regional Surface Layer Wind Field Characteristics Using a Three-Layer Mesoscale Model.
Beitr. Phys. Atmosph., 59, 1986, p. 518 - 537

/6/ Kolb, H.
Vergleich verschiedener Methoden der Übertragung von Statistiken der Ausbreitungsverhältnisse in orographisch modifiziertem Gelände.
Arch. Met. Geoph. Biokl., Ser. B, 24, 1976, p. 57 - 68

/7/ Mc. Nider, R. T.; Pielke, R. A.
Numerical Simulation of Slope and Mountain Flows.
J. Clim. Appl. Meteorol., 23, 1984, 1441 - 1453

/8/ Richter, C.-J.
Ein Vorschlag zur Berücksichtigung lokaler orographischer Gegebenheiten bei der Standortplanung.
Ann. Meteorol., 22, 1985, p. 124 - 125

Synthetic Wind Climatology evaluated by the Non-Hydrostatic Numerical Mesoscale Model KAMM.

Gerhard Adrian
Institut für Meteorologie und Klimaforschung
Universität Karlsruhe /
Kernforschungszentrum Karlsruhe
Kaiserstraße 12
7500 Karlsruhe 1

ABSTRACT: Statistical informations about the wind field, like distributions of the wind direction or the wind velocity, are often needed for regional planning purposes. Thereby a high spatial resolution is required. But the density of the station network is usually poor, because only few measurements of few stations are available. But these measurements show, that the wind field is influenced by local processes, so that an interpolation of the measurements seems not to be wise without any additional information.

It will be demonstrated, that a mesoscale model can be used to construct synthetic maps of statistical parameters of the wind field with high resolution. These statistical parameters are the distribution of the mean wind velocity and the wind direction. Also the spatial correlation function can be evaluated. This function can be helpful to decide a station to be representative for a special location.

1. AVAILABLE INFORMATION ABOUT WIND STATISTICS

Statistical informations about the wind field like distributions of the wind direction or the wind velocity are often needed for regional planning purposes or pollution problems. Another application of wind statistics is to find suitable locations for wind converters. The classical method to get this information for a special place is to interpolate or extrapolate the statistical parameters, evaluated from observations. These observations must be carried out for a

K. Grefen and J. Löbel (eds.), Environmental Meteorology, 397–411.

long time to assure, that the statistical parameters are stable. Therefore this information must be taken from operational stations.

One must presuppose, that the observations, taken at different stations, are carried out under comparable conditions. These conditions are the use of similar instruments at the same height above the ground, an undisturbed surrounding and periodical observations at the same time. An important requirement is the sufficient long period of observation.

Most of these conditions can only be fulfilled by operational weather stations. In Germany only about 90 stations of this kind are working (Benesch et. al. 1978) In 1975 the scale of measuring the wind direction was changed, so that only the data of the wind direction taken before this time or after this year can be used to evaluate statistical climatological parameters. Between 1969 and 1974 only the data of 74 stations were useful for this purpose (Benesch et. al.)

Some of these stations had different heights of anemometers for different reasons. To use these data for interpolation purposes, they must be reduced to the same height of reference, otherwise they would not be comparable. The known methods of reduction require homogeneous and stationary conditions, because the theory of the Prandtl-layer is used. Additionally information about the stability and roughness is needed. It seems to be very difficult to get this information. The requirements of the Prandtl-theory are normally violated in the typical inhomogeneous terrain.

The wind field is strongly influenced by the topography and the required information is needed at special locations. An interpolation is only possible if the resolution of the station network is sufficient or if the interpolation scheme includes physical information additionally. An example of such a scheme is the "optimum interpolation" of Gandin (1965). But this scheme needs correlation functions of the interpolated variables with a sufficient high resolution at least in the mesoscale. Both requirements can not be satisfied by a station network of 90 stations in Germany for all applications.

2. MESOSCALE MODELS IN CLIMATOLOGY

For these reasons additional information about the wind field is required in the mesoscale. It will be shown here, that mesoscale models are able to give such information and that they can be an important tool for climatological purposes.

The most obvious applications of mesoscale models are simulations of single, typical events. These simulations can help to explain or to analyse observed features of the wind field. An example is given in (Adrian, Fiedler 1987). There the model KAMM was used to analyse the wind field in the upper Rhine valley. In most of all cases with a geostrophic wind from south east to south a north-westerly wind is observed in Freiburg in the upper Rhine valley. This phenomenon can not be explained by operational observations only. The result of the mesoscale model of such a case shows a large eddy at the lee side of the black forest mountains inside the Rhine valley. This eddy is caused by a flow splitting due to the southern part of the black forest. The simulation shows further, that this phenomenon is connected with the so called "Moehlin Jet" observed at the southern slope of the black forest in the Rhine valley.

This kind of application of mesoscale models is important for analysing observations or for grasping special processes, which may be relevant for the local climate. The example of the wind direction in Freiburg is observed. But the observation at one station gives no idea, whether this phenomenon is local or not. But the model results show, that this phenomenon has a length scale of 60 km.

An other application of mesoscale models is to construct maps of statistical parameters artificially to complete the poor informations, taken from observations.

A lot of informations are available about the large scale atmospheric fields from the weather stations and the daily weather forecasts. And this kind of information is needed to get the external forcing parameters of mesoscale models. One thinkable way is to use these external parameters to force the mesoscale model and to run it for several years of model time.

In the case of the model KAMM the amount of computing time is one sixth of the model time on the vector computer available at the University of Karlsruhe. Today there exist vector computers, which are ten times faster. But such a simulation on a modern computer would still need several weeks of computing time. Therefore the use of a multi-layer, mesoscale, non-hydrostatic model, like KAMM, is not practicable in this way.

Heimann (1986) used a three layer, hydrostatic mesoscale model for such a simulation of the local wind climate of the Taunus mountains. A lot of features of the flow field can be described by such a simplified model. But if there are steep slopes and high mountains in the model area or, if the wanted resolution is very high, an improved model must be used.

Therefore the disadvantage of these models, to consume to much computing time, must be avoided. It seems only to be possible to simulate typical situations, which can be classified by external parameters like the geostrophic wind, and to infer the statistical parameters from the features of the chosen external parameters. Wippermann and Groß (1981) used this method to construct an artificial frequency distribution of the wind direction at the weather station Mannheim and demonstrated a good agreement between the observed and synthetic windrose. They used a two dimensional version of their model FITNAH. Here the three dimensional model KAMM is used.

2.1 The Karlsruhe Atmospheric mesoscale model KAMM

The non hydrostatic model KAMM has been developed in recent years at the University of Karlsruhe (Prenosil 1980, Dorwarth 1986). Here the three dimensional version is used. It solves the equations of momentum and heat using an anelastic form to filter sound waves. It has a terrain following coordinate system (Adrian 1987a) and it is specially suitable for simulation flow fields in the mesoscale and in the microscale . It is forced by a large scale basic state which is geostrophic and hydrostatic (Adrian 1987b). Therefore the geostrophic wind and the large scale temperature field can be chosen externally to classify the simulation. But they can also be determined by the operational weather analysis or forecast (Adrian 1987b) as Vogel, Adrian and Fiedler (1987) demonstrate in their paper here.

The subgrid fluxes are parameterized by a first order closure which evaluates the turbulent fluxes by a gradient relationship using Prandtl"s theory and mixing length. This seems to be sufficient in those cases, which are presented here.

The shown simulations are done with fixed external parameters, like the geostrophic wind and soil temperatures. The simulation time was chosen, that the results are stationary, typically after three hours simulation time. The horizontal grid size was 5 km, in the vertical direction the grid size varies between 40 m near the surface and 400 m at the top of the model area.

3. MEAN FEATURES OF THE WIND FIELD IN SOUTH - WEST GERMANY

As discussed in the introduction the knowledge of the mean features of the wind field is very poor in the mesoscale . It is demonstrated here, how a mesoscale model can be used to get more information. The chosen model area is the south western part of Germany, the state of Baden-Württemberg and its surrounding. Figure 1 shows a contour plot of this area. the Vosges, Black Forest and the Schwäbische Alb can be recognized. The height at the bottom of the Rhine valley is between 150 m and 250 m above sea level, the tops of the Vosges and Black Forest between 1000 m and 1500 m. The model area can be seen as typical to the German highlands.

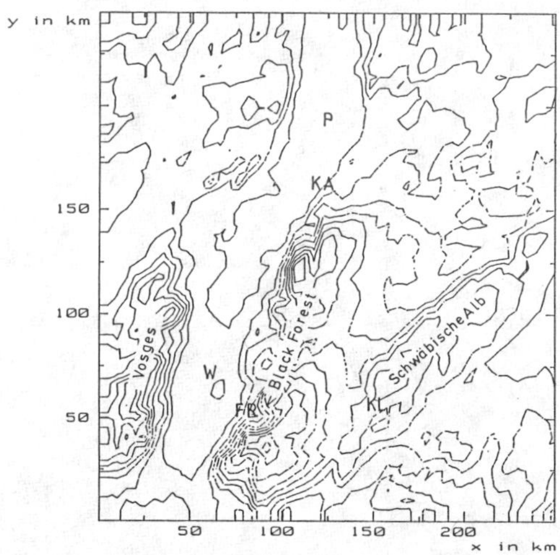

Figure 1: The orography of the model area. The contour lines differ of 100 m. The signed locations are used in the text:KA=Karlsruhe, KL=Klippeneck, FR=Freiburg, W=Wyhl, P=Philippsburg.

3.1 Synthetic Windroses

For this model area synthetic wind roses are constructed. Thereby Wippermann's and Groß's (1981) method was used. Firstly a distribution of the geostrophic wind direction was taken. Here the distribution of the station Karlsruhe in the middle of the area was chosen, which was evaluated by Malberg, Bökens and Frattesi (1980). Afterwards the wind field was computed for eight different geostrophic wind directions, every 45 degrees, using the model KAMM. Only a stable stratification was considered. The frequency of the geostrophic wind direction was attached to the wind direction, simulated by the model. So a synthetic wind rose was got at every grid point. Before the results will be discussed, two synthetic windroses are compared to observations to validate the method.

Figure 2 and 3 show the comparison. The observations are taken at the meteorological towers in Philippsburg and Wyhl. They represent mean values at a height of 40 m above the ground. The meaning time is 30 minutes. All values of five years (1976-1981) are taken in account. Although the simulated windroses describe only stable cases, the agreement between the simulated and observed windroses is quite good.

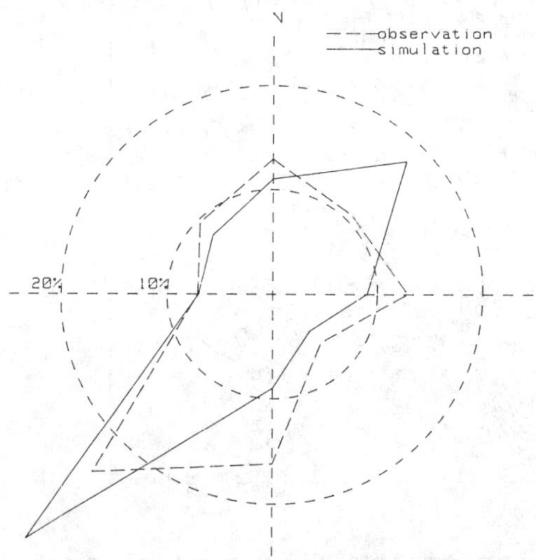

Figure 2: Comparison between observed (dashed line) and synthetic (full line) windrose at the meteorological tower of Philippsburg.

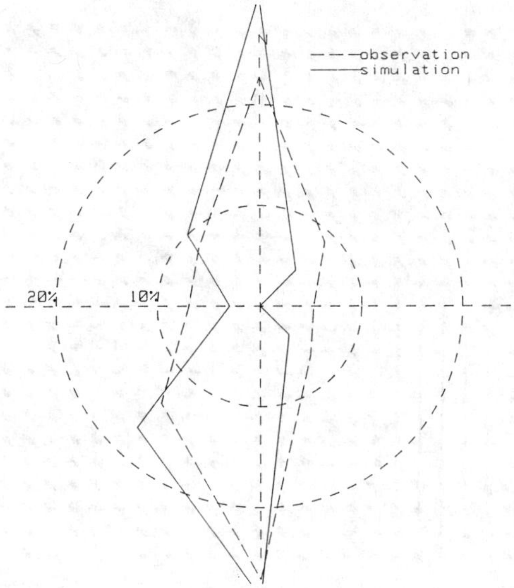

Figure 3: Comparison between observed (dashed line) and synthetic (full line) windrose at the meteorological tower of Wyhl.

Because of these good agreement the result of the hole model area is shown in Figure 4. Only every second wind rose is plotted, because of the clearness of the picture. The information is available at every grid point, every five kilometers. Obviously the distribution of the wind direction is strongly influenced by the orography. The channeling effect can be recognized in the Rhine valley, in the Kinzig valley between the northern and southern part of the Black Forest, and also at the slopes of the Schwäbische Alb.

The local structures of the distribution of the wind direction demonstrate, that they can not determined by observations alone, and that the structures of the mean wind field is complicated and strongly influenced by the orography. But they can be described by mesoscale models and validated by a comparison with observations at single stations. Further tests show, that the windroses are not sensitive to variations of the geostrophic wind velocity. This is at least true for values between 5 m/s and 15 m/s.

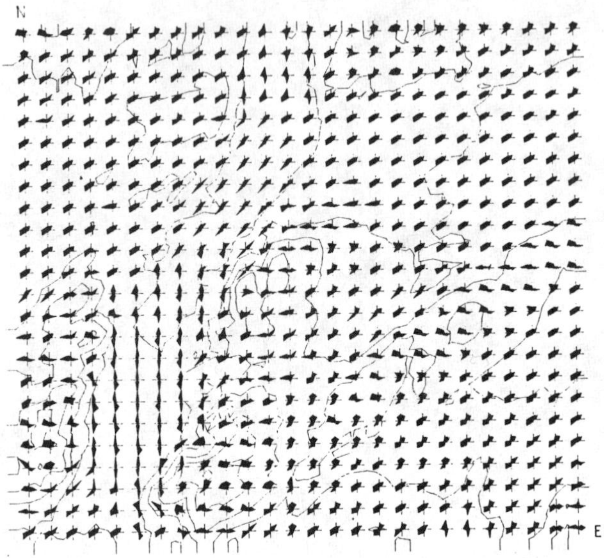

Figure 4: Map of synthetic wind roses in the model area. Only every second one is shown.

3.2 The distribution of the mean wind velocity

Models of the flow field do not give only the wind direction but also the wind velocity. Of course the local wind velocity does not only depend in the orography but also strongly on the geostrophic wind velocity. Therefor a dimensionless wind velocity is introduced:

$$c = u/u_g$$

u is the local, u_g the geostrophic wind velocity. One result of the simulations is, that this drag coefficient is not very sensitive to the geostrophic wind velocity. This was also tested for geostrophic wind velocities between 5 m/s and 15 m/s.

The figures 5 to 7 show the distribution of the drag coefficient c at different geostrophic wind directions from south , south-west and west. The areas of low values can be observed at the lee sides of the large mountains. They turn into the same direction as the geostrophic wind. The highest values occure at the Schwäbische Alb, if the geostrophic wind direction is south, or if the surface wind is south-east because of the wind shift in the Ekman - Layer.

In the upper Rhine valley the wind velocity is almost low.

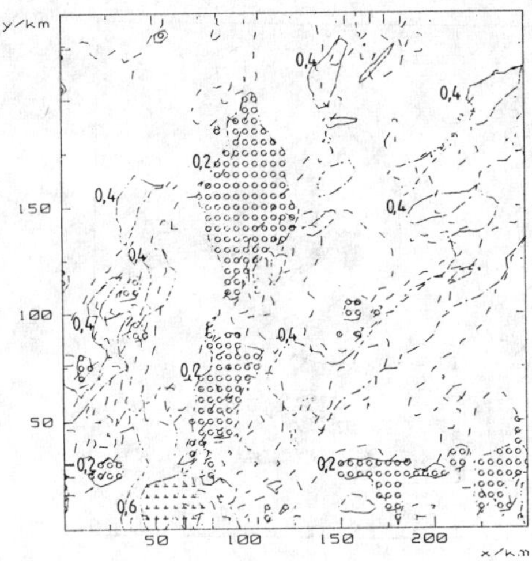

Figure 5: The distribution of the drag coefficient at a geostrophic wind from south. Areas with values less than 0.2 are signed by circles, areas with values greater than 0.6 are signed by crosses.

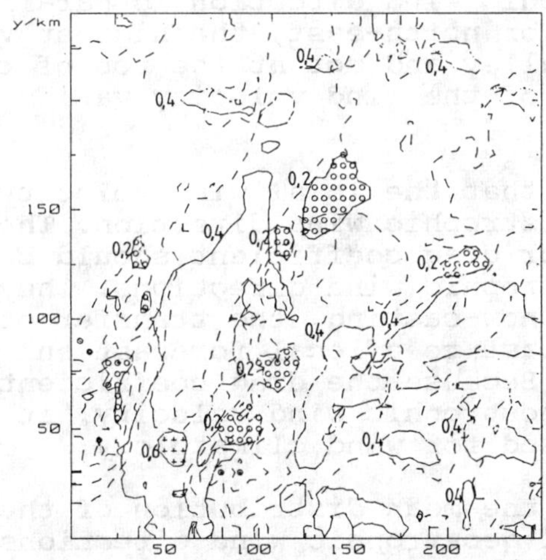

Figure 6: The distribution of the drag coefficient at a geostrophic wind from south-west. Areas with values less than 0.2 are signed by circles, areas with values greater than 0.6 are signed by crosses.

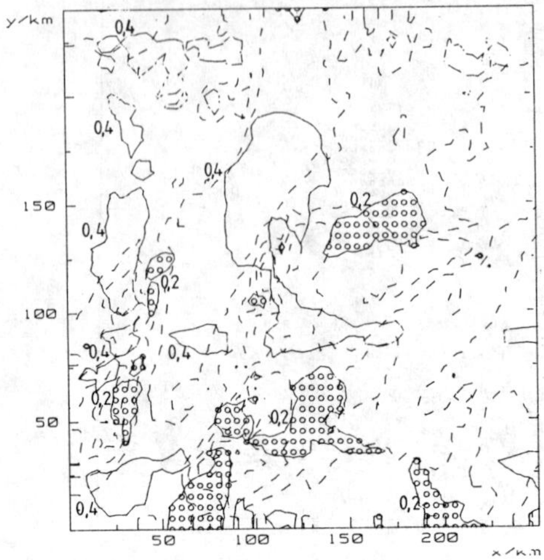

Figure 7: The distribution of the drag coefficient at a geostrophic wind from west. Areas with values less than 0.2 are signed by circles, areas with values greater than 0.6 are signed by crosses.

But if the geostrophic wind direction is parallel to the valley, south-west or north-east, the highest values of c are found in the valley and not at the top of the mountains. Thereby the values of the wind velocity varies of a factor of five.

The conclusion is, that the local wind velocity depends strongly on the geostrophic wind direction. Therefore maps of the wind velocity or drag coefficient should be constructed for different geostrophic wind directions. They can be an important tool for now-casting, the transfer of the large scale weather forecast to a local forecast and for analysing the local climate. Because the drag coefficient is not sensitive to the geostrophic wind velocity, it is a suitable number to be analysed for wind climatology.

The figure 8 shows the mean distribution of the drag coefficient for all geostrophic wind directions. It is evaluated in the same way as the wind roses, as a weighted mean of the results of the different directions. The weights are the frequencies of the geostrophic wind directions. Only few structures can be seen. Near Freiburg the wind velocity is significantly small and large near the saddle between the northern and the southern part of the Black Forest and at the top of the Schwäbische Alb. So the conclusion is, that such a map of the mean velocity like Figure 8 is not so important than other specific maps for different directions.

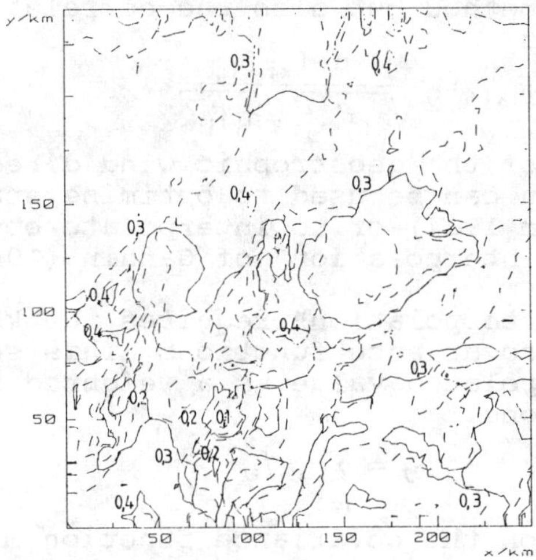

Figure 8: Mean distribution of the drag coefficient.

The model results show, that the geostrophic drag coefficient is an interesting and useful number at least for the now-casting problem. It is simple to evaluate this number from model results, but it take pains to evaluate it from observations. Therefore no analysis is available to validate these results. But the validation of the distribution of the wind direction and of single events shown in (Adrian, Fiedler 1987, Vogel 1987, Vogel, Nester, Baer 1987) is possible. Therefore these results of the mean velocity distribution seem to be correct.

It is also a climatologic application of mesoscale models to test new methods of analysing, because to use a model is cheaper than to do measurements for a long time.

4. SPATIAL CORRELATION OF THE WIND VELOCITY - OPTIMIZING STATION NETWORKS

Not only the mean values of the wind velocity can be determined by this method but also the correlation function. It can be defined as

$$m_{kl} = \frac{n_i f'_{ki} f'_{li}}{\sqrt{n_i f'^2_{ki}} \sqrt{n_i f'^2_{li}}}$$

n is the frequency of the geostrophic wind direction. This correlation function can be used to optimize station networks (Grimm-Luczak, Grimm 1984) or to interpolate observations by using the "optimum interpolation" of Gandin (1964).

Gandin's "optimum interpolation" requires the knowledge of the correlation or covariance functions. This scheme determines an interpolated value as a weighted mean of all available observations.

$$f_{0j} = \sum_{i=1}^{N} p_i f_{ij}$$

The weights depend on the covariance function m :

$$\sum_{i=1}^{N} p_i m_{ij} = m_{0j}, j = 1, 2, ..., N$$

m is the covariance between two observations. Also a mean interpolation error can be determined. It depends also on the covariance.

The advantage of using this scheme is, that it includes additional physical information given by the covariance. If the correlation between a station and the point, where an interpolated value is required, is high, the weight p would be also large. If there is no correlation, the station would not be taken in consideration by the interpolation scheme. The disadvantage of the optimum interpolation is, that these covariance function can only be determined, if the station network is dense. But then a simpler method can be used. In the numerical weather forecast there is assumed, that these function are isotrope. Using this assumption, sufficient data would be available to determine the correlation functions. But in the mesoscale this assumptions seems not to be correct. But as discussed these functions can be determined by a mesoscale model.

If the special case of one station is assumed, the weight is of course p=1. The interpolation error is more interesting in this case,

$$E = (1 - \mu) m_{00}$$

or the relative interpolation error is

$$1-\mu$$

This error can be used to decide, whether a station can be seen as representative or not. A station is representative for an area, if the relative interpolation error is smaller than a critical value.

In the figures 9 and 10 the correlation of the wind velocity between the stations Karlsruhe in the Rhine valley and Klippeneck at the top of the Schwaebische Alb is shown as examples. The station Karlsruhe is representative for the area inside the Rhine valley. But the station Klippeneck is only representative for a small area near the tops of the mountains of the Schwaebische Alb. This analysis can be repeated for all stations in this area. Afterwards one can see, whether additional stations are required.

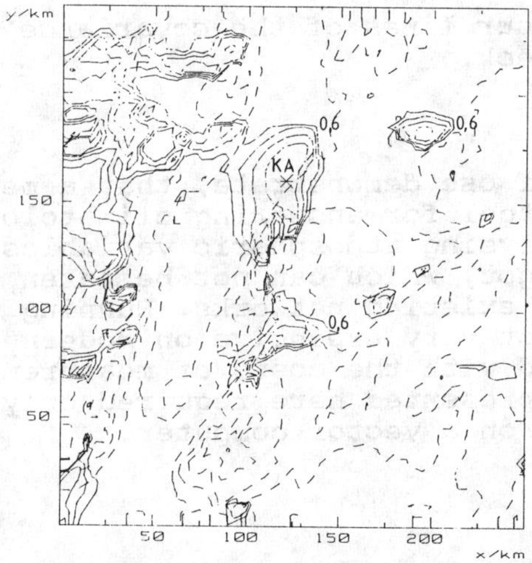

Figure 9: Contour lines of the covariance for the Station Karlsruhe

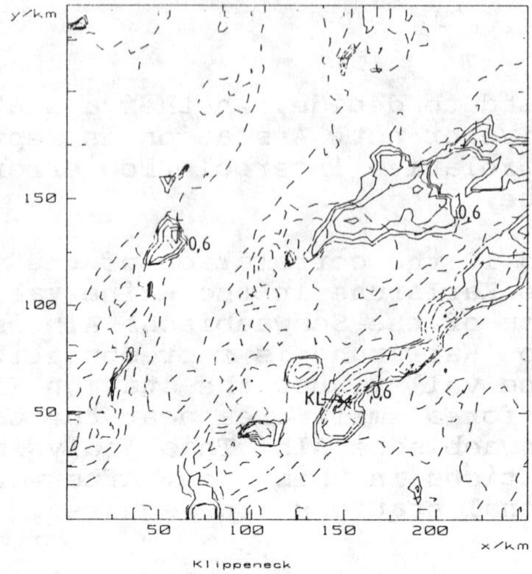

Figure 10: Contour lines of the covariance for the Station Klippeneck.

5. CONCLUSIONS

Different examples of use demonstrate, that a mesoscale model can be an important tool for answering climatological problems and for analysing atmospheric variables. Thereby informations can be got, which can not be taken from observations done by existing networks. Running such a mesoscale model is not very expensive on modern computers, if the price is compared with the costs of measurements. For example the results presented here required only about five hours computing time on a vector computer.

References:

/1/ Adrian,G., Fiedler,F., 1987: Mesoscale eddies induced by topography analysed with a non hydrostatic model. Preprints 16th International Technical Meeting on Air Pollution Modeling and its Applications. Lindau 1987.

/2/ Adrian, G., 1987a: Zur Berechnung horizontaler Differenzen in geländefolgenden Koordinatensystemen unter Berücksichtigung variabler Gitterweiten. Met. Rdsch. 40, 19-24.

/3/ Adrian, G., 1987b: Determination of the Basic State of a Numerical Mesoscale Model from Operational Numerical Weather Forecasts. Beitr. Phys. Atmosph. 60, 361-370.

/4/ W. Benesch, G. Duensing, G. Jurksch, R. Zöllner, 1978: Die Windverhältnisse in der Bundesrepublik Deutschland im Hinblick auf die Nutzung der Windkraft. Berichte des Deutschen Wetterdienstes Nr. 147.

/5/ Dorwarth, G., 1986: Numerische Berechnung des Druckwiderstandes typischer Geländeformen. Wiss. Ber. Inst. Met. Klimaf. Univ. Karlsruhe, Nr. 6.

/6/ Gandin,L., 1965: Objective analysis of meteorological fields. Translated from Russian. Jerusalem: Israel Program for Scientific Translations.

/7/ Grimm-Luczak, E., Luczak, H., 1984: Die optimale räumliche Dichte von klimatologischen Meßnetzen. Berichte des Deutschen Wetterdienstes Nr. 166.

/8/ Heimann, D., 1986: Estimation of Regional Surface Layer Wind Field Characteristics Using a Three-Layer Mesoscale Model. Beitr. Phys. Atmosph. 59, 518-537.

/9/ Malberg, H., Bökens, G., Frattesi, G.:, 1980: Mittlere geostrophische und beobachtete Strömungsverhältnisse im Oberrheingraben. Annal. Meteorol. (N.F.) 16, 85-89.

/10/ Prenosil,Th., 1980: Anwendungen eines numerischen Modells für die /berströmung mesoskaliger Geländeformen. Annal. Meteorol. (N.F.) 16, 106-108.

/11/ Vogel,B., Nester,K., Baer,M., 1987: Comparison between simulated and measured concentration and dry deposition of SO2. Preprints 16th International Technical Meeting on Air Pollution Modeling and its Applications. Lindau.

/12/ Vogel, B., Adrian, G., Fiedler, F., 1987: The Sandoz Accident - Numerical Simulations of the Transport and the Dispersion of the Airborne Releases. Proceedings International Symposium Eviromental Meteorology Würzburg.

/13/ Wippermann, F., Gross, G., 1981: On the Construction of Orographically Influenced Wind Roses for given Distributions of the Large-Scale Wind. Beitr. Phys. Atmosph. 54, 492-501.

Airborne Measurements to Investigate the Transport and the Chemistry of Polluted Air Leeside of the City of Munich

F.M. Rösler, D. Paffrath, W. Peters
Institut für Physik der Atmosphäre
DFVLR Oberpfaffenhofen
D-8031 Weßling, F.R.G.

Abstract. Aircraft measurements of NO_x, NO, O_3, SO_2 were carried out leeside of the city of Munich during weak-wind inversion situations. Such meteorological situations occur in the Munich area mainly during anticyclonic weather or 'Föhn'-situations. In both cases low easterly winds and an inversion layer lead to highly polluted urban plumes with impact on the so called "Fünf-Seen-Land" west of Munich.
NO_x concentrations exceeding 190 ppb and O_3 concentrations exceeding 140 ppb were observed in some cases. The NO/NO_x ratio was decreased and the O_3 concentrations were significantly enhanced in the downwind areas.
Massflux estimations showed a total NO_x emission in Munich of 2.1 kg/s. Due to the emission situation in Munich it can be assumed that the main emitter groups are the car traffic, domestic heating systems and power plants.
The SO_2 measurements indicate that there are no significant SO_2 sources present in the Munich area. As a result of recent measurements one can conclude, that mainly long-range transport of SO_2 from our eastern neighbouring countries (CSSR, GDR) is responsible for high SO_2 concentrations in the Munich area.
Additional computer simulations over the complex terrain of interest show a strong influence of topography on transport and dispersion.

1. Introduction

Urban agglomeration areas can be regarded as one of the main sources of anthropogenic pollution. Especially inversion situations result in high concentration levels in the lower atmosphere. In such meteorological conditions very often a weak advection occurs, which transports the urban immissions to the leeside of the urban area forming a highly concentrated urban plume. The measured horizontal concentration distributions at ground-level, leeside of the urban areas result from 3-dimensional transport and mixing processes. In addition, very complex chemical reactions determine the qualitative and quantitative composition of the pollutants. In particular, the photochemical production of agressive and toxic substances like ozone, PAN (Peroxiacetylnitrate) and other oxidants in highly concentrated urban plumes is of increasing importance due to questions arising from their effects on human health or

K. Grefen and J. Löbel (eds.), Environmental Meteorology, 413–424.

from the so called novel forest deseases. The understanding of the above mentioned processes and reactions is important in order to answer questions in regional planning and clean air policy. This study intends to present airborne measurement results during weak-wind inversion situations leeside of the city of Munich. Also computer simulation experiments which examine the influence of topography on transport and dispersion will be presented.
The measurements were carried out by the Institute of Atmospheric Physics (DFVLR) from 1983 to 1986 using an aircraft measuring system described by Paffrath D. (1985). The horizontal and vertical concentration distributions of the pollutants NO, NO_x, O_3, SO_2, the meteorological parameters temperature and humidity, the concentration of Aitken nuclei and the scattering coefficient were recorded. Furthermore, windspeed and -direction were measured by pilot baloons.
The computer simulations were made using the hydrostatic three-layer mesoscale model REWIMET (Heimann, 1986). This model allows the time-dependent simulation of the mesoscale airflow and the transport of pollutants over complex topography. The used version did not consider any chemical reaction mechanism. The results presented here try to answer the following selected questions:

- What are the typical pollution distributions in highly concentrated urban plumes and which areas are mostly affected by them?
- What kind of horizontal and vertical pollution concentration profiles are found in the Munich plumes?
- Can the production of photooxidants (*e.g.*O_3) in urban plumes be detected? What are the maximum O_3-concentrations found?
- What is the influence of the topography on transport and dispersion?

2. Experimental Design

Inversions near ground level and weak winds below the main inversion are the basic meteorological conditions leading to an accumulation of pollutants in the mixing layer. These conditions are typically connected with the occurence of anticyclonic weather. Because of the special location of Munich near the Alps, the 'Föhn'-situations, leading to strong inversions near ground level must be considered (Hoinka, Rösler 1987). In both cases one observes mainly weak easterly flow below the main inversion in the Munich area.
The frequency distribution of the mean wind direction and -speed in Munich-Airport shows a maximum of low easterly winds between July and November. This behaviour is confirmed by wind recordings at Oberpfaffenhofen and Fürstenfeldbruck. Thus, this time period is predisposed to the occurence of highly polluted urban plumes. Therefore, we selected an 80km x 80km area west of Munich as the region of interest . This area was investigated by measurements and computer simulations.
The inclined suface of the Alpine-Foreland forces thermal circulations which superimpose the weak easterly flow. As a consequence, changing wind speeds and directions cause a very complex behaviour of the urban plumes.

In addition, the temperature stratification below the main inversion sometimes indicates small inversions or isothermals leading to several vertically separated pollution layers. Vertical concentration changes of 40 - 50 % were observed. Thus, small changes in the flight altitude during horizontal flight may cause significant changes of the observed pollution concentrations. To avoid misinterpretations, it was necessary to execute vertical soundings by the aircraft before and during horizontal flight. Also the horizontal flight pattern was planned in flight with regard to the measurements.
Usually the airplane was flown at constant altitudes above sea-level and thus at changing altitudes above ground (150 -500 m GND).

3. Case Studies

3.1. Föhn-Case

Down-slope air flow occuring north of the Alps is called 'Föhn' when it is characterized by strong gusty winds and an increase in temperature. The Föhn-air which has crossed the Alps from the south is warmed up by the downward movement north of the Alps. The advection of this warm air over the colder surface air builds up a very strong sloping inversion with temperature changes of up to 15 Kelvin across this boundary layer within 100 m (Nater et al. 1979). In some of the Föhn events Munich itself remains in the cold air with prevailing low easterly winds. Inversion heights above Munich area are typically 300-700 m above ground (Hoinka, Rösler 1987).
Such conditions were present on October 3. 1984. In the low-level cold air the immissions from Munich were transported, due to the easterly advection, to its western surroundings forming an urban plume. The vertical pot.-temperature profile, taken in the middle of the plume 60 km leeside of Munich, shows a stable stratification up to 1400 MSL (fig. 1). A significant inversion with a temperature change of 4.5 K / 150 m builds up a seperating layer between the ground-level cold air and the warmer 'Föhn' air. Pilot baloon wind data indicate a windshear in 1000 MSL. In the free troposhere above the main inversion layer, strong southwesterly winds (25 m/s) prevailed, while low easterly flow was observed in the ground layer (4 m/s).
The main inversion is an effective barrier against the vertical mixing of pollutants as can be seen by the strong concentration gradients.
Due to the fast reaction:

$$O_3 + NO \rightarrow NO_2 + O_2$$

NO is transformed to NO_2 reducing the O_3 concentration. Due to 7/8 cloudiness over the area of interest on that day no significant photochemical ozone production could be observed. The vertical profile of NO shows an excess of more than 20 ppb below the inversion layer with a consequently reduced O_3 concentration of only 6 ppb compared with a tropospheric background of 37 ppb. Because there are no significant NO sources between Munich and the area where the vertical sounding took place one can conclude, that the horizontal and vertical entrainment of air with background O_3 concentration of 35 ppb was insufficient to convert NO totally to NO_2. This consideration is also confirmed by the distribution of the NO_x concentration and the NO/NO_x ratio

(fig. 2/3). These distributions were derived from measurements recorded from 11:00 to 12:50 UTC along the indicated flight tracks. The flight altitude was 800 MSL or 250 - 400 m GND. The highest concentration of NO_x (60 ppb) at the flightlevel were found about 60 km west of Munich, which indicates the low vertical diffusion of the pollutants. The constant value of the NO/NO_x ratio along the axis of the plume and the gradients at the edges of the plume show that there was few horizontal entrainment also.
One flight-track was flown in two different heights below the inversion. From massflux calculations the NO_x emission strength for Munich city was estimated to 2.1 kg/s.
The SO_2 contribution to pollution is negligible. It shows an increase of only 2 ppb above background-level over the urban area.

3.2. Anticyclonic Case

Subsident inversions near groundlevel and weak easterly winds appear during stagnant high pressure influence in the area of interest. Frequently, one observes no-wind conditions at night and beginning weak advection with sun rise. This is a typical feature leading to a pollution enrichment in the air over the city of Munich by night, as well as over other densely populated areas and to a slow leeward movement of the highly polluted air-package after the sunrise forming a highly polluted urban plume.
Such conditions were present on September 13. 1985. The measurements were carried out from 12:00 to 14:00 UTC. The figures 4/5 show the distributions of the NO_x and O_3 concentrations.
During midnight a no-wind condition resulted in a pollution enrichment in the air over the city. Before the measurements the polluted air package was shifted by the weak winds (1-2 m/s) to the western border of Munich.
The obtained background concentrations were: NO_x: 6 ppb, O_3: 40 ppb, SO_2: 5 ppb. NO_x and O_3 concentrations of more than 190 ppb and 70 ppb respectivly were measured inside the plume. Also the NO concentration reached values of partly 75 ppb which reduced O_3 concentrations to less than 20 ppb.
In connection with the low advection velocity, thermal circulations and topographic conditions have a strong influence on the movement of the polluted air mass. Computer simulations show that the air package was first moved by katabatic surface winds to the north (4:00-7:00 UTC). After insolation started to heat the ABL the wind direction changed from southerly to north easterly direction and thus the polluted air moves slowly back towards the city (8:00-13:00 UTC). Since the NO/NO_x ratio depends on the travel time of the polluted air different NO/NO_x ratios were found close together in the air over Munich. The O_3 concentration showed changes of more than 25 ppb in a range of 4 km.
Some flight tracks crossed the highway from Munich to Stuttgart. NO_x concentrations of 95 ppb besides the highway were observed. This is nearly 50% of the maximum values found in the urban plume. Due to the emission situation in Munich, the main sources of NO_x can be assumed to be the car exhaust, domestic heating systems and power-plants. The production of photooxidants strongly depends on the presence of hydrocarbons and NO_x. Both species are

found in the exhaust fumes of vehicles. Therefore, questions on transport and chemical reaction of car exhausts must consider the highways as well as the urban agglomeration areas.
The photochemical production of O_3 inside the urban plume leads to O_3 concentrations ecxeeding the background of more than 30 ppb. Under similar meteorological conditions O_3 concentrations of 135 ppb were found inside the urban plume in Munich on August 10. 1983.
The SO_2 measurements again indicate that there are no significant SO_2 sources present in Munich. The SO_2 concentration inside the polluted air mass is less than 5 ppb. The mesoscale variation of the SO_2 concentration in the area of interest frequently shows an increase from south to north. This observation becomes clear from recent measurements which investigated the long-range transport of SO_2 under weak-wind inversion situations (Paffrath, Peters, Rösler, Baumbach 1987). These measurements show that the predominant part of high SO_2 immissions in Germany during the mentioned meteorological conditions can be due to long-range transport from our eastern neighbouring countries, CSSR and GDR. The concentration gradient from south to north is caused by the SO_2 source distribution in the CSSR and the GDR.

4. Simulation Experiments

Numerical modelling of transport and dispersion of pollutants over complex terrain is nowadays an important instrument not only for theoretical work, but also for interpreting measurements and for planning new experiments. Such models can show pollution movements continously in time and space and they allow simulation experiments by changing the boundary conditions. Figure 6 shows the simulation of the distribution of a non-reacting pollutant by the model REWIMET for the anticyclonic case on August 13. 1985 at 12 UTC. In order to compare the measurement with the simulation, the NO_x is assumed to be an inert quantity. The agreement between the simulation result and the measured distribution is rather good (fig. 4/6).
Figure 7 shows the result of simulation after the following 3-hours time intervall at 15 UTC. This distribution is interesting because of the sharp left turn of the plume indicating some possible topographic effect on the transport and dispersion mechanism. To investigate this question we ran the model twice with the same input parameters, except the ground-level elevation inventory data and the surface roughness data. To eliminate the influence of orography the groundlevel elevation was set constant over the whole area during the first run (fig. 8). In the second run, we set the surface roughness to be constant over the whole area considering only the groundlevel elevation data (fig. 9).
It is evident that the results in figures 8 and 9 deviate significantly from the calculated distribution in figure 7, where the whole topographic influence was considered. These results were not verified by corresponding measurements. The simulation results indicate a considerable influence of topography on transport and dispersion under the mentioned weather situations. These results are partly verified (Rösler 1986) and are to be taken as a basis for new experimental design.

5. Conclusions

The measurements during weak-wind inversion situations leeside of the city of Munich helps to understand the three-dimensional mixing and transport processes relevant for immissions released in urban agglomeration areas mainly by cars and domestic heating. With the used aircraft and measuring technique, it was possible to gather concentration distributions in an acceptable time periode. The measurements show that in urban plumes of Munich NO_x and O_3 concentrations of 190 ppb and 140 ppb respectivly can occure. It can be assumed that under extreme photochemical conditions O_3 concentrations of 150 ppb or more may be generated in the plume. SO_2 pollution obviously plays a minor role in the area of Munich. Further efforts should be made in investigating the impacts of the increasing O_3 concentrations in the biosphere, particularly on forest ecosystems?

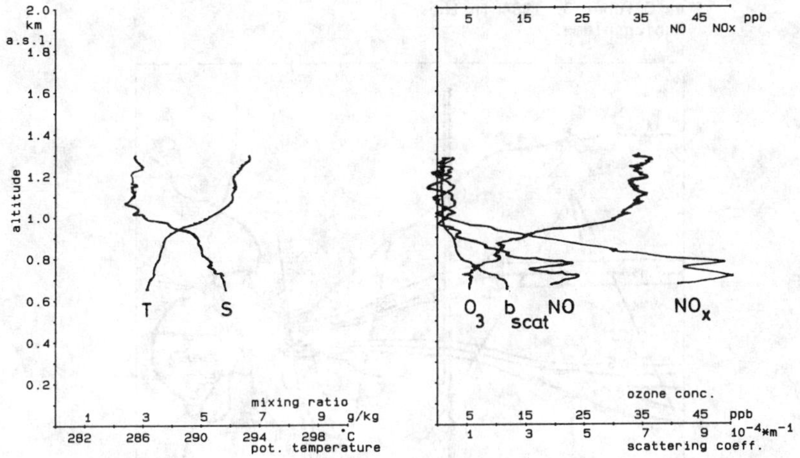

Fig. 1: Vertikal sounding 60 km west of the city Munich in the middle of the plume.

Fig. 2: NOx concentration distribution (in ppb) west of Munich measured on October 3. 1984 during a Föhn event along the indicated flight tracks

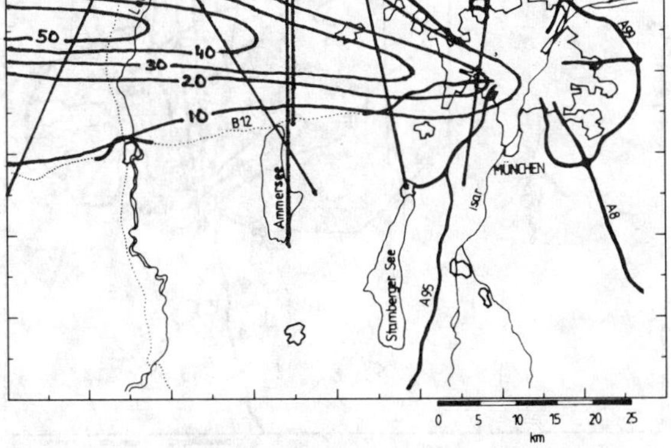

Fig. 3: Distribution of NO/NOx ratio on October 3. 1984 in the Munich plume.

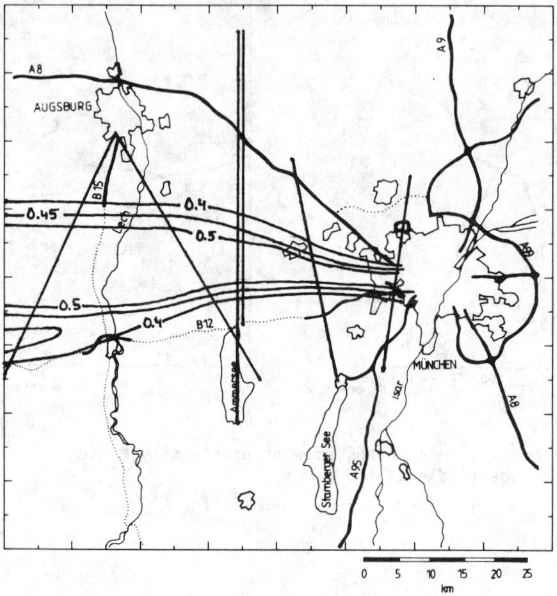

Fig. 4: NOx concentration distrubution (in ppb) west of Munich measured along the indicated flight tracks on September 13. 1985 during an anticyclonic situation.

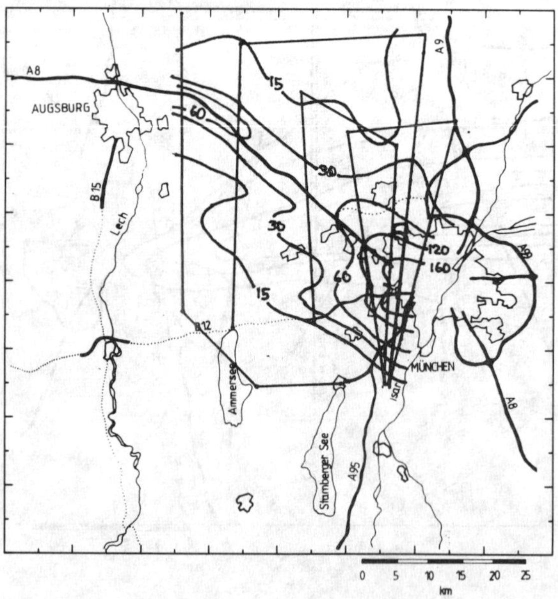

Fig. 5: Ozone concentration distribution measured on September13.1985 in the Munich plume (in ppb)

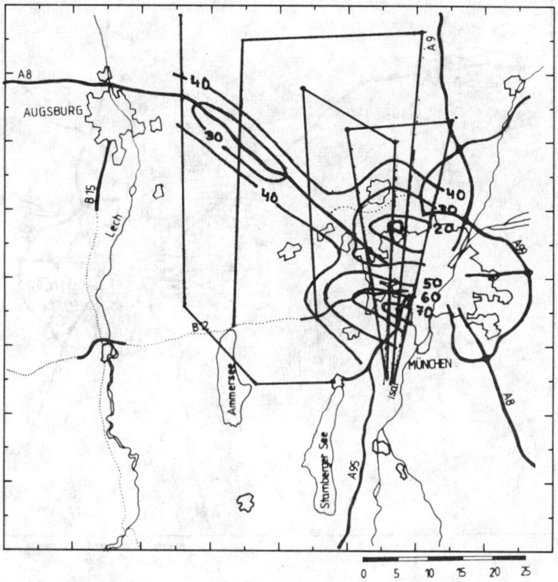

Fig. 6: Numerical simulation of the distribution for a non reacting pollutant for the September 13. 1985 (12 UTC). This simulation result is qualitativly comparable with the measurement of fig. 4

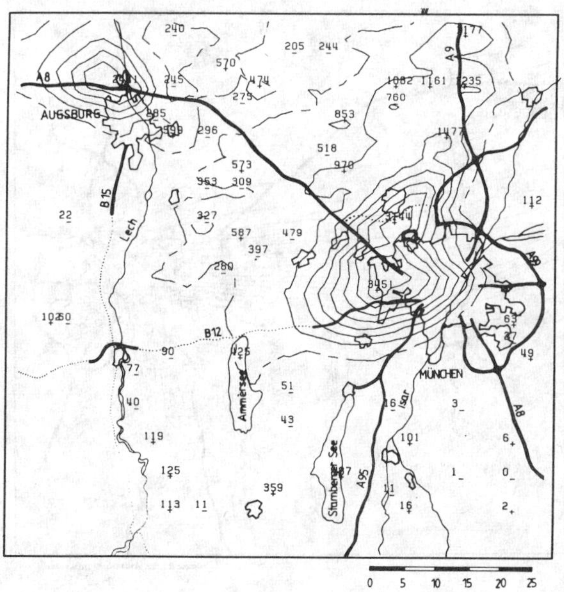

Fig. 7: Modelprediction for the distribution after the following 3-hours time intervall (15 UTC).

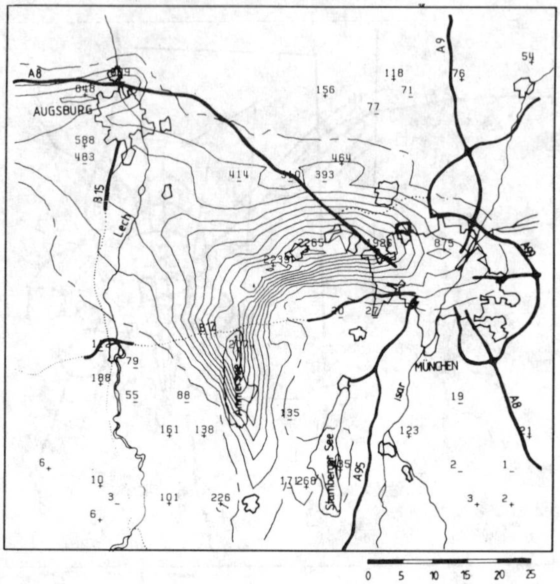

Fig. 8: Simulation using the same input parameters as in fig. 7, except the groundlevel elevation which was set constant in the modelling area. The different surface roughness was considered.

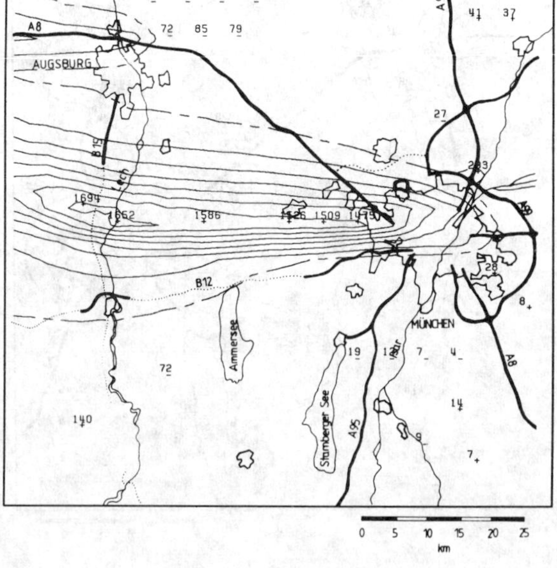

Fig. 9: Simulation using the same input parameters as in fig. 7. The surface roughness was set constant in the whole area and the groundlevel elevation inventory was considered.

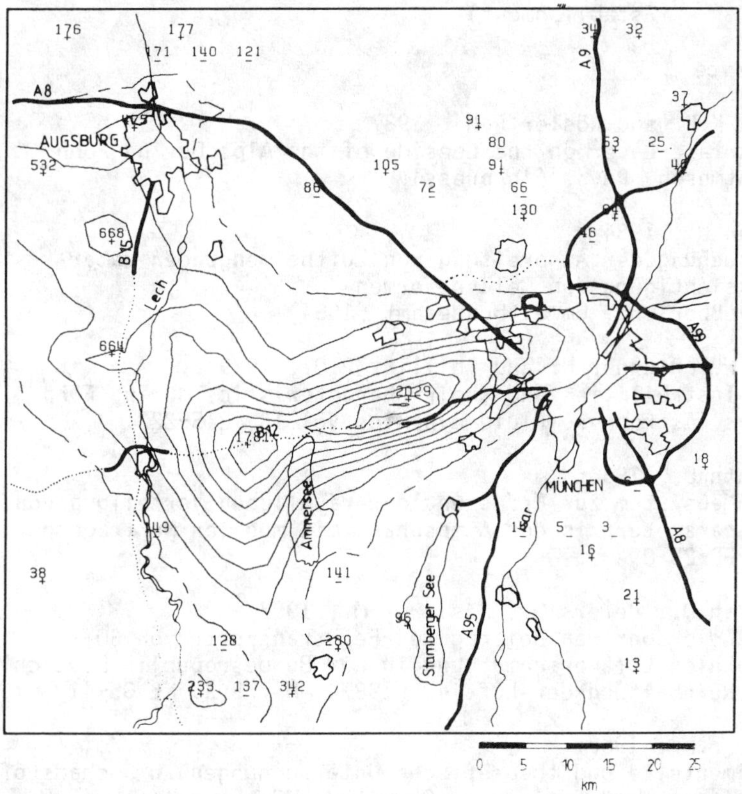

Abbreveations

GND	height above ground
MSL	height above mean sea level
ABL	atmospheric boundary layer
UTC	universal time coordinated
DFVLR	Deutsche Forschungs- und Versuchsanstalt für Luft- und Raumfahrt (German Aerospace Research Establishment)

Reference

Hoinka K.P. and Rösler F.M., 1987
The Surface Layer on the Leeside of the Alps During Föhn
Met. Atmosph. Phys. (in press)

Heimann D., 1984
Untersuchung der Ausbreitung von Luftbeimengungen unter Berücksichtigung von Geländeformen
Abschlußbericht, Umweltbundesamt (1984)

Nater W., Richner H. and Phillips P.D., 1979
Shear Instabilities and their Characteristics during Föhn
Geophys. Astrophys. Fluid Dynamics Vol.13, 215-223

Paffrath D., 1985:
DFVLR-Meßsystem zur Erfassung der räumlichen Verteilung von Umweltparametern in der Atmosphäre mit mobilen Meßträgern
DFVLR-FB 85-09

Paffrath D., Peters W., Rösler F.M., 1987
Fallstudie über den Beitrag des Ferntransports von SO_2 zur lokalen Luftverschmutzung in der Bundesrepublik Deutschland
Staub-Reinhaltung der Luft 47 (1987), H 7/8 S. ML 35-41

Rösler F.M., 1986:
Experimentelle und theoretische Untersuchungen zur Schadstoffausbreitung am Beispiel der Großstadt München
Diplomarbeit, TU-München, Physik Department

THE DETERMINATION OF MIXING HEIGHTS BY SODAR IN AN URBAN ENVIRONMENT

M. Piringer
Central Institute for Meteorology and Geodynamics
Hohe Warte 38
1190 Wien
Austria

ABSTRACT. During July and August 1987, the project "VISAS", short for "Vienna Summer Aerosol Study", was carried out. Its purpose was to determine the chemical and meteorological conditions in the summertime boundary layer over the city of Vienna esp. during anticyclonic conditions. For the meteorological measurements, a SODAR and a tethersonde supplemented the routine network. Three examples of the SODAR's performance in detecting mixing heights are shown; these are compared with mixing heights derived from the tethersonde and from radiosonde temperature profiles. Often, there is a good agreement between the three systems; differences and failures can be explained meteorologically.

1. INTRODUCTION

The producers of SODARS offer convenient and easy-to-handle software tools which greatly facilitate the interpretation of the large amount of data delivered by the instrument. Nevertheless, the detection of mixing heights by a single criterion (e.g. σ_w) can give only very rough estimates if the many different physical mechanisms leading to mixing layers of various depths in the boundary layer are kept in mind.

The purpose of this paper is to show that analysis of SODAR data can still be a very delicate task and needs careful interpretation by the meteorologist and sometimes additional meteorological information.

The examples shown were derived from the project "VISAS", short for "Vienna Summer Aerosol Study" (Puxbaum et al., in preparation).

During July and August 1987, the Institutes of Analytical

K. Grefen and J. Löbel (eds.), Environmental Meteorology, 425–444.

Chemistry, Experimental Physics and Meteorology of the University of Vienna carried out the "VISAS" field study sponsored by the community of Vienna to determine the chemical and meteorological conditions in the summertime boundary layer over the city of Vienna. For the meteorological measurements, a SODAR and a tethered balloon-system were used; interpretation of SODAR echoes was facilitated by radiosonde data twice daily at 0 and 12 GMT, and additional wind profiles at 6 and 18 GMT.

2. AIMS OF THE "VISAS" PROJECT

The purpose of the enhanced meteorological network was the following:

1. to get time-height-variations of meteorological parameters,especially of wind and temperature, and to get information on onset, prevailing, mean heights and disappearance of inversion layers and mixing heights under various synoptic situations;
2. to construct statistics of mixing heights in summer over Vienna under anticyclonic conditions.

This kind of information should allow for the interpretation of transport mechanisms,especially the inflow of clean or polluted air and the outflow of city-generated pollution through the city boundaries. These results are not available yet. The meteorological measuring campaign partly missed its aims for two reasons:

1. Meteorological conditions were seldom anticyclonic during the extensive field programme between July 15 and August 15, 1987 and the 10 days-extention added. Available data is not sufficient to create the mixing height statistics mentioned before.
2. Because of a hardware defect, the SODAR was out of function from July 23 on. Until the end of the field programme, there remained only the tethersonde to measure atmospheric stability whose operation was limited by wind or rain.

3. PRINCIPLES OF SODAR OPERATION; THE MEASURING SITE

The SODAR operates on a backscatter principle very similar to that of radar. Pulses of sound are transmitted into the atmosphere by three antennas alternatively. One of the antennas is oriented vertically, the others are tilted 18° to the vertical; the angle between the tilted antennas is 90°.

Acoustic backscattering is caused by small-scale temperature inhomogeneities, which are produced by turbulence in regions of larger-scale potential temperature gradients. The backscattered signal, the echoe, undergoes a Doppler-shift because of the mean and turbulent flow in the scattering volume.

Besides the analysis of the value of the backscattered signal, the three-dimensional configuration of the SODAR allows for measurements of the three wind components and the standard deviations of wind direction and vertical velocity from the frequency shifts.

An extended description of acoustic measurement in general is found in Brown and Hall jr. (1978).

In Austria, a REMTECH SODAR is in operation since the beginning of 1986. Data are displayed by means of REMTECH software on a colour screen and a colour printer.

The operational mode chosen was 10 to 15 minutes of measurements per halfhour. Vertical averaging was 40 meters, begining at 60 m and extending up to 820 m (20 layers).

The SODAR was in operation on the grounds of the Vienna International Fair near the center of Vienna. The tethersonde was positioned in a large park in the central area 3 km apart from the SODAR.

4. MIXING HEIGHT DETERMINATION BY SODAR

Definitions and concepts of mixing heights vary (e.g. Ball (1960), Benkley et al. (1979), Gamo et al. (1979)), but they are all related to turbulence. Turbulence in the boundary layer is either generated by the upward directed transport of heat ("thermal turbulence" in the daytime boundary layer heated by the sun) or by wind shear ("mechanical turbulence" in the night-time shear-driven boundary layer).

In the case of thermal turbulence, it is commonly accepted that the top of the ground-based convective layer with an almost adiabatic temperature lapse rate and strong vertical mixing, capped by a stable layer or an inversion, is identical with the top of the mixing layer. In the coloured SODAR outputs the top of the daytime mixing layer is assumed to be where echoe intensities aloft increase again.

The mixing height concept was extended later on to the nocturnal ground-based shear layer. The layer is stably stratified, nevertheless there is some turbulence because of strong vertical wind shear. The mixing height is then considered to be the maximum height to which pollutants actually mix (by turbulence) during the period of concern (Olsson et al. (1974)). In the colour SODAR-outputs, the layer showing a strong vertical gradient in echo intensities is considered to be the top of the nighttime mixing layer. A detailed description of SODAR data processing is found in Russell and Uthe (1978); the method outlined there was adjusted to colour prints.

5. RESULTS

5.1 Combined SODAR-tethersonde measurements

The combined measurement period lasted from July 15 to July 17, but only the first four hours giving the most interesting results will be discussed.

On July 15, there was a high pressure area over Scandinavia and Eastern Europe, Austria lying in an easterly current. Skies were clear over Vienna during the night.

The radiosonde profile (fig. 1 a) showed a stable layer between 200 and 500 m above ground at 0 GMT on July 15. The shallow ground-based inversion detected by the radiosonde is assumed to be site-specific, because it is shown neither by the tethersonde nor by the SODAR. At this time, the echo intensities (fig. 2) had high values near the ground and decreased gradually with increasing height, not giving any indication of an inversion aloft. The vertical directional wind shear was very pronounced (fig. 1 a). Though the layers below 200 and above 500 m showed a temperature decrease, the lapse rate was slightly stable in the whole first km from the ground, which might have been the cause of weak turbulence everywhere giving rise to only poor SODAR echoes.

The detailed comparison between the tethersonde temperature profiles and the SODAR echoes shows an echo maximum at the inversion base at about 2 a.m. (fig. 3 a), but there were several echo maxima half an hour later below and throughout the inversion (fig. 3 b). The highest echoe values always occured at the lowest level. This is not a very encouraging result for echo interpretation. It is again postulated that if the atmosphere is stably stratified through a layer deeper than the SODAR range without pronounced potential temperature gradients, echoes are weak in general and higher values are not restricted to shallow inversions amidst the stable layer.

This was true also for the two following nights of the tethersonde measurements where there was thermal stability throughout the first km of the atmosphere but without any larger potential temperature gradients.

5.2 A frontal passage

A cold front associated with thunderstorms crossed Austria from July 8 to July 9 and interrupted a prolonged anticyclonic period. In the Vienna area, there was typically only little rain, but a marked jump in wind direction (fig. 4) and a change in the sign of vertical velocity at about 10.30 p.m. (fig. 5). Positive vertical velocities occured again with thundershowers in the morning. Echo intensities showed no significances associated with the frontal passage.

At the time of the passage, the SODAR performance was poor, possibly because of the rain and the neutrally stratified atmosphere; wind measurements left out completely for one measuring cycle.

5.3 Anticyclonic conditions, no ground-based inversion

Normally one would expect that under anticyclonic conditions in summer with clear nights, a ground-based radioative inversion would develop. In the Vienna area however, the build-up of a nocturnal ground-based inversion under high atmospheric pressure seems to depend on the location of the high pressure centre. If the centre is located to the west or northwest of Austria, a northerly airflow transporting well-mixed cool and moist air in all layers is likely to prevent the build-up of a ground-based inversion in many cases.

There was an anticyclonic period from July 2 to July 6 where the high pressure centre always remained to the northwest of Austria (over the North Sea). During this time period, no nighttime ground-based temperature inversion occured in Vienna. Results of July 3 are discussed. The radiosonde profile of Vienna at 0 GMT on July 3 (fig. 1 b) shows a ground-based adiabatic layer and a stable layer about 300 m aloft, extending to about 1 km and containing two inversions. There is considerable wind shear up to the first inversion base. The stable layer aloft can also be seen from the echo profiles (fig. 6); the increase in intensity coincides well with the inversion base. The inversion base remained in 250 m above ground during the second half of the night. Because of the negative vertical velocities throughout all layers (fig. 7), it is suggested that the stable layer was caused by continued subsidence.

Because of the heating of the ground by radiation after 8 a.m., the inversion base was lifted continually to greater heights, as can be seen from the echoes (fig. 6). The midday radiosonde temperature profile shows the inversion base at about 1 km above ground (fig. 1 c). The development of the wellmixed convective cells extend up to 200 m.

The end of ground-based convection was at about 19.00; a stable layer aloft developed again after 22.00 with the establishment of marked subsidence (fig. 6, fig. 7).

5.4 Anticyclonic conditions, ground-based inversion

On July 6, the high pressure centre crossed Austria and moved to Eastern Europe, creating a southerly airflow over Austria. During the night from July 6 to July 7, in the warm southerly airflow a ground-based temperature inversion developed (fig. 1 d).

The air was drier than on July 3 throughout the whole layer. There was considerable wind shear (in both direction and velocity) in the inversion layer. The radiosonde profile gives the top of the stable layer in 350 m above ground at 0 GMT.

The SODAR echo and wind profiles (fig. 8, 9 a - 9 d) show the situation in more detail. The time scale ranges from 12 noon on July 6 to 12 noon on July 7. The daytime convective layer can easily be recognized, also its complete breakdown at about 7 p.m. Vertical velocities were more upward directed than on July 3 (fig. 10). In all layers, a weak easterly airflow was dominant. Soon after sunset (before 8 p.m.), a ground-based inversion layer developed. At the time of the midnight radiosonde temperature profile, the SODAR shows the inversion height only in about 180 m above ground (instead of 350 m as by the radiosonde temperature profile) (fig. 1 d). The difference can be explained by the fact that the strong potential temperature gradient was restricted to these heights, whereas in the isothermal layer above, echoe intensities remained low because of weak potential temperature gradients. Thus, the height of a ground-based stable layer might be underestimated by the SODAR under such conditions.

During the second half of the night, the inversion deepened and the potential temperature gradients became sharp especially around 6 in the morning (fig. 8). There was a pronounced wind shear in the inversion layer.

With the onset of insolation, the inversion base rose successively to greater heights (fig. 8). Because the SODAR reached only heights of about 300 m (perhaps due to ambient noise), the development of the daytime turbulent mixing layer cannot be followed in detail.

6. CONCLUSIONS

be an eccellent tool in continously detecting either the daytime convective boundary layer or the nighttime shear-driven boundary layer and the associated mixing heights. The importance of the mixing height for the concentration of air pollution in the atmosphere has not been stressed so far. The results gained during the VISAS field study are not available in full detail yet, but there is some indication that there are big differences in the ozone and nitrogen oxide contents between the ground-based nighttime shear layer containing low ozone and high nitrogen oxide concentrations and the neutral layer above, with undistrubed high ozone values and very low NO_x content. During the episode, a chemical measuring site situated in the Vienna surrounding forests to the west of the city 300 m above city level was overflown by polluted city air when the ground-based inversion grew higher than 300 m.

REFERENCES:

Ball, F.K. (1960): Control of inversion height by surface heating. Quart. J. R. Met. Soc. 86, 483-494.

Benkley, C.W. and L.L. Schulman (1979): Estimating hourly mixing depths from historical meteorological data. J. Appl. Meteor. 18, 772-780.

Brown, E.H. and F.F. Hall jr. (1978): Advances in atmospheric acoustics. Rev. Geophys. Space Phys. 16, No. 1, 47-110.

Gamo, M. and O. Yokoyama (1979): Growth of the mixing depth and the diurnal variation of vertical profiles of temperature and turbulence characteristics in the mixing layer. J. Met. Soc. Japan, 57, Nr. 2, 159-171.

Olsson, L.E., M.P. McCormick, W.P. Elliott, S.H. Melfi (1974): An observational study of the mixing layer in western Oregon. Atm. Environment 8, 241-252.

Puxbaum, H., et al.: The Vienna Summer aerosol study of 1987 (in preparation).

Russell, P.B. and E.E. Uthe (1978): Regional patterns of mixing depth and stability. Sodar network measurements for input to air quality models. Bull. Am. Met. Soc. 59, No. 10, 1275-1287.

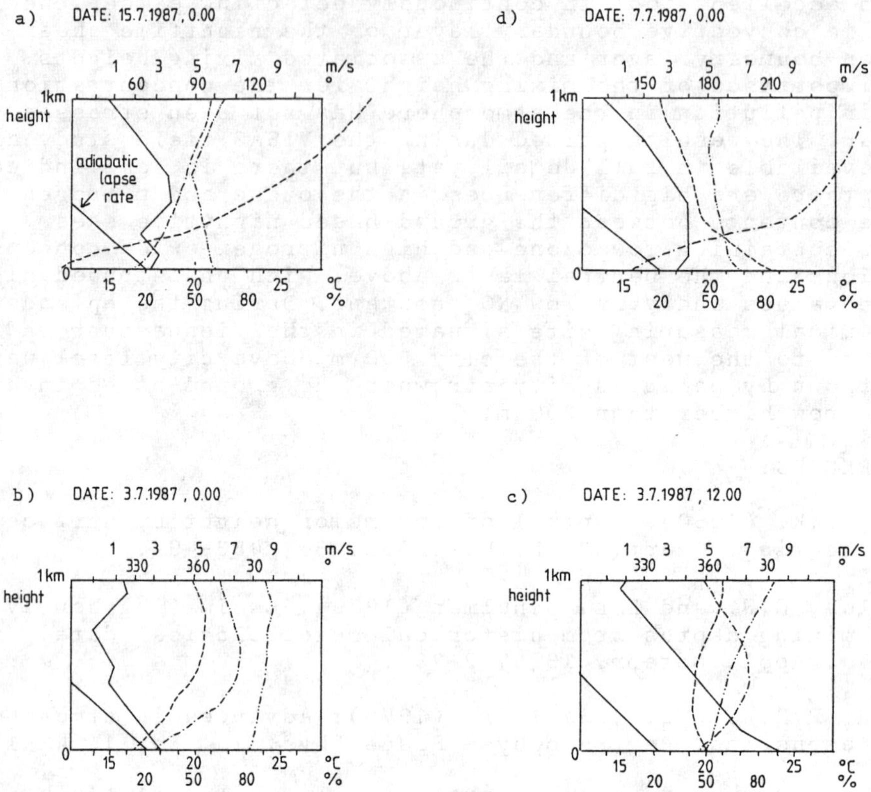

FIG. 1: RADIOSONDE PROFILES VIENNA

——— TEMPERATURE
-··-··- REL. HUMIDITY
– – – – WIND DIRECTION
-·-·-·- WIND VELOCITY

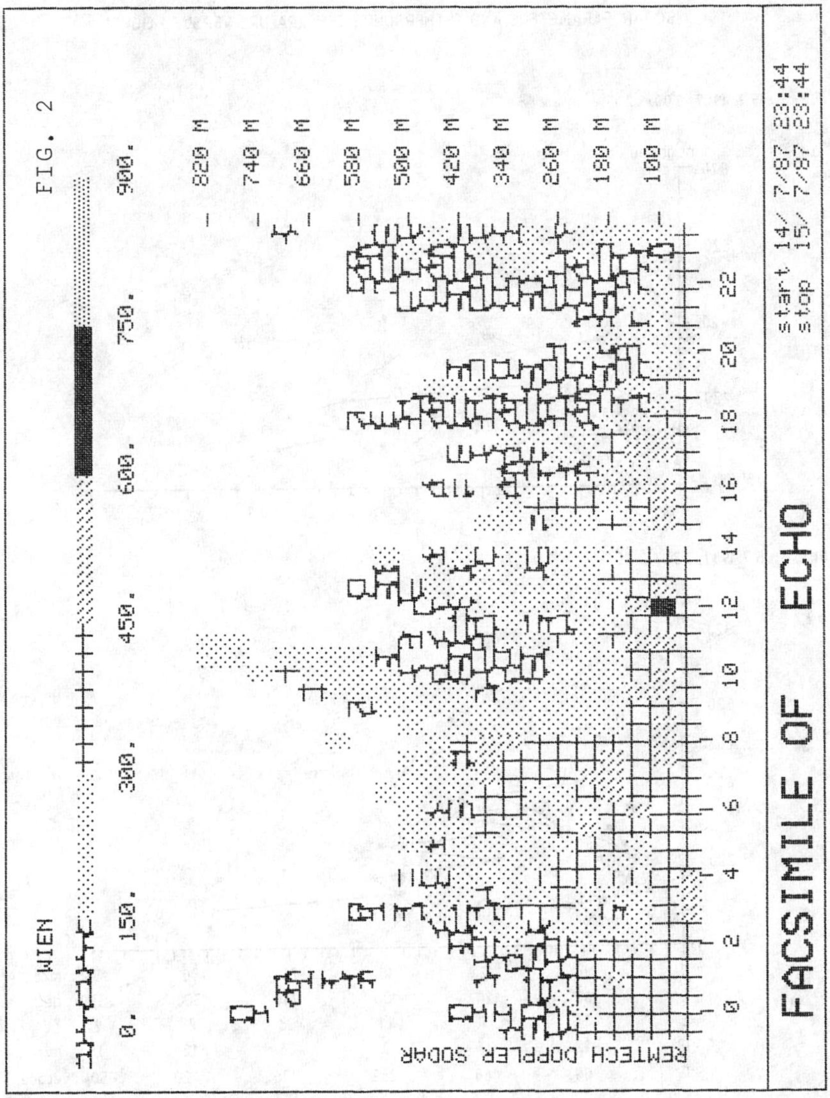
WIEN
0. 150. 300. 450. 600. 750. 900.
820 M
740 M
660 M
580 M
500 M
420 M
340 M
260 M
180 M
100 M
REMTECH DOPPLER SODAR
0 2 4 6 8 10 12 14 16 18 20 22
start 14/ 7/87 23:44
stop 15/ 7/87 23:44
FACSIMILE OF ECHO

FIG. 2

SODAR PARAMETERS AND TETHERSONDE TEMPERATURE VERSUS HEIGHT

FIG. 3 : 15.7. 1987, 2.00 – 2.15

a)

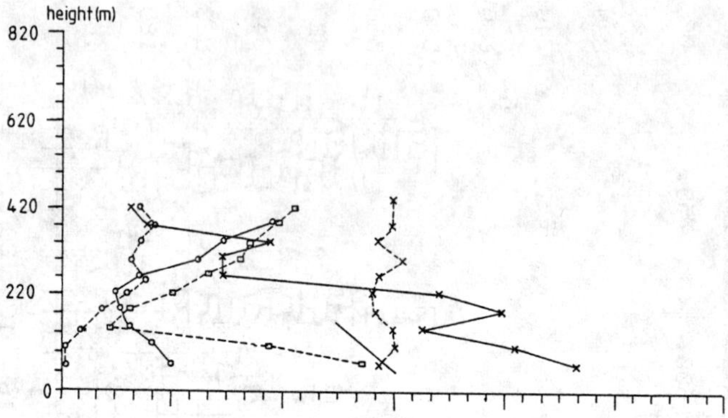

FIG. 3: 15.7.1987, 2.30 – 2.45

b)

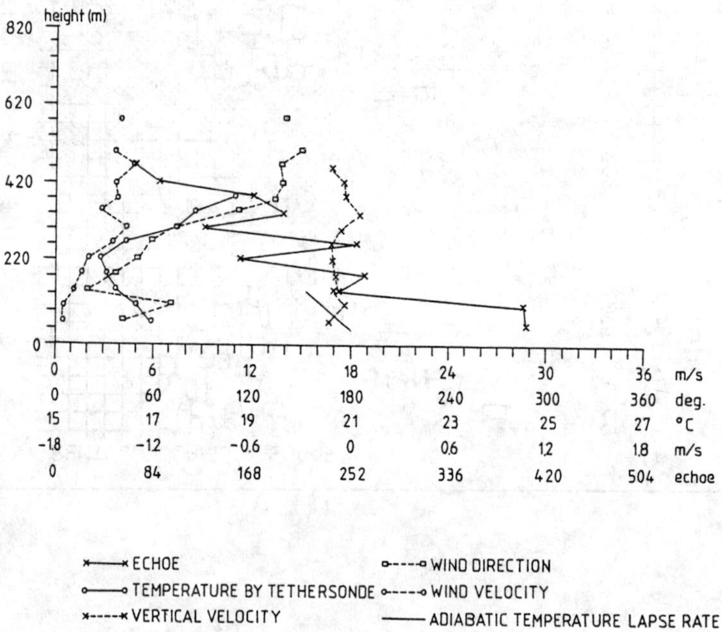

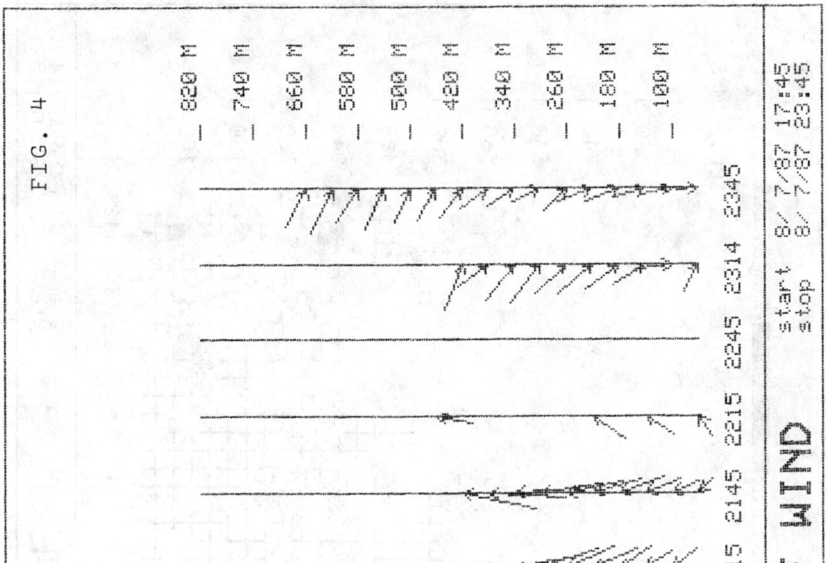
WIEN
1000.CM/S SCALE
REMTECH DOPPLER SODAR
820 M
740 M
660 M
580 M
500 M
420 M
340 M
260 M
180 M
100 M
1814 1844 1914 1945 2015 2045 2115 2145 2215 2245 2314 2345
VECTORIAL PLOT OF WIND
start 8/ 7/87 17:45
stop 8/ 7/87 23:45

FIG. 4

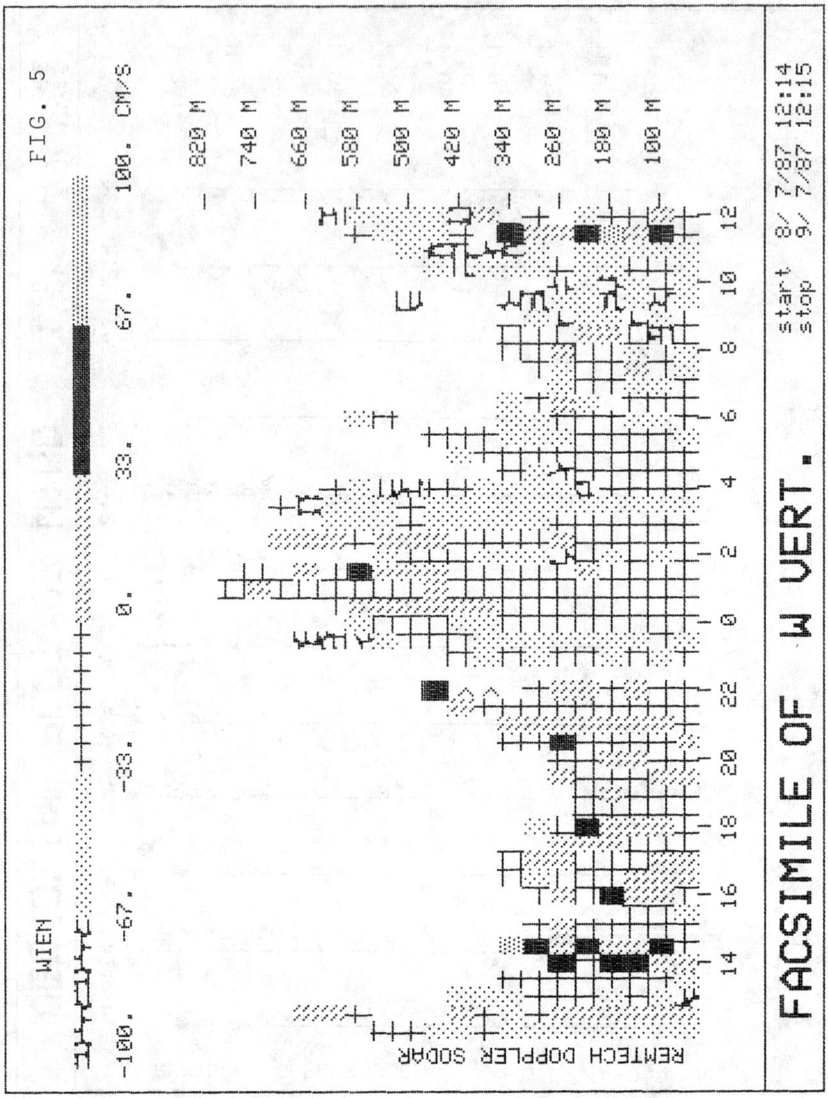
FIG. 5
WIEN
-100. -67. -33. 0. 33. 67. 100. CM/S
820 M
740 M
660 M
580 M
500 M
420 M
340 M
260 M
180 M
100 M
REMTECH DOPPLER SODAR
14 16 18 20 22 0 2 4 6 8 10 12
FACSIMILE OF W VERT.
start 8/ 7/87 12:14
stop 9/ 7/87 12:15

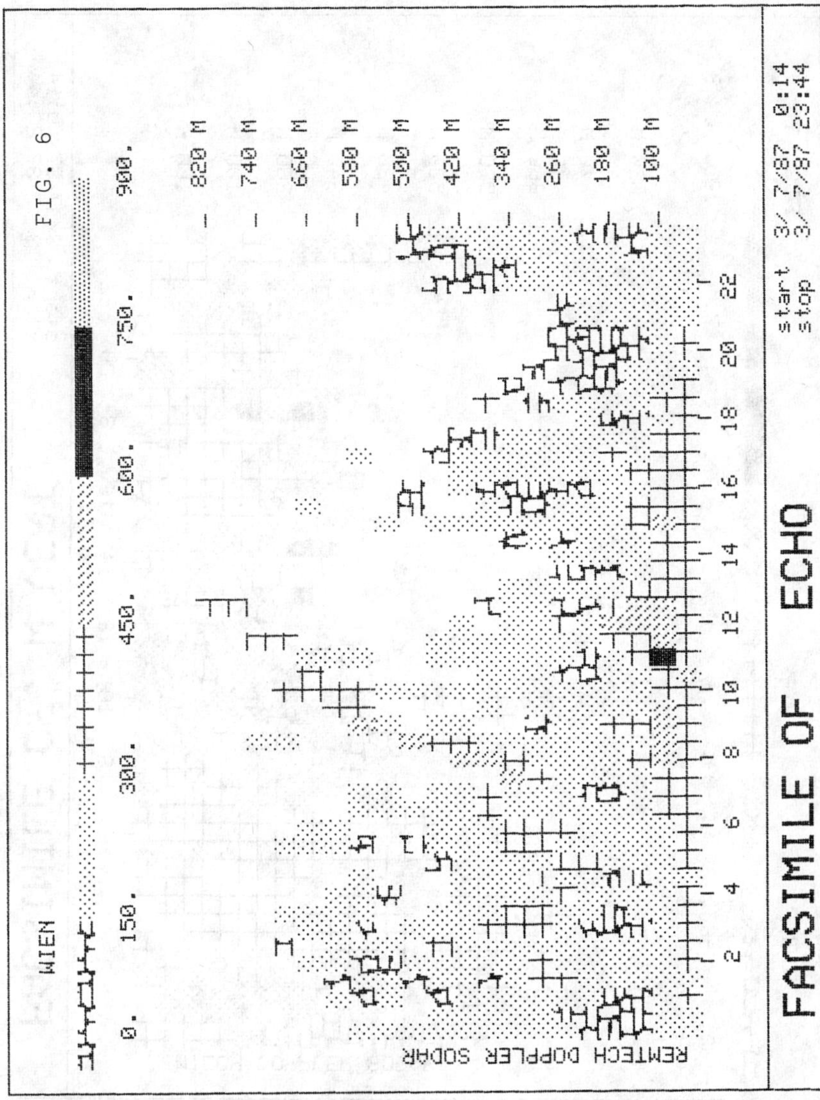
FIG. 6
WIEN
0.
150.
300.
450.
600.
750.
900.
820 M
740 M
660 M
580 M
500 M
420 M
340 M
260 M
180 M
100 M
REMTECH DOPPLER SODAR
2
4
6
8
10
12
14
16
18
20
22
FACSIMILE OF ECHO
start 3/ 7/87 0:14
stop 3/ 7/87 23:44

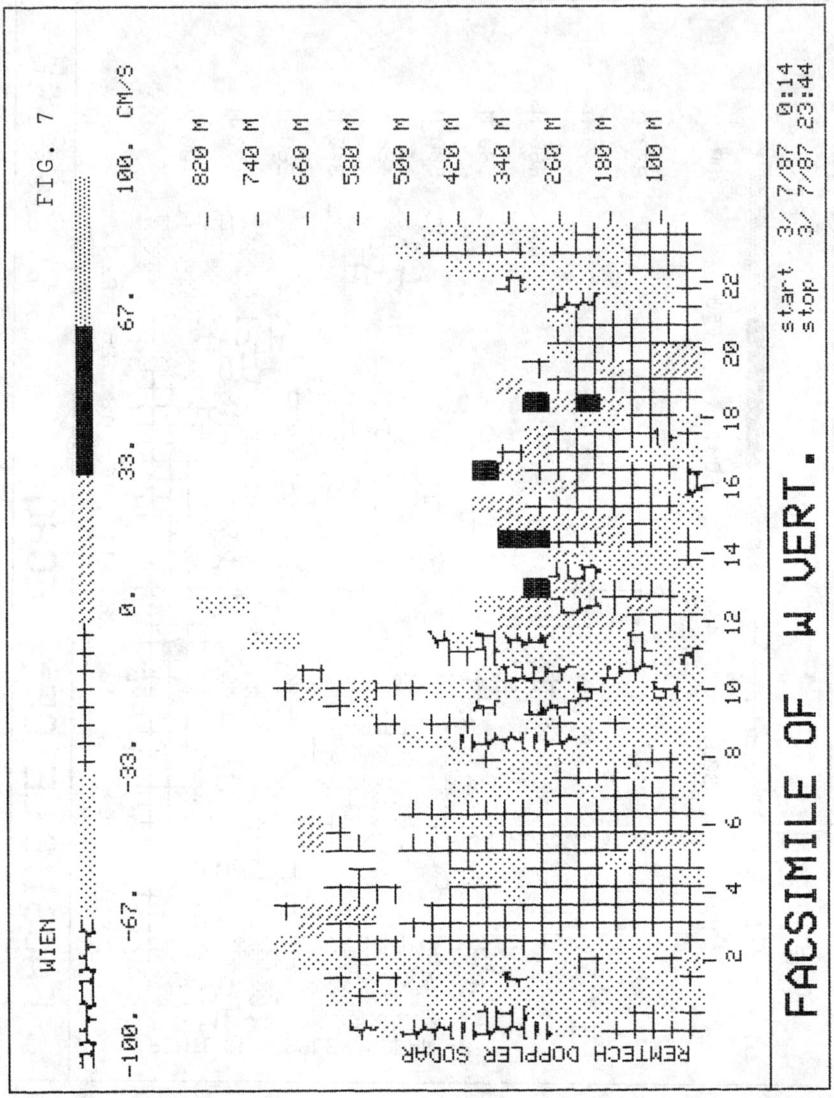
FIG. 7
WIEN
-100. -67. -33. 0. 33. 67. 100. CM/S
820 M
740 M
660 M
580 M
500 M
420 M
340 M
260 M
180 M
100 M
2 4 6 8 10 12 14 16 18 20 22
REMTECH DOPPLER SODAR
FACSIMILE OF W VERT.
start 3/ 7/87 0:14
stop 3/ 7/87 23:44

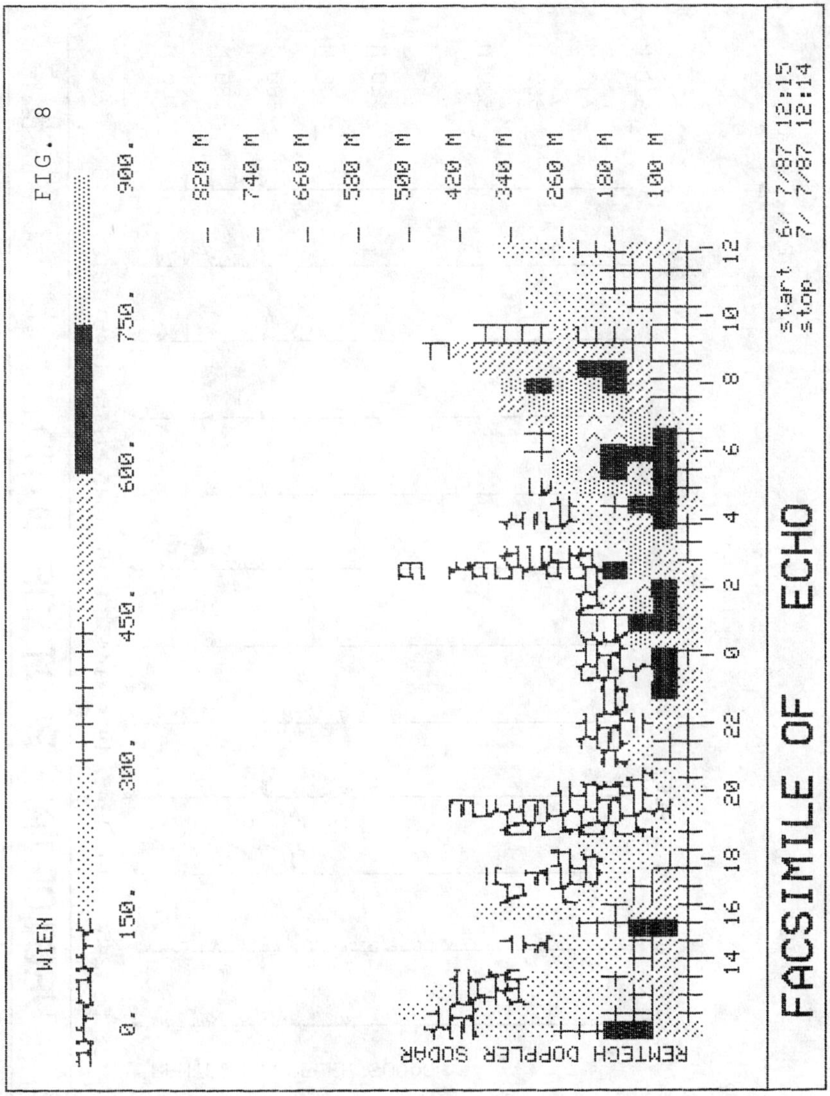
WIEN
FIG. 8
0. 150. 300. 450. 600. 750. 900.
820 M
740 M
660 M
580 M
500 M
420 M
340 M
260 M
180 M
100 M
14 16 18 20 22 0 2 4 6 8 10 12
REMTECH DOPPLER SODAR
FACSIMILE OF ECHO
start 6/ 7/87 12:15
stop 7/ 7/87 12:14

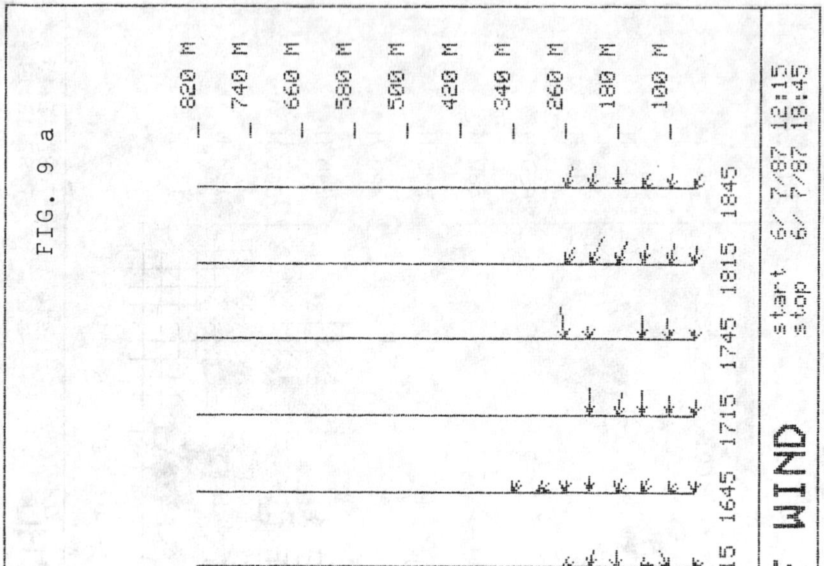
FIG. 9 a
WIEN
1000.CM/S SCALE
REMTECH DOPPLER SODAR
820 M
740 M
660 M
580 M
500 M
420 M
340 M
260 M
180 M
100 M
1215 1245 1314 1345 1414 1445 1515 1645 1715 1745 1815 1845
VECTORIAL PLOT OF WIND
start 6/ 7/87 12:15
stop 6/ 7/87 18:45

WIEN

FIG. 9 b

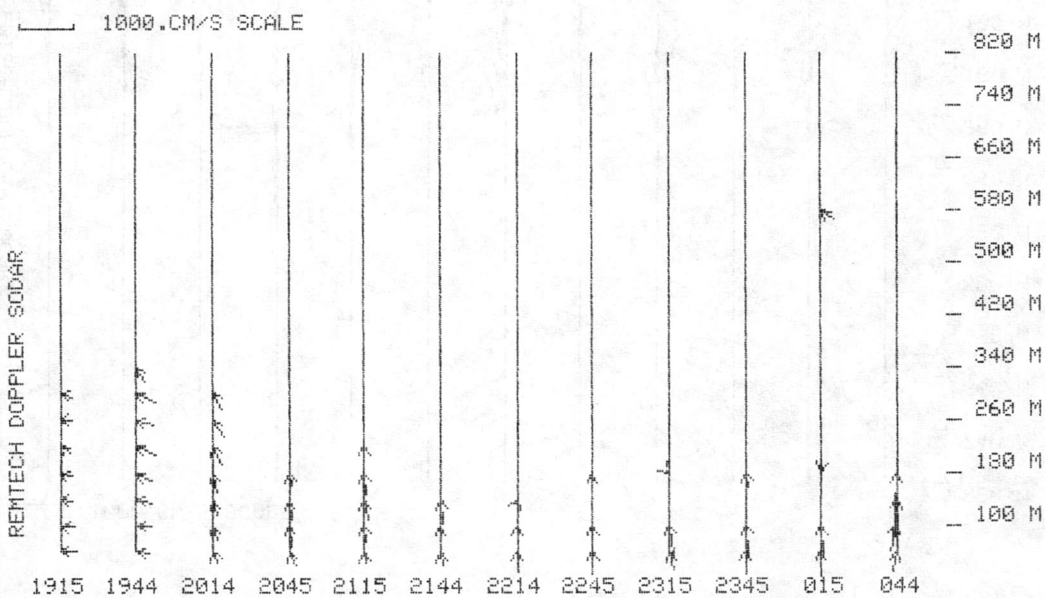

VECTORIAL PLOT OF WIND

start 6/ 7/87 18:45
stop 7/ 7/87 0:44

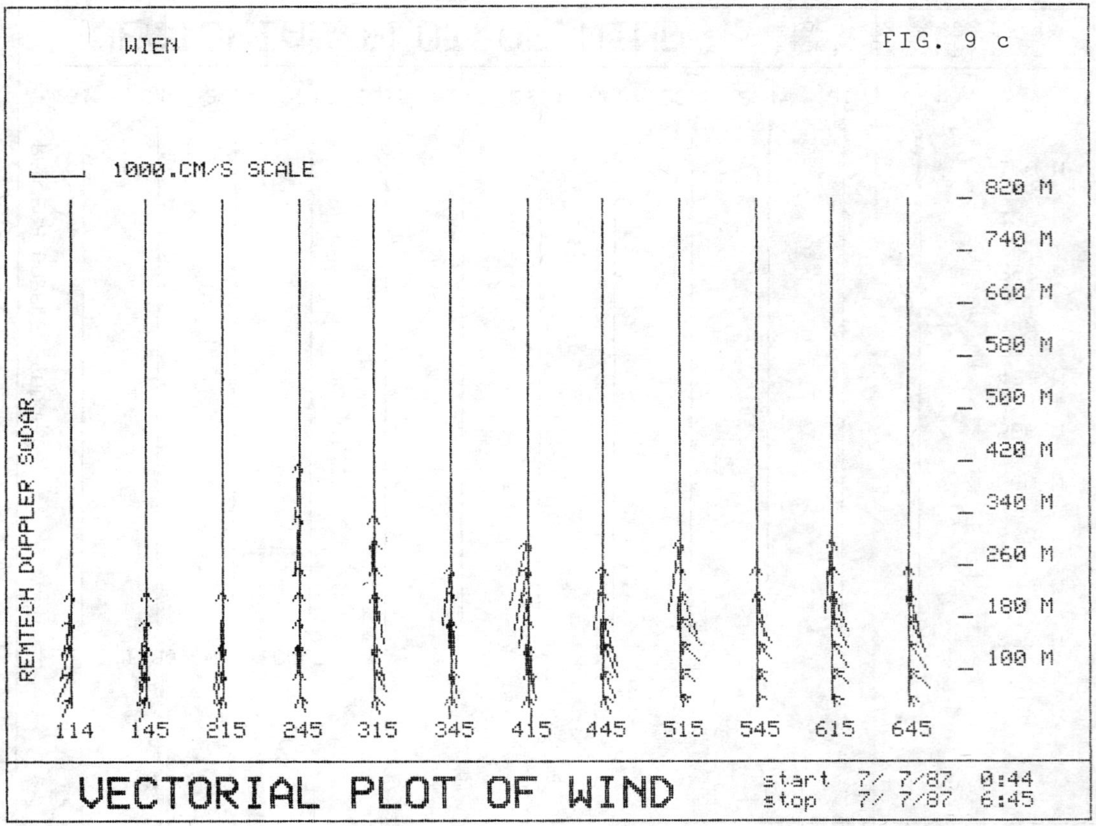
WIEN
1000.CM/S SCALE
REMTECH DOPPLER SODAR
820 M
740 M
660 M
580 M
500 M
420 M
340 M
260 M
180 M
100 M
114 145 215 245 315 345 415 445 515 545 615 645
VECTORIAL PLOT OF WIND
start 7/ 7/87 0:44
stop 7/ 7/87 6:45

FIG. 9 c

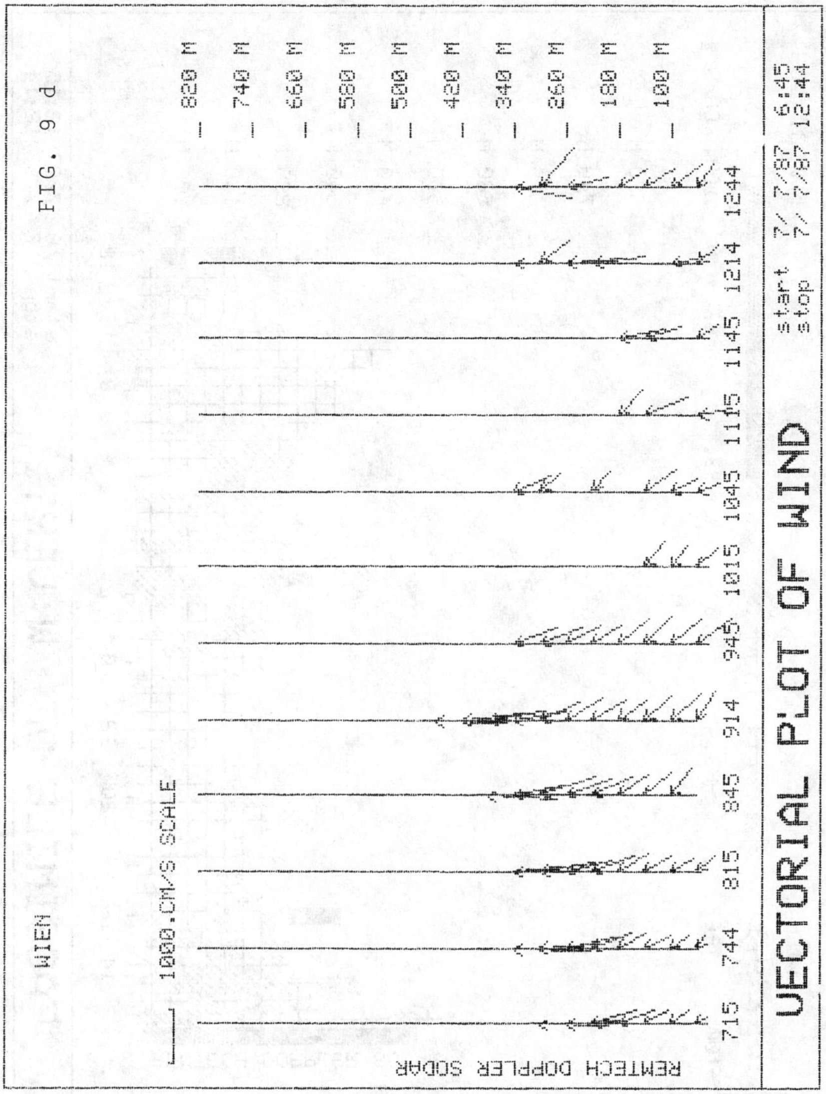
FIG. 9 d
WIEN
1000.CM/S SCALE
REMTECH DOPPLER SODAR
820 M
740 M
660 M
580 M
500 M
420 M
340 M
260 M
180 M
100 M
715 744 815 845 914 945 1015 1045 1115 1145 1214 1244
VECTORIAL PLOT OF WIND
start 7/ 7/87 6:45
stop 7/ 7/87 12:44

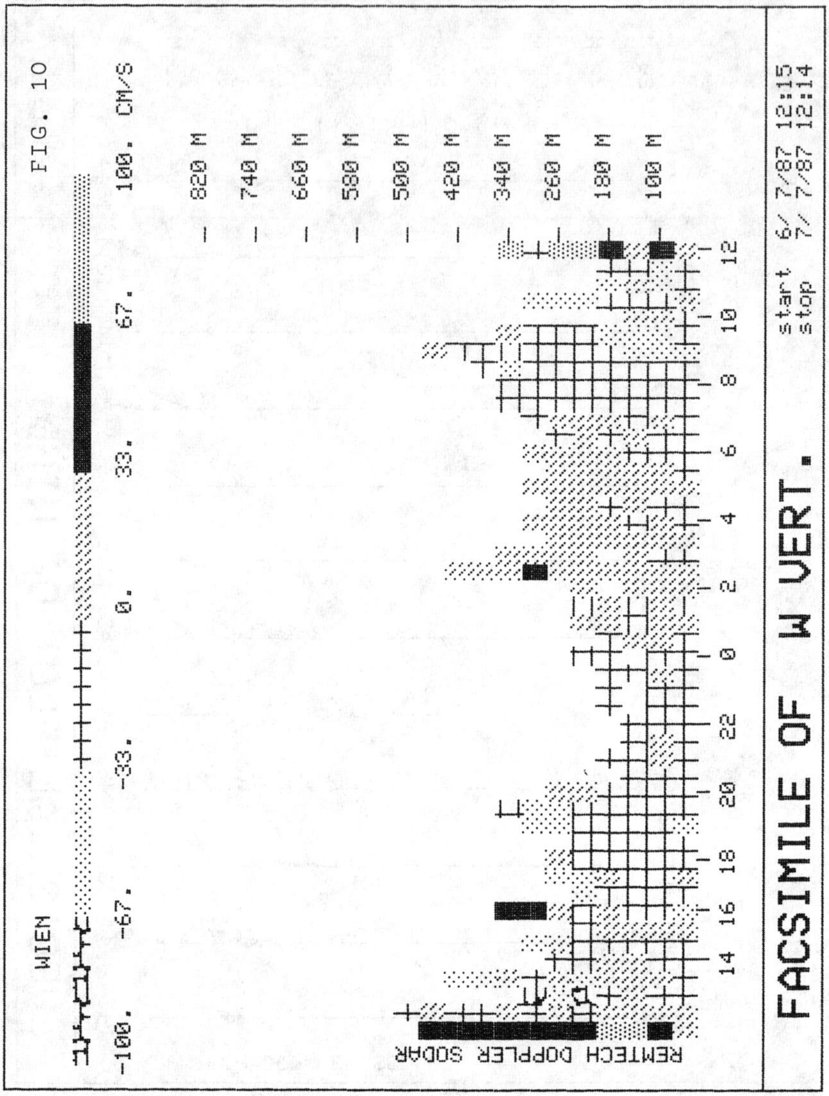

WIEN
FIG. 10
-100. -67. -33. 0. 33. 67. 100. CM/S
820 M
740 M
660 M
580 M
500 M
420 M
340 M
260 M
180 M
100 M
REMTECH DOPPLER SODAR
14 16 18 20 22 0 2 4 6 8 10 12
start 6/ 7/87 12:15
stop 7/ 7/87 12:14
FACSIMILE OF W VERT.

SO_2 CALCULATIONS IN SEVERAL DANISH CITIES.

Per J. Løfstrøm, Lisbeth Stenfalk
Air Pollution Laboratory
Danish National Agency of Environmental Protection
Risø, 4000 Roskilde
Denmark

ABSTRACT. The Danish National Agency of Environmental Protection has in collaboration with local and regional authorities conducted a nationwide Air Quality Evaluation Program for seven Danish cities, with populations ranging from 30.000 to 1.000.000. A mapping was undertaken of SO_2 emissions from larger Point Sources, and from smaller sources grouped together in Area Sources (500 X 500 m²). The emissions from this mapping have been used in model calculations of the SO_2 concentrations in the cities. The OML model, which is a Gaussian model, was applied. Calculations were performed for every hour over a two-year period. A major result is that, in all cities, a few "large" sources (i.e. either source with large emission or sources where the plume is under influence of the wake from surrounding buildings) dominate the SO_2 levels for short time periods. Small and large sources are important for long time periods. The levels of SO_2 concentrations and their geographical distribution do not vary considerably from one year to another.

1. INTRODUCTION

The Danish National Agency of Environmental Protection has in collaboration with local and regional authorities conducted a nationwide Air Quality Evaluation Program (abbreviated to LMP in Danish) for several Danish cities. The purpose of the program was fourfold: to gain knowledge on the air quality in selected Danish cities, to test meteorological air quality models, to monitor the development of the air quality, and to form the initial basis for a mapping of the air quality in Denmark. Seven cities were selected. These were Copenhagen, Næstved, Odense, Fredericia, Randers, Esbjerg and Aalborg, with populations ranging from 30.000 to 1.000.000. In each of the seven cities several monitors were set up to measure SO_2 (and other gases). In the following we will describe the method used in Denmark to calculate the SO_2 concentrations in the cities. Such calculations have been concluded for all cities except Copenhagen.

K. Grefen and J. Löbel (eds.), Environmental Meteorology, 445–460.

2. EMISSION

In the planning stage of the LMP program it was decided to collect information on all sources within the municipalities of the seven mentioned cities and to divide them into two categories: point sources and area sources. Point sources were defined as sources with a maximum fuel consumption of more than 50 kg fuel per hour. The remaining sources were grouped into area sources, which were squares of 500x500 m². Emission from traffic and remote sources was not taken into consideration.

The sources contributing to the area sources were assumed to have a uniform distribution within the area. With respect to variation in time, the emission was assumed to be constant during any specific month. The variation from month to month followed the typical consumption due to domestic heating. The relative variation from January to December was thus described by 12 so-called emission factors: 1.00, 0.93, 0.88, 0.60, 0.25, 0.12, 0.12, 0.12, 0.19, 0.53, 0.76, 0.95.

In practice the emission from area sources was determined by first calculating the floor space of all buildings except those heated by the point sources. (This was done by using a preexisting database with information on every address.) The following categories of floor space were used: residential houses, apartment complexes, offices (including trade and service), night-and-day institutions and summer houses. These groups were divided further into subgroups depending on the heating system, age of heating system and type of fuel.

The subgrouping was used to describe the efficiency of the heating systems and the sulphur content in the fuel. Additionaly, for each subgroup the yearly energy consumption due to heating was calculated by use of standard yearly needs per square meter. Using this information on needs of energy, efficiency of heating system and sulphur content the SO_2 emission from each group was estimated. The emission from all of the subgroups was added up to yield the total emission from the square.

TABLE I. Emission and population statistics.

City or municipality	SO_2 emission over one year: All point and area sources (1000*Tons)	Contribution to total emission (%): All area sources	The 5 largest point sources: 1.	2.	3.	4.	5.	Population: City proper (1000)	Municipality (1000)	Number in order*
Naestved	1.65	15.	44	8	6	6	5	38	45	12
Randers	4.06	4.	45	14	9	9	5	56	61	6
Fredericia	13.0	1.0	48	37	9	2	2	29	46	16
Odense	21.5	1.0	74	5	1	1	0.8	138	171	3
Esbjerg	21.9	0.3	92	4	2	0.5	0.3	71	81	5
Aalborg	44.1	0.8	56	24	8	2	0.8	114	155	4

Emission from traffic is excluded. * Number in the order of magnitude of all danish cities.

With respect to point sources, they were all visited by participants in the LMP program. Forms were filled in with information on fuel consumption, types of fuel, sulphur content in each of these, stack height, diameter of stack top, height of nearby buildings, gas temperature, percent of oxygen in gas and other data. Also the typical seasonal, weekly and diurnal variation of fuel consumption was determined. This was done by obtaining the fuel consumption for the individual months over a year. Furthermore, the consumption for the individual days over a typical week during the Winter was provided. The same was done for a typical Summer week. Finally, the consumption for the eight individual 3-hour-periods during a typical Winter day and during a typical Summer day was discribed. In this way the hourly emission was determined throughout the year.

Some statistics on emission of SO_2 and on population in the six cities are listed in Table I.

3. METEOROLOGY

The micrometeorological parameters used in the dispersion model were in any particular case derived from either standard synoptic observations at airports or specially equipped meteorological masts located near the cities (so-called LMP masts). Furthermore, nearby radiosonde data were used.

When using standard synoptic data, the heat flux, friction velocity and Monin-Obukhov length were calculated applying the so-called resistance method (Berkowicz et al. 1982a). This method requires net radiation as input, which is computed by using solar elevation, cloud observations (type and amount), precipitation, ground albedo and ground heating coefficient (Nielsen et al. 1981). When using data from the meteorological masts, where temperature gradients were measured, the micrometeorological parameters were computed applying the profile method (Berkowicz et al. 1982b).

4. MODEL

The so-called OML Model was used to estimate the SO_2 concentrations (Berkowicz et al. 1985, 1986). OML is an abbreviation of Danish words, which translated are Operational Meteorological Air Quality Models. The OML model has been developed at the Air Pollution Laboratory of the Danish EPA.

A brief description follows. The OML model is a Gaussian model. It calculates hourly mean concentrations at fixed receptor points assuming stationary conditions. The dispersion parameters σ_y and σ_z are continuous funtions of both micrometeorological parameters (friction velocity, heat flux, Monin-Obukhov length, large changes in wind direction and others) and source parameters (source height, plume buoyancy and others). Furthermore, the model takes into account boundary layer height, partial penetration of the plume through the boundary layer top, turbulence in wake of buildings and stack downwash.

The OML model has been tested on data from large power plants in both Denmark and USA as well as on data from tracer experiments (Berkowicz at al. 1985, 1986). These tests have led to continuous improvements.

Beyond the present report, an evaluation of the model performance by R. Berkowiczis is found in the present proceedings: "Evaluation of a Model Performance with Data from a Large Power Plant".

5. RESULTS

The OML model has so far been applied to six Danish cities: Esbjerg, Odense, Aalborg, Randers, Fredericia and Næstved. For each city calculations have been performed for every hour over a two-year period. In the first three cities the bulk of the domestic heating was excess heat from power plants. Test runs of the model showed that the contributions from area sources to the SO_2 concentrations at the receptor points was negligible. Area sources were therefore omitted in the computations.

As an example of computer expendituies the time consumpton on a IBM 3081 computer for one year of simulation for the city of Odense with 228 point sources and 218 receptor points is 3 1/2 CPU hours.

Output from the model was yearly and monthly mean values, 99-percentiles of hourly mean values on a monthly basis, and for daily averages on a yearly basis 98-, 50-, 50-Winter(Oct.-March)-percentiles and the largest concentrations exceeded in 3 consecutive days ("3-day-max"). Maximum values of some of these statistics for all of the cities are compared with different Limit Values and Guide Values in Chapter 6.

The statistical output was plotted as contour maps (Løfstrøm et al. 1987). Additionally, maps were produced of the so-called maximum monthly 99-percentiles during a year, i.e. for each receptor the maximum of the 12 monthly 99-percentiles during a year. Thereby values from different months are compacted in to the same map. In the following only a few examples of these maps for the city of Odense are shown and discussed.

In Fig. 1 the location of point sources and receptor points are indicated for the city of Odense. Odense is the third largest city in Denmark with a population of 171000 in the municipality of which 138000 live in the city proper. In Table I population statistics for the other five cities are listed. Calculations were made at the receptor points and interpolated to points on a square grid with 250 m spacing. Contour maps could thus be drawn.

Fig. 2 shows a contour map of the maximum monthly 99-percentiles for the city of Odense. In the municipality 228 point sources of which 147 are heated greenhouses contribute to the SO_2 distribution. As mentioned earlier, area sources and traffic are not included in the calculations for Odense. Likewise, no long range transport is included. The SO_2 emission is dominated by a single combined power and heating plant emitting 74% of the total yearly emission. The second largest source emits only 5%. This distribution in order of magnitude is not unusual for the six cities, as is seen in Table I. The maximum hourly SO_2 emission from the two sources is 995 and 47 g/s respectively and the stack heights are 102-149 m and 39 m respectively. More detailed information on the five sources with the largest yearly emission is listed in Table II.

The largest maximum monthly 99-percentiles are found near these two sources: more than 500 μg/m³ near the second largest source and 400 μg/m³

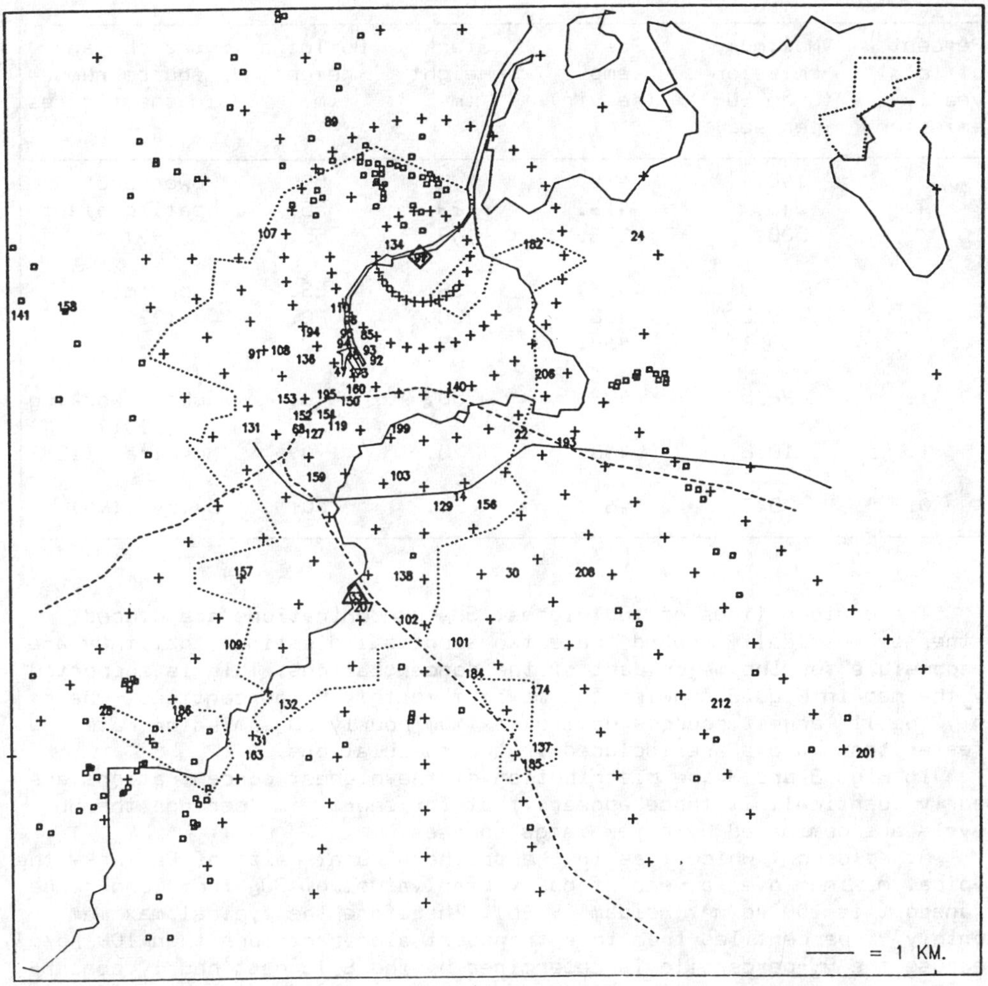

Figure 1. Map of the city of Udense with locations of receptor points and point sources. ····: limits of built-up area; ——: main road, coastline or stream; — — — : railway; "number": point source; □ : greenhouse; +: receptor point; ◇: the largest source emitting 74% of the total yearly SO_2 emission in the municipality and △: the second largest source, which emits 5%.

near the largest. The reason for the highest SO_2 concentration near the second largest source is, that a relative tall building is located next to the stack, and the plume often enter the wake of the building and is immediately dissipated to the ground.

TABLE II. Information on the 5 point sources with the largest yearly SO_2 emission in Odense.

Percent of total yearly emission	Maximum emission (gram SO2 per sec)	Gas temp. (Kelvin)	Stack height (m)	Building height (m)	Branch and source number in the figures
	142.	418.	102.	55.	Power and
74.	345.	413.	123.	55.	heating plant
	508.	383.	149.	55.	(97)
	8.3	468.	38.	23.	Paper mill
5.	8.3	468.	38.	23.	(33)
	30.1	468.	39.	23.	
1.	22.5	473.	50.	0.	Rubber working (101)
1.	18.2	468.	75.	0.	Hospital (159)
0.8	5.5	468.	30.	0.	Dairy (194)

The contour lines of the highest SO_2 concentrations are placed rather symmetrically around those two sources indicating, that they are responsible for the major part of the concentrations. This is supported by the map in Fig. 3 showing the maximum monthly 99-percentiles, where only the 11 largest sources having maximum hourly SO_2 emission rate greater than 10 g/s are included in the calculations.

In Fig. 2 and 3 the distribution of the highest concentrations are nearly identical. It those appear, that for short time periods the SO_2 levels are dominated by a few large sources.

At Keldsnor, which lies in the southern rural parts of Denmark, the typical maximum over a year of daily mean values of SO_2 from long-range transport is 100 $\mu g/m^3$ (Heidam 1986b). Therefore the typical maximum monthly 99-percentiles from this transport alone are more than 100 $\mu g/m^3$, because the 99-percentile is determined by the 8 largest hourly concentrations in a month. It is not possible to include the long-range transport in calculation of the 99-percentiles.

With respect to long time averages two maps of yearly means of SO_2 concentrations are shown in Fig. 4 and 5. As with the 99-percentiles all point sources are included in calculations in Fig. 4 and only the 11 largest in Fig. 5. In Fig. 4 concentrations in Odense are typical 10 $\mu g/m^3$ with a local maximum up to 25-30 $\mu g/m^3$. In Fig. 5 the SO_2 levels in the city reduces to around 5 μg/m³, and several local maxima disappear. Longrange transport of SO_2 adds 8-10 μg/m³ to the yearly mean values (Heidam 1986a).

It is seen that for long time averages it is not enough to include the largest sources. All sources in the area contribute to the concentrations. The reason is, that the plume from a single source over a year is dissipated in all directions, so in mean a receptor point is only in

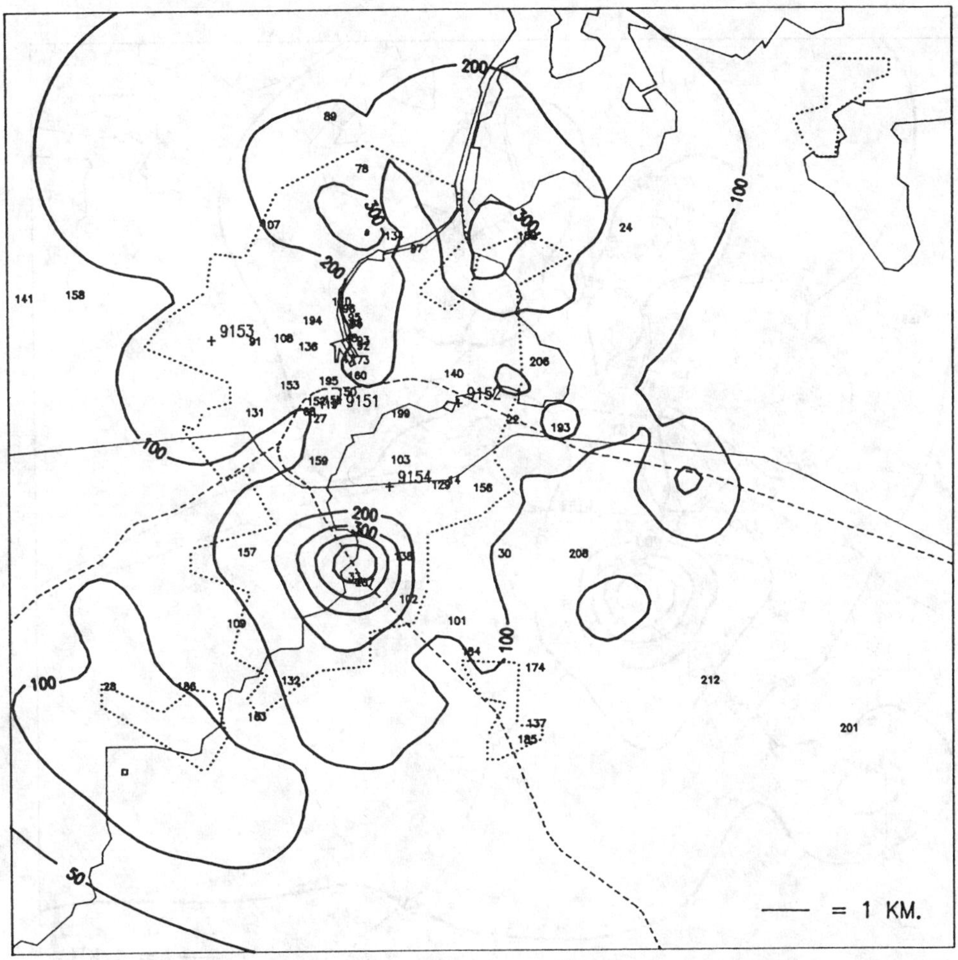

Figure 2. Maximum of monthly 99-percentiles (μg/m³) of hourly averages of SO_2 for each receptor point during 1982 in Odense . All point sources are included in the calculations. + and 4-digit numbers are locations and identifications of monitor stations. See Figure 1 for a further list of signs.

say the 10 degree plume angle in 3% of the time, and the average values will be low. In opposite for the 99-percentiles the plume in extreme only have to blow in one direction in 8 hours in a month to produce a large value.

Fig. 6 and 7 show the maximum 99-percentiles and the yearly means of SO_2 concentrations in Odense for the following year, 1983. The difference between these maps and those shown in Fig. 2 and 4 for 1982 are minor. It thuse appeare, that the geographical distribution and the

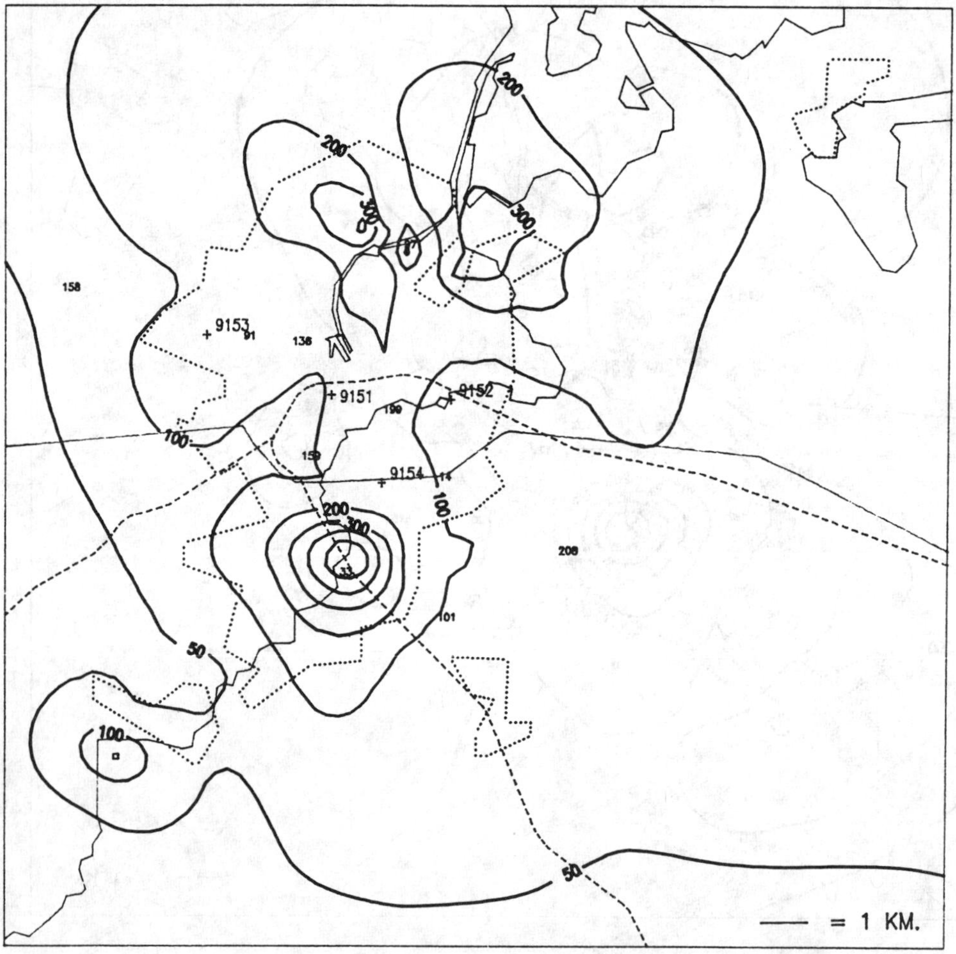

Figure 3. Maximum of monthly 99-percentiles (μg/m³) of hourly averages of SO_2 for each receptor point during 1982 in Odense. Only point sources with maximum emission rate > 10 g/s of SO2 are included in the calculations. See Figure 1 and 2 for a list of signs.

levels of SO_2 do not vary considerably from one year to another.

These conclusions on source contribution to yearly means, maximum monthly 99-percentile and on geographical distribution and levels, are repeated for all of the five other cities, where meteorology from other years and locations have been used.

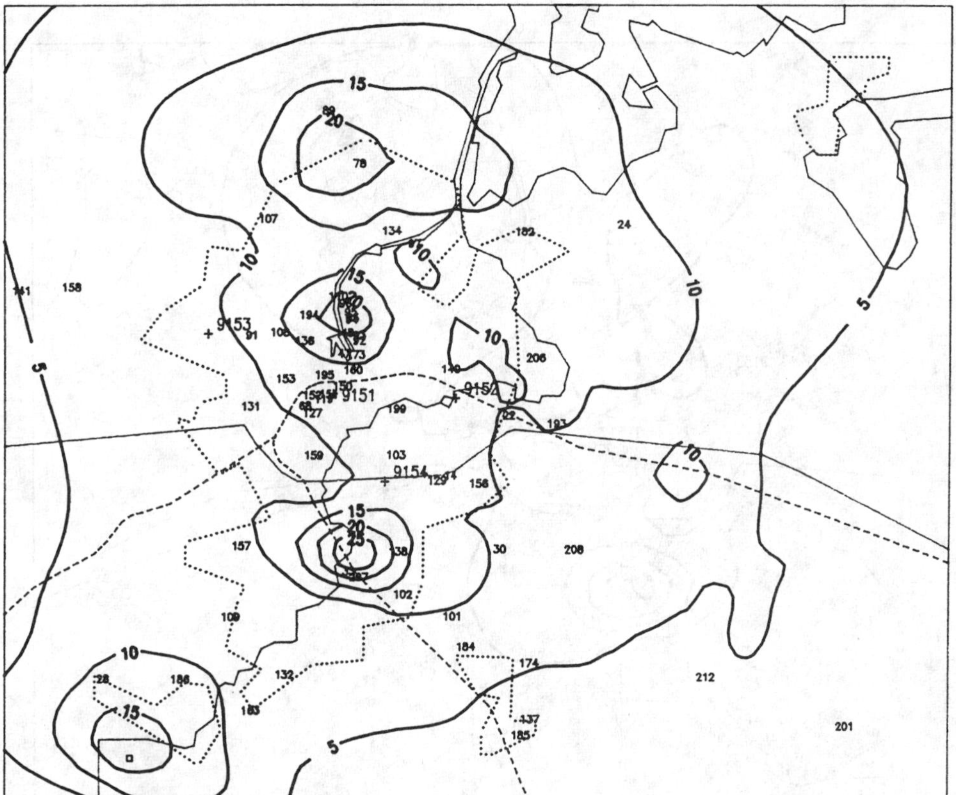

Figure 4. Yearly average of SO_2 ($\mu g/m^3$) in Odense for 1982. All point sources are included in the calculations. See Figure 1 and 2 for a list of signs.

6. LIMIT VALUES

As mentioned other types of statistics were produced by the model. Among this were the following for daily averages on an yearly basis: 98-, 50-, 50-Winter-percentiles and 3-day-max. In Table III the largest value of these statistics in each of the cities are compared with Limit Values and Guide Values. It is seen that there are no violations of the EEC Limit Values. The 98-percentiles and 3-day-max are nearest to the Limit Values. The 50-percentiles are well below the Limit Values. When background concentrations of SO_2 from long-range transport, which are of

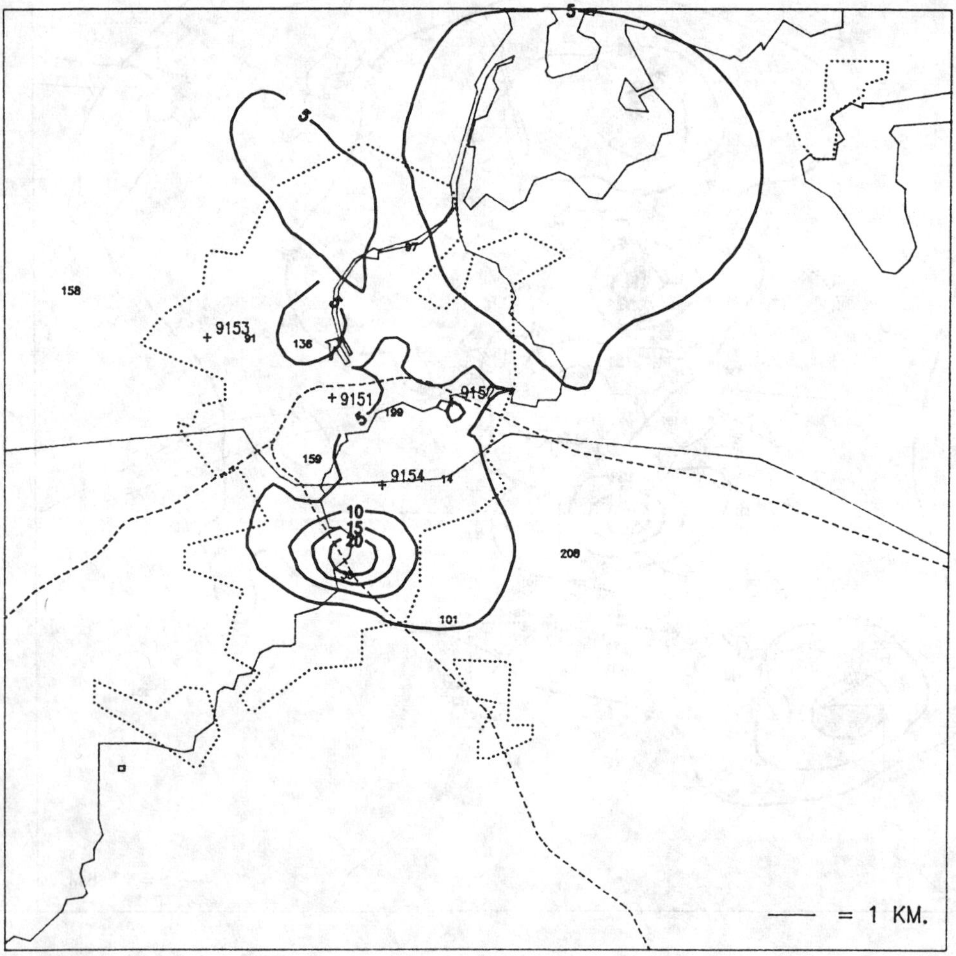

Figure 5. Yearly average of SO_2 (µg/m³) in Odense for 1982. Only point sources with maximum emission rate > 10 gram SO2/sec are included in the calculations. See Figure 1 and 2 for a list of signs.

magnitude 8-10 µg/m³, are added to the yearly averages, they approximate the EEC and WHO Guide Values in the city of Aalborg and are slightly below in most of the other cities.

The pattern of SO_2 concentrations for 98-percentiles and 3-day-max are similar to the pattern of the maximum 99-percentiles. The pattern of 50-percentiles corresponds to the yaerly averages.

TABLE III. Comparison of maximum calculated SO_2 concentration with Limit Values and Guide Values (µg/m³). Long-range transport is not included.

EEC Limit Values of daily averages on a yearly basis	Næstved	Randers	Fredericia	Odense	Esbjerg	Aalborg
250 98-percentile	150	150	100	150	200	225
80 50-percentile	15	10	10	15	10	25
130 50-Winter-perc.	25	15	15	25	10	35
250 "3-day-max"	100	125	125	125	175	200
EEC and WHO Guide Values						
40-60 yearly mean	25	25	15	25	25	35

"3-day-max" is the largest concentration exceeded in 3 consecutive days.

7. COMPARISON WITH MEASUREMENTS

In Odense four monitoring stations were located in part of the calculation period (Palmgren 1983). The locations of the stations are shown in the figures with contour lines. All of the stations monitored daily means of SO_2 concentrations. Additionally the one in the city center monitored half-hour means.

Before comparing of calculated concentrations with measurements, it is important to point out, that this is very difficult for various reasons. First, the emissions used are not exact for any given hour but only typical values, and detailed values from at least the largest sources are needed to expect good agreement. Second, the monitors were not working all the time. The monitor measuring half-hour mean values was working 80% of the time. Third, small errors in wind directions can move the pattern of concentrations and make great changes at the individual receptor points.

These limitations must be considered when comparing calculations and measurements in the city center (station no. 9151). This is done for monthly mean values and for the 99-percentiles in Table IV. The influence from long-range transport is taken into account with regard to the monthly mean values by adding the average background concentrations measured at Keldsnor and Tange (in central Jylland) to the calculated values. It is seen that the calculated and measured SO_2 values of both 99-percentiles and monthly means correspond quite good, but with slightly lower calculated values.

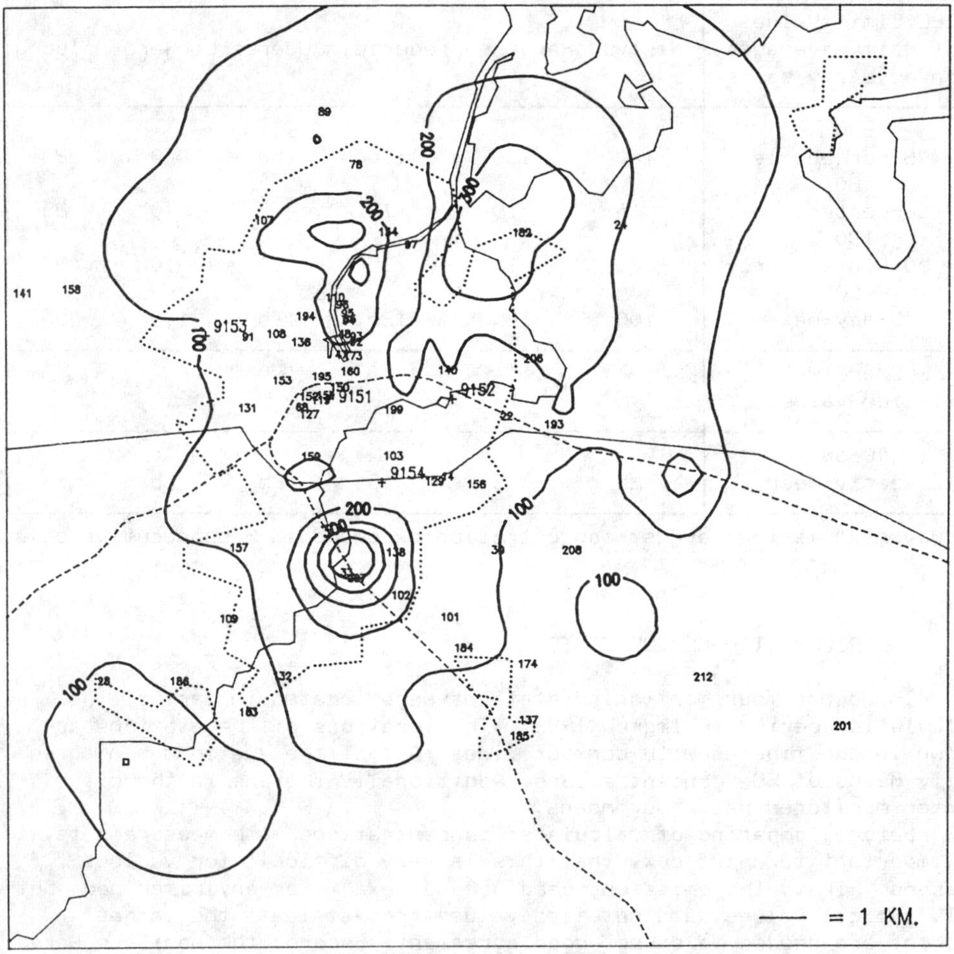

Figure 6. Maximum of monthly 99-percentiles ($\mu g/m^3$) of hourly averages of SO_2 for each receptor point during 1983 in Odense . All point sources are included in the calculations. See Figure 1 and 2 for a further list of signs.

In Fig. 8 the 10 monthly means in 1982 and the means of all 10 months from calculations and measurements are shown for all of the 4 stations. The values for the city center station are from the monitor which measures daily means, and they differs a litle from the values in Table IV. Except from the southernmost station, 9154, the values correspond reasonable well.

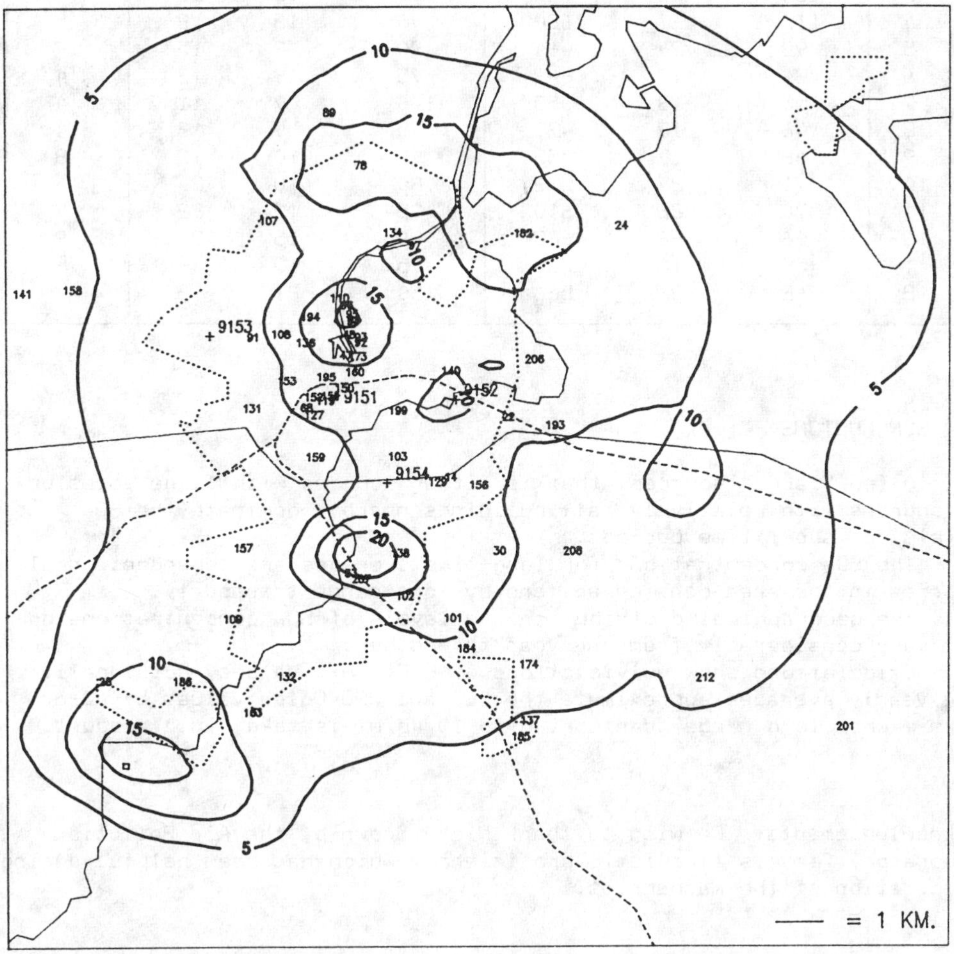

Figure 7. Yearly average of SO_2 ($\mu g/m^3$) in Odense for 1983. All point sources are included in the calculations. See Figure 1 and 2 for a list of signs.

TABLE IV. Comparison of calculated SO_2 concentration with measurements at station no. 9151 in the city center (μg/m³).

Month during 1982/83	Measurement			Calculation			Background
	99-percentile of hourly averages	Mean	Number of observations	99-percentile of hourly mean	Mean	Mean + background	Mean
3	120	32	736	100	16	26	10
4	112	27	620	90	13	18	5
5	98	24	464	90	8	15	7
6	84	17	643	75	9	13	4
7	155	18	555	100	7	11	4
8	46	12	471	75	9	13	4
9	68	17	553	90	10	18	8
10	71	27	538	80	10	24	14
11	70	26	610	75	13	22	9
12	96	22	668	90	15	24	9
1	59	16	735	75	13	-	-
2	162	32	566	150	21	-	-

8. CONCLUSIONS

A few "large" sources, that is either sources with large emission or sources with relatively tall buildings nearby, dominate the SO_2 levels for short time periods.

The SO_2 concentrations for long-time averages are dependent on all sources in the area considered (and by long-range transport).

The geographical distribution and levels of SO_2 concentrations do not vary considerably from one year to another.

Calculations show no violations of EEC Limit Values, but locally the yearly averages approximate the EEC and WHO Guide Values of 40-60 μg/m³, when long-range transport of 8-10 μg/m³ is taken into account.

Acknowledgements. We wish to thank Nicky Brown of the Air Pollution Laboratory for his linguistic proficiency, which has been helpful during preparation of the manuscript.

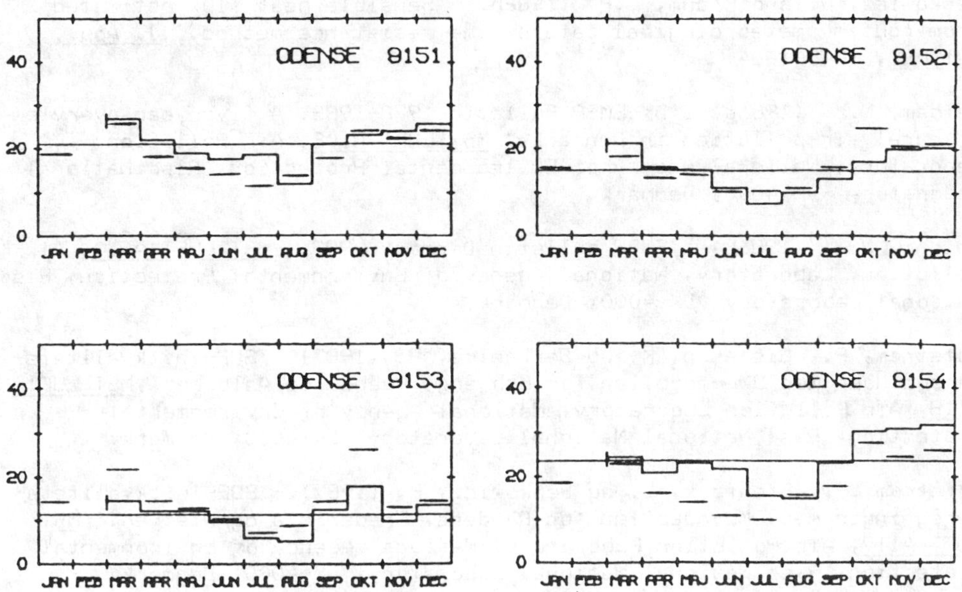

Figure 8. Calculated and measured monthly averages of SO_2 concentrations ($\mu g/m^3$) in Odense for 1982. The numbers of the monitor stations are indicated and equals the numbers in Figures 2-7. Continuous lines are measurements of the individual months and the 10-month average. The short lines are calculated values to which the background values have been added.

REFERENCES

Berkowicz, R., Olesen, H.R. and Torp, U. (1985). 'The Danish gaussian air pollution model (OML): Discription, test and sensitivity analysis in view of regulatory applications.' Proceedings of the 15th international technical meeting on air pollution modelling and its applications. St. Louis, USA, 1985.

Berkowicz, R., Olesen, H.R. and Gislason, K.B. (1986). 'Modelling air pollution from industrial and urban sources.' WHO conference: Air pollution modelling and its applications. Leningrad, USSR. MST LUFT-F30.

Berkowicz, R. and Prahm, L.P. (1982a). 'Evaluation of the profile method for estimation of surface fluxes of momentum and heat.' Atm. Env. **16**, 2809-2819.

Berkowicz, R. and Prahm, L.P. (1982b). 'Sensible heat flux estimated from routine meteorological data by the resistance method.' J. Appl. Met. 21, 1845-1864.

Heidam, N.Z. (1986a). 'DK-EMEP Bulletin 1978-1982. A five-year overview of rural air pollution in Denmark.' MST LUFT-A102, Air Pollution Laboratory, National Agency of Environmental Protection, Risø National Laboratory, DK-4000, Denmark.

Heidam, N.Z. (1986b). 'EMEP bulletin Denmark 1983.' MST LUFT-A110, Air Pollution, Laboratory, National Agency of Environmental Protection, Risø National Laboratory, DK-4000, Denmark.

Løfstrøm, P., Gislason, K. og Berkowicz, R. (1987). 'SO2-luftkvalitets-beregninger med OML-modellen for Esbjerg, Odense og Aalborg.' MST LUFT-A118, Air Pollution Laboratory, National Agency of Environmental Protection, Risø National National Laboratory, DK-4000, Denmark.

Løfstrøm, P., Stenfalk, L. og Berkowicz, R. (1987). 'SO2-luftkvalitets-beregninger med OML-modellen for Randers, Fredericia og Næstved.' MST LUFT-A119, Air Pollution Laboratory, National Agency of Environmental Protection, Risø National National Laboratory, DK-4000, Denmark.

Nielsen, L.B., Prahm, L.P., Berkowicz, R. and Conradsen, K. (1981). 'Net incoming radiation estimated from hourly global radiation and/or cloud observation.' J. Climat. 1, 255-272.

Palmgren Jensen, F. og Kemp, K. (1983). 'Det landsdækkende luftkvalitets-måleprogram. Luftkvalitetsmålinger, Årsrapport 1982.' MST LUFT-A78, Air Pollution, Laboratory, National Agency of Environmental Protection, Risø National Laboratory, DK-4000, Denmark.

Palmgren Jensen, F. og Kemp, K. (1987). 'Det landsdækkende luftkvalitets-måleprogram. Luftkvalitetsmålinger, Årsrapport 1985.' MST LUFT-A113, Air Pollution, Laboratory, National Agency of Environmental Protection, Risø National Laboratory, DK-4000, Denmark.

WHO (1979). 'Sulphur oxide and suspended particulate matter.' Enviromental Health Criteria 8.

URBAN REDEVELOPMENT UNDER CONSIDERATION OF CLIMATOLOGICAL ASPECTS

Lutz Katzschner
Gesamthochschule Kassel, Stadt-und Landschaftsplanung
Henschelstr. 2
35 Kassel
B.R.D.

ABSTRACT. The renewal of the inner city in Kassel should take place with knowledge of the ecological data as background to improve the living conditions of the people in respect to bioclimate and air pollution. Within the inner city of Kassel exist quite different micriclimatic conditions in dependence of the city structure. From this proposals could be conducted how open spaces can be reconstructed to improve the urban climate. One result of this investigation was the fact, that improvements of the climatic conditions can be done on a very small scale, as specially the daily and annual variation differs much within a small area.

1. Urban Climate and Planning

Man reacts not directly but very sensitive on the climatic conditions in a city. The behaviour and the usage of free spaces is dependent from the microclimatic conditions. In dependence of the city structure however the different microclimatic condition can be formed through surface materials and vegetation in general.
To improve the living conditions different climatic situations have to be available so that the usage of the same place can be possible during various weather conditions.

Climate can be an indicator for the environmental situation of a city. The variaty of different climate divided by the scale in which they occur can be one criteria. This also garantees a good possible uasge if that places, more activities can be done. In order to discuss the connection between climate and the behaviour of man as planning tool it is important to evaluate not only the mean values of the meteorological parameters, but discuss also case studies and the daily variations.

This can be shown by a study of the behaviour of people in dependence of the climatic situations and the frequency of the usage at

K. Grefen and J. Löbel (eds.), Environmental Meteorology, 461–472.

different places (Geiger, 1961). So an answer can be given, what sort of climate do we aim at.

For an investigation to study the connection between the behaviour of people and climate to much time is needed. Therefore the present climatic situation was investigated to conduct from that planning proposals which can effect the behaviour of man. This had to be integrated into the urban planning system, so that the advices derived from the meteorological data could be implemented to the general urban renewal plans. The meteorologist can descripe the bioclimatical and the disrtibution of air pollution in the city. The places of the maximum heat island, worse bioclimatic conditions as well as bad polluted areas can be found by meteorological measurements. This can be included in the urban renewal programme and the planner can add its own priorities to that, which implements the meteorological situation as well. Through such a handling the existing conflict can be solved. A bioclimatic improvement can effect more stable situations,which lead to higher air pollution if trees for example disturb the vertical exchange.

It could shown bey by different authors (Eriksen 1964,Sperber 1974), that the influence of green is often restricted to the area itself. Especially in a densly built up area the climatic effect of green cannot infiltrate into the housing areas. Therefore the structure and the vegetation of the areas itself become more important than the park in the neighbourhood. To improve the ecological conditions in the city much can be done on a very small scale.

2. Urban Climate and City Structure

The often desriped heat island (Baumüller, 1986) is very much depending from the city size. But this does not mean, that the heat island is to be seen as homogenous increasing towards the city centre. The urban structure and the surface conditions lead to sdmall scale variations. Meteorological investigation have to consider this. Especially the working methods must have this in mind. The regular net of weather stations does not cover the urban situation, therefore new stations had to be set up and mobile measurements had to be carried out.

From Figure 1 the investigation area and the city structure can be seen. One can seperate into a buisiness area, the pedestrian zone and a living quarter with some small industries. Important for the climatic situation are the green areas around the churches and the Friedrichsplatz. As seen later all these areas have their own climatic characteristic. In the buisiness centre the percentage of green is very low. All backyards are filled with parking places. In the living quarters the structure is more inhomogenous as many places are covered with vegetation.

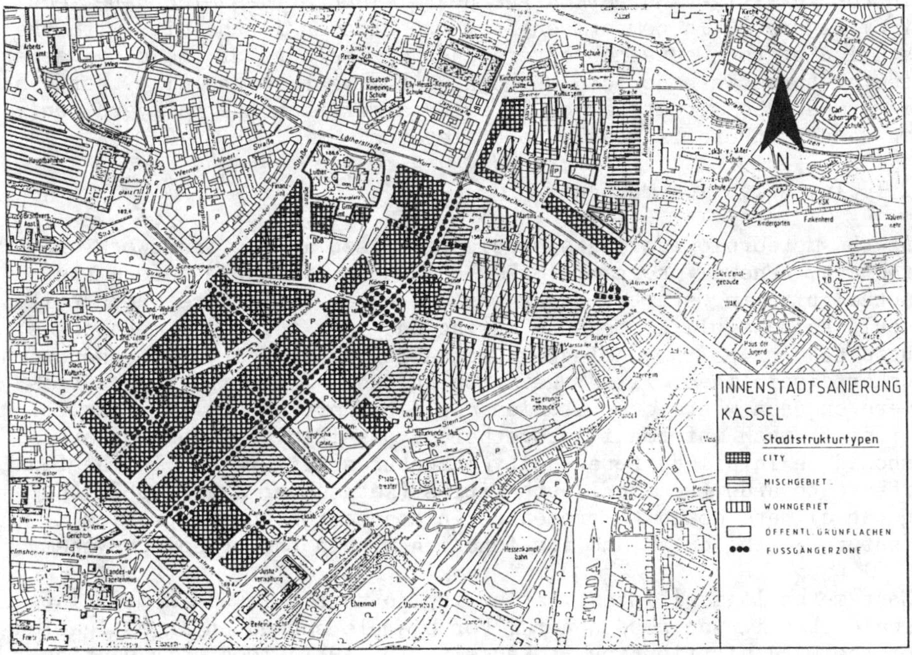

Figure 1. The city structure of the investigation area

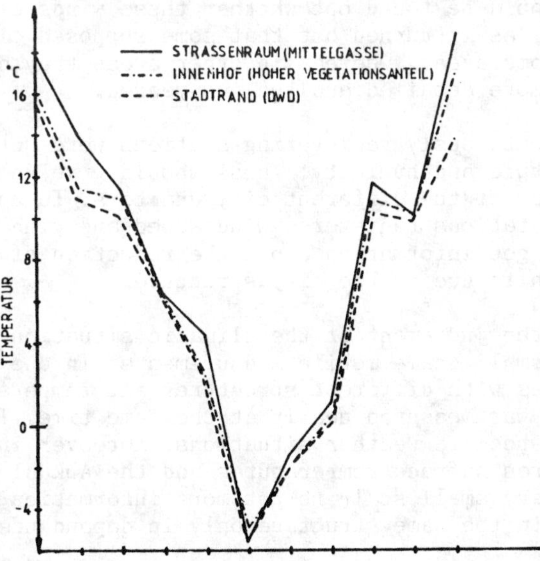

Figure 2. Annual variation of the air temperature in different city structures

In the investigation period the heat island of Kassel can be seen from figure 2. In comparison from two station in the city with the station of the Deutsche Wetterdienst at the edge of the city warmer temperatures were noticed throughout the year in the city. Especially in summer the street canyon is even warmer than the backyard, only in January there is nearly no difference.

2.1. Investigation Methods

As the meteorological data should be raised in the framework of the city renewal programme, they had to be available before the actual planning starts. As it was not clear in advance whether it would be possible to measure during one year following method was used to desripe the ecological situation of the city. Together with other disciplines, like vegetation and soil investigations, areas were choosen where case studies could be carried out in order to bring as much information together as possible. The choosen areas should be in all different city structures (figure 1). In the final report towards the local authorities (Katzschner u.a., 1987) the urban climate could be presentet in connection to the problem "nature in the city" and its function within the urban ecosystem.

Most meteorlogical investigations up to that time had a much larger scale than it was needed here. For example showed an investigation of the wind distribution in Kassel (DWD, 1982) local circulation systems, which could suppose a urban rural circulation. For the problems here it should be found out whether these winds can penetrate into the city, as it turned out that some supposed circulation effects only some areas, whereas in other areas they did not exist at all. So a more detailed analyse was needed.

For one year five continously registering stations were set up to measure air temperature and humidity. These should give knowledge about the mean values of the different city quarters. In addition to that three wind stations registered wind speed and wind direction in the city to get information about the reduction of the windspeed and windshift due to the city structure.

Very important for the judgement of the climatic situation in the area have been the small scale mobile measurements. In one meter hight above 40 places with different structures air temperatures as well as humidity was measured nearly at the same time. This was carried out during specific weather situations. Moreover in different city structures sufrace temperatures and the Abkühlungsgröße was observed in a very small scale to get more information about the variations within the same structure only in dependence of the surface.

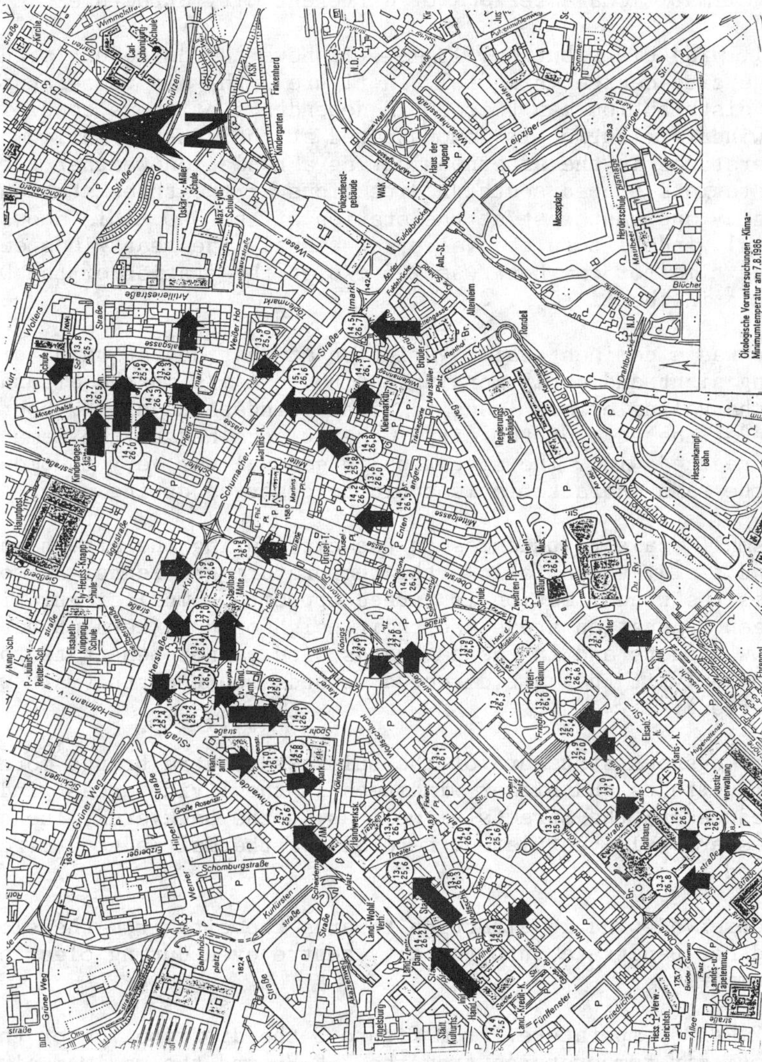

Figure 3. Minimum (above), maximum (below) air temperatures and and winddistribution in the investigation area

2.2. Climatic Characteristic of different City Structures

2.2.1. Wind Distribution. Figure 1 shows the different structures of the city centre in Kassel. In figure 3 the air temperatures and wind distribution can be seen in dependence of these conditions. The winddistribution is demonstrated at the case of maximum air temperatures in the afternoon during a radiation weather type. The windstations showed southern directions at the three stations in the city and southwest at the station of the DWD. This shows the general windshift in the valley of Kassel. The mean windspeed at the time for one hour was 2,5 m/s at the station of the DWD and 1,8 m/s in the city.

There was a dominant daily variation at that day, with no wind during night and early morning and up to 4 m/s in the afternoon. The channeling effect leads to higher windspeeds as the hourly mean value at the edge of the city. Other places had at the same time very reduced wind. The situation in general was very inhomogenous, but the same characteristic could be observed during other days.

The small scale turbulences in the city during the afternoon create a wind field, which follows the temperature gradients. At some places the winddirection is turned opposite. Small circulation between vegetation areas and their neighbourhood were obvserved. In narrow streets and in the backyards wind cannot penetrate to the bottom only if they are directed towards south-north (main wind-direction).

The homogenous structure of the buisiness centre brings more channeling effects and higher windspeed in the living areas wind speed is reduced in the streets as well as in the backyards. The open places are well ventilated, here also a thermal induced circulation occured.

The observed wind in direction towards the city from the slopes could not penetrate into the city centre even during clear nights in summer.

2.2.2. Air Temperature. From figure 3 the distribution of the minimum air temperatures from the 7.8.86 and the maximum air temperature from 14.8.86 is drwan. The highest values are in the buisiness centre and at southwards oriented streets. Here the cooling during the night is very low. The vegetation areas have naturally the lowest temperatures no matter how big they are. Even the greens in the buisiness centre show this cooling effect. In the street canyons without trees very high temperatures occur up to the late evening. Here, in the living area, the differences between the air temperatures of the street and in the backyard was big, while backyards in the buisiness centre, which have mostly no vegetation, tend to stay very warm.

2.2.3. Abkühlungsgrößen. This complex number can descripe the bioclimatologicak situation in the city. Different city structures were compared (figure 4). It can be seen that the pedestrian zone has the highest daily variation and high values in the morning. The living area and the street have a moderate variation. This is mainly caused through the higher windspeeds in the morning. The surface roughness effect very low values in small streets and backyards. The influence of air temperature and humidity is negletable against wind.

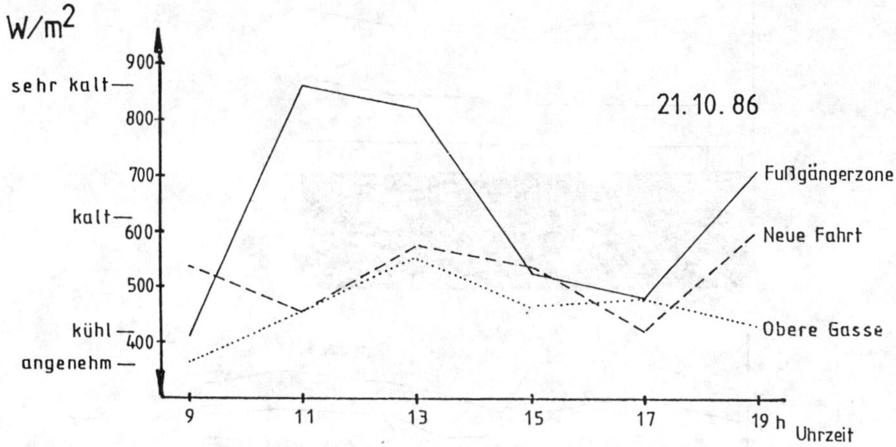

Figure 4. Daily variation of the Abkühlungsgröße in different city structures in W/m^2

The situation in the pedestrian zone tends to cold in the winter, autumn and spring and only moderate in summer. To increase the surface roughness would bring here a better bioclimatological conditions.

3. Open Spaces and Microclimate

In figure 5 two profiles if the air temperatures are drawn in dependence of the city structure. In the left the daily variation in the living area is shown. The close backyards with a high percentage of green devlopes quite another situation than in the streets. The differences increase during the course of the day.The climatic situation of the backyards seems to have nothing to do with the streets. In the closed backyard (B 14 and B 16) have quite different daily variations than in the streets.

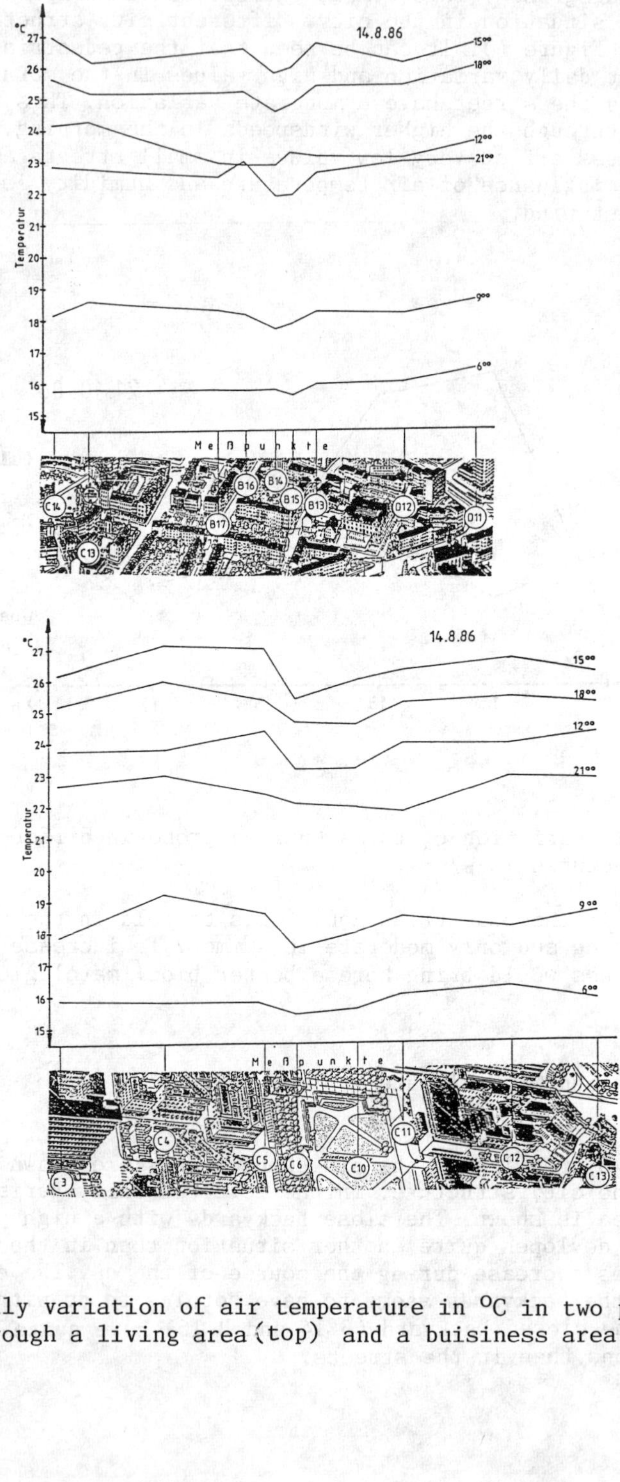

Figure 5. Daily variation of air temperature in °C in two profiles through a living area (top) and a buisiness area (below)

Another situation is observed at the Friedrichsplatz (C 10) As only one side of the place has a clear border to the surrounding streets the change from the open place to the streets is more smoothly. But already the trees can bring again this sharp division.

It can be seen that open spaces have their own climate, which is more dominant if the boarders to the neighbourhood prevent an exchange.

How big the variatiy of the small scale climatic conditions can be is shown in fugure 6. The surface temperatures as well as the air temperatures demonstrate the influence of the surface material and the vertical building structure. Asphalt surfaces have warmer values than the grass in the direct neighbourhood. This effect is also in 1 m height present. So it can be seen that a garage or a single tree can bring up to one degree differences even in october.

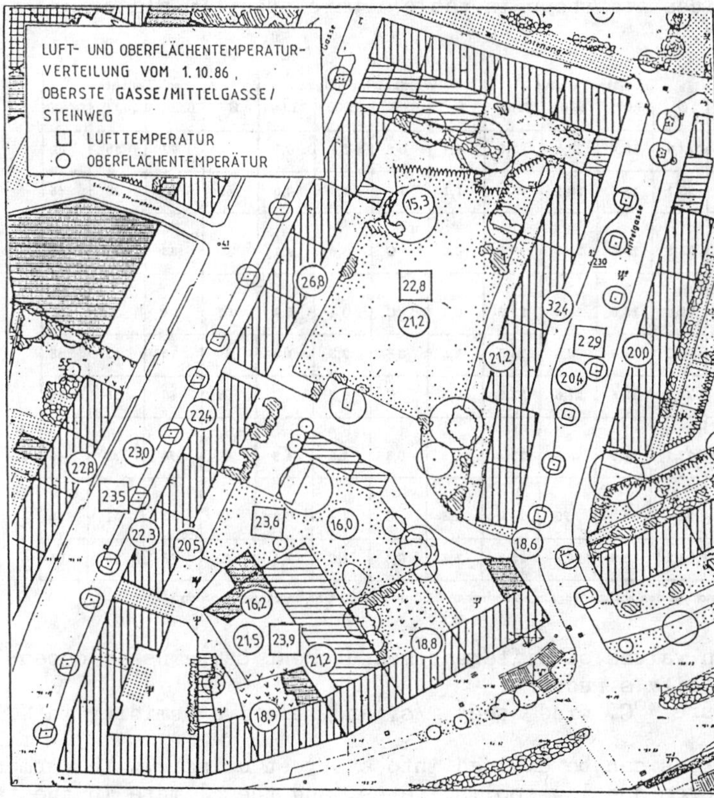

Figure 6. Surface and air temperatures in ^{o}C in a backyard

From figure 6 another instresting information can be observed. Air temperatures are warmer that the surface temperatures. This is the case in the living areas in afternoon only. At that time warming semms to be an effect from radiation of the urban boundary layer. This means that in the afternoon from autumn to spring stable situations occur, in contrary to the buisiness centre, where the surface temperatures are higher also in the evening.(Figure 7)

4. Conclusions and Planning Aspects

With this method to deduct results mainly from case studies have the problem that these results should be now generalised for the whole year. The case studies descripe specific situations during a certain weather type. But on the other han for planning processes small scale investigation were needed. In a conclusion results of some mean values are shown in table 1 were specific places are characterised.

	Lufttemperatur				Windgeschwindigkeit				rel. Feuchte			
	Ø	6.00	15.00	21.00	Ø	6.00	15.00	21.00	Ø	6.00	15.00	21.00
City Straße	22,8	13,7	28,3	23,5	1,4	0,6	2,1	0,5	50	77	35	49
City Innenhof	23,1	13,5	28,1	23,9	1,1	1,1	1,0	0,5	49	80	35	45
Asphaltierte offene Plätze	23,3	13,3	29,2	23,5	0,5	C	0,4	1,1	50	83	31	51
Friedrichs-platz	22,9	13,0	28,4	23,2	0,5	C	0,4	0,9	52	84	36	55
Fußgängerzone	23,1	13,6	28,8	23,5	1,3	0,5	2,3	1,3	49	76	32	48
Lutherplatz	22,3	13,5	27,6	22,6	0,3	C	0,2	C	53	80	35	54
Wohngebiet Innenhöfe	23,1	13,9	28,3	23,4	1,2	0,5	1,5	1,3	48	76	33	49
Wohngebiet Straßen	23,4	14,3	29	23,9	1,3	C	1,4	0,9	48	73	31	50
Randstraßen	23,2	13,9	28,7	23,3	1,2	0,6	1,6	1,1	49	78	33	51

Die Ø-Werte sind aus den sechs Meßgängen zwischen 6.00 und 21.00 Uhr berechnet worden.

Table 1. Mean values of different measurement campains in dependence of the city structure.
Air temperature ^{o}C, wind speed m/s, and relative humidity in %

The living areas can be divided into a street climate with warmer air temperatures and low cooling rates, and the climate in the back yards with a very inhomogenous structure low air temperatures and a decent bioclimate. Here no proposals for a change in structure could be made, while in the streets more green was recommended. Backyards tend to stable situations during nights.

In the buisiness centre the heat island is most deveolped. The streets as well as the backyards have warm temperatures and low cooling rates. The traffic emissions lead to higher CO concentrations than at the official stations. In rush hours up to 7 mg/m^3 was measured against 3,5 mg/m^3 at the station of the HLfU (Katzschner u.a.,1987). Here the wind speed is beside of some channeling effects reduced.

In the pedestrian zone a high diurnal variation of all meteorological paramters occur, which lead to bad bioclimatological conditions. This should be improved by an higher percentage of vegetation.

The variaty of climates in the city can be recomended to improve the climatical situation. It garantees a better usage of the open spaces during every time of the year. Here it seems that one conflict between economy and ecology can be solved as it is in general intrest that people visit a city if there are decent environmental conditions.

With such an investigation the city is descriped in view of the climatological situation. This can be used as background for further planning. But the scientistic analyses cannot take away the political decision about the solution of conflict areas between social, economical and ecological problems. The scientistic data answer only the question what impact a certain development may have.

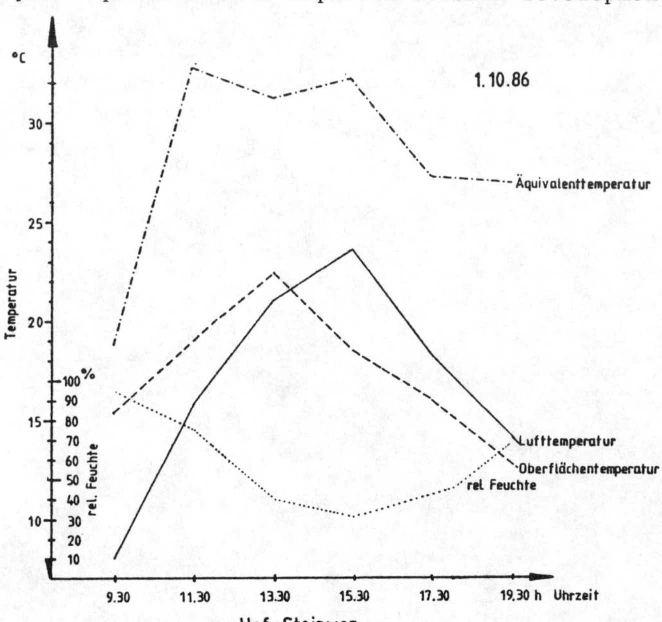

Figure 7. Daily variation of different parameters in a backyard of the living area

References:

/1/ Baumüller,J;Hoffmann U:
Verbesserung der Infrastruktur des Stadtklimas,VDI Bericht Nr.605, Düsseldorf, 1986

/2/ Deutscher Wetterdienst Frankfurt
Klimatologische Untersuchung zum Landschaftsplan Kassel, Die Windverhältnisse bei austauscharmen Wetterlagen, Offenbach, 1982

/3/ Eriksen,W:
Beiträge zum Stadtklima Kiel, Schriftenreihe des Geographischen Insituts der Universität Kiel,22, Kiel, 1964

/4/ Geiger,R:
Das Klima der bodennahen Luftschicht, Friedrich Vieweg Verlage Braunschweig, 1961

/5/ Katzschner,L; Kunick,W; Mayer,R; v.Reuß,J:
Untersuchungen zur ökologischen Situation der Kasseler Innenstadt, Gutachten im Auftrag der Stadt Kassel, GhK, 1987

/6/ Sperber,H:
Mikroklimatisch-ökologische Untersuchungen an Grünanlagen in Bonn, Dissertation am Geograpgischen Institut der Universität Bonn, 1974

FIELD EVALUATION OF A FLUCTUATING PLUME MODEL FOR ODOURS WITH SNIFFING TEAMS

Frans de Bree and Hendrik Harssema
Department of Air Pollution
Agricultural University
P.O. Box 8129
NL 6700 EV Wageningen.
THE NETHERLANDS

SUMMARY. In the Netherlands, up till now a preliminary air quality standard for odours is established. It is based on an hourly averaged concentration level, which is calculated with a generally accepted gaussian dispersion model in the Netherlands. For odours, hourly averaged concentration levels are of little use,however. Because of the turbulent character of the dispersion process, high concentrations may occur within an hour in which the average concentration is well below the detection limit. For this reason a dispersion model was developed, based on the Högström odour dispersion model. This is a fluctuating plume model and it yields estimates of the perceived odour frequency downwind from the source. The dispersion parameters in the model were adapted to the generally accepted ones in the Netherlands.

The model was evaluated around three different source types;a high point source, a low source and an area source. Measured odour frequencies were compared with model calculations with a bigaussian model and with the modified Högström model. The relation between observed odour frequencies and calculated hourly averaged concentrations was poor, especially for the high source, and different for each source type. The relation between the observed and calculated odour frequencies was much better and the same for all sources.

The conclusion is that the modified Högström model is a better tool for estimating odour exposures than gaussian models based on hourly averages, specially for high sources. This model has been tested for relatively short travel times. From a theoretical point of view it is likely, that for large distances the applicability is less. Since most odour problems are on a local scale, this is no serious limitation in practice.

1. INTRODUCTION.

Recently, governments of some countries in Europe are preparing national air quality standards for odours. One of the instruments for regulating odour emissions is the use of dispersion models to relate emissions of odorous substances to exposure of the population to odours. Exposure

K. Grefen and J. Löbel (eds.), Environmental Meteorology, 473–486.

may be expressed as the amount of time that a certain concentration is exceeded (Harssema 1987).
In the Netherlands, up till now a preliminary air quality standard for odours is established, based on a concentration level for populated areas. For new sources, the 99.5 percentile of hourly averaged concentrations shall not exceed 1 odour unit per cubic metre (the odour detection limit for 50 percent of the population) (Wijnen,1985). This level is calculated by means of the generally accepted dispersion model for the Netherlands (Werkgroep Verspreiding, 1983). This model is based on the Gaussian plume concept and yields estimates of the 1-hourly averaged concentration levels, with a total observation period of one year.

For odours, however, hourly averaged values are of little use. The turbulent character of the dispersion process causes large variations in concentration levels, specially in the near vicinity of sources. Within an hour with an average concentration well below the detection threshold, short periods of high concentrations may occur, probably causing nuisance.

For that reason the Department of Air Pollution of the Dutch Agricultural University developed a dispersion model, which accounts for short averaging times (Voerman and Harssema,1984). This model, proposed by Högström(1972), is based on the fluctuating gaussian plume model, the principles of which were published by Gifford (1959). The dispersion parameters of this model were adapted to be in line with the generally accepted parameters in the Dutch National Model. The model was evaluated around three source types: an elevated point source with a physical height of 70 metres, a complex source consisting of a number of vents on the roof of a building at a height of about 10 metres and an area source at ground level, consisting of a sewage water facility. For all three sources emission estimates were available or made. Odour detectability downwind of the sources was determined by the use of sniffing teams, consisting of eight trained panel members. The experimental results were compared with model calculations with a bi-gaussian plume model and with the modified Högström model.

In this paper the theoretical backgrounds of the dispersion of and the exposure to odours are shortly discussed in section 2. In section 3 the modified Högströmmodel is described, in section 4 the experimental setting wil be given, the results are presented and discussed in section 5 and conclusions are formulated in section 6.

2. THEORETICAL BACKGROUNDS

For the measurement and modelling of the detectability of an odour at a downwind position of the source it is important to know the nature of the fluctuations of the odour detectability. The odour detectability is determined by the odour threshold. This threshold is defined as the odour concentration which can be detected by 50 % of the population. This odour concentration is, by definition set equal to 1 odour-unit per cubic meter (1 ou/m3) and is comparable with the commonly used concentration unit expressed in terms of g/m^3. Thus fluctuations in the detectability are determined by fluctuations in the odour concentration.

To determine the nature of concentration fluctuations we look at the plume from a fixed position downwind from the source. It appears that the plume shape is very irregular and that the plume axis has a variable spatial position. All turbulent eddies are effective in the dispersion of the plume. The concentration distribution within the plume is dominated by eddies which are smaller than the dimensions of the plume. This dispersion within the plume is called relative diffusion. Eddies which are much larger than the plume dimensions are responsible for displacements of the axis of the plume, the plume-meandering.

Both relative diffusion and plume-meandering occur in horizontal as well as in vertical direction. They are more or less independent turbulent processes. As a result of this turbulent character of the dispersion process, the momentary concentration downwind of a point source shows a strong variability.

Wilson, Robins & Fackrell(1982) concluded from windtunnel experiments that most of the variance of the concentration fluctuations is produced at near distances from the source. This variance is dissipated during the downwind transport of the plume. During this downwind dispersion the dimensions of the plume are increasing. The meandering of the plume is the dominant factor in the dispersion process for travelling times smaller than the Langrangian timescale (Hanna 1984). The influence of the relative diffusion is increasing with increasing traveling times. The effect of the meandering on the concentration fluctuations, however, may still be significant at large traveling times because of the influence of mesoscale effects and local turbulences (Hanna 1986).

The stack height has a strong influence on the vertical meandering of the plume. For low sources the vertical meandering process is negligible, for high stack sources it is the dominant factor in the generation of concentration fluctuations (Högström 1972). The initial dimensions of the plume are determined by the dimension of the source. A plume arising from an area source will have larger dimensions than the one emitted from a point source and relative diffusion will be more important for area sources compared with point sources.

In general the stability of the atmosphere may be expected to influence the concentration fluctuations. Under unstable conditions fluctuations will be more pronounced.

Ramsdell and Hinds(1971) and Hanna(1986) found that concentration fluctuations are the largest at the centre of the plume and they decrease in the direction perpendicular to the wind direction.

Resuming, concentration fluctuations and thus fluctuations in the detectability of odours in a fixed position downwind form the source, are determined by the horizontal dimensions of the source, the downwind distance to the source (traveling time), the stability of the atmosphere and the position of the observer in the cross wind section of the plume.

3. MODELLING OF THE EXPOSURE TO ODOURS.

The exposure to odours downwind of a source can be described by models which account for the fluctuations in the detectability of the odour.

From the preceeding chapter it is concluded that these fluctuations have a small timescale and are produced by two turbulent processes, the relative diffusion and the meandering of the plume. Thus dispersion models with small averaging times and which take account for both turbulent processes are needed. Most dispersion models, however, only describe one of the two processes (Hanna 1984). Only statistical models (Monto Carlo Models) and large eddy simulation models can describe both processes. They are large computer time consumers, however, and not all required meteorological data are available on a routine basis.

As most odour problems are on a small scale at which the meandering of the plume is probably the most dominant factor in the dispersion of the plume, models that only describe plume meandering may be sufficient for calculating the odour exposure. From a literature survey Voerman and Harssema (1985) concluded that the Högströmmodel (Högström 1972) was at that time the only available model. This model is based on the fluctuating gaussian plume model, the principles of which were published by Gifford(1959). The model yields estimates of the odour frequency at a certain point downwind of the source. The odour frequency is defined as the percentage of the time within the observation period, mostly one hour, that the odour exceeds the threshold concentration.

In the model, the momentary plume is assumed to have fixed dimensions and the concentration distribution within the plume, in both the horizontal and vertical direction is assumed to be gaussian, with parameters $\sigma_r(y)$ and $\sigma_r(z)$. Within an hour, this momentary plume meanders around the hourly averaged plume axis. The spatial distribution of the position of the plume axis in the horizontal and vertical direction is also assumed to be gaussian, with parameters $\sigma_m(y)$ and $\sigma_m(z)$. The σ_r and σ_m parameters are related to the 'normal' hourly averaged σ-values by:

$$\sigma^2 = \sqrt{(\sigma_r^2 + \sigma_m^2)} \qquad (1)$$

The dispersion parameters σ, σ_r, and σ_m, used by Högström in his model, are somewhat different from the σ-values generally accepted in the Netherlands (Anon 1984). Moreover, they are based on different stability definitions. In order to get parameters which are in line with the accepted ones in the Netherlands, the ratios FACY and FACZ of σ_r to σ (the hourly averaged parameters $\sigma(y)$ and $\sigma(z)$)) according to Högström(1972) were applied to the accepted σ-values in the Netherlands, as follows:

$$\begin{aligned} \mathrm{FACY} &= \{\sigma_r(y)/\sigma(y)\}_{\text{Högström(1972)}} \\ &= \{\sigma_r(y)/\sigma(y)\}_{\text{Anon(1984L)}} \end{aligned} \qquad (2)$$

$$\begin{aligned} \mathrm{FACZ} &= \{\sigma_r(z)/\sigma(z)\}_{\text{Högström(1972)}} \\ &= \{\sigma_r(z)/\sigma(z)\}_{\text{Anon(1984L)}} \end{aligned} \qquad (3)$$

The ratio FACY was made dependant on stability class only, the ratio FACZ was made dependant on stability class and stack height. It is assumed that the values of FACY and FACZ are independent of travel time, which, from a theoretical point of view is only true for travel times that are not much greater then the Langrangian timescale(Csanady 1973).

The central point in Högström's approach is the determination of the weighted odorous width of the plume at ground level and its variation due to the meandering of the plume. The odorous width at a certain distance from the source is that part of the instantaneous plume within which the odour threshold (1 o.u./m3) is exceeded at ground level. This is illustrated in figure 1.

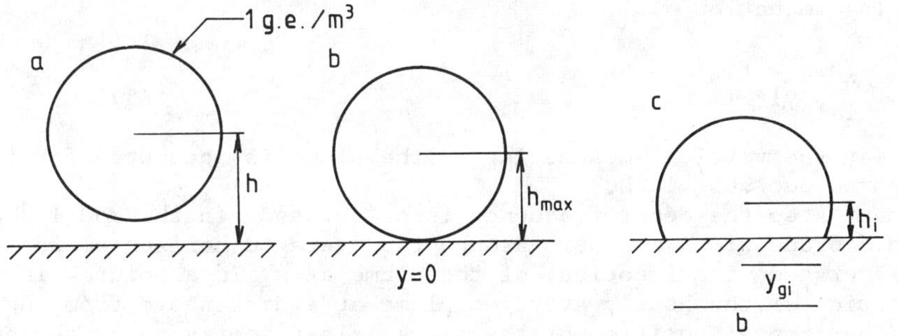

Figure 1. Schematic diagrams of cross sections of the detectable part of the instantaneous plume, at different heights (h_i) and the associated instantaneous odourwidth y_i.

The odour concentration within the cross sections in figure 1 is greater or equal to 1 ou/m^3. The local plume-centroid is at an ever differing height(h_i) above the ground. When it is close to the ground (figure 1a), the odorous width (y_i) at ground level is large. When the plume centroid is heigher, y_i decreases to zero in figure 1b. For a certain meteorological condition, defined by stability class and windspeed (u), the odorous width of the plume at groundlevel (at distance x), for each height of the plumecentroid, can be determined from the emisson strength (Q) and a threshold concentration (C_t) = 1 o.u./m3 with the aid of the gaussian plume model with (instantaneous) $\sigma_r(y)$ and $\sigma_r(z)$ values, according to equation 4:

$$C_t = \frac{Q}{\pi * \sigma_r(y) * \sigma_r(z) * u} * \exp(-h_i^2/2\sigma_r^2(z)) * \exp(-y_i^2/2\sigma_r^2(y)) \quad (4)$$

For this meteorological condition, a weighted odorous width is determined. First the value of h is calculated for which y_i=0 (in other words: the "odour-cone" just touches the ground - see figure 1b); this height is called h_{max}.
Then the value of y_i is calculated for 10 equally spaced values of h between h = 0 and h=h_{max}. With the aid of the probability distribution of the values of h, given by the normal distribution defined by the hourly averaged vertical position of the plume axis (the effective stack height) and the value of $\sigma_m(z)$ and taking into account the reflection of the plume at ground level, the fraction of the time (P_i) that each value of h occurs is calculated. The weighted odour width Y_w is determined as the sum of the products of each y_i-value with its probability P_i (see equation 5).

$$Y_w = \sum_{i=1}^{i=10} yi*Pi \quad (5)$$

In this way the vertical meandering of the plume is accounted for in the weighted odorous width.
In the next step the odour frequency is calculated. In the model the weighted odorous width meanders around the averaged horizontal position (hourly averaged winddirection) of the plume axis. In a point W in the cross section of the hourly averaged plume at a distance d from the averaged horizontal position of the plume axis (see figure 2) the odour frequency is equal to the percentage of the time that the weighted odour width (the distance between z_1 and z_2 in the figure) reaches as far as this point. This percentage is equal to the surface beneath the normal distribution in figure 2. This normal distribution is the probability distribution of the horizontal position of the plume axis, defined by its hourly averaged position and the value of $\sigma_m(y)$.

The first version of this modified Högströmmodel was published in 1984 (Voerman and Harssema,1984). Since then the model has undergone some changes in the definition of the dispersion parameters and in the calculation of the weighted odorous width. The first change was necessary because of the fact that the dispersion parameters $\sigma(y)$ and $\sigma(z)$ of the Dutch National Model were changed (Anon 1984). In the former version (Werkgroep Verspreiding, 1983) the value of $\sigma(z)$ was dependent on stack height and via equation (3) $\sigma_m(z)$ was also dependent on stack height. In the later version of the Dutch National Model the dependancy of $\sigma(z)$ and thus $\sigma_m(z)$ on stack height disappeared, which, from a theoretical point of view is incorrect for the vertical meandering of the plume. For this reason the value of $\sigma_m(z)$ in the latest version of the modified Högströmmodel is made dependent on stackheight. According to the original experimental data of Högström(1964) a lineair relation between the logarithm of the effective stack height and FACZ was calculated and implemented in the model.

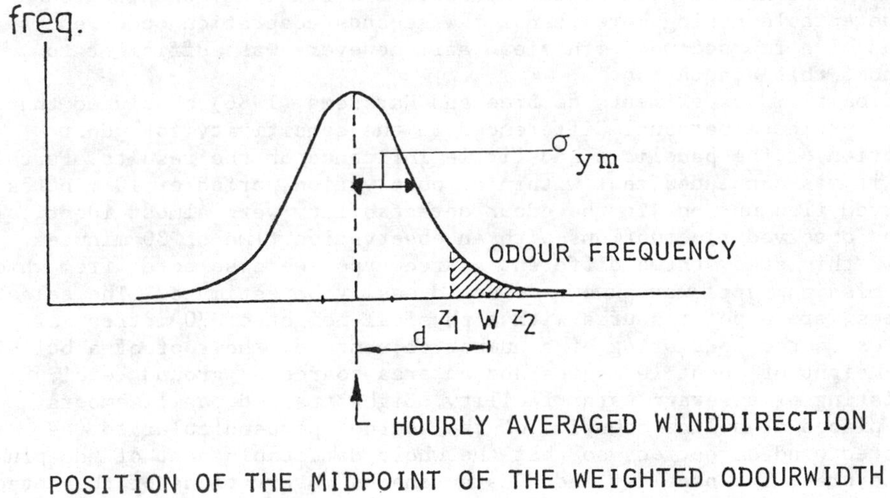

Figure 2. Schematic presentation of the probability that the odour can be detected by observer W in the horizontal cross wind section of the plume.

The second change, the calculation of the weighted odorous width was made to improve the accuracy of the model. In the original version Y_w was calculated in steps of h_i of the size of $0.1*\sigma_m(z)$. For large values of $\sigma_m(z)$ in combination with small values of h_{max}, only one or a few steps of h_i would be performed. For that reason the model has been changed as described above.

4. THE EXPERIMENTAL SETTING.

In order to evaluate the validity of the Högströmmodel an experimental program was set up, in which odour frequencies were measured in the cross section of the plume with trained observers, in dependency of the source dimensions, the travel time and the stability of the atmosphere. The averaging time and the observation time are important factors in the experimental setting. The averaging time determines the high frequency edge of the concentration fluctuations. The observation time is important for the determination of the fluctuations which are caused by eddies at a low frequency (Högström 1968). To measure short time fluctuations in the odour detectability the averaging time must be very small. From simultaneous sensoric and instrumental measurements Barynin and Wilson(1972) concluded that the signal of the concentration measuring device was followed well by the sensoric observations. When the odour was detectable during more then a few seconds adaptation occurred; a period of a few seconds with clean air, however, was sufficient to overcome this adaptation.

From field experiments de Bree and Harssema (1986) concluded that inter- or intra-personal differences in the sensitivity for odour detection of the panelists had little influence on the results. Furthermore it was concluded that within an observation period of 10 minutes the observed fluctuations in the odour detectability were almost identical to the observed fluctuations with an observation time of 30 minutes.

In this study three different source types were selected from which the emission rates were kwown or could easily be estimated. The selected sources were a point source with a physical height of 70 metres, a complex source consisting of a number of vents on the roof of a building at a height of about 10 metres and an area source at ground level, consisting of a sewage water facility. Eight trained panel members were placed at a line downwind of the source, perpendicular to the expected wind direction, so that the whole detectable part of the plume was covered. The panelist made instantaneous observations of the detectability of the odour (with an 'averaging' time of a few seconds), every 10 seconds during 10 minutes. For expressing their odour judgements, they used a four-point odourintensity scale (score 0:no odour detected, score 1: a weak odour detected, score 2: the odour can easily be detected and score 3: a strong odour detected). All judgements were written on computer readable paper.

At a certain distance three consecutive runs of 10 minutes were performed within one hour, with a period between two measurements of 5 minutes. The stability of the atmosphere was determined by the visually measured cloud coverage during the measurements and by windspeed recordings obtained from meteorological stations in the near vicinity of the sources.

5. RESULTS AND DISCUSSION.

For each source, 60 to 100 10-minutes runs were obtained, covering various distances to the source and unstable to near-neutral stability conditions. The results will be handled with in two different ways:

1. The fluctuations of the detectability of the odour have been examined.
2. Model performances are compared with the experimental results.

For examining the fluctuations of the detectability of the odour typical results of runs at the three sources under unstable conditions are presented in figure 3. As these observations are made in fixed positions, this is a recording of the instantaneous fluctuations of the detectability of the odour in a Eulerian framework.

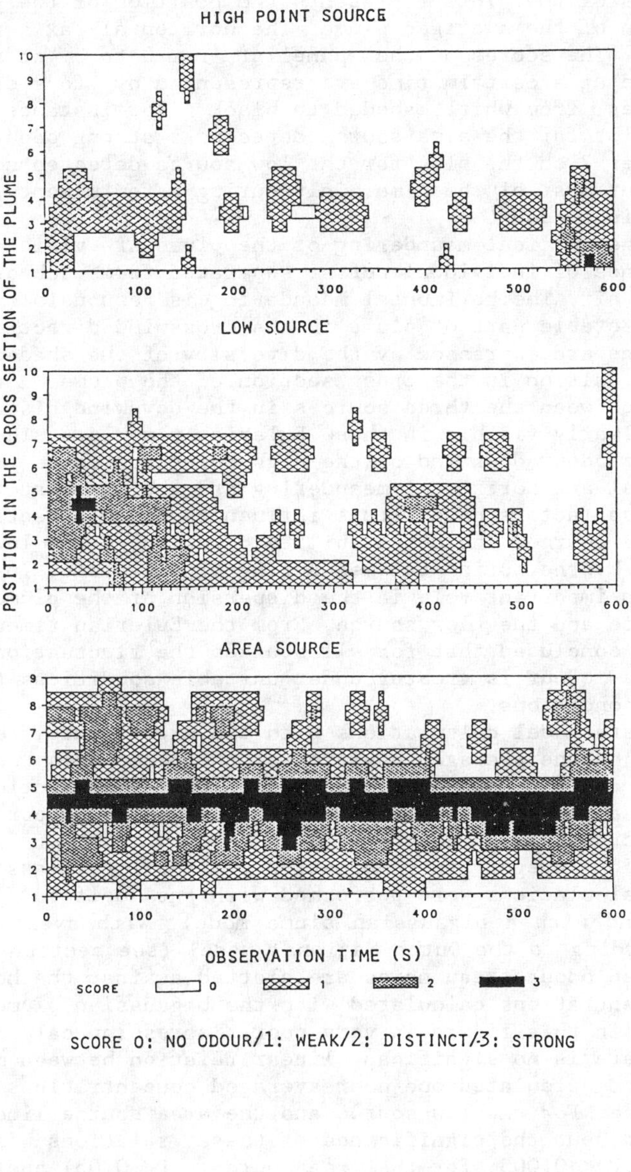

Figure 3. Eulerian timeplots of the observed odour intensity in the cross section of the plume for the three sources under unstable atmospheric conditions.

The vertical axis of the plots represents the position of the panelist in the cross section of the averaged plume. The horizontal axis represents the timescale. The scores of the panelist from 0 to 3 at a certain position and at a certain time are represented by four different "coloured" rectangles, from white, shaded to black. For instance observer 4 on the plot for the area source detected a strong odour most of the time. Observer 7 on the plot for the low source detected no odour or a weak odour most of the time. Only during a few moments he or she detected a distinct odour.

The effects of the vertical meandering of the plume are visible in the plots in the occurence of individual puffs, separated from each other by periods of clean air. The horizontal meandering is responsible for the sway of the detectable part of plume in the crosswind direction. In-plume fluctuations are expressed by the diversity of the shaded rectangles at one position in the cross section of the plume.

The differences between the three sources in the downwind dispersion of the odour are clearly visible in these Eulerian timeplots. The detectability of the odour downwind of the high point source is very fluctuating. Vertical and horizontal meandering are the dominant turbulent processes. These factors have little influence on the detectability downwind of the area source. The detectability downwind of the low source is dominated by the horizontal meandering of the plume. In-plume fluctuations play an important role in the dispersion of the plume from the high point source and the low source. From the Eulerian timeplots of all runs, it was concluded that for all sources the fluctuation of the detectability of the odour is greater under unstable conditions than under near neutral conditions.

In order to compare model calculations with the experimental results the odour frequency on the averaged position of the plume axis was determined from the odour frequencies observed on the cross section of the plume, assuming a gaussian distribution of the observed odour frequencies in the cross wind direction.

For each dispersion category (a combination of stability class and traveltime) model calculations were performed with the modified Högströmmodel and with a bigaussian plume model, with averaged σ_y and σ_z values according to the Dutch National model (see section 3). In figure 4 the measured odour frequencies are plotted against the hourly averaged odour concentrations calculated with the bigaussian plume model. The relation in this figure is very poor. Regression calculations show that there is no significant linear relation between measured odour frequencies and calculated one-hour averaged concentrations for the high point source. For the low source and the area source linear relations were found, but the significance of these relations was poor (for the low source: $P < 0.005$; for the area source: $P < 0.05$) and moreover, the lineair relations of the two sources differed significantly ($P < 0.025$).

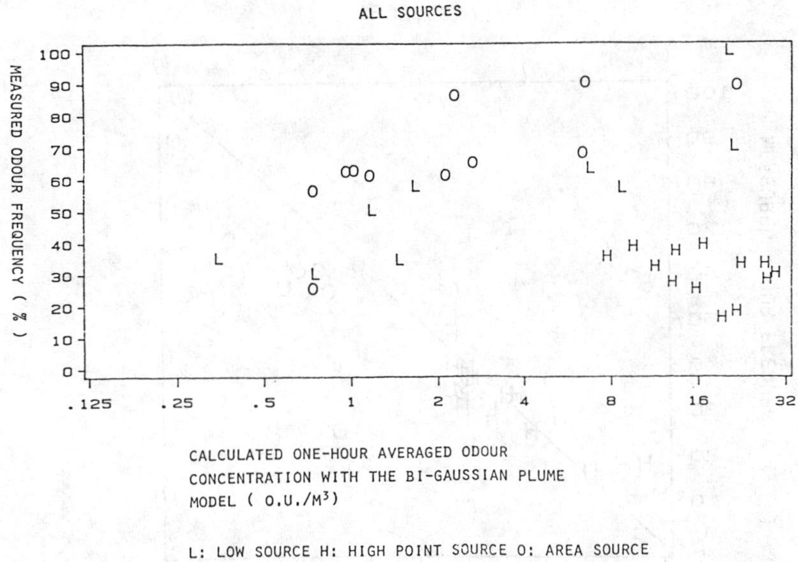

Figure 4. Relation between the observed odour frequencies at the plume axis and the one-hour averaged odour concentrations calculated with the bi-gaussian plume model, for all sources and for unstable and near neutral atmospheric conditions.

In figure 5 the measured odour frequencies are plotted against the odour frequencies calculated with the Högströmmodel.

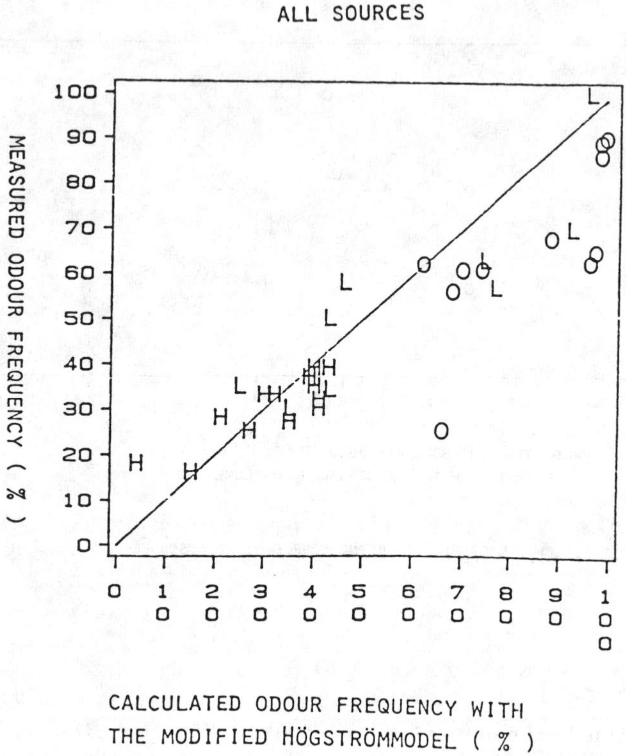

Figure 5. Relation between the observed odour frequencies at the plume axis and the calculated odour frequencies with the modified Högströmmodel, for all sources and for unstable and near neutral atmospheric conditions.

From this figure it is clear that there is a good agreement between the measured and the calculated odourfrequencies. Lineair regression calculations were performed between measured and calculated odour frequencies for all three sources individually and for all sources together. It was found that there was no difference in the relation between measured and calculated odour frequencies between different stability classes and between the three sources ($P<0.001$). The relation for all sources together was significant ($P<0.001$), with a coefficient of correlation of $r=0.90$.

This relation has the following form:

Measured odour frequency = 0.70 * calculated frequency + 9.

The fact that the regression coefficient of 0.70 is less then 1 can be contributed to the fact that the calculated odour frequencies of the area source were over prediction of the measured odour frequencies (see figure 5). The uncertainty about the emissionrate of this source may be credit to this fact.

6. CONCLUSIONS.

It can be concluded that, at least under near neutral and unstable atmospheric conditions, there is a good agreement for all sources between the measured odour frequencies and those calculated with the modified Högströmmodel. The agreement between measured odour frequencies and calculated hourly-averaged odour concentrations is poor for all sources and under all atmospheric conditions.

The meandering of the plume has a great influence on the detectability of the odour downwind from the high point source. Thus for the modelling of the detectability of odours downwind from (high) point sources, the use of a fluctuating plume model, like the modified Högströmmodel is necessary. For these sources a bi-gaussian plume model is insufficient to give a proper description of the exposure to odours.

Finally, since it is concluded that the modified Högströmmodel is a better tool for estimating odour exposure than gaussian models based on hourly averaged concentrations, a recommendation is made for changing the Dutch preliminary concentration standards for odours in exposure standards that are based on the detectability of odours within an hour.

The modified Högströmmodel was tested for relatively short traveltimes only, at wich it was found that relative diffusion played a less important role in generating concentration fluctuations as plume meandering did. Under these conditions the application of the modified Högströmmodel is recommended. From a theoretical point of view it is likely that for large distances the applicability is less, because then relative diffusion will be more dominant than plume meandering. Since most odour problems are on a local scale, this is no serious limitation in practice.

ACKNOWLEDGEMENTS

This study was financed by and conducted for the Dutch Ministry of Housing, Planning and Environment (VROM).

REFERENCES

Anon.(1984):"Parameters in the long-term model of air pollution dispersion. New recommendations". Study and Information Centre TNO for Environmental Research, Delft, The Netherlands.

Barynin,J.A.M. Wilson,M.J.G. (1972):"Outdoor experiments on smell". Atm. Env. vol 6 97-207.

de Bree,F.B.H. Harssema,H. (1986):"Vervolgonderzoek verspreidingsmodellen voor stank. Eindverslag Deel 1". Report R-149, Dept. of Air Pollution, Agricultural University Wageningen, The Netherlans (published in Dutch).

Csanady,G.T.(1973):"Turbulent diffusion in the environment". J. Atmos. Science, 24, 21-25.

Gifford,F.(1959):"Statistical properties of a fluctuating plume dispersion model". Adv. in Geophys.,6, 117-138.

Hanna,S.R.(1984):"Concentration fluctuations in a smoke plume". Atm. Env. 18, 1091-1106.

Hanna,S.R.(1986):"Spectra of concentration fluctuations: the two time scales of the meandering plume". Atm. Env. 20, 1131-1137.

Harssema,H.(1987):"Characterization of exposure in odour annoyance situations". In: Koelega,H.S., Environmental Annoyance: Characterisation, Measurement and Control, pp 96-104.

Högström,U.(1964):"An experimental study on atmospheric diffusion". Tellus 16, 205-251.

Högström,U(1968):"A statistical approach to the airpollution problem of chimney emissions". Atm. Env. 6.

Högström,U.(1972):"A method for predicting odour frequencies from a point source". Atm. Env. 6, 103.

Ramsdell Jr,J.V. Hinds,W.T.(1971):"Concentration fluctuations and peak-to- mean concentration ratios in plumes from a groundlevel continious point source". Atm. Env. 5,483-495.

Voerman,J.W.Th., Harssema,H.(1984):"Development of a dispersion model for odours". In: Characterization and control of Odoriferous Pollutants in Process Industries. Louvain La Neuve, pp 231-244.

Voerman,J.W.Th., Harssema,H.(1985):"Inventarisatie verspreidingsmodellen voor stank". Documentatie-Lucht 1985-1, Ministerie VROM, Leidschendam (in Dutch).

Werkgroep Verspreiding Luchtverontreiniging(1983):"Frequency distributions of air pollution concentrations; recommendations for a method of calculation". Delft, The Netherlands

Wijnen,H.(1985):"Air quality standards on odours in The Netherlands". VDI-Berichte 561, 365-385.

Wilson,D.J. Robins,A.G. Fackrell,J.E. (1982):"Predicting the spatial distribution of concentration fluctuations from a groundlevel source". Atm. Env. 16, 497-504.

THE INFLUENCE OF A BUILDING ON THE DISPERSION OF EMISSIONS FROM LOW CHIMNEYS
The Dutch recommendation for a calculation procedure

A. van Melle/H.P. Baars
MT-TNO
Department of Fluid Dynamics
Department of Environmental Chemistry
respectively
P.O. Box 342, 7300 AH Apeldoorn, The Netherlands

ABSTRACT. A new model has recently been included in the Dutch National Model for the dispersion of air pollutants. It includes a new method for calculating the concentrations of inert substances emitted from a stack close to a rectangular building. This method allows concentrations to be calculated near the building as well as further downwind. The article summarises the research that has led to the new model. It discusses basic flow characteristics around buildings, describes the windtunnel experiments and presents the major results. Finally it explains the basic concepts of the new model.

1. INTRODUCTION

1.1 The National Model

A concensus exists in the Netherlands with regard to the calculation of air pollution concentrations by means of what is known as the National Model. In 1976, a method for the calculation of average annual concentrations of air pollutant components was published (1). This was added to in 1984 to include high point sources. A booklet was issued in 1981 giving the method for the calculation of percentage values, i.e. concentrations exceeded for a particular percentage of the time (2).

The Blue Book (1) sets a limit of 100 metres within which the model is not considered accurate enough since within this distance from the source obstacles such as buildings may exercise a major influence on the local dispersion of air pollutants. Even at greater distances, the influence of buildings may be significant.

This prompted the Ministry of Housing, Physical Planning and Environment to commission research to find a model-based formula. This research was carried out by the TNO division of Technology for Society (MT-TNO) and the Centre for Energy and Environmental Studies of the

K. Grefen and J. Löbel (eds.), Environmental Meteorology, 487–495.

University of Groningen [3]. In collaboration with the Working Party on Dispersion of Air Pollution, a new recommendation was formulated [4]. In this article, we begin by giving some of the characteristics of the flow around buildings. Next, we shall deal with the windtunnel research and the results obtained. Finally, we shall discuss the chief aspects of the new recommendations for a method of calculation.

1.2 Flow round a building

Before going into the windtunnel experiments and recommended calculation method, it will clarify matters to describe in general terms the flows in the vicinity of a building. The diagrams in Figure 1 show a simplified version of the flow lines, whirls and wind profile near a cubiform building. In the left-hand diagrams the wind flow is at right angles to the wall. On the right, the cube is turned 45°. Figures a1 and b1 give a view from above, the four figures at the bottom give the view from the side. We shall first discuss certain interesting aspects of these diagrams.

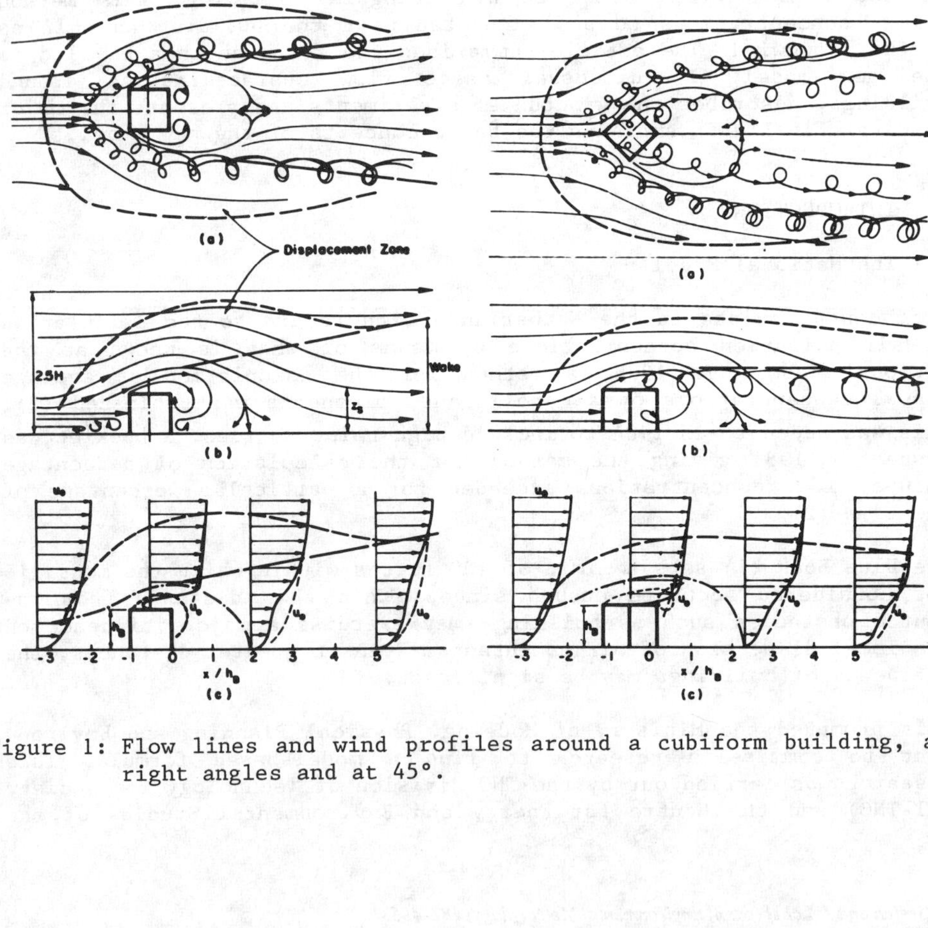

Figure 1: Flow lines and wind profiles around a cubiform building, at right angles and at 45°.

A windflow over a building or any obstacle creates an upward pressure on the upwind side and a downward pressure on the downwind side. The flow lines diverge. Secondary flows and increased turbulence are created. On the downwind side a leeward whirl is created with increased turbulence and even a backflow to the ground. The pollution emitted by such a leeward whirl can momentarily cause extremely high concentrations locally but these are also very quickly remixed with the total volume of the leeward whirl. Depending on the size of the building, the leeward whirl may extend downwind to a distance of at least 2 and at most 6 times the height of the building. Its height varies from 1.2 to 1.5 times the height of the building. A leeward whirl is not a static occurrence. Small changes in the velocity and direction of the wind quickly create different patterns of turbulence. The 'dimensions' of the leeward whirl also fluctuate as a result.

When the windflow is from a certain angle (45° in the right-hand figures), strong spiral whirls are created from the corners of the roof. These can keep going for quite a long distance. They reduce the leeward whirl and increase mingling. In this situation the hourly average concentration in the leeward whirl can, as the result of a source at that point, be multiplied by a maximum factor of 5 compared with a flow at right angles to the wall. Between the leeward whirl and the undisturbed flow (above approx. 2.5 times the height of the building) is an area in which the flow turbulence is not increased. The flow is sucked downwards (including any smoke plumes containing pollutants) as the result of the lower pressure behind the building. Above 2.5 times the height of the building it is assumed that the influence of the building is no longer significant.

It should be clear that Figure 1 describes an idealised model. Where building size varies or extra obstacles are present in the surrounding area, the flow profiles may change dramatically and, as a result, the dispersion of air pollution.

2. INTERNATIONAL EXPERIENCE

Internationally speaking, there is hardly any concensus with regard to a generally valid model for dispersion around buildings, in contrast for example to the Gaussian Plume Model for undisturbed situations. This is so in spite of the fact that since the Fifties a great deal of research has been done, including the use of windtunnels, both on scale models (nuclear power stations) and on simplified situations with a view to developing models. Because of the complexity of the flow around obstacles, an unambiguous mathematical description is extremely difficult. The main directions of empirical theories are a little more united on some of the more general, straightforward aspects. Sometimes for the lack of better ones. For example, on the

subject of long-term average concentrations in leeward whirls resulting from a source which is also in the leeward whirl, with the wind in certain directions. The leeward whirl is assumed to be homogeously mixed. The concentration can simply be deduced from the 'replacement' of the leeward whirl volume by the wind. There is agreement also on the way to calculate concentrations at a greater distance from the obstacle, when the flow is more or less undisturbed. The effect of a building on dispersion may then be described by means of a simple broadening of the plume ($\sigma_y + \sigma_{y0}$). The greater the distance downwind, the smaller is the influence exerted by a building on the concentration in any case. σ_y becomes greater with distance but the extra dispersion term σ_{y0} remains constant.
Thirty years of research have not resulted in a generally valid model, either on the basis of the Gaussian plume concept or with the help of diffusion comparisons.

3. WINDTUNNEL TESTS

TNO's PIA (Pollution Industrial Aerodynamics) windtunnel was used for the research into the disperion of chimney plumes in the vicinity of buildings. A scale model can be placed inside a boundary-layer windtunnel of this type and all the processes which take place in the open air can be simulated. The profile of the airflow inside the windtunnel corresponds to the wind field of the earth, both in velocity profile and in the level of turbulence. In addition to the dispersion of pollutants, wind hindrances for pedestrians and the pressure distribution for example, around large building can be studied.

A number of scale rules must be observed in order to arrive at a realistic simulation. These are dimensionless groups which must have the same value in the windtunnel as in the open air. The most obvious of these is that the ratios for the scale model must correspond with actuality. In addition, the Froude number, for example, must remain constant. This gives the ratio between the upwards forces due to differences in density and the resistance experienced by the light gas when it rises.

In order to formulate a model, the idealised situations in the windtunnel were examined: a cube and a chimney stack were placed in different relative configurations, and the height of the stack also was varied. The stack was always placed against the wall or corner of the building. The emission of 'smoke' from the stack was simulated, with the addition of a tracer gas, in order to determine the dilution at various points downwind from the model.

The effect of the type of plume was also studied in detail. As we shall see, the type of plume may have a major influence on dispersion in the vicinity of a building. A plume with a low emission impulse (i.e. velocity), one with a high impulse and a warm smoke plume with a low impulse were compared. The light gas was used to simulate a warm plume (which has a lower density than the air at the surrounding temperature) to model scale. For the light gas, a mixture of air and helium was used.

Plume profiles were arrived at from the various configurations, from which calculation rules were then adduced for the purpose of formulating the model.

4. RESULTS

4.1 Building influence

One of the most important results of the windtunnel experiments was the difference which was observed in the way the plume rose for the different types of plume. A 'cold' impulse plume very quickly (within a few metres) reached the final plume height, while the warm/light plume rose slowly and asymptotically. Figure 2 illustrates this. This phenomenon is of great importance to the dispersion behind the building. The warm plume without much impulse can, due to the small amount of rise in the first 20 or 30 metres, become incoporated in the leeward whirl and be transported straight to the ground. There localised and momentarily very high concentrations may be created at points where people may be present. A cold plume with some degree of impulse, however, can just clear the immediate area of influence of the building (the leeward whirl) so that pollution is not noticeable close to the building.

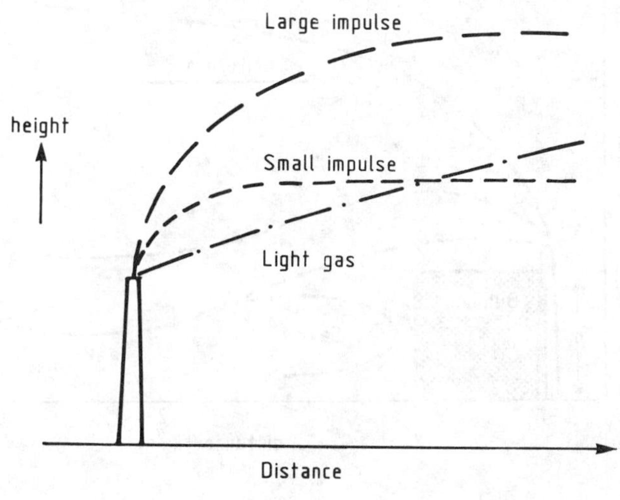

Figure 2: Plume rise as a function of the distance transported for three types of plume.

It was also observed during the experiments that there are in fact three systems by which the building exerts an influence. Depending on the point where the plume axis is located on the leeward whirl, the plume may be caught by and incorporated in the leeward whirl, or the plume axis may fall without influencing dispersion, or the plume may not be influenced at all. The background to the windtunnel experiments, the results and the experience in other countries are described, accompanied by over a hundred references, in the booklet (3).

4.2 New recommendation

In formulating the new recommendation, an attempt was made to keep as near as possible to the Gaussian Plume concept of the National Model. In many cases it was possible, by the addition of a parameter or the upscaling of a value, to keep the basic algorithm. However, we did not succeed in describing it all in one simple formula, one reason being that the various systems must be differentiated. The booklet (4) will serve the user as a guide.

As already mentioned, there are three systems of influence by the building. The critical dimension for deciding into which system the plume falls is the height of the plume at the highest point of the leeward whirl (see Figure 3).

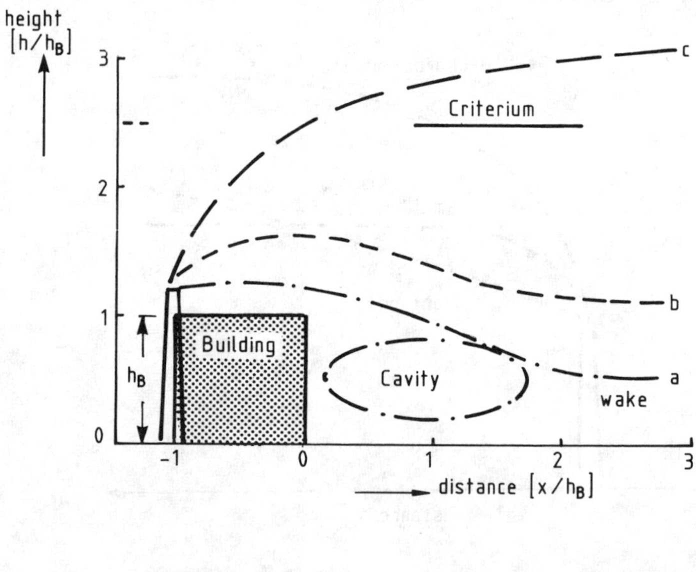

Figure 3: Three systems of influence exerted by a building on a plume.

If the plume is lower at that point than the height of the leeward whirl, the pollution is assumed to be homogeneously mixed with it (a). Above this height, the plume merely arches downwards (b) or is even completely uninfluenced (c). We have already seen that the emission impulse, and not the heat content, is important to the rise of the plume in the first few metres after discharge. Unfortunately, the plume rise formula in the present National Model only describes the contribution made by the heat content (buoyancy) to this rise. And in any case only the final height of the plume rise is calculated. The new recommendation makes use of a formula of Briggs upon which the National Model was based but which describes the rise caused by impulse as well as the buoyancy. In addition, Briggs gives the rise as a function of the distance transported. The rise resulting from buoyancy and impulse is described by this formula in the same way as was observed in the windtunnel. The final plume rise is approximately the same by both formulae.

Using this plume rise formula, it may thus be determined to which system of influence the plume belongs. The calculations for these cases are briefly summarised below:

(a) The plume is completely incorporated in the leeward whirl and asumed to be mixed homogeneously with it. The concentration is then in proportion to the emission and in inverse proportion to the 'exchange' of the leeward whirl by the wind. In order to calculate the concentration at a greater distance downwind, the leeward whirl may be regarded as a kind of volume source or a point source which already has some initial dispersion. This is then discounted in the dispersion coefficients σ_y and σ_z of the Gaussian plume formula.

(b) The plume axis follows the flow lines and falls as a result of the down pressure behind the building. The dispersion of the plume (σ_y and σ_z) remains unchanged. The degree of fall from the effective plume height is determined by the calculated height above the leeward whirl.

(c) The plume is entirely uninfluenced. The National Model may be applied in the normal way.

This recommendation must be seen as a method of estimation to be used for rectangular buildings of a fairly simple shape, the ratio of whose height, width and depth is not more than a factor of 2. The chimney stack should be in the immediate vicinity of the building, on it or against it. More complex situations require more careful attention and/or further windtunnel research.

4.3 The model on floppy disc

As already observed, we were not successful, due to the complexity of the material, in keeping the prescribed calculation down to a single formula. Calculations of an hourly average concentration may be made using the method in the booklet (4). The calculation of an annual average concentration in accordance with the distribution of climatological frequencies, however, demands a good deal of calculation and patience on the part of the user. The degree of building influence (the system) is, for example, dependent on wind velocity and wind direction. It was therefore obvious that this calculating should be computerised. The TNO division of Technology for Society (MT-TNO) is engaged at present on making the National Model 'user-friendly' and placing it on disc. Any user in possession of an IBM (or compatible) PC will in due course be able to purchase this package and carry out his own dispersion calculations. An attempt will also be made, although at a slightly later date (mid-1988), to implement this new recommendation as well.

Acknowledgements

This research was subsidized by the Air Directorate of the Ministry of Housing, Physical Planning and Environment. The authors would like to thank Dr. P.J.H. Builtjes (TNO division of Technology for Society MT-TNO) and Dr. M.P.J. Pulles (Centre for Energy and Environmental Studies of the University of Groningen) for their contributions to this research.

Bibliography

1. Working Party on the Dispersion of air pollution.
   Models for the calculation of the dispersion of air pollution. The long-term model.
   Staatsuitgeverij, 1976 (in Dutch).
   Obtainable in English from SCMO-TNO,
   P.O. Box 186, 2600 AB Delft.

2. Working Party on the Dispersion of air pollution.
   Frequency distributions of air pollution concentrations.
   Staatsuitgeverij, 1981 (in Dutch).
   Obtainable in English from SCMO-TNO,
   P.O. Box 186, 2600 AB Delft.

3. Baars, H.P., A. van Melle.
   The influence of a building on the dispersion of emissions from chimneys.
   IVEM Report No. 4, August 1985. Also published in the MT-TNO Report CMP 85/10, October 1985 (in Dutch).

4. Working Party on the Dispersion of air pollution.
   The influence of a building on the dispersion of emissions from chimneys; recommendations for a calculation procedure.
   Vereniging Lucht, October 1986. Obtainable from SCMO-TNO,
   P.O. Box 186, 2600 AB Delft (in Dutch).

Explanation of the symbols

σ_y is the lateral dispersion coefficient of the formula according to the Gaussian plume model;

σ_{y0} is the term for the initial dispersion;

σ_z is the vertical dispersion coefficient.

* H.P. Baars was at the time of the research a member of the Centre for Energy and Environmental Studies of the University of Groningen; he is currently working as a member of the TNO division of Technology for Society (MT-TNO) in Delft. A. van Melle is attached to the TNO division of Technology for Society (MT-TNO) in Apeldoorn.

MODELLING DISPERSION OF AIR POLLUTANTS EMITTED BY POWER STATIONS USING FLUCTUATIONS OF WIND DIRECTION

H. van Duuren and J.J. Erbrink
N.V. KEMA[1)]
Environmental Research Department
P.O. Box 9035
6800 ET ARNHEM
The Netherlands

ABSTRACT. To calculate the effects at ground level of emissions of air pollutants from large power stations, a dispersion model has been developed by which hourly concentrations can be calculated. The model is based on the Gaussian dispersion model of which the parameters σ_y and σ_z are calculated from the fluctuations of the horizontal wind direction. To do so, the fast stability-dependent fluctuations are separated from the slower stability-independent ones. The fast fluctuations are characterized by the standard deviation σ_v of the lateral wind velocity and by the Eulerian time-scale T_e. The values of σ_v and T_e are used to calculate σ_z. Parameter σ_y is also calculated from σ_v and T_e, by which a contribution of the stability-independent fluctuations is taken into account.
To test the model, two sets of dispersion experiments were carried out; SF_6 measurements near two non-buoyant sources and lidar observations of a smoke plume. From these experiments values of the hourly-mean concentration and of σ_y and σ_z were derived, which are compared with values calculated by applying the wind fluctuation model and three other dispersion models. It is concluded that σ_v and T_e can easily be obtained by measuring wind direction and wind speed, and that dispersion may well be modelled through direct conversion of these parameters into σ_y and σ_z without the use of a stability parameter.

1. INTRODUCTION

In the Netherlands investigations into the dispersion of air pollutants emitted by power plants started in the early sixties. The experiments were set up and carried out by KEMA.

In the period up to 1970 measurements have been performed around several power plants to determine the deposition of fly-ash and soot

---

1) Joint Laboratories and Other Services of the Dutch Electricity Supply Undertakings

K. Grefen and J. Löbel (eds.), Environmental Meteorology, 497–513.

and to gain insight into the concentration levels of sulphur dioxide and oxides of nitrogen up to a distance of 5 km. The data were used, among other things, to test guidelines on dispersion calculations, as for instance given in VDI-directive 2289 (Elshout and Duuren, van, 1967). The knowledge obtained also made it possible to predict concentration levels around new power stations to be built.

After 1970 fundamental research was started on the atmospheric dispersion of flue gases emitted by tall stacks of 150 m and higher. This research was part of co-operation with the National Institute of Public Health and Environmental Hygiene (RIVM) and with the Royal Netherlands Meteorological Institute (KNMI). A measuring method was developed using sulphur hexafluoride (SF_6) as a tracer (Duuren, van, et al., 1975). The method was employed in a large number of dispersion experiments from the 213 m high meteorological mast of KNMI (Agterberg et al., 1983). The aim of this study was to test whether the values of the dispersion coefficients σ_y and σ_z as given in the Dutch National Model for long-term average concentrations (TNO, 1976, 1981) could also be used for very high sources (Duuren, van, and Nieuwstadt, 1980).

In 1978 the development of a continuous SF_6 monitor was started. A prototype came into operation in 1982; a second improved type is now in operation.

For many years it has been common practice to calculate the yearly-averaged ground-level concentrations of air pollutants around power stations as a measure for the environmental impact. However, the increased interest in short-term exposures that even recently in the Netherlands led to the formulation of air quality standards for SO_2 and NO_2 as 1-hour-average concentrations, made it necessary to develop a dispersion model by which hourly concentrations can be calculated.

This model is based on the widely used Gaussian dispersion model; the dispersion parameters σ_y and σ_z are calculated from the fluctuations of the horizontal wind direction.

In chapter 2 the backgrounds which led to the development of the model and the model itself are described. The model was tested by carrying out a series of dispersion experiments using SF_6 and a series of lidar measurements. The experimental design, materials and methods are given in chapter 3.

The results of the experiments, the comparison of these results with calculations of the model described and with other dispersion models, are given in chapter 4 and discussed in chapter 5.

2. MODEL DESCRIPTION

The starting-point of the development of a dispersion model by which hourly-mean concentrations can be calculated was to make use of an operational model which can easily be applied to tall stacks, with input parameters which can be measured with available meteorological instruments or can be derived from such measurements in a relatively simple manner. An extensive literature survey showed that most of the operational models are based on the well-known Gaussian distribution function to describe plume dispersion. Furthermore, it was concluded that the dispersion parameters σ_y and σ_z can be calculated succesfully if the statistical dispersion theory of Taylor (1921) for a continuous source is applied (Irwin, 1983). This has the advantage that the use of stability classes can be avoided, because as a rule a reliable determination of stability is difficult.

In the next paragraphs only a brief summary of the model developed is given. A detailed description of the model and of the theoretical backgrounds will be published elsewhere (Erbrink, 1988).

Dispersion can be described using Taylor's dispersion formulae:

$$\sigma_y^2 = 2\sigma_v^2 \, T_{Ly}^2[t/T_{Ly} + \exp(-t/T_{Ly}) - 1] \tag{1}$$

and

$$\sigma_z^2 = 2\sigma_w^2 \, T_{Lz}^2[t/T_{Lz} + \exp(-t/T_{Lz}) - 1] \tag{2}$$

σ_y and σ_z are the horizontal and vertical dispersion parameters, σ_v and σ_w the standard deviations of the horizontal and vertical wind velocity fluctuations, T_{Ly} and T_{Lz} the Lagrangian timescales of the horizontal and vertical wind components and t is the travelling time.

The values of σ_v and σ_w can be derived from near-surface measurements of wind direction fluctuations and wind speed. This is based on the following considerations. First, the assumption is made that σ_v equals σ_w over the height interval $0.1\ z_i < z < 0.9\ z_i$, z_i being the height of the mixing layer (Wilczak and Phillips, 1986).

The value of σ_v is independent of height within the mixing layer. This is concluded from an analysis of fast wind direction fluctuations as measured at heights of 10 m and 200 m along the meteorological mast of KNMI at Cabauw. The wind direction fluctuations measured within a period of an hour can be separated into a fast and a slow component using the method of the moving average. The average signal is a measure for the slow component and the residual signal, the difference between the original and the average signal, for the fast component. From wind measurements performed at an aircraft base near Arnhem during several days in winter and in summer it was concluded that the fast fluctuations depend on stability. This is not the case for the slow fluctuations.

The value of the fast fluctuations is used directly to calculate the vertical dispersion. The horizontal dispersion is obtained by combining the fast and slow fluctuations.

The Eulerian timescale T_e of the fast fluctuations is calculated from the autocorrelation of the time series of the wind direction data. A negative exponential autocorrelation function with timescale T_e is assumed:

$$R(\tau) = \exp(-\tau/T_e) \tag{3}$$

$R(\tau)$ being the autocorrelation function for a time step τ. The value of T_e depends on stability; typical values are 5 s for stable, 20 s for neutral and 30 s for unstable conditions. Conversion of the Eulerian timescale T_e into the Lagrangian one T_L is done using a proportionality factor β equal to 3 (Moore et al., 1985). The timescale of the slow fluctuations is assumed to be infinite.

Summarizing, the values of T_e and σ_v are directly derived from data of wind direction and wind speed, after separation of the wind direction signals into stability-dependent (fast) and stability-independent (slow) fluctuations.

σ_{yf}, due to the fast fluctuations, is calculated using equation (1) with $\sigma_y = \sigma_{yf}$ and $T_{Ly} = \beta T_e$. The value of σ_z is equal to that of σ_{yf}. σ_y is obtained using the equation:

$$\sigma_y^2 = \sigma_{yf}^2 + \sigma_{ys}^2 \tag{4}$$

with $$\sigma_{ys} = \sigma_{vs} t \tag{5}$$

σ_{vs} is the standard deviation of the slow horizontal wind velocity fluctuations.

In view of the assumption of isotropy within the mixing layer, this method can only be applied to tall stacks with an effective height between 0.1 z_i and 0.9 z_i.

The method of calculating fast and slow fluctuations is not available from routine measurements at meteorological stations. Therefore, measurements were done during a great many of hours (about 500) at an aircraft base, as mentioned before.

For each hour σ_z-values as a function of distance were calculated. The stability class (A to F) according to Pasquill-Gifford-Turner was also obtained using the scheme as defined in the Dutch National Model (TNO, 1976, 1984). In this model the stability class is obtained combining cloud cover and wind speed, dependent on season and time of the day (day and night). Using the stability class and Pasquill's equations to calculate σ_z (TNO, 1984), σ_z values as a function of distance were calculated too.

As an example in Figure 1 σ_z-curves are given for a number of hours on December 3 and December 4, 1984. During stable conditions at night (stability classes E and F) and neutral conditions in the day-time (class D) the agreement between both curves is reasonably good. Under so-called unstable conditions (class B) in the day-time the curves differ considerably. From σ_z-curves obtained for other days it is seen that especially during the transitional stages between night and day there is a gradual change in σ_z calculated from wind fluctuations as distinct from the Pasquill-values. In the latter cases for instance the stability class suddenly changes from F to A, which also results in a sudden change in the σ_z-curve.

When on the basis of wind fluctuations the average σ_z-curves are calculated for all neutral cases (class D), separated into day-time and night-time, there is a large difference between both

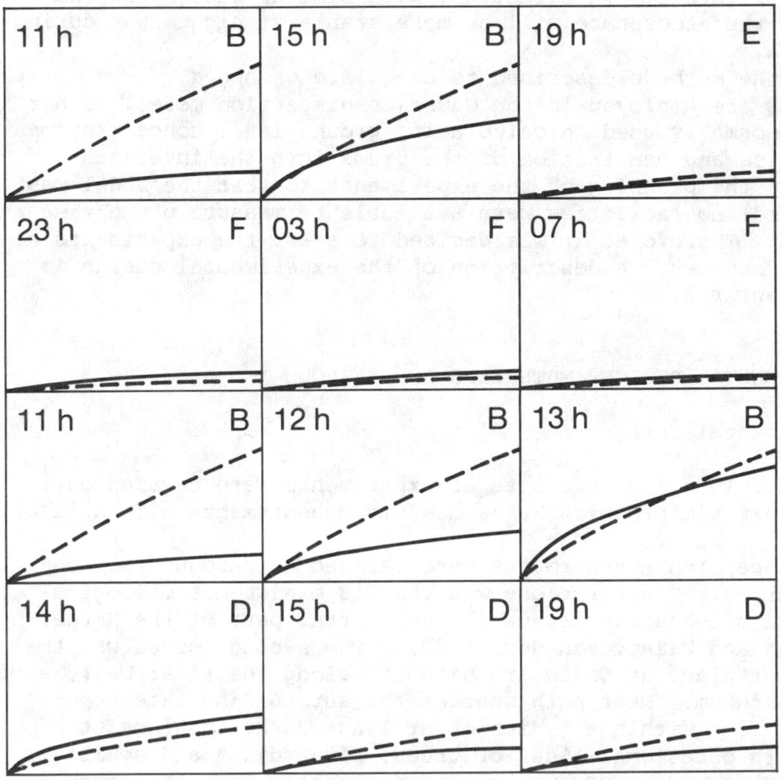

Figure 1. σ_z-curves calculated from wind fluctuation measurements (unbroken lines) and according to Pasquill (broken lines) for the period December 3, 1984, 11h00, to December 4, 1984, 19h00, (B, D, E, F: stability class; x-axis: 0-15 km, y-axis: 0-1 km).

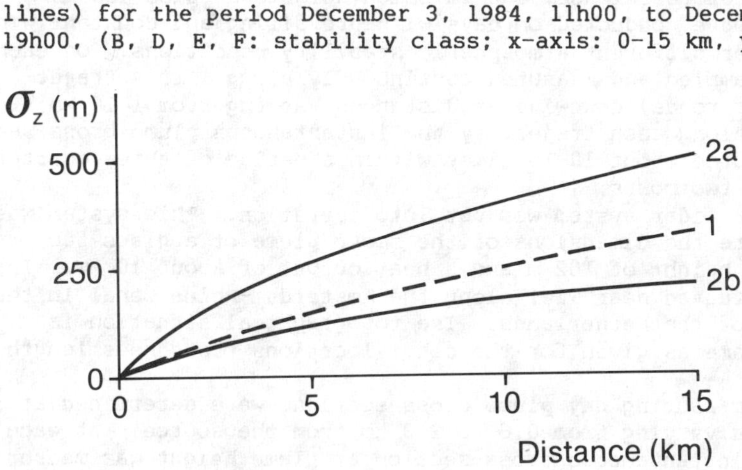

Figure 2. Average σ_z-curves calculated from wind fluctuation measurements for stability class D during day-time (2a) and night-time (2b) as compared with Pasquill's σ_z-curve (1).

curves; see Figure 2. In situations with similar wind speed and cloud cover the atmosphere will be more stable at night than during the day-time.

Using the methods described to calculate σ_y and σ_z, these values are employed in the Gaussian dispersion model. Other parameters normally used in calculating ground-level concentrations are plume rise and penetration of the plume into the inversion layer. When the planning of the experiments to test the model was first started, no facilities were available to measure plume rise and penetration. Therefore, it was decided to start the experiments at non-buoyant sources. A description of the experimental design is given in chapter 3.

3. EXPERIMENTAL DESIGN, MATERIALS AND METHODS

3.1. Experimental design

In the period 1985-1987 two sets of experiments were carried out; dispersion experiments with SF_6 and plume measurements with a lidar system.

The dispersion measurements were carried out around two non-buoyant sources. The first one was the 213 m high meteorological mast of KNMI situated at Cabauw in the central part of the Netherlands (Monna and Vliet, van der, 1987). The second source was the nuclear power plant at Dodewaard situated along the river Waal, about 20 km from Arnhem. Near both sources the surroundings are topographically flat within a radius of at least 20 km and consist of meadows, with occasional lines of trees, river dikes and small villages. At Cabauw pure SF_6 was released directly into the atmosphere at a height of 100 m. At Dodewaard SF_6 was added to the ventilation air before release; the emission height was also 100 m. Measurements were conducted on days with prevailing and constant wind direction under different atmospheric stability conditions. On each day SF_6 was sampled and measured continuously along 1 to 4 trajectories (public roads) downwind at distances varying from 1-10 km from the source. Along each trajectory the instantaneous plume cross-section was measured about 10-15 times within a period of three quarters of an hour to two hours.

In 1986 a lidar system was put into operation. This system was used to measure the dimensions of the smoke plume of a glass factory with a source height of 102 m and a heat output of about 10 MW. The factory is situated near Tiel along the Amsterdam-Rhine Canal in the central part of the Netherlands. The topographical situation is roughly the same as given for the other locations (roughness length 0.25 m).

On each measuring day plume cross-sections were determined at 1 to 4 distances varying from 0.6 to 2.3 km from the source. At each distance the instantaneous cross-section at plume height was measured about 15 to 20 times within a period of one hour.

Each day, both during the SF_6 experiments and during the lidar measurements, meteorological measurements were carried out:

- near-surface measurements (10 m) of wind speed, wind direction, relative humidity, temperature and global radiation
- periodic radiosonde observations of temperature and relative humidity, wind direction and wind speed.

During the measurements near Tiel also a doppler sodar system of KNMI was used to obtain profiles of wind speed and wind direction. At Cabauw wind speed and wind direction were measured at several heights along the mast.

3.2. Materials

- SF_6 monitoring system: this system is based on the continuous measurement of the concentration of SF_6 in a gas flow with an electron capture detector. The oxygen in the air sampled is removed by means of catalytic combustion with hydrogen over palladium and by drying the gas after condensation with a perma-pure dryer. The detection limit of the monitor is about 4 ppt(v/v) at a 90% response time of 6 s. The monitor is placed in a measuring van of which speed and direction are measured. All data are stored on magnetic tape.
- Lidar system: this system is constructed around a pulsed Nd-YAG laser (Quanta Ray) with a wavelength of 1064 nm. The maximum pulse energy is 300 mJ; the pulse duration is 8-9 ns and the pulse frequency is 10 Hz. The intensity of the reflected light is detected with an avalanche photo-diode with a sensitivity of 34 A/W and is recorded by means of a transient recorder. All data are stored on hard disk and on magnetic tape. Both the scanning system and the detection system are computer-controlled.
- Meteorological instruments: the near-surface measurements are carried out with a Kaymond weather system. Global radiation is measured with a Kipp solarimeter. The data are sampled with a computer and stored on magnetic tape. The radiosonde system is from AIR (Atmospheric Intrumentation Research, Inc.) and consists of an ADAS model AIR-3C atmospheric data acquisition system for receiving and processing telemetered data of signals emitted by radiosondes (model AS-3A series). Wind speed and wind direction are obtained by tracking the balloon with an optical theodolite (Warren-Knight, Inc.) with shaft encoders.

3.3. Methods

From the data measured a number of parameters are derived. From the instantaneous SF_6 plume cross-sections measured along one trajectory the 1-hour mean concentration value and the hourly-mean value of the lateral dispersion parameter σ_y are calculated. The 1-hour mean concentration is obtained by calculating the average cross-section. After correction for the background concentration the highest average concentration value is taken as the 1-hour mean value.

The σ_y-value is obtained by calculating the centre of gravity and the standard deviation or second moment of each individual cross-section. From the centres of gravity the standard deviation is calculated, which is a measure for meandering of the plume. Combining this value with the mean value of the standard deviations of the cross-sections, by taking the square root of the sum of the squares, gives the hourly σ_y-value, which has to be corrected for the angle between the trajectory and the plume axis.

The plume cross-sections measured by the lidar system at one distance from the stack are averaged. After subtracting background reflection, from the 1-hour mean cross-section thus obtained the standard deviations of the lateral and vertical concentration distributions are calculated. σ_y is obtained after correction for the angle between the mean cross-section and the plume axis. σ_z is calculated by fitting a Gaussian distribution function, taking into account the reflection of the plume against the top of the mixing layer and the ground surface.

Values of wind direction and wind speed measured near the surface are processed to calculate the values for σ_y and σ_z as described in chapter 2.

From the profiles of temperature and relative humidity as obtained from the radiosonde data the value of the mixing height is deduced. The mixing height is also derived from the lidar measurements, because above the mixing height there are usually less scattering aerosols.

The radiosonde data are also used to calculate the wind speed at plume height by averaging over an interval of 50-200 m height.

Values for the meteorological parameters measured are used to calculate or to derive input parameters needed for the various dispersion models; see chapter 4.

4 RESULTS

The model developed on the basis of wind fluctuations is tested by doing dispersion experiments as described in the previous chapter. On 20 days, mainly in summer (April-October), SF_6 measurements were carried out. A total of 52 values both σ_y and 1-hour mean concentrations are available. The lidar measurements were done mainly in winter (October-April). On 11 days 33 hourly-mean cross-sections were determined from which σ_y and σ_z values are derived. All measured data will be collected in a special report (Duuren, van, et al., 1988).

On the basis of the meteorological measurements and other data such as stack height and emission rate the values for σ_y, σ_z and hourly-mean concentration are calculated and compared with the measured values. This was not only done for the wind fluctuation model, but also for three other dispersion models.

Before giving the results of the comparison between observed and estimated values, a summary is presented of the various models tested and of the several parameters used for comparison.

4.1. Model descriptions

4.1.1. Dutch National Model. This model was developed to calculate long-term average concentrations (TNO, 1984). However, it is also frequently used in the Netherlands to calculate hourly-mean concentrations.

The main parameters are:

- stability class (A-F), derived from cloud cover and wind speed
- wind speed at 10 m height, at stack height and at plume height
- height of the mixing layer
- roughness length.

σ_z is calculated using Pasquill's curves (Pasquill, 1974); for calculating σ_y the formulae of Singer and Smith (1966) are applied (TNO, 1981). Plume rise is calculated according to Briggs (1971) and inversion penetration according to Manins (1979, 1984).

4.1.2. PPSP-model. This model was developed by Weil and Brower (1984), especially for unstable conditions. The main input parameters are:

- heat flux in the day-time to derive stability class. The heat flux is estimated using global radiation, wind speed and cloud cover (Holtslag and Ulden, van, 1983; Ulden, van, and Holtslag, 1985). At night stability is derived from cloud cover and wind speed
- wind speed at 10 m height, at stack height and at plume height
- mixing height
- roughness length.

σ_y, σ_z, plume rise and inversion penetration are calculated according to Briggs (1975, 1984).

4.1.3. RIVM puff model. This model, developed by the National Institute of Public Health and Environmental Hygiene, is mainly used to describe long-range transport of pollutants emitted in large industrial areas. But it can also be used to calculate dispersion from individual point sources (Egmond, van, and Kesseboom, 1983a, b).

The parameters are:

- Obukhov-length (L) for determining stability according to Pasquill (Golder, 1972). L is estimated using global radiation, wind speed and cloud cover
- wind speed at 10 m height, at stack height and at plume height
- mixing height
- roughness length.

σ_y is calculated from the standard deviation of the horizontal wind direction fluctuations using a constant Lagrangian timescale of 5 minutes. A contribution due to wind-shear can be taken into account.

σ_z is based on Pasquill's curves (Pasquill, 1974). The plume rise is calculated according to Briggs (1971) and penetration of the inversion layer according to Manins (1979, 1984).

4.1.4. Wind fluctuation model. The model description has already been presented in chapter 2, so that only the input parameters are

given here:
- σ_v, T_e and σ_{vs} to calculate σ_y and σ_z
- wind speed at 10 m height, at stack height and at plume height
- height of the mixing layer
- roughness length.

Plume rise can be calculated according to Briggs (1971) and inversion penetration according to Manins (1979, 1984).

4.2. Test parameters

To compare the observed and estimated values a number of parameters are used:
- the mean ($\bar{x}_o$) of the observed values and the mean ($\bar{x}_e$) of the estimated values
- the linear correlation coefficient (r)
- the number N2 of observed values between x_o = 0.5 x_e and x_o = 2.0 x_e
- the mean (MQ) of the ratios between estimated (x_e) and observed (x_o) values and the standard deviation (s_{MQ}) of these ratios:

$$MQ = \frac{1}{n} \Sigma \left(\frac{x_e}{x_o}\right) \tag{6}$$

n is the number of observations
- the mean (MM) of the differences between estimated and observed values and the standard deviation (s_{MM}) of these differences:

$$MM = \frac{1}{n} \Sigma (x_e - x_o) \tag{7}$$

- the mean fractional error (FE) and the standard deviation (s_{FE}). FE is calculated by means of:

$$FE = \frac{2}{n} \Sigma \left(\frac{x_e - x_o}{x_e + x_o}\right) \tag{8}$$

- the index of agreement (Willmot, 1982):

$$IA = 1 - \frac{\Sigma(\Delta x_e - \Delta x_o)^2}{\Sigma(|\Delta x_e| + |\Delta x_o|)^2} \tag{9}$$

with $\Delta x_e = x_e - \frac{1}{n} \Sigma x_e$ (10)

and $\Delta x_o = x_o - \frac{1}{n} \Sigma x_o$ (11)

When the agreement between observed and estimated values is perfect, the value for IA equals one. If there is no agreement at all, IA will be equal to zero.

TABLE I SF_6 experiments. Comparison of observed and estimated σ_y-values (for explanation see text)

parameter	WFM	DNM	PPSP	RPM
n	47	47	39	47
$\bar{x}_o$ (m)	335	335	359	335
$\bar{x}_e$ (m)	338	367	345	293
r	0.89	0.75	0.72	0.80
N2	46	43	38	44
MQ	0.98	1.20	1.01	0.90
s_{MQ}	0.32	0.42	0.32	0.30
MM (m)	2	31	-14	-43
s_{MM} (m)	121	141	152	122
FE	-0.07	0.11	-0.04	-0.15
s_{FE}	0.32	0.40	0.34	0.32
IA	0.93	0.86	0.84	0.87

TABLE II SF_6 experiments. Comparison of observed and estimated 1-hour mean concentrations (for explanation see text)

parameter	WFM	DNM	PPSP	RPM
n	52	52	44	52
$\bar{x}_o$ (ppt)	186	186	174	186
$\bar{x}_e$ (ppt)	149	114	118	145
r	0.73	0.19	0.68	0.53
N2	36	27	26	41
MQ	0.94	1.76	1.07	1.35
s_{MQ}	0.80	4.41	1.71	1.63
MM (ppt)	-37	-72	-56	-41
s_{MM} (ppt)	112	166	123	138
FE	-0.27	-0.30	-0.39	-0.04
s_{FE}	0.56	0.79	0.70	0.64
IA	0.82	0.32	0.74	0.64

TABLE III Lidar measurements. Comparison of observed and estimated σ_y-values (for explanation see text)

parameter	WFM	DNM	PPSP	RPM
n	32	32	29	32
$\bar{x}_o$ (m)	116	116	121	116
$\bar{x}_e$ (m)	125	207	181	90
r	0.67	0.17	0.38	0.45
N2	29	20	24	27
MQ	1.05	1.91	1.54	0.78
s_{MQ}	0.46	0.90	0.58	0.33
MM (m)	8	91	78	46
s_{MM} (m)	55	91	75	43
FE	-0.04	0.52	0.34	-0.31
s_{FE}	0.44	0.40	0.39	0.39
IA	0.74	-0.22	0.23	0.52

TABLE IV Lidar measurements. Comparison of observed and estimated σ_z-values (for explanation see text)

parameter	WFM	DNM	PPSP	RPM
n	33	33	30	33
$\bar{x}_o$ (m)	70	70	73	70
$\bar{x}_e$ (m)	76	85	146	75
r	0.58	0.18	0.13	0.11
N2	28	19	18	20
MQ	1.15	1.60	2.44	1.33
s_{MQ}	0.56	1.30	1.80	0.91
MM (m)	6	17	73	6
s_{MM} (m)	36	57	89	43
FE	0.03	0.18	0.55	0.06
s_{FE}	0.47	0.66	0.61	0.61
IA	0.73	0.24	-0.11	0.43

4.3. Comparison of observed and estimated values

The results of the SF_6 experiments and the lidar measurements are given separately. The different models are abbreviated as follows: WFM for the wind fluctuation model, DNM for the Dutch National Model, PPSP for the model developed by Weil and Brower and RPM for the RIVM puff model. The number of observations (n) as given in the various tables is always lower for the PPSP-model, because of the fact that for some cases the determination of stability class was not possible. In Table I the comparison is given between observed and estimated σ_y-values as derived from the SF_6 experiments.

All models estimate the σ_y-value with a fair degree of accuracy. In view of the values for the several standard deviations, the differences between the four models are not statistically significant.

In Table II the results of the comparison of the 1-hour mean concentration values are given. All models underestimate the concentration (except RPM) as follows from the negative values for MM and FE. The Dutch National Model gives the worst results. The index of agreement is best for the wind fluctuation model.

The comparison between observed and estimated values as derived from the lidar measurements is given in Tables III and IV. In Table III the parameter values for σ_y are mentioned. From these values it can be concluded that the differences between the models are large. In general the wind fluctuation model gives better results than the other ones.

The results of the comparison concerning σ_z are given in Table IV. Here there are also large differences between the four models, with the best results on the whole for the wind fluctuation model and the worst for the PPSP model.

5. DISCUSSION

One of the starting points in developing a new dispersion model was to avoid the use of stability classes. Determination of stability is difficult, as can be concluded from the data presented in Table V. In this table the frequencies of stability classes are given for the SF_6 and lidar datasets as determined from the stability parameters used in the three dispersion models (see chapter 4). As mentioned before, the SF_6 experiments were carried out in summer and the lidar observations in winter. The differences between the frequency distributions of the stability classes are large, which is reflected in the differences between observed and estimated values as given in Tables I to IV.

From the data presented in these tables it is concluded that plume dispersion can be described better using the model based on wind fluctuations than with the three other models. This holds especially for dispersion during the winter period. In that period small differences in the energy balance near the surface will result in large differences in stability and turbulence.

TABLE V Frequencies (%) of stability classes as deduced from stability parameters of three dispersion models

stability class	SF_6 experiments			lidar measurements		
	DNM	PPSP	RPM	DNM	PPSP	RPM
A	0	9	0	0	9	0
B	17	29	6	30	37	24
C	17	31	50	40	39	58
D	54	8	34	30	6	12
E	4	4	6	0	0	0
F	8	6	4	0	0	6
R	0	13	0	0	9	0

R : stability class not determined
SF_6: n = 52; lidar: n = 33

Also at night the determination of stability is difficult; stability is then derived only from cloud cover and wind speed. As discussed by Draxler (1987) large differences between measured and modelled values occur especially at night. In general, the amount of cloud cover is overestimated resulting in a shift in stability towards more neutral conditions. Hence, turbulence is underestimated during the day-time and overestimated at night.

It will be clear that determining turbulence directly from measurements of wind direction and wind speed, followed by direct conversion into σ_y and σ_z is a far better method to describe dispersion than estimating the stability class in combination with average curves of σ_y and σ_z measured elsewhere.

The 1-hour-average plumes, as calculated from the measurements of the instantaneous plume cross-sections, can be described well by the Gaussian distribution function. This holds for the lateral as well as for the vertical distribution.

An important point of interest is the uncertainty of the results of the model developed. A measure of this uncertainty is the standard deviation, s_{FE}, of the fractional error FE, and the number N2 of observed values differing not more than a factor two from the estimated values (see chapter 4). The value of s_{FE} is 0.32 to 0.56; 31% of the observed concentration values and 2-15% of the observed σ_y- and σ_z-values differ more than a factor two from the estimated ones (Tables I to IV). The deviation between observed and measured values is still rather great. Is it in fact possible to reduce this deviation by making the model more complex, taking into account more physical processes which affect the result? According to Carson (1986), Venkatram (1984) and Benarie (1987), there will always be a deviation between modelled and measured values due to the stochastic nature of turbulence. Besides, uncertainty is increased by inaccuracies in the input parameters. The more input parameters are needed, the more the model becomes a kind of random generator.

The most important advantage of the wind fluctuation model is that a good estimation of plume dimensions is obtained using simple and reliable measuring techniques. Moreover, it is possible to estimate the variance around hourly-mean values, so that frequency distributions of concentrations can be calculated. However, to do this one needs accurate data of mixing height and profiles of wind direction and wind speed with height. For this purpose the use of lidar techniques will be investigated in the near future.

Model evaluation will be extended in 1987 and 1988 doing dispersion experiments at night. Up till now the model has only been tested near non-buoyant sources; in 1988 a number of measuring campaigns is scheduled to test the model at large power stations with tall stacks. Also, in 1988 research will start into the penetration of plumes into the inversion layer.

6. ACKNOWLEDGEMENTS

The authors wish to thank the Boards of Directors of N.V. GKN (Joint Nuclear Power Station of the Netherlands), KNMI and "Machinale Glasfabriek De Maas B.V." for their kind permission to do the experiments, and for the technical assistance rendered by these organizations. They are much indebted to Messrs. L. van den Beld, P.H.J. Gamelkoorn, and R.A. de Vries who carried out the SF_6 experiments and to Messrs. J.E. Jaspers and R.D.A. Scholten for doing the lidar observations and the meteorological measurements.

They also wish to express their gratitude to Mr. A.A.M. Holtslag of KNMI and Messrs. N.D. van Egmond and D. Onderdelinden of RIVM for their stimulating discussions.

7. REFERENCES

AGTERBERG, R., NIEUWSTADT, F.T.M., DUUREN, H. VAN, HASSELTON, A.J. and KRIJT, G.D., 1983. 'Dispersion experiments with sulphur hexafluoride for the 213 m high meteorological mast at Cabauw in the Netherlands'. De Bilt, KNMI; Arnhem, N.V. KEMA.

BENARIE, M.M., 1987. 'The limits of air pollution modelling'. In: Atmospheric Environment 21, pp. 1-5.

BRIGGS, G.A., 1971. 'Some recent analyses of plume rise observation'. In: Proceedings 2nd Intern. Clean Air Congress, 6-11 December, 1970, New York, Academic Press, pp. 1029-1032.

BRIGGS, G.A., 1975. 'Plume rise predictions'. In: Lectures on air pollution and environmental impact analysis, Boston MA, Am. Meteor. Soc., pp. 59-111.

BRIGGS, G.A., 1984. 'Plume rise and buoyancy effects'. In: Atmospheric Science and Power Production, Office of Scientific and Technical Information, US Dept. of Energy, NTIS DE 84005177 (DOE/TIC-27601), pp. 327-366.

CARSON, D.J., 1986. 'A report on the symposium on uncertainty in modelling atmospheric dispersion'. In: Atmospheric Environment 20, pp. 1047-1049.
DRAXLER, R.R., 1987. 'Accuracy of various diffusion and stability schemes over Washington D.C.'. In: Atmospheric Environment 21, pp. 491-499.
DUUREN, H. VAN, ERBRINK, J.J., SCHOLTEN, R.D.A., 1988. 'Dispersion experiments with sulphur hexafluoride and lidar observations. A data-base for testing dispersion models'. (In preparation) Arnhem, N.V. KEMA.
DUUREN, H. VAN, KRIJT, G.D., ELSHOUT, A.J., 1975. 'Zwavelhexafluoride als tracer voor de verspreiding van luchtverontreinigende componenten vanuit puntbronnen'. In: Elektrotechniek 53, pp. 135-143.
DUUREN, H. VAN, NIEUWSTADT, F.T.M., 1980. 'Dispersion experiments from the 213 m high meteorological mast at Cabauw in the Netherlands'. In: Atmospheric Pollution 1980, Proceedings of the 14th International Colloquium (Paris, May, 5-8, 1980), M.M. Benarie (ed.), Studies in Environmental Science, Volume 8, Amsterdam, Elsevier Scientific Publishing Company, pp. 77-90.
EGMOND, N.D. VAN, KESSEBOOM, H., 1983a. 'Mesoscale air pollution dispersion models - I. Eulerian grid model'. In: Atmospheric Environment 17, pp. 257-265.
EGMOND N.D. VAN, KESSEBOOM, H., 1983b. 'Mesoscale air pollution dispersion models - II. Lagrangian puff model and comparison with Eulerian grid model'. In: Atmospheric Environment 17, pp. 267-274.
ELSHOUT, A.J., DUUREN, H. VAN, 1967. 'SO_2-Konzentrationen in der Umgebung von Kraftwerken'. In: Mitteilungen der VGB 107, pp. 119-126.
ERBRINK, J.J., 1988. 'A dispersion model, based on wind fluctuations, for calculating short-term ground-level concentrations'. (To be published) Arnhem, N.V. KEMA.
GOLDER, D., 1972. 'Relations among stability parameters in the surface layer'. In: Boundary-Layer Meteor. 3, pp. 47-58.
HOLTSLAG, A.A.M., ULDEN, A.P. VAN, 1983. 'A simple scheme for daytime estimates of the surface fluxes from routine weather data'. In: J. Climate Appl. Meteor. 22, pp. 517-529.
IRWIN, J.S., 1983. 'Estimating plume dispersion. A comparison of several schemes'. In: J. Climate Appl. Meteor. 22, pp. 92-114.
MANINS. P.C., 1979. 'Partial penetration of an elevated inversion layer by chimney plumes'. In: Atmospheric Environment 13, pp. 733-741.
MANINS, P.C., 1984. 'Chimney plume penetration of the sea-breeze inversion'. In: Atmospheric Environment 15, pp. 2339-2344.
MONNA, W.A.A., VLIET, J.G. VAN DER, 1987. 'Facilities for research and weather observations on the 213 m tower at Cabauw and at remote locations'. De Bilt, KNMI, Scientific reports WR - No. 87-5
MOORE, G.E., MEI-HAO LIU, LU-HUAI SHI, 1985. 'Estimation of interface time scales from a 100 m meteorological tower at a plains site'. In: Boundary-Layer Meteor. 31, pp. 349-368.
PASQUILL, F., 1974. Atmospheric diffusion, Chichester, John Wiley and Sons.

SINGER, I.A., SMITH, M.E., 1966. 'Atmospheric dispersion at Brookhaven National Laboratory'. In: Int. J. Air Water Poll. 10, pp. 125-135.

TAYLOR, G.I., 1921. 'Diffusion by continuous movements'. In: Proc. London Meth. Soc. 20, pp. 196-202.

TNO, 1976. Modellen voor de berekening van de verspreiding van luchtverontreiniging inclusief aanbevelingen voor de parameters in het lange-termijnmodel. Werkgroep Verspreiding Luchtverontreiniging, 's-Gravenhage, Staatsuitgeverij.

TNO, 1981. Frequentieverdelingen van luchtverontreinigingsconcentraties. Een aanbeveling voor een rekenmethode. Werkgroep Verspreiding Luchtverontreiniging, 's-Gravenhage, Staatsuitgeverij.

TNO, 1984. Parameters in het lange-termijnmodel verspreiding luchtverontreiniging. Commissie Onderzoek Luchtverontreiniging TNO, Delft.

ULDEN, A.P. VAN, HOLTSLAG, A.A.M., 1985. 'Estimation of atmospheric boundary layer parameters for diffusion applications'. In: J. Climate Appl. Meteor. 24, pp. 1196-1207.

VENKATRAM, A., 1984. 'The uncertainty in estimating dispersion in the convective boundary layer'. In: Atmospheric Environment 18, pp. 307-310.

WEIL, J.C. AND BROWER, R.P., 1984. 'An updated gaussian plume model for tall stacks'. In: J. Air Poll. Contr. Ass. 34, pp. 815-827.

WILCZAK, J.M., PHILLIPS, M.S., 1986. 'An indirect estimation of convective boundary layer structure for use in air pollution dispersion models'. In: J. Climate Appl. Meteor. 25, pp. 1609-1624.

WILLMOT, C.J., 1982. 'Some comments on the evaluation of model performance'. In: Bull. Am. Meteor. Soc. 63, pp. 1309-1313.

PERFORMANCE EVALUATION OF AIR POLLUTION IMPACT ASSESSMENT MODELS : FIFTEEN YEARS OF PRACTICAL EXPERIENCE

J.G. Kretzschmar and G. Cosemans
Nuclear Energy Research Centre
Boeretang 200
B-2400 Mol
Belgium

ABSTRACT. Despite enormous efforts over the past years in the field of air pollution transport, dispersion and deposition modelling, no consensus on a unique set of procedures for model performance evaluations has been reached. After some comments on the consequences of this fact, a systematic review is given of how the Belgian Nuclear Energy Research Centre got involved in the modelling field and what the major steps forwards were as a function of time and available resources. The final conclusion is that specifically computer hardware and environmental monitoring improved significantly over the past fifteen years, with as major consequence an enhanced strategic position for the human expert.

1. INTRODUCTION

A physical or mathematical model basically aims at the simulation of some physical, chemical or physico-chemical phenomena in such a way that it gives a reliable, handy and flexible tool to freely manipulate and combine all the variables that are important (input parameters) for the processes under investigation. Whatever model, simple or complex, naive or sophisticated, is nevertheless useless for practical applications if no proofs can be given of the fundamental correctness of its formulation and of its ability to replicate known conditions within acceptable limits of accuracy and precision. These statements look rather trivial, but they aren't when it comes to translating them in practice. Even if we only focus on the second requirement, being the specific topic of this paper, there seems to be more confusion than agreement despite the fact that, for air pollution modelling, the problem dates back to the late sixties at least.

Over the past years many different attempts with respect to the development and evaluation of atmospheric transport and dispersion models have been noted. The pretty long, but nevertheless very incomplete list of references, given at the end of this paper, illustrates only a minor part of the efforts with respect to model evalu-

K. Grefen and J. Löbel (eds.), Environmental Meteorology, 515–530.

ations. A fast screening of the different titles in that list immediately reveals a striking confusion in the terminology. Validation, verification, evaluation, calibration, performance assessment, testing, accuracy assessment and so on, are all somewhat different terms used, or still in use to refer to basically the same bottom line question : How well does the model predict actually measured ground-level concentration and deposition values ? Some notable experts even tried to arrive at definitions of terms used in evaluation of models that are consistent with normal dictionary usage (Turner, 1979; Munn, 1987). Although not entirely successful, it must be pointed out that there is a tendency to use (model) performance evaluation as standard term nowadays, to denote the ensemble of procedures used to systematically compare observed and calculated values as a function of time and space.

While the divergence in terminology is rather a "fait divers", the absence of a consensus on a unique set of procedures is certainly a much more important handicap. To overcome this problem, many proposals have been formulated and eventually illustrated by means of small or larger scale examples. The long list of references is certainly illustrative for this. Once again it has to be remembered that it was not the purpose of authors to make a complete or in-depth review of the hundreds of papers and reports that do actually exist. As an illustration for the incompleteness of the survey it can be pointed out that the proceedings of several national and international conferences, symposia and workshops do contain many more interesting papers. An example here is undoubtedly the collection of the proceedings of the, up to now sixteen, International Technical Meetings on Air Pollution Modelling and its Application, organised since 1970 by the NATO Committee on the Challenges to Modern Society.

2. PERFORMANCE EVALUATION IN GENERAL

As already mentioned in what precedes, no consensus has been reached over the past years on a unique set of procedures for model performance evaluation. As well the research community, as the industrial sectors and the regulatory bodies, do nevertheless agree upon the fact that no single quantitative index for performance assessment exists. Furthermore there seems to be a consensus on the fact that performance assessments will always be less or more heavily biased by the objectives of the modelling effort, or the regulatory framework wherein it is or will be applied. (Bencala et Seinfeld, 1979; Fox, 1981; Hilst, 1981a; Weber and Garrett, 1985; Munn et al., 1987). As a conclusion it could be stated that it will always be necessary to form a qualitative overall evaluation of the results of various quantitative procedures based on performance criteria on one hand and on statistical measures on the other one.

A less or more general flow chart summarising the different steps to be considered when starting a model performance evaluation is sketched in Figure 1. Under "Problem Definition" various things have to be done such as defining the pollutants under investigation, the region and its vicinity where the study will be carried out, the period of time taken into account and the averaging time(s) of the individual concentration and deposition values. The regulatory framework has of course its influence and this is summarised by the "Air Quality Standards" block and its links. Once the problem has been specified with some detail, the model has to be chosen together with the required emission and meteorological input data. Problems of representativeness, completeness, accuracy and resolution as a function of time and space do normally appear here. The same is true for the right side of the flow chart where the available observed levels (concentration and/or deposition values) come into the picture. Apart from the spatial and temporal coverage of the data, the importance of background levels, due to sources external to the region under investigation, must be stressed. Before comparing calculated and observed values the latter must be corrected for (eventually time and space dependent) background values not calculated by the model by definition. Once sets of comparable calculated and (corrected) observed values for (a) given averaging time(s) have been obtained the intercomparison can go two ways, namely as a function of time at a given location or as a function of space at a given moment. In both branches different sets or subsets of data can be made as a function of e.g. specific meteorological conditions, combination of sites, specific time intervals, etc... The final analysis of whatever (sub)set is then or an "analysis of residuals" or what is labelled "other statistics" for the sake of simplicity. As the former is an almost standard technique it doesn't require any further explanation. This is possibly not the case for "other statistics" as this box obviously contains quite a variety of techniques. Among those we mention : regression and correlation techniques, scatter diagrams, spectral analysis, histograms, cumulative frequency distributions, different types of spatial analysis, iso-lines, pollution roses, analyses of extremes, resampling techniques and so on.

The final impression from what precedes is that indeed a rather large and diversified toolbox has been built-up by modellers and model users. The specific choices one will make, when confronted with a model performance evaluation, will undoubtedly be significantly influenced by the objectives of the model and its application. If it concerns an impact assessment study the regulatory framework will already specify what parameters have to be checked as a minimum. If an air quality standard sets limits for e.g. the yearly arithmetic mean, the 98th-percentile of the daily averages and the maximum hourly average values in that year, it is obvious that at least these statistical parameters will have to be studied with a sufficient high degree of spatial resolution in the region under investigation. It has nevertheless not to be forgotten that acceptable agreement of statistics does not necessarily imply acceptable agreement of the raw data, namely the simultaneously observed and calculated levels. Therefore it has to be repeated that the regulatory framework sets the minimum requirements but not the model performance evaluation in globo.

3. IMPACT ASSESSMENT MODELLING AT THE SCK/CEN

The Belgian Nuclear Energy Research Centre (SCK/CEN) has been involved in atmospheric transport and dispersion modelling since the mid sixties. This was largely due to the fact that the institute, with its three on-site research reactors and several laboratories where radioactive materials were processed, had to evaluate its possible impact upon the local environment by means of the most up-to-date techniques and methodologies. This knowledge, primarily developed for internal purposes, was later on used for impact assessment studies for other nuclear installations in Belgium and abroad. By the early seventies the modelling activities were extended into the non-nuclear field, as many questions with respect to the actual and future impact of especially SO_2 sources were waiting for an appropriate answer. For the modellers there were obviously some fundamental changes inherent to this broadening of the application field of their models. The most important change was possibly the switching from an almost single source configuration - with well-known characteristics, mostly neglectable or no emissions at all under normal conditions, but eventually extremely large ones under accidental conditions - to multiple source configurations, with continuous and continuously changing emissions of significant magnitude. The only way to tackle those new problems in a feasible way seemed to be by means of some kind of statistical approach.

First of all the bi-Gaussian approach was chosen since the very beginning for SCK/CEN's impact assessment modelling efforts. Secondly it was felt that a specific source's impact within a multiple source configuration, had to be evaluated by means of its (absolute and relative) influence upon the most relevant part of the cumulative frequency distributions of the (short) time averaged ground-level concentrations or immission values. The name of the model IFDM or Immission Frequency Distribution Model clearly refers to this basic philosophy. It must be pointed out that this philosophy proved it's soundness as it was never changed during the almost fifteen years that the code is in operational use in Belgium for many different impact assessment studies, as well for the authorities as for the private sector.

4. PERFORMANCE EVALUATIONS OF IFDM

Around the mid seventies the first larger scale attempt to evaluate IFDM's performances was carried out on a 15 km x 20 km test region in Antwerpen, Belgium's main harbour and one of its most important industrialised urban areas (Kretzschmar et al., 1976). Available emission data was as follows : calculated yearly average emissions for ≈150 industrial stacks totalising 126.000 t SO_2/yr, and estimated space heating emissions of 14.000 t SO_2/yr for a population of

≈ 800.000. Based on the degree day approach space heating emissions were divided over the year on a daily basis. Hourly averages for wind speed, wind direction by sectors of 20 degrees and temperature were available at the 51 m and 154 m level on a local meteorological tower along the Schelde. Observed daily averages of the SO_2 concentration levels came from 6 local monitoring sites (3 city centre, 1 suburban, 2 industrial) of the national Sulphur-Smoke network. Different combinations of these sites were used for checks on three successive years, namely 1971, 1972 and 1973. Calculated values were hourly values in the grid points of a 1 km by 1 km mesh, combined subsequently into daily averages. Only days with available observed SO_2-values were taken into account. Due to computer resource restrictions IFDM basically calculated standard pollution maps, respectively for point and area sources, for each of the combinations of 7 stability classes, 4 wind speed classes and 18 wind direction sectors. The uniform background SO_2-level was set at 20 $\mu g\ SO_2/m^3$, the mixing height was constant at 550 m, wind speed at the effective heights of the point sources was derived from the exponential wind profile law with m = 0.28 and no deposition phenomena were included. Performance evaluation was based on the following three outputs for simultaneously observed and calculated daily averages for each site-year combination : time series plots, scatter plots, inclusive regression analysis and cumulative frequency distribution plots on log-probit paper, supplemented by intercomparisons for yearly averages, medians, 95-percentiles and maxima. The main conclusions were : IFDM seems to be an acceptable tool for impact assessment studies but performance evaluation has to be continued taking into account that better input data and observed data will gradually become available as a function of time. The model as such has to get away from the "standard pollution map" approach based on too coarsely subdivided meteo situations.

In the next phase some new output facilities were added to IFDM. First of all calculation of the absolute and relative differences between selectable statistical parameters of sets of respectively observed and calculated concentrations was included. Secondly some subroutines were added to generate three dimensional plots of the spatial patterns of emissions, concentration fields, residuals and the previously mentioned relative or absolute differences for representative statistical parameters (Kretzschmar et al., 1977 and 1978).

By the end of the seventies, our air pollution knowledge base had increased significantly, automatic monitoring networks had become fully operational and emission inventories were refined. This resulted in more reliable and more detailed input and control data, which together with the availability of increased computer power and facilities, triggered some new performance evaluation exercises. Within the context of a National R & D Programme Environment Air, the industrialised urban area of Gent-Terneuzen was chosen as a multiple source test case for SO_2 (Cosemans et al., 1981), while a Sb-aerosol emitting factory in Beerse almost ideally fitted the requirements of a single source configuration (Vanderborght et al., 1983).

Apart from a faster IFDM code - with enhanced time resolution and output facilities - and more detailed emission, meteorological and immission data as a function of time and space, the Gent-Terneuzen study was characterised by two innovations in the performance assessment package. With the availability of measured half hourly averages for the local meteorological parameters and immission data it became feasible to use the concept of air pollution roses to statistically test the model as a function of wind direction in each of the twelve SO_2-monitoring sites of the local automatic network. As already pointed out by H. Fortak in his introductory remarks to a 1973 NATO/CCMS Panel Meeting on Modelling (Fortak, 1973), this was a rather critical but very instructive test. Figures 2a and 2b highlight this by means of one of the many possibilities namely for 1978 the 95th-percentiles of the half-hourly SO_2-averages as a function of wind direction, subdivided into 10 degree sectors. When comparing the observed and calculated results, two phenomena are striking. First of all in the southern part the presence of an important CS_2-source, interfering upon the SO_2-measurements, even at greater distances. Secondly for easterly wind direction, and more pronounced towards the north of the test region, the very important influence of SO_2 advected from outside the test region (background). This clearly illustrates the importance of the "background-levels" box in the general flow chart of Figure 1, especially for countries like Belgium with its major agglomerations being so closely spaced and its industrial areas scattered all over the place. The obvious question "how to cure this" has no trivial answer. For yearly averages wind direction dependent background levels can be derived from the detailed monitoring results, and accounted for, certainly for spatial lay-outs like the one in the Gent-Terneuzen area. For percentiles the only way out is to correct, before any further treatment, the individual measurements at a given moment (with e.g. t_{av} = 30 min) for the background contribution under the given meteorological condition. The latter has obviously to be derived too from the ensemble of available measurements, preferentially from several sites, as well inside as outside the test region. This was one of the major lessons learnt from the 1978 Gent-Terneuzen exercise. The second innovation originated from the increase in temporal resolution that was realised by using half-hourly averages as input and output data. This obviously generated an amazing amount of data on a yearly basis, possibly confronting the modeller with a not always required data avalanche and subsequent information glut. Was it not possible to carry out major parts of the performance assessment package by means of a more limited number of randomly chosen data out of the total population ? As the data set would then be limited, the (human) efforts could be more focused on increasing the accuracy of time and space varying emissions, meteorological conditions and background levels. Several studies (Juda, 1968; Junker, 1976; Kretzschmar and Cosemans, 1981a; Ott and Mage, 1981) have indeed pointed out that for lognormal populations of half hourly concentration values with geometric standard deviations below 4.5, as was the case in Gent, random samples of size 400 to 600 performed pretty well, even at the 98th-percentile level. A test based on a sample size of 450 showed that, as expected, the differences with the cumulative frequency distributions based on all data were quite acceptable as they even didn't exceed the 10 percent at the 98th-percentile level for the 12 sites taken into account. These findings were considered important, as they open the way to a faster way of checking the (preliminary) results of the many different approaches, and corresponding multitude of computer runs, that model evaluators tend to continuously generate during their assessments.

The Beerse test case (Vanderborght et al., 1983) was in so far different from the other model performance assessment studies that it concerned a specific aerosol normally only present at background levels in ambient air. The air pollution roses approach was consequently not hampered anymore by the previously mentioned background problems. As there are no ideal test cases, a spectrum of other problems was noted this time. First of all the Sb-emissions were a combination of less or more known point source emissions through three high stacks, and very variable and completely unknown process linked fugitive emissions generated at different spots in the factory. Activity related emission factors for the chimney releases were determined by an extensive, but discontinuous monitoring in the stacks, while SF_6 tracer releases were used to achieve the same for the fugitive emissions. Second problem in this study was the longer averaging times required for the ambient measurements. The applicability and possibilities of the Sb-pollution roses approach were certainly limited by the fact that ground-level concentration values had to be daily averages. As for deposition only monthly means were available, and as the degree of uncertainty in such a measurements is well-known, it was not possible to check the performances of IFDM's deposition part at the level the investigators would like to do. Despite those limiting boundary conditions the exercise was very successful, as well for the modellers as for the other R & D teams and certainly for the plant under investigation.

5. CONCLUSIONS

What precedes is undoubtedly a very incomplete recapitulation of what the SCK/CEN modellers team learned and developed over the past fifteen years. Apart from warnings and best wishes to the newcomers in the modelling community the following summary could be forwarded. For industrialised and/or urban areas covering some 20 km by 20 km in a region with not too complicated topography, bi-Gaussian transport and dispersion models are still the preferred approach within the context of impact assessment studies. No real fundamental changes did occur over the past years. Resolution and accuracy in time and space of the basic input data is still the problem especially for emission data, but also for meteorological data. Ambient measurements did significantly improve as a function of time as well with respect to quantity as quality, although whatever concerns deposition is far behind what has been achieved for ground level concentrations. Tracer techniques and specific monitoring networks for model performance assessment studies have proven their usefulness. With respect to hardware, software and computer peripherals the improvements have to be expressed in orders of magnitude. One of the main driving forces for models, namely the regulatory framework and legislation in general, have become much stricter and more demanding with respect to model performances and number of different scenarios and outputs. With (much) faster computers and many more runs, each time with a wide spectrum of outputs, the next bottleneck seems to be the human expert(s) trying to find his (their) way through the abundance of data and corresponding information glut. The time has possibly come to switch from data based to knowledge based systems in the modellers world too.

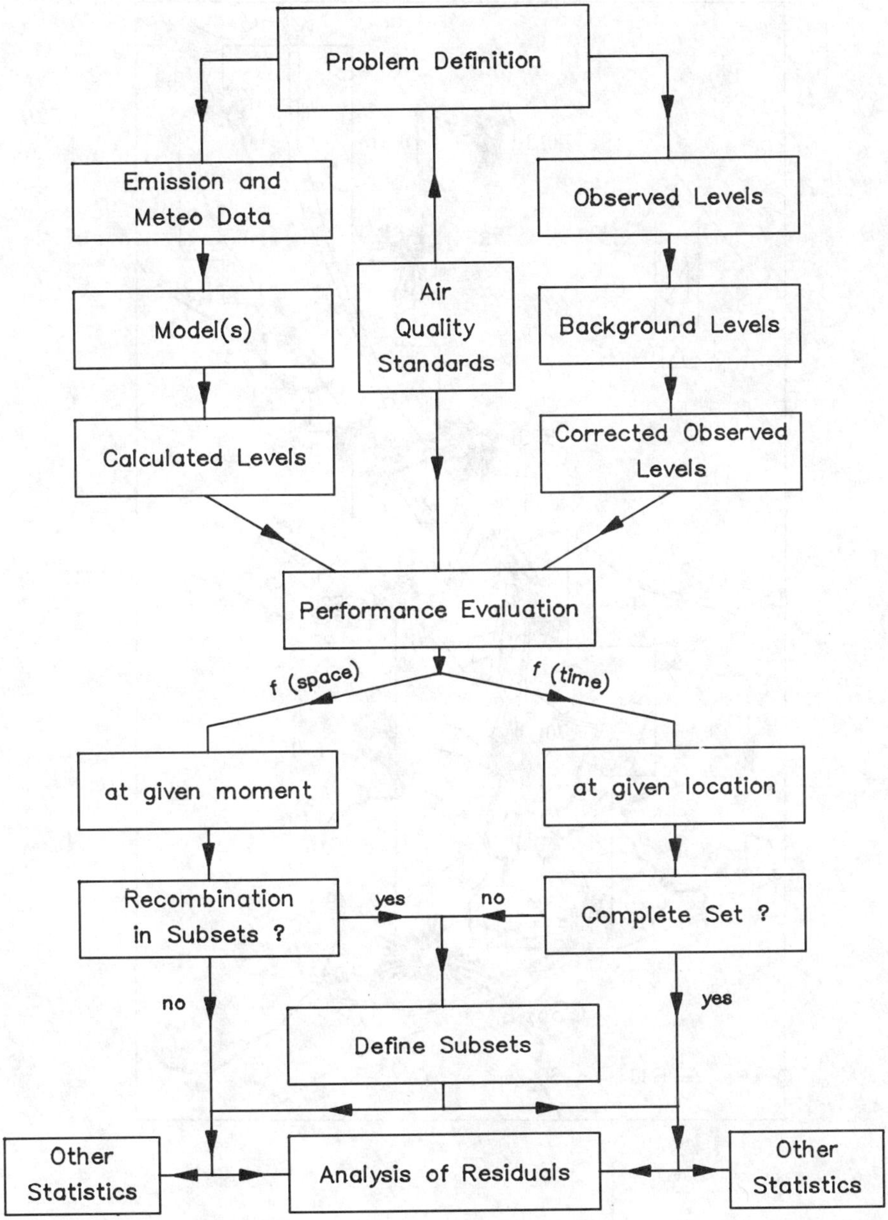

Figure 1 : General flow chart for model performance evaluations.

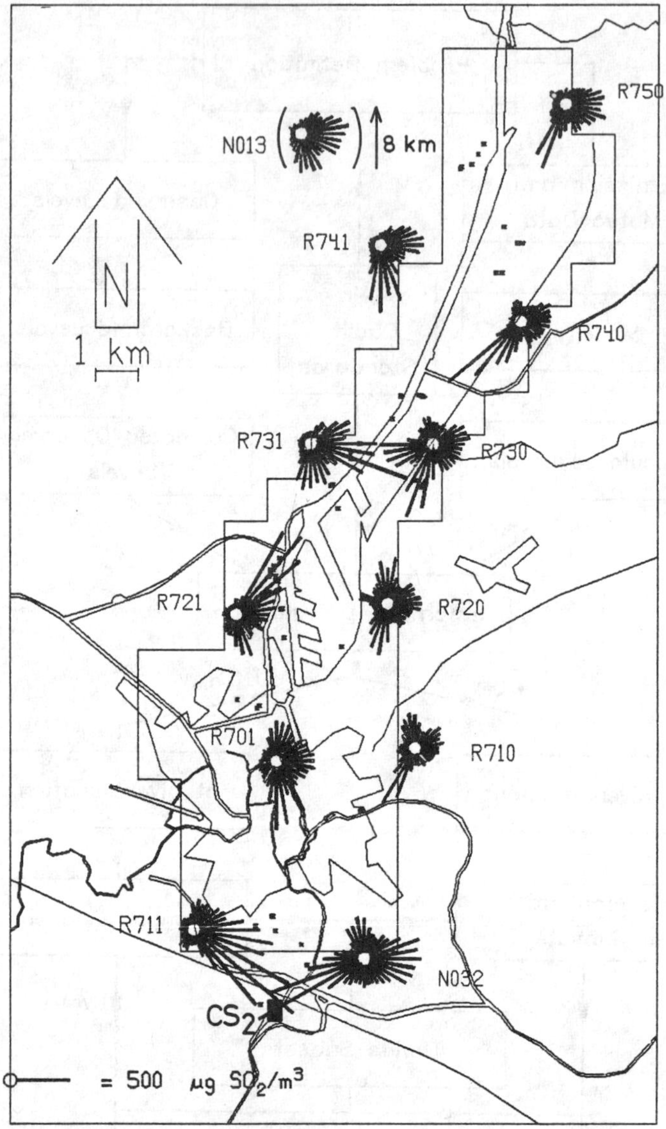

Figure 2a : SO_2–roses of the 95–percentiles of the measured half hourly averaged SO_2–concentrations during 1978 in the Gent–Terneuzen region.

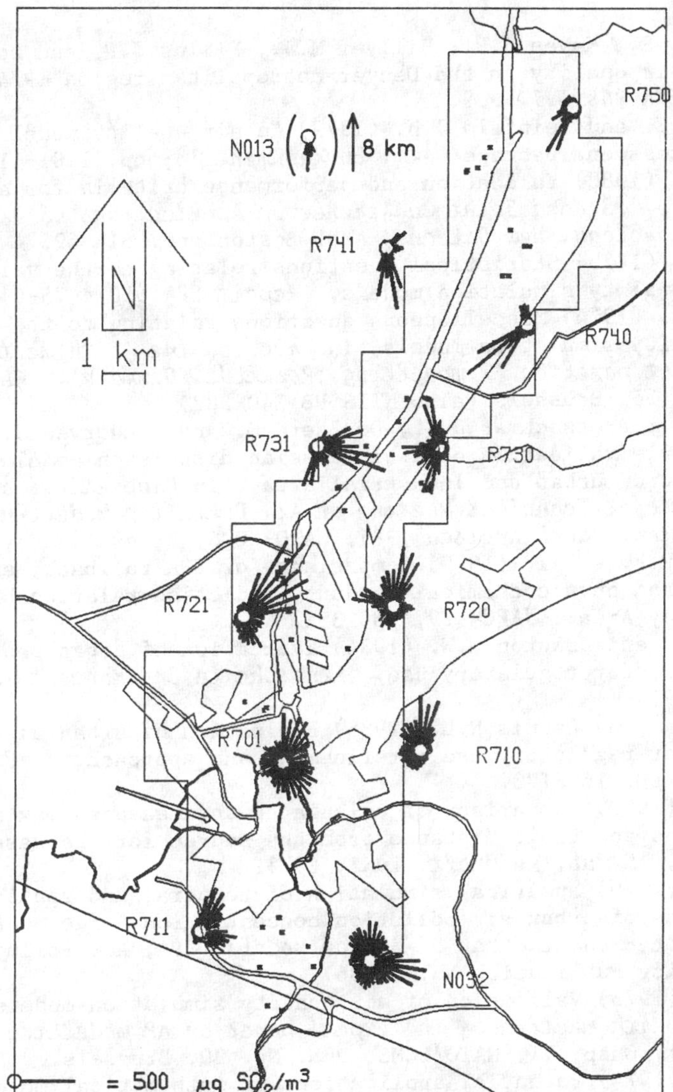

Figure 2b : SO_2–roses of the 95–percentiles of the calculated half hourly averaged SO_2–concentrations during 1978 in the Gent–Terneuzen region.

REFERENCES

Anderson G.E., Hayes S.R., Hillyer M.J., Killus J.P. and Mundkur P.V. (1977) Air quality in the Denver metropolitan region 1974 - 2000. *Report EPA 908/1-77-002.*

Bencala K.E. and Seinfeld J.H. (1979) An air quality model performance assessment package. *Atm. Environment 13*, pp. 1181-1185.

Bowne N.E. (1980) Validation and performance criteria for air quality models. *Second Joint Conference on Applications of Air Pollution Meteorology, New Orleans*, AMS, Boston, pp. 614-626.

Brier G.W. (1975) Statistical questions relating to the validation of air quality simulation models. *Report EPA 650/4-75-010*, 22 pp.

Calder K.L. (1974) Miscellaneous questions relating to the use of air quality simulation models, in *Proc. of the fifth meeting of the expert panel on AP modelling (Roskilde, Denmark)*. NATO/CCMS, Pub. No. 35, Brussels (also NTIS PB 240-582).

Cosemans G., Kretzschmar J.G., De Baere G. and Vandervee J. (1981) Large scale validation of a bi-Gaussian dispersion model in a multiple source urban and industrial area. In *Proceedings of the 12th International Technical Meeting on Air Pollution MOdelling and Its Application, San Francisco 1981*, NATO-CCMS, Brussel.

De Mandel R.E. et al. (1979) Comparisons of EPA rollback, empirical, kinetic and physicochemical oxidant prediction relationships in the SF Bay Area. *JAPCA 29*, pp. 352.

Dennis R.L. and Downton M.W. (1984) Evaluation of urban photochemical models for regulatory use. *Atmospheric Environment 18*, pp. 2055-2069.

Downton M.W. and Dennis R.L. (1985) Evaluation of urban air quality models for regulatory use : refinement and approach. *J. Clim. Appl. Met. 24*, pp. 161-173.

Fortak H. (1969) Comparison of calculated and measured maximum ground immissions and their distance from the source for the case of large emitters. *Staub. Reinhalt. Luft, 29-12.*

Fortak G. (1970) Numerical simulation of temporal and spatial distributions of urban air pollution concentration. *Proc. of Symp. on Multiple-Source Urban Diffusion Models*. EPA, Air Pollution Control Office Publication No. AP-86.

Fortak H. (1973) Validation of air quality simulation models, in *Proc. of the fourth meeting of the expert panel on AP modelling (Oberursel, FRG)*, Chap. VI, NATO/CCMS, Pub. No. 30, Brussels.

Fortak H. (1974) Potential applications of mathematical-meteorological diffusion models to the solution of problems of air quality maintenance, in *Proc. of the fifth meeting of the expert panel on AP modelling (Roskilde, Denmark)*. NATO/CCMS, Pub. No. 35, Brussels (also NTIS PB 240-582).

Fox D.G. (1981) Judging air quality performance. *Bulletin American Meteorological Society 62, No. 5*, pp. 599-609.

Gibson L.V. and Peters L.K. (1977) A short-term air quality model for sulfur dioxide in Louisville, Kentucky. *JAPCA 27, No. 3*, pp.218-223.

Guldberg P.H. and Kern Ch.W. (1978) A comparative validation of the RAM and PTMTP models for short-term SO_2-concentrations in two urban areas. *JAPCA 28, No. 9*, pp. 907-910.

Gutfreund P.D., Liu C.S., Nicholson B.R. and Roberts E.M. (1983) COMPLEX I and II model performance evaluation in Nevada and New Mexico. *J. Air Pollut. Control Ass.* 33, pp. 864 -

Guzewich D.C. and Pringle W.J.B. (1977) Validation of the EPA-PTMTP short-term Gaussian dispersion model. *JAPCA 27, No. 6*, pp. 540-542.

Hayes S.R. (1979) Performance measures and standards for air quality simulation models. *Report EPA-450/4-79-032*, 160 pp.

Hayes S.R. and Moore G.E. (1986) Air quality performance : a comparative analysis of 15 model evaluation studies. *Atmospheric Environment 20*, pp. 1897-1911.

Hillyer M.J., Reynolds S.D. and Roth P.M. (1979) Procedures for evaluating the performance of air quality simulation models. *Report EPA-450/4-79-033*, 146 pp.

Hilst G.R. (1981a) Validation of plume models statistical methods and criteria. *EPRI EA-1673-SY, Project 1616-1*, pp. 27.

Hilst G.R. et al. (1981b) Preliminary Results from the EPRI plume model validation project - Plains Site. *EPRI EA-1788, Project 1616*, pp. 103.

Irwin J. and Smith M. (1984) Potentially useful additions to the rural model performance evaluation. *Bull.Am.Met.Soc. 65*, pp.559.

Johnson W.B. (1972) Validation of air quality simulation models, in *Proc. of the third meeting of the expert panel on AP modelling (Paris, France)*. NATO/CCMS, Pub. No. 14, Brussels (also NTIS PB 240-574).

Johnson W.B., Sklarew R.C. and Turner D.B. (1976) Urban air quality simulation modelling. In *Air Pollution*, Vol. I, pp. 503-562 (ed. by Stern A.C.). Academic Press, New-York.

Juda J. (1968) Planung und Auswertung von Messungen der Veruntreinigungen in der Luft, *Staub. Reinhalt. Luft 28*, 186.

Junker A. (1976) Statistische Auswertung von Messwertkollektiven zur Ermitlung von Kenngrössen-Perzentil-Schätzung, *Staub Reinhalt. Luft 36*, p. 253.

Koch R.C. and Thayer S.D. (1971) Validation and sensitivity analysis of the Gaussian plume multiple-source urban diffusion model. *NTIS, PB 206-951*.

Koch R.C. and Fisher G.E. (1973) Evaluation of the multiple-source Gaussian plume diffusion model - Phase 1. *EPA-650/4-75-018a*.

Kretzschmar J.G., De Baere G. and Vandervee J. (1976) Validation of the Immission Frequency Distribution Model in the region of Antwerpen, Belgium. In *Air Pollution N. 51 : Proceedings of the Seventh International Technical Meeting on Air Pollution Modelling and its Application*, NATO-CCMS, Brussels, pp. 235-270.

Kretzschmar J.G., Cosemans G., De Baere G., Mertens I. and Vandervee J. (1977) Some practical examples of the impact of individual sources upon the cumulative frequency distributions of the daily SO_2-concentrations in an urban and industrial area. In *Air Pollution N. 80 : Proceedings of the Eight International Technical Meeting on Air Pollution Modelling and its Application*, NATO-CCMS, Brussels, pp. 2-23.

Kretzschmar J.G., De Baere G. and Vandervee J. (1978) The Immission Frequency Distribution Model of the SCK/CEN, Mol, Belgium. In *Modelling, Identification and Control in Environmental Systems* (edited by Vansteenkiste Gh.), IFIP, North-Holland Publishing Company, pp. 379-391.

Kretzschmar J.G. and Cosemans G. (1981a) Random- and Minimax-Campaigns for the determination of the actual air pollution levels in an unknown region. *Atmospheric Environment 15* : p. 1047.

Ku J.Y., Rao S.T. and Rao K.S. (1987) Numerical simulation of air pollution in urban areas : model performance. *Atmospheric Environment 21*, pp. 213-232.

Kummler R.H. et al. (1978) A comparative validation of the RAM and modified SAI models for short-term SO_2-concentrations in Detroit. *JAPCA 28*, pp. 907.

Kurzeja R.J. and Weber A.H. (1985) Volume 3 : Summary, Conclusions and Recommendations. *Proc. of the DOE/AMS Air Pollution Model Evaluation Workshop*. Kiawah, South Carolina, Oct. 23-25, 1985. DP-1701-3, E.I. Dupont de Nemours & Co., Savannah River Laboratory, Aitken, SC 29808.

Lee R.F., Mills M.T. and Stern R.W. (1975) Validation of a single source dispersion model. *Proc. Sixth Int. Technical Meeting on Air Pollution Modelling and its Application, Frankfurt a/m, FRG*. NATO/-CCMS Pub. No. 42.

Liu M.K., Whitney D.C., Seinfeld J.H. and Roth P.M. (1976) Continued research in mesoscale air pollution simulation modelling : analysis of prior evaluation studies. *Report EPA 600/4-76/016a.*

Londegren R.J. (1980) Validation of plume models, statistical methods and criteria. *TRC Environmental Consultants Inc.*, Wethersfield, Conn., 24 pp.

Londergan R.J. et al. (1982) Evaluation of rural air quality simulation models. *Report EPA-450/4-83-003.*

Londergan R.J., Minott D., Wachter D. and Fizz R. (1983) Evaluation of urban air quality simulation models. *Report EPA-450/4-83-020.*

Lott R.A. (1984) Performance of a time dependent dispersion model in elevated terrain. *Atmospheric Environment 18*, pp. 2367-2376.

Maldonado C. and Bullin J.A. (1977) Modelling carbon monoxide dispersion from roadways. *Envir. Sci. Technol. 11*, pp. 1071-1076.

Manier G. (1970) Die Fehler bei der rechnerischen Erfassung der Ausbreitung atmosphärischer Spurenstoffe. *Staub. Reinhalt. Luft, 30-1*, pp. 11-16.

Marziano G.L., Shir C.C., Shieh L.J., Sutera A., Gianolio L. and Ciprian M. (1979) Study of the SO_2-distribution in Venice by means of an air quality simulation model. *Atmospheric Environment 13*, pp. 477-487.

Mills M. and Record F.A. (1975) Comprehensive analysis of time-concentration relationships and the validation of a single source dispersion model. *Report EPA-450/3-75-083.*

Mills M. and Stern R. (1975) Model validation and time-concentration analysis of three power plants. *Report EPA-450/3-76-002.*

Munn R.E., Alcamo J. and Fedorov V. (1987) Evaluating the performance of air quality models in a policy framework. *Preprints of the 16th ITM on Air Pollution Modelling and its Applications*, April 6-10, 1987, Lindau, FRG.

Nappo C.J. (1974) A method for evaluating the accuracy of air pollution models. *Preprints of the Symposium on Atmospheric Diffusion and Pollution*, Santa Barbara, CA, Am. Met. Soc., pp. 325-329.

Noll K.E. (1978) A comparison of three highway line source dispersion models. *Atm. Env. 12*, pp. 1323.

Okamoto S. and Shiozawa K. (1978) Validation of an air pollution model for the Keihin area. *Atmospheric Environment 12*, pp. 2139-2149.

Ott W.R. and Mage D.T. (1981) Measuring air quality levels inexpensively at multiple locations by random sampling. *JAPCA, Vol. 31*, No. 4, pp. 365-369.

Prahm L.P. and Christensen M. (1977) Validation of a multiple source Gaussian air quality model. *Atmospheric Environment 11*, pp. 791-795.

Rao S.T., Sistla G., Pagnotti V., Petersen W.B., Irwin J.S. and Turner D.B. (1985) Evaluation of the performance of RAM with the regional air pollution study data base. *Atmospheric Environment 19*, pp. 1503-1518.

Rao S.T., Sistla G., Pagnotti V., Petersen W.B., Irwin J.S. and Turner D.B. (1985) Resampling and extreme value statistics in air quality model performance evaluation. *Atmospheric Environment 19*, pp. 1503-1518.

Smith M.E. (1984) Review of the attributes and performance of 10 rural diffusion models. *Bull. Am. Met. Soc. 65*, pp. 554-558.

Tirabassi T., Tagliazucca M. and Zannetti P. (1986) Kappa-G, a non-Gaussian plume dispersion model: description and evaluation against tracer measurements. *JAPCA 36*, pp. 592-596.

Turner D.B., Zimmerman J.R. and Busse A.D. (1972) An evaluation of some climatological dispersion models in *Proc. of the third meeting of the expert panel on AP modelling (Paris, France)*. NATO/CCMS, Pub. No. 14, Brussels (also NTIS PB 240-574).

Turner D.B. (1974) Air quality frequency distributions from dispersion models compared with measurements. *Proc. Symp. on Statistical Aspects of Air Quality Data*, Chapel Hill, NC, EPA-650/4-74-038.

Turner D.B. (1979) Atmospheric dispersion modelling : a critical review. *JAPCA 29*, pp. 502-519.

Turner D.B. and Irwin J.S. (1981) Comparison of sulfur dioxide estimates from the model RAM with St. Louis RAPS measurements. *Proc. Twelfth Int. Meeting on Air Pollution Modelling and Its Applications, August 1981, Ca*, NATO/CCMS.

Turner D.B. and Irwin J.S. (1982) Extreme value statistics related to performance of a standard air quality simulation model using data of seven power plant sites. *Atmospheric Environment 16*, pp. 1907-1914.

Vanderborght B., Mertens I. and Kretzschmar J.G. (1983) Comparing the calculated and measured aerosol concentrations and depositions around a metallurgical plant. *Atmospheric Environment* 17, pp. 1687-1701.

Venkatram A. (1979) A note on the measurements and modelling of pollutant concentrations associated with point sources. *Boundary Layer Meteorol.* 17, pp. 523-536.

Weber A.H. and Garrett A.J. (1985) *Proceedings of the DOE/AMS Air Pollution Model Evaluation Workshop*. Savannah River Laboratory DP-1701-1, NTIS.

Wilmott C.J. (1982) Some comments on the evaluation of model performance. *Bull. Am. Meteor. Soc.* 63, pp. 1309-1313.

Wong T.S., Heidorn K. and Yap D. (1976) Modelling sulphur dioxide levels in the Sarnia area. *Water, Air and Soil Pollution* 5, pp. 407-414.

NUMERICAL CASE STUDIES OF AIR POLLUTION IN AND AROUND URBAN AGGLOMERATION AREAS DURING INVERSION SITUATIONS

D. Heimann , F. Rösler
Institut für Physik der Atmosphäre
DFVLR Oberpfaffenhofen
D-8031 Weßling , F.R.G.

M. Baltrusch
Hessische Landesanstalt für Umwelt
Unter den Eichen 7
D-6200 Wiesbaden , F.R.G.

ABSTRACT. A numerical three-layer mesoscale model is used to simulate the air flow and the transport of pollutants during typical inversion episodes in and around Frankfurt and München. The model considers the complex topography which causes inhomogeneous and instationary meteorological fields. The use of several concentration arrays in the computer program makes it possible to distinguish the impact of distinct species of pollutants, of diverse types of sources, and of different time intervals of emission. Thus, the model achieves estimates on source-receptor relationships and allows certain puffs of impurities to be traced. The computed results support the interpretation of available measurements and can contribute to regional air pollution control strategies and decision making.

1. INTRODUCTION

Monitoring networks of air pollution are presently established in most of the urban agglomeration areas of the Federal Republic of Germany. The recordings of these stations allow a continuous control of the temporal and spatial distribution of the concentration of various species of pollutants. Nevertheless, for many questions which arise in regional planning and clean air policy, additional knowledge is required which cannot be extracted from operational network data. Hereto we count for example:

- How much do specific emittors or groups of emittors contribute to local concentration ?
- What time has passed between the release of a pollutant and its local impact somewhere away from the source ?

K. Grefen and J. Löbel (eds.), Environmental Meteorology, 531–544.

The second question is especially important if chemical processes may modify the pollutants. The answer to such questions requires either special observations which enable one to trace certain plumes, e.g. airborne measurements, or the use of a dispersion model which allows to 'mark' the exhausts of different sources or different times of release.

The present study intends to demonstrate the advantage of combined mesoscale air flow and dispersion models for applications during weak-wind inversion situations. Under such meteorological conditions the regional wind field, in general, is neither stationary in time nor homogeneous in space, particulary over topographically structured areas. Therefore it is necessary to apply a mesoscale model which simulates the time depending regional wind field and turbulence parameters according to the large scale forcing and the regional distribution of topography. The method of the dispersion calculation must be appropriate to inhomogeneous and instationary flows.

2. THE MODEL

The simulations have been carried out using the hydrostatic three-layer mesoscale model REWIMET (Heimann, 1986). This model allows the time dependent simulation of the mesoscale air flow and the corresponding transport of pollutants over complex topography. The model atmosphere is dynamically forced by a prescribed horizontal supra-scale pressure gradient (geostrophic wind). The thermal stability is controlled by prescribing the initial stratification and the temporal variation of the temperature at the earth surface.

All model layers are assumed to be vertically mixed. Wind shear and thermal stratification is expressed by zero-order discontinuities of wind and potential temperature at the interfaces between the layers.

The two lower layers represent the turbulent atmospheric boundary layer (ABL). They are coupled by advective and tubulent fluxes of momentum, heat, and concentrations. While the lowest layer ('surface layer') is defined to have a constant thickness of 50 m, the middle layer ('mixed layer') always adapts to the actual local height of the ABL which is controlled by both advection and entrainment processes. The third model layer is established above the ABL. It is capped by a free and impermeable surface which acts like an upper level inversion and limits the growth of the ABL.

Air pollutants are emitted into one of the model layers depending on their height of release. They are vertically mixed throughout this layer immediately after their emission.

The numerical procedure provides foreward differences in time, upstream differences for the advective terms, and central differences else. The horizontal advection of concentrations is handled with the more advanced technique after Smolarkiewicz (1984) to minimize the artificial diffusion.

3. CASE STUDIES

3.1. Case 1: A smog episode in Frankfurt

A five days episode of high pollution was simulated over the Frankfurt area. The dense population and the high industrialization lead to a remarkable impact on air quality. The Taunus range northwest of Frankfurt often channelizes the flow and causes rather complex wind fields. The aim of the simulation was to investigate how the exhausts of the regional sources are distributed during a typical smog episode, i.e. a weak-wind situation with a low inversion level and a duration of five days.

The meteorological input data have been choosen close to the real situation between 16 and 20 January 1982. This episode was characterized by a strong inversion (up to 15 K), weak winds, low surface temperatures (-12 °C to -2 °C), and considerabely high concentrations. As a simplification the geostrophic wind was set to a constant value of 3 m/s from 180° during the entire episode.

The emission data were taken from the official inventory of the Untermain region. The time dependency of the emission was considered. Two species (SO_2 and NO_x) and three groups of sources (heating, traffic, industry), i.e. six classes of emissions, were treated seperately in the model.

Figures 1 and 2 show snapshots of the episode. The nightly situation on 19 Jan. 1982 at 01 CET is characterized by very weak down hill flows (Fig. 1a). The simulated wind field is completely inhomogeneous. The calculated NO_x-concentration (Fig. 1b) has maximal values in the center of Frankfurt and the western periphery. The most important originators of the NO_x pollution in the towns are the heating facilities. Near the freeways the contribution of traffic is dominant. Only in some industrialized locations west of Frankfurt the high stacks attain a major role in the pollution. Although the industry has the greatest part of the emission in this region its plumes seldom penetrate down to the surface layer due to the stable stratification with reduced turbulent exchange.

Twelve hours later (at 13 CET) the wind speed has slightly increased (Fig. 2a), but the flow field is still rather inhomogeneous. Especially noticable is the channeling along the Taunus range with northeasterly winds despite of the geostrophic wind from the south. This effect is known as a 'counter-current' (Wippermann, 1984). In spite of the increased wind speed the concentration is higher than it was during the night. This is a consequence of the stronger emission during the busy hours. The traffic has become the most important originator of the NO_x concentration in most parts of the domain.

The temporal variations of the simulated concentrations at a location southwest of Frankfurt (black triangle in Fig. 1a/2a) are illustrated in Fig. 3b/c. They show that the importance of each types of sources varies during the course of a day. In the daytime the SO_2 impact is dominated by the heating facilities while the NO_x concentration is mainly caused by traffic. The height of the ABL (Fig. 3a) increases during the day as usual, and shrinks during the night. Higher surface temperatures on the last two days yield a expansion of the ABL and lowers the concentration level.

3.2. Case 2: An urban NO_x plume leeside of München

Inversion situations in the Alpine foreland which appear during stagnant high pressure influence or foehn conditions (Hoinka and Rösler, 1987) are often connected with a weak easterly flow which transports the trapped pollution to the western surroundings of the town. NO and NO_2 emissions by traffic are dominant in München where heavy industry only plays a minor role. During the transport NO is oxidated to NO_2. Furthermore, solar radiation yields photochemical processes which generate ozone and other toxic substances.

The inclined surface of the Alpine Foreland forces thermal circulations which superimpose the weak easterly flow. As a consequence, changing wind speeds and directions cause a very complex behaviour of the urban plume.

Such conditions were present on 11 September 1985. This case is documented by aircraft measurements accomplished by the DFVLR (Rösler, 1986). Figure 4a/b show the patterns of the NO_x concentration and the NO/NO_2 ratio. These fields are derived from aircraft data recorded from 1150 to 1430 UTC along the flight tracks indicated in Fig. 4a. The altitude of the airplane was 900 m MSL or 250-400 m GND, respectively.

Besides the plume southwest of München with high concentrations of 60 ppb and more a second maximum was found 40 km away from München west of the Ammersee (Fig. 4a). The minimum of the NO/NO_2 ratio (Fig. 4b) suggests that these pollutants as well as those over the Starnberger See have been emitted already some hours ago where most of the NO has been oxidated to NO_2 during the transport. Small NO/NO_2 ratios normally are correlated with a high content of ozone. O_3 concentrations of up to 70 ppb were found over these areas during the measurements. Since O_3 is one of the trace gases suspected to be responsible for damages on forests the answers to the following questions may be valuable for an improvement of that problem:

- Are the pollutants found over the Starnberger See and west of the Ammersee originated in München ?
- On what paths were these pollutants transported ?
- How old are the pollutants when they arrive in these areas ?

However, these questions cannot be answered by interpretating the measurements alone. Therefore the REWIMET model is used to supplement the aircraft data of this particular case which is typical for weak wind inversion weather in the Alpine foreland.

The geostrophic wind was derived from weather maps and set to 2.5 m/s from 80°. The surface temperature was prescribed and rises from 7 °C at 03 UTC (05 CEST) to 22 °C at 15 UTC (17 CEST). The ABL is stablely stratified until 0830 UTC and becomes more and more unstable afterwards. The upper level inversion was observed at 1300 m MSL. It is represented in the model by the free and impermeable model lid which was initialized in that altitude.

Unfortuately there does not exist a NO_x emission inventory for München. As a substitute the spatial distribution of the emission rate was deduced from the land use data. The level of NO_x release was choosen to 2.4 kg/s within the town area of München. This value was estimated from aircraft measurements of the NO_x plume downstream of München during homogeneous and stationary wind conditions. Since the overwhelming part of the NO_x output is caused by traffic, the emission was injected into the lowest model layer.

Figures 5a/b illustrate the surface layer flow field at 06 UTC and 12 UTC, respectively. As long as the ABL is stablely stratified the surface layer wind field is very complex and characterized by catabatic flows (Fig. 5a). This is evident over München where the sloped surface yields a southerly wind component. After insolation has started to heat the ABL the wind speeds up and the flow field becomes more homogeneous (Fig. 5b).

For a better comparability to the measurements the simulated NO_x patterns of the middle model layer (mixed layer above 50 m GND) at 12 UTC is displayed in Figure 6. It shows structures similar to the observations. By using different arrays to store the concentration values emitted either inside or outside München it is possible to estimate the origin of the local NO_x impact. The result suggests that only 50 % of the total concentration west of the Ammersee is shared by pollutants originated in München while the plume between München and the Starnberger See is produced almost entirely by sources in München.

What is the history of transport which brought about the situation at noon ? To answer this the concentration was stored by the model in different arrays depending on the 3-hour interval of emission. The shape of the area polluted by NO_x emissions in München between 03 UTC and 06 UTC can be traced in Figure 7. In the morning the southerly wind in the surface layer and the easterly flow above significantly deform the plume. After 08 UTC the wind shifts to a more uniform northeasterly direction in all layers. The western branch of the plume now moves toward the southwest and reaches the Ammersee by 12 UTC. The eastern branch is first carried as far as 20 km to the north of München and is then blown back. It moves over the town again where new emissions charge the already polluted air before it reaches the Starnberger See at 12 UTC.

Figure 8 indicates the age of the urban pollutants at 12 UTC. The air west of the Ammersee was loaden with NO_x already 6 to 12 hours ago. Around the Starnberger See the NO_x impact consists partly of younger pollutants transported here on a direct way. But also older impurities which at first were moved northward in the morning affect the air quality in this area.

4. CONCLUSION

The computations show that the application of a mesoscale model helps to understand the air pollution transport mechanisms in and around urban agglomeration areas even under the complex conditions of inhomogeneous and instationary wind fields.

Other than monitoring networks the model can distinguish different originators and can estimate their importance to local concentrations. The simulation of a typical smog episode in greater Frankfurt exhibits the possibility to support decisions on preventive measures.

The numerical imitation of a weak wind inversion situation in the Alpine foreland gives an impression how the exhaust-gases of the München town area are transported to the rural surroundings, and what portion of the impact is originated in the agglomeration area. The indirect estimation of the travel time of the pollutants by evaluating the NO/NO_2 ratio is confirmed by the simulations. In addition the model achieves a possible history of the situation found at noon by the measurement flights.

Abbreviations.

GND	height above ground
MSL	height above mean sea level
ABL	atmospheric boundary layer
UTC	universal time coordinated
CET	Central European time (UTC+1h)
CEST	Central European summer (daylight saving) time (UTC+2h)
DFVLR	Deutsche Forschungs- und Versuchsanstalt für Luft- und Raumfahrt (German Aerospace Research Establishment)

References.

Heimann, D., 1986:
'Estimation of Regional Surface Wind Field Characteristics Using a Three-Layer Mesoscale Model'
Beitr.Phys.Atmosph., 59 ,518-537

Hoinka,K.P., Rösler,F., 1987:
'The Surface Layer on the Leeside of the Alps during Foehn'
To be printed in: *Meteorol. and Atmosph. Phys.*

Rösler, F., 1986:
'Experimentelle und theoretische Untersuchungen zur Schadstoffausbreitung am Beispiel der Groβstadt München'
Diploma thesis, Techn.Univ. München, Physics Department

Smolarkiewicz,P.K., 1984:
'A Fully Multidimensional Positive Definite Advection Transport Algorithm with Small Implicite Diffusion'
J.Comp.Phys., **54** ,325-362

Wippermann,F., 1984:
'Do Flat Mountain Ranges also Channel the Air Flow ?'
Beitr.Phys.Atmosph., 57 ,282-284

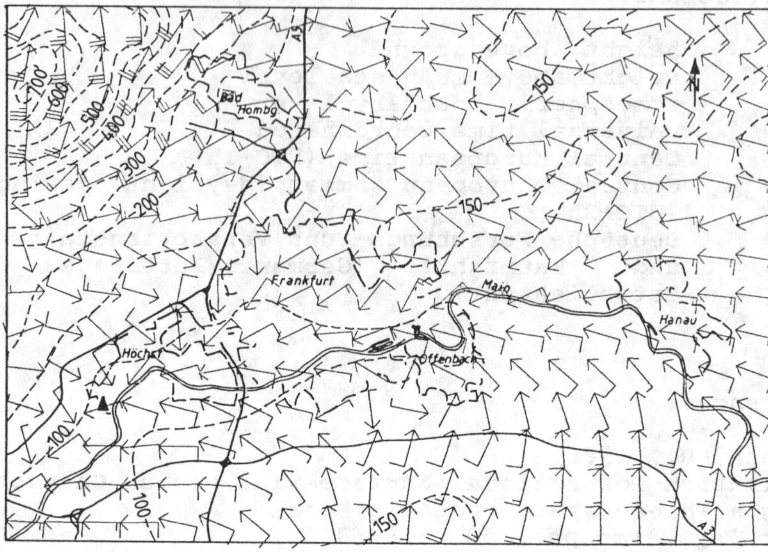

Figure 1a. Calculated surface layer wind field for 19 January 1982 at 01 CET. The short and long feathers denote wind speeds of 0.5 and 1 m/s, respectively.

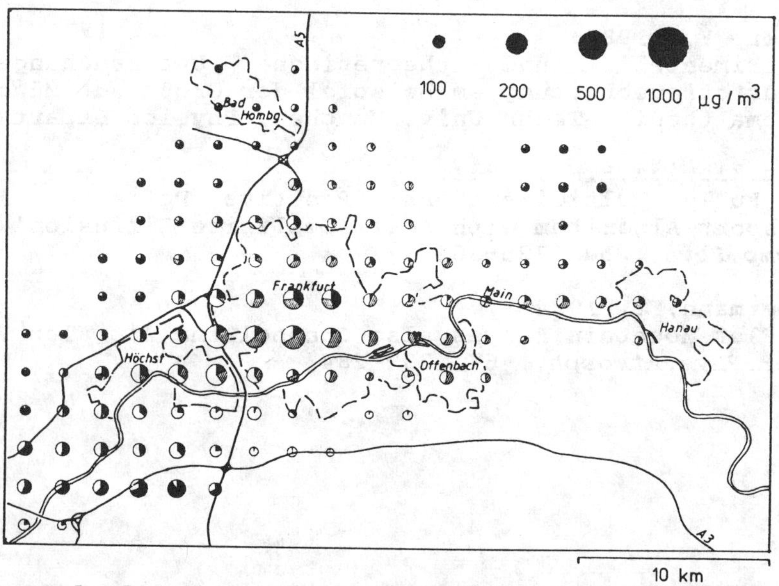

Figure 1b. Calculated surface layer NO_x concentration at 01 CET. The origin of the pollution is indicated by black segments (industry), grey segments (heating facilities), and white segments (traffic). The segment areas are proportional to the share of concentration.

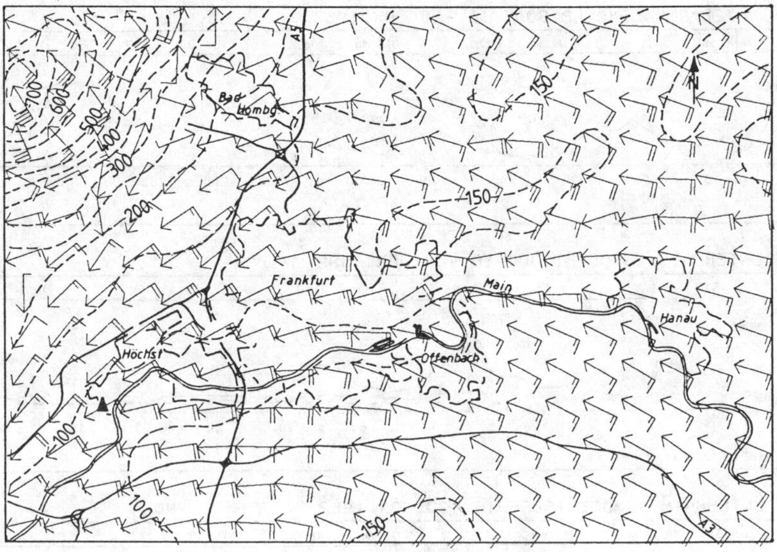

Figure 2a. Same as Figure 1a but for 19 January 1982 at 13 CET.

Figure 2b. Same as Figure 1b but for 19 January 1982 at 13 CET.

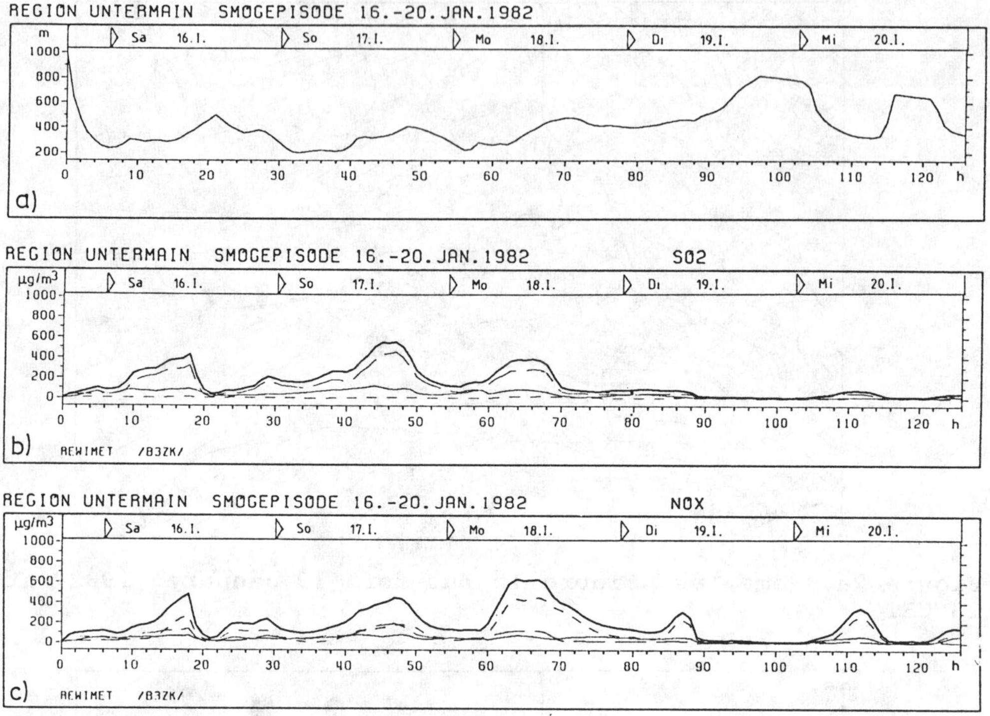

Figure 3a,b,c. Temporal variation during the five days episode at Sindlingen southwest of Frankfurt. This location is indicated by a black triangle in Figure 1a and 2a.
a) Thickness of the calculated atmospheric boundary layer (ABL)
b) Calculated SO_2 concentration.
c) Calculated NO_x concentration.
The thick solid line indicate the total concentration. The partial concentrations originated by industry, heating facilities, and traffic are plotted with a thin solid line, and long and short dashes, respectively.

Figure 4a. Measured concentration (ppb) of NO_x on 11 September 1985 from 1150 to 1430 UTC. The thin straight lines represent the flight tracks along which the concentrations were recorded in an altitude of 900 m MSL.

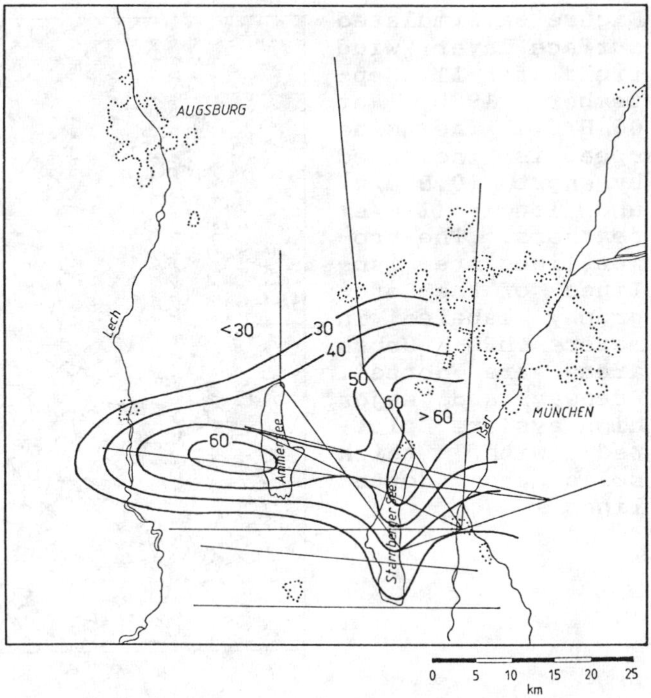

Figure 4b. NO/NO_2-ratio derived from the aircraft measurements on 11 September 1985 from 1150 to 1430 UTC in an altitude of 900 m MSL.

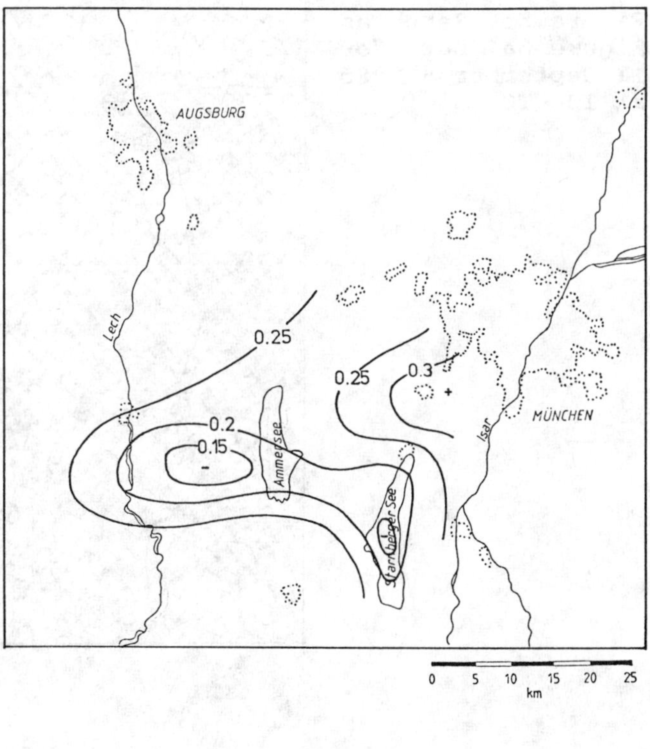

Figure 5a. Simulated surface layer wind field for 11 September 1985 at 06 UTC. The wind speed is indicated by short (0.5 m/s) and long (1 m/s) feathers. The broken lines are isolines of the orography labeled in meters MSL. Urban areas are dotted. Freeways and major highways are plotted with thick solid and dotted lines.

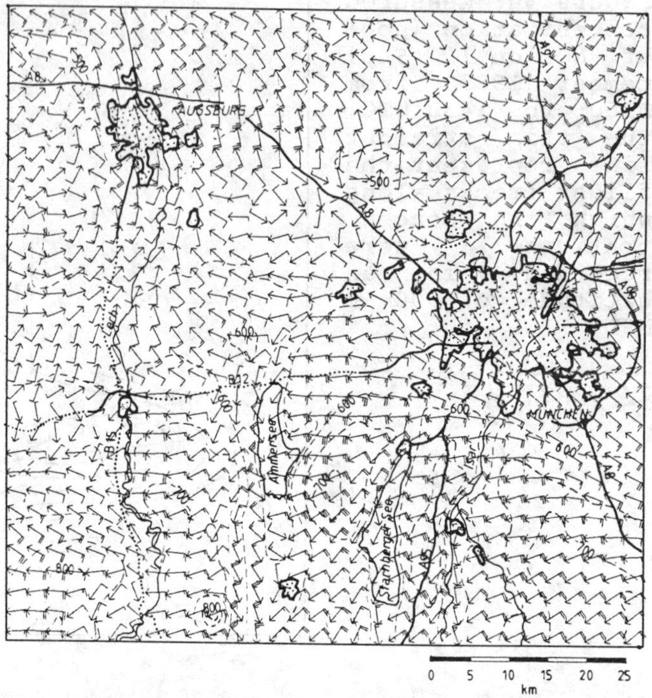

Figure 5b. Same as Figure 5a but for 11 September 1985 at 12 UTC.

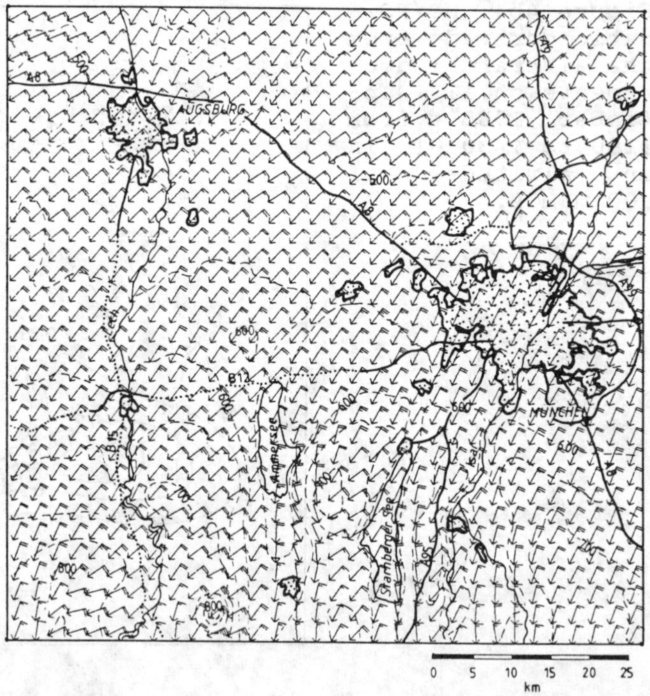

Figure 6. Simulated NO_x concentration for 11 September 1985 at 12 UTC in the ABL above 50 m GND. The circular areas are proportional to the total concentration (plotting threshold is 10 ppb). Regions with a concentration of 50 ppb and more are enclosed by the broken line. The black segments shows the partial concentration originated in München (i.e. inside the rectangular frame).

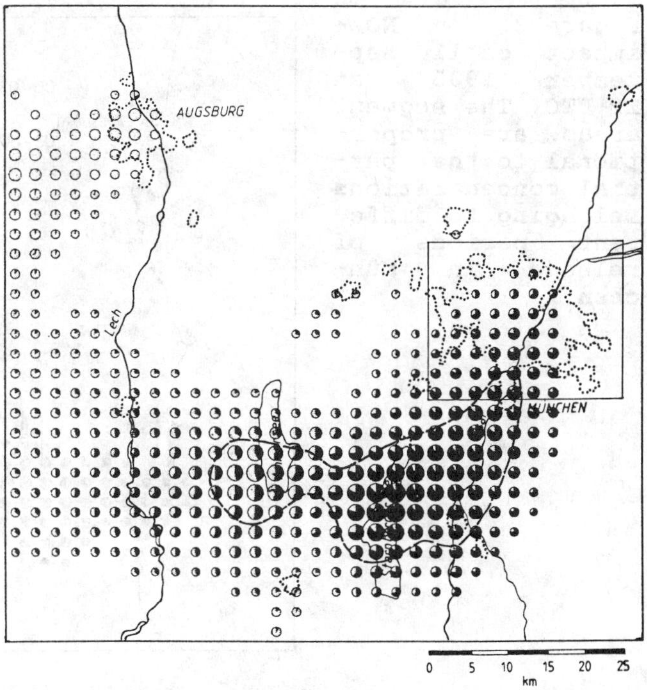

Figure 7. Locations of the 10 ppb isolines of a simulated NO_x puff at different point of times on 11 September 1985.
The puff was released in München between 03 and 06 UTC.

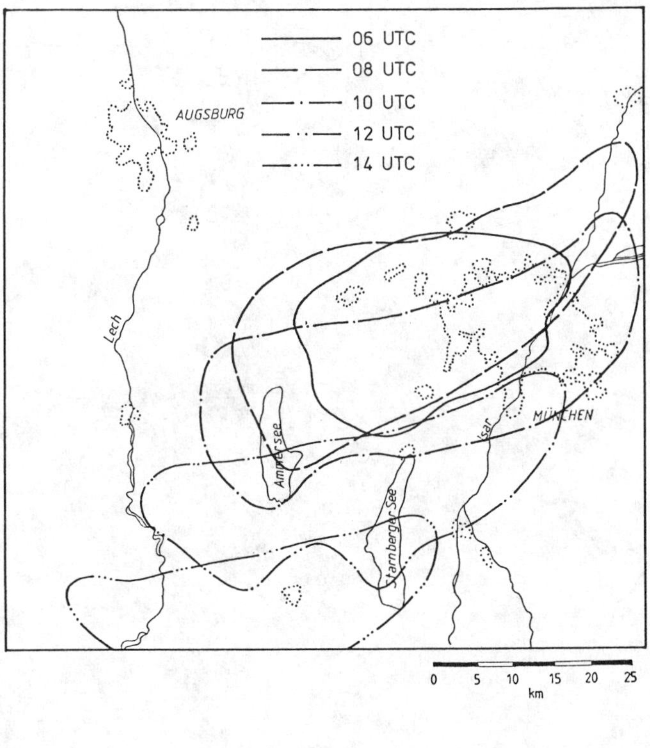

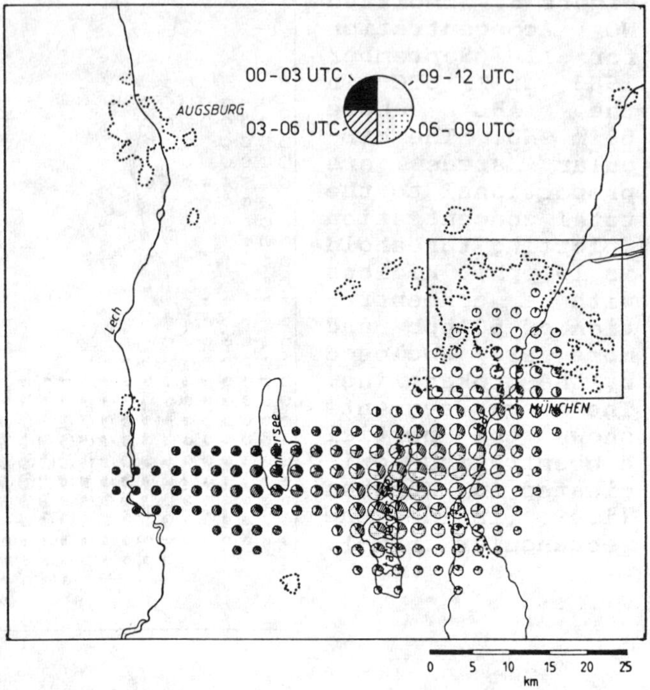

Figure 8. NO_x-impact on 11 September 1985 at 12 UTC. The segment areas are proportional to the partial concentrations belonging to different periods of release in München.

MODELLING OF THE SO_2 DISPERSION OVER BADEN-WÜRTTEMBERG

K. Nester, F. Fiedler, M. Baer
Institut für Meteorologie und Klimaforschung
Kernforschungszentrum Karlsruhe/Universität Karlsruhe
Postfach 3640, Weberstr. 5
7500 Karlsruhe 1
Germany

ABSTRACT. The model DRAIS was developed to simulate the transport and dispersion of pollutants in a mesoscale area. Concentration and deposition calculations over Baden-Württemberg have been carried out using the emission data and the input flux of the TULLA experiment from March 25, 1985.

Further calculations for March 25 were performed taking into account seperately the input flux and the emissions from point and area sources. The results show that the contribution of the different sources to the total dry deposition is 50 %, 22 % and 28 % respectively.

Calculations of the SO_2 concentrations and dry deposition with geostrophic winds from different directions have been carried out. The results, weighted with the frequency distribution of the wind direction, are drawn as mean values and as roses. The roses show at each grid point of the model area which sources from Baden-Württemberg contribute mainly to the SO_2 concentration and dry deposition. It is a first estimate of a climatology of the SO_2 burden over Baden-Württemberg.

1. INTRODUCTION

There is still a substantial lack of knowledge about the transport, deposition and chemical transformation of airborne pollutants in mesoscale areas, eg. several hundred square kilometers. In order to improve this knowledge, experimental and theoretical investigations have been carried out for the area of Baden-Württemberg.

On March 1985 the TULLA experiment was carried out. During intensive measuring phases concentrations of SO_2, NO, NO_2, O_3 and other species were determined. Additionally meteorological measurements were carried out in the lower troposphere /1/.

The DRAIS model was developed to simulate the dispersion of pollutants over irregular terrain in a mesoscale area. It was first applied to the TULLA episodes.

K. Grefen and J. Löbel (eds.), Environmental Meteorology, 545–557.

Further calculations of the SO_2 concentration and deposition over Baden-Württemberg were based on wind fields calculated for different geostrophic wind directions. These results together with the frequency distribution of the geostrophic wind provides a first estimate of a SO_2 statistic in Baden-Württemberg.

2. THE DRAIS MODEL

The model DRAIS solves the diffusion equation on an Eulerian grid.

$$\frac{\partial \bar{c}_i}{\partial t} + \nabla(\vec{\bar{v}}\bar{c}_i) = \nabla\cdot(K\cdot\nabla\bar{c}_i) + f_i\ (\bar{c}_i \ldots \bar{c}_p) \tag{1}$$

$\bar{c}_i$: mean concentration of chemical species i
$\vec{\bar{v}}$: mean advective velocity field $\vec{\bar{v}}=(u,v,w)$
K : second order diffusivity tensor
f_i: includes chemical formation (or depletion) rate of species i.

with the lower boundary condition:

$$-K\ \nabla\bar{c}_i\ \vec{n}_h = v_{D,i}\bar{c}_i \tag{2}$$

$\vec{n}_h$: unit vector normal to the ground
$v_{D,i}$: dry deposition velocity of the chemical species i.

To simulate the diffusion over irregular terrain, a terrain following coordinate system is used. The vertical coordinate in the terrain following coordinate system in DRAIS is given by:

$$a(1-\eta)^2 + b(1-\eta) = \frac{h(x,y)-z}{h(x,y)-H} \tag{3}$$

H is the top of the modeled layer. By choosing appropriate values for the constants a and b, an improved resolution near ground level can be achieved.

3. PARAMETERIZATION OF THE DRY DEPOSITION VELOCITY

In most models the deposition rate is described by a single quantity, the pollutant deposition velocity v_D. The flux of material, F, directed towards the lower boundary surface, is defined by:

$$F = v_D \cdot \bar{c}(z_r) \tag{4}$$

where $\bar{c}(z_r)$ is the concentration of the material at a reference height z_r. The dry deposition velocity involves a complex linkage between turbulent diffusion in the surface boundary layer, molecular scale motion at the air-ground interface and interaction of the material with the surface.

To parameterize the dry deposition velocity of gases to vegetated surfaces a 'big-leaf' multiple resistance model is used, see Fig. 1.

$$v_D = \frac{1}{r_a + r_b + r_c} \tag{5}$$

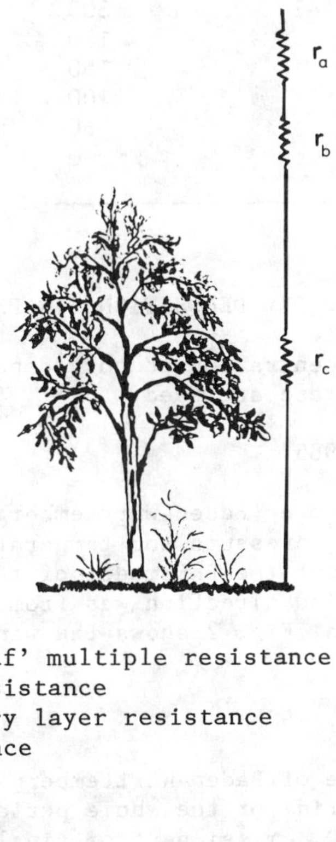

Figure 1. 'Big-leaf' multiple resistance model.
r_a: aerodynamic resistance
r_b: surface boundary layer resistance
r_c: canopy resistance

The aerodynamic resistance r_a associated with atmospheric turbulence depends on the friction velocity u_*, the Monin Obukhov stability length L_* and the roughness lenght z_o. The quasi-laminar boundary resistance r_b, which takes into account the resistance to mass transfer through the quasi-laminar layer of air in contact with the surface elements, is influenced by the diffusivity of the material being transferred and depends on the friction velocity. The formulas for r_a and r_b are given in /2/. The canopy resistance r_c, which is dominated by biological surface factors, includes the stomatal, mesophyll and cuticular uptake resistances. However, the stomatal resistance tends to dominate for gases. The canopy resistances for our calculations are summarized in Table I.

TABLE I. SO_2 canopy resistances.

ground cover/land use	r_c (s/m)
high density residential	1000
low density residential	300
forest	150
scrub	150
soil dry	100
soil wet	60
water surface	0

4. SIMULATED CONCENTRATION AND DRY DEPOSITION FIELDS FOR MARCH 25, 1985

For the calculation of the concentration and dry deposition fields of SO_2 the following input parameters are used.

4.1. Wind Field on March 25, 1985

The wind and turbulence fields over Baden-Württemberg were simulated by the KAMM model, using the wind, pressure and temperature field from the operational numerical analysis of the BKF model of the German Weather Service at noon. The general wind direction was from southwest, the most frequent direction in this area. Fig. 2 shows the wind field at ground level.

4.2. SO_2 Emissions on March 25, 1985

The SO_2 emissions for the state of Baden-Württemberg are available as hourly averages in a 1x1 km^2 grid for the whole period of the TULLA experiment /3/. The relevant SO_2 emissions from single sources on March 25, are given in Figure 3. The SO_2 emission from single sources on that day was about 18 t/h.

Since Figure 3 contains also the topography, it can be seen that the sources are concentrated in the Upper Rhine and Neckar Valley.

Figure 4 shows the distribution of the area sources, such as traffic, household and small industries. The main sources are located again in the northern part of Baden-Württemberg. The emissions of the area sources add up to about 12 t/h.

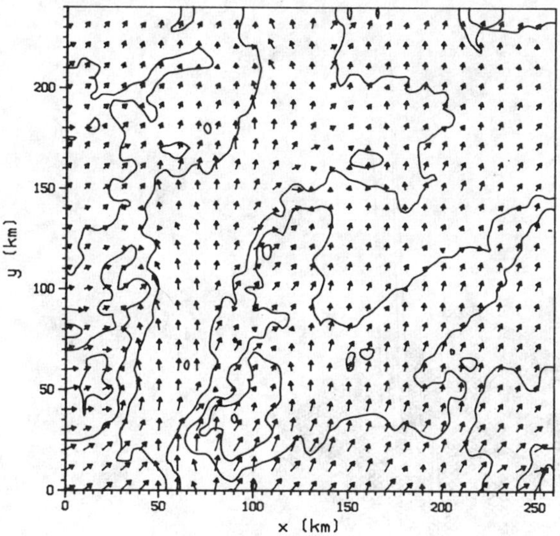

Figure 2. Wind field at ground level (March 25, 1985).

4.3. Determination of the SO_2 Flux into Baden-Württemberg

On March 25, 1985 four aircrafts measured SO_2 concentration profiles on a flight pattern around Baden-Württemberg. The lowest flight level was about 200 m above ground level, varying between 100 and 300 m. The other aircrafts were flying at levels about 100 m distant from each other.

From the measured concentrations a profile could be derived at each grid point along the flight route. At the inflow boundary these profiles are used as the boundary conditions for the DRAIS model.

4.4. Model Results

A calculation of the dispersion of SO_2 over Baden-Württemberg has been carried out with the DRAIS model using the emissions from all sources and the input flux of SO_2. The results are shown inside of the flight route of the aircrafts around Baden-Württemberg (Figure 5). Higher concentrations occur in the area east of Karlsruhe caused by emissions from the area around Rastatt. Another area with high concentrations extends along the Neckar Valley and then to the north and north-east of Baden-Württemberg. Low concentrations are calculated in the higher regions of the Black Forest and in the south of Baden-Württemberg. The corresponding dry deposition field (Fig. 6) shows a similar distribution. The total dry deposition inside of the flight route was calculated to about 6 t/h which emphasizes the importance of this removal process.

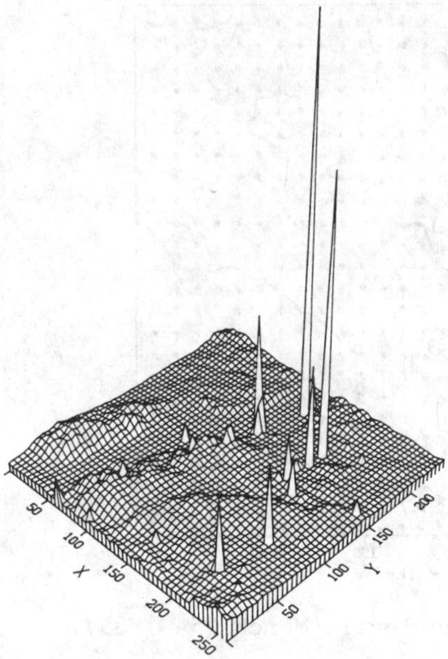

Figure 3. SO_2 emissions from point sources (March 25, 1985, maximum 4.6 t/h).

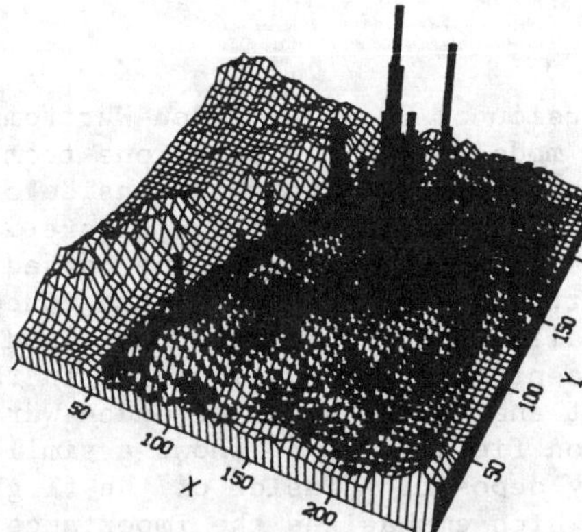

Figure 4. SO_2 emissions from area sources (March 25, 1985, maximum 152 kg/h).

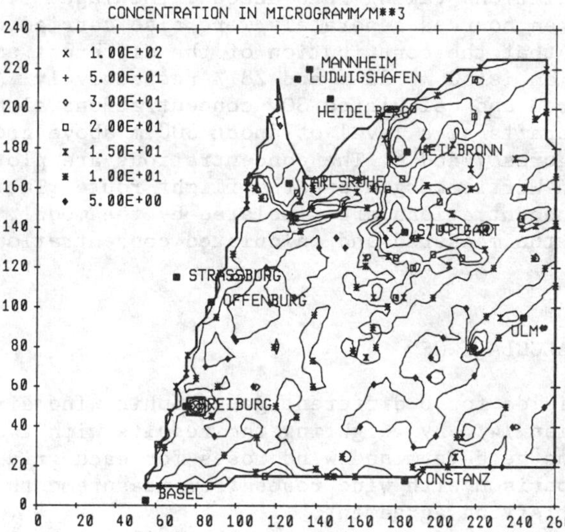

Figure 5. SO_2 ground level concentration in Baden-Württemberg (March 25,1985).

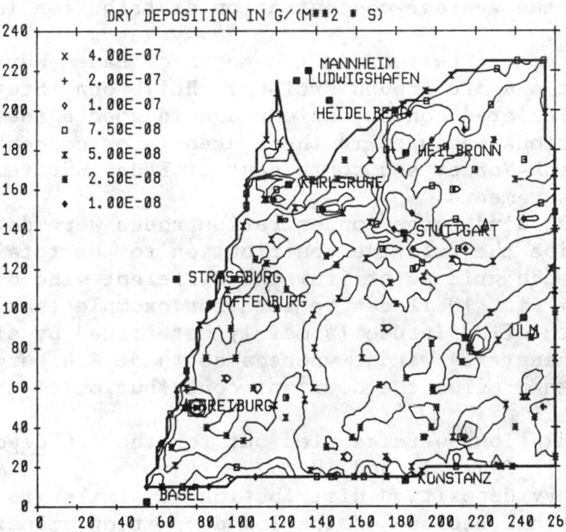

Figure 6. SO_2 dry deposition in Baden-Württemberg (March 25, 1985).

Several calculations taking into account the input flux, the point sources and the area sources seperately were also carried out. These calculations show that the contribution of the different sources to the total dry deposition is 50 %, 22 % and 28 % respectively. A comparison between the measured and calculated SO_2 concentrations along the flight routes of the aircrafts at a level of about 300 m above ground is plotted in the Figures 7 and 8. The concentrations are plotted over the orography of Baden-Württemberg along the flight route. The main features of the measured concentrations are simulated by the model. The agreement between the measured and calculated concentrations is quite satisfactory.

5. STATISTICAL CALCULATIONS

Calculated wind fields for 8 different geostrophic wind directions were calculated by Adrian /4/. By weighting the results with the frequency of the wind directions he determined wind roses for each grid point of our model area. A comparison with wind roses from measurements near ground level showed a satisfying agreement.

This result encouraged us to carry out similar calculations for the dispersion of SO_2. We used as input parameters for the DRAIS model

- the wind fields calculated by Adrian /4/,
- the point and area source distribution from the TULLA-experiment,
- the input flux of SO_2 from the TULLA-experiment.

Eight calculations have been carried out and the results were weighted with the frequency of the different geostrophic wind directions.

Fig. 9 shows the average concentration distribution inside of the input flux pattern.

We find high concentrations in the north of Baden-Württemberg. The highest concentrations are around Freiburg, Heilbronn, Stuttgart und Karlsruhe. The calculated concentrations are in good agreement with average concentrations measured in these areas. Low concentrations are located in the Black-Forest and south east of Baden-Württemberg, as expected from measurements.

Similar to the wind roses concentration roses were determined. These roses describe the relative contribution to the total concentration at each grid point from the different wind directions near ground level. From Fig. 10 it can be seen for example that the concentration north of Karlsruhe is mainly determined by situations with westerly and northeasterly winds, whereas south of Freiburg winds from southwest and north provide the dominant contribution to the SO_2 concentration.

Similar calculations were carried out for the dry deposition. The results are shown in Figs. 11 and 12.

The average dry deposition distribution is similar to those for the average concentration. Because of the low deposition velocity in the dense populated area of Stuttgart the dry deposition is lower than expected from the high concentration.

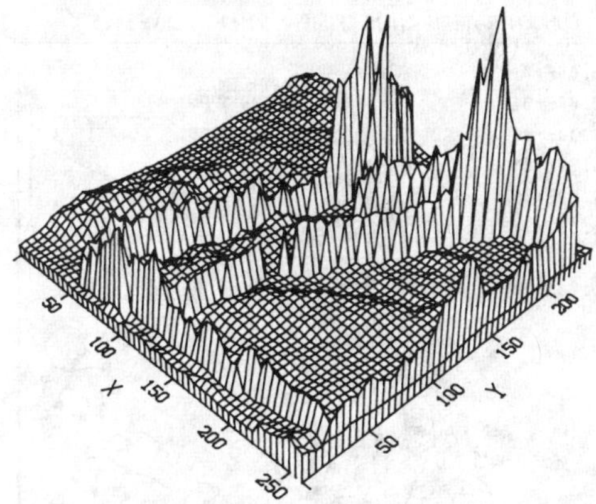

Figure 7. Measured SO_2 concentrations along the flight route at a level of about 300 m above ground (March 25, 1985, maximum: 35 μg/m^3).

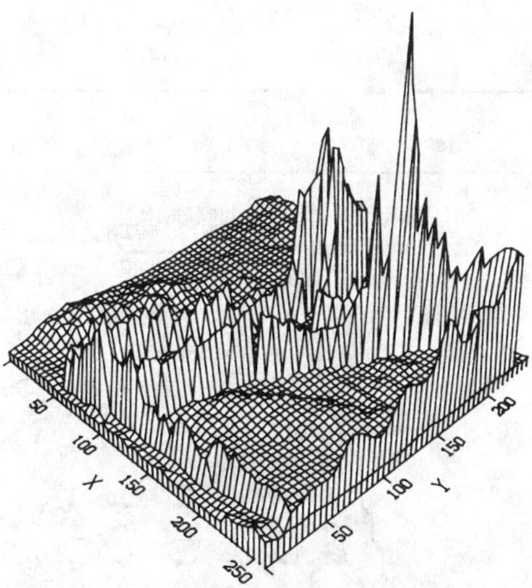

Figure 8. Calculated SO_2 concentration along the flight route at a level of about 300 m above ground (March 25, 1985, maximum: 54 μg/m^3).

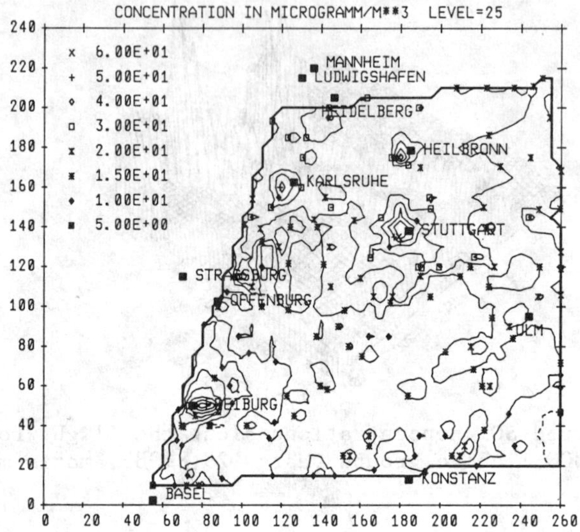

Figure 9. Mean SO_2 ground level concentration in Baden-Württemberg.

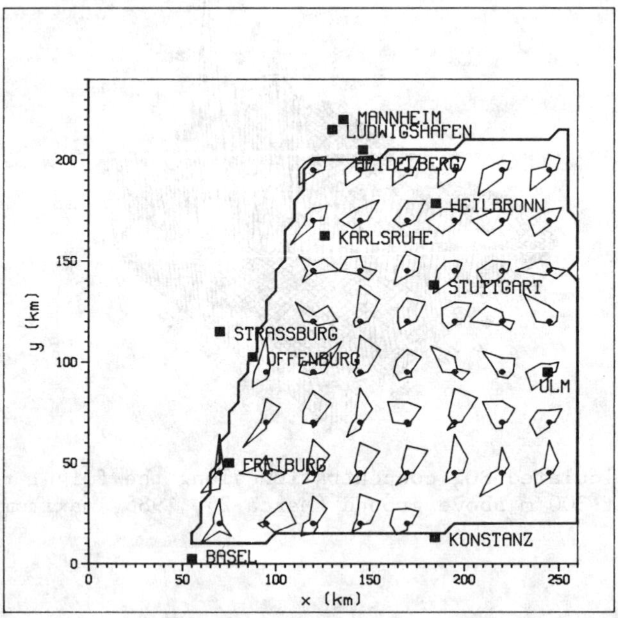

Figure 10. Concentration roses in Baden-Württemberg.

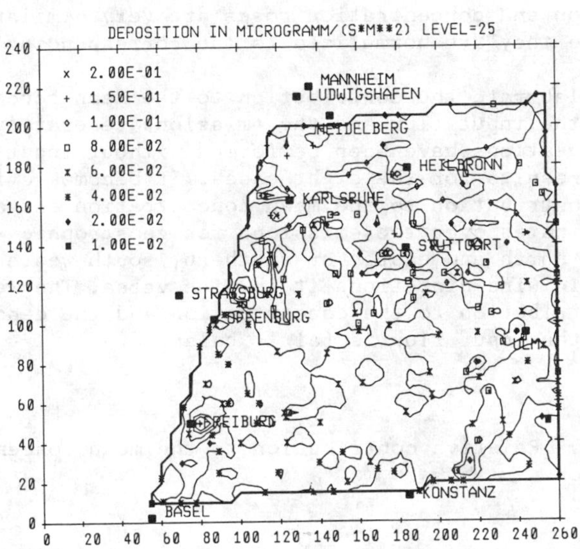

Figure 11. Mean dry deposition in Baden-Württemberg.

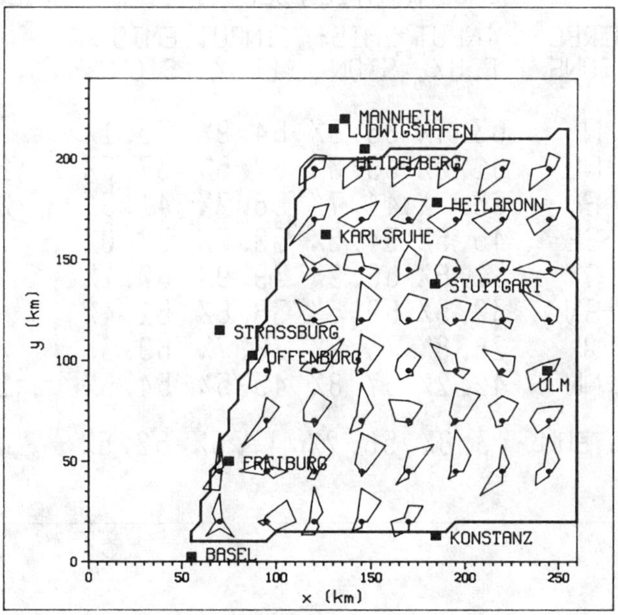

Figure 12. Dry deposition roses in Baden-Württemberg.

The deposition and concentration roses are very similar to each other since they are normalized to the corresponding averages at each grid point.

In order to elaborate the contribution to the mean concentration and deposition by the input flux and the emissions seperately, corresponding calculations have been performed without input flux. Tab. II summarizes the results for the eight cases. It becomes evident from Tab. II that the contribution to the mean concentration and deposition by the input flux dominates over the emissions for geostrophic wind directions between north and east. For southern, south-westerly and westerly geostrophic wind directions it is vice versa. The weighted sum shows that the contribution to the concentration and the deposition by the emissions and the input flux is fairly balanced.

TABLE II. Relative contribution to the mean concentration and deposition.

CASES	DIRECTIONS	CONTRIBUTION TO THE MEAN CONCENTRATION		CONTRIBUTION TO THE MEAN DEPOSITION		FREQUENCY OF GEOSTR. WIND DIRECTION
		INPUT FLUX	EMISSION	INPUT FLUX	EMISSION	
1	N	63.4%	36.6%	64.9%	35.1%	9%
2	NE	61.0%	39.0%	62.5%	37.5%	10%
3	E	55.3%	44.7%	58.7%	41.3%	21%
4	SE	45.5%	54.5%	50.0%	50.0%	7%
5	S	31.8%	68.2%	39.9%	60.1%	5%
6	SW	30.5%	69.5%	38.6%	61.4%	4%
7	W	26.8%	73.2%	30.7%	69.3%	27%
8	NW	42.2%	57.8%	45.5%	54.5%	17%
WEIGHTED SUM		43.8%	56.2%	47.5%	52.5%	

CONCLUSIONS

The calculated SO_2 concentration field for March 25, 1985 has shown that the DRAIS model is able to simulate the dispersion of SO_2 over irregular terrain. A first estimate of a climatology of the SO_2 burden over Baden-Württemberg shows the possibilities of such a model also for the determination of statistical features. The resolution in space is always better than it can be achieved by a network of measuring stations. A further improvement of the model results can be expected if non steady state simulations are carried out.

REFERENCES

/1/ S. Vogt, F. Fiedler:
Measuring Activities of the TULLA-Experiment (Field Phase Report).
KfK **4167**, Februar 1987

/2/ M. Baer, K. Nester:
Simulation of Concentration, dry Deposition and Fluxes of SO_2 during TULLA.
Annales Geophysicae, 1987, **5B**, (5), 497-502

/3/ B. Boysen, F. Friedrich, Th. Müller, N. Scheirle, A. Voss:
Feinmaschiges Kataster der SO_2- und NO_x-Emissionen in Baden-Württemberg für die Zeit der TULLA-Meßkampagne.
2. Statuskolloquium 4.-7.3.1986, Projekt Europäisches Forschungszentrum für Maßnahmen zur Luftreinhaltung im Kernforschungszentrum Karlsruhe, KfK-PEF **4**, April 1986

/4/ G. Adrian:
Synthetic Wind Climatology evaluated by the Non-hydrostatic Numerical Mesoscale Model KAMM.
Proceedings of the international Symposium on Environmental Meteorology, Würzburg, 29.9.-1.10.1987

MODELING OF THE WIND FIELD AND THE DISPERSION OF REACTIVE POLLUTANTS

Th. Flassak and N. Moussiopoulos
Institut für Technische Thermodynamik
Universität Karlsruhe
D-7500 Karlsruhe 1
F.R.G.

ABSTRACT. In the first two sections of this paper two models providing wind fields over complex terrain are presented: A diagnostic wind model and a non-hydrostatic dynamic mesoscale model. The former is used to reconstruct the flow field in the Upper Rhine valley, the latter is applied to simulate the inviscid flow over a single mountain. The third section of the paper deals with simulations of the dispersion of reactive pollutants. As an application, the influence of the diurnal variation of emission rates on the formation of ozone is investigated.

1. THE DIAGNOSTIC WIND MODEL CONDOR

1.1. Model description

A simple approach for analyses of the air flow over complex terrain consists in generating mass-consistent (i.e. non-divergent) wind fields. An initial wind velocity field $\underline{u}^\circ$ ("observed" field) is first generated by an interpolation technique on the basis of surface measurements and upper air soundings. The "observed" field in general does not satisfy continuity. With $\underline{u}$ as the mass-consistent wind field to be calculated ("adjusted" field), the incompressible form of the continuity equation $\nabla\underline{u} = 0$ yields for the velocity difference $\underline{c} = \underline{u} - \underline{u}^\circ$ the equation $\nabla\underline{c} = - \nabla\underline{u}^\circ$. For the determination of $\underline{c}$ it is assumed that the vorticity calculated on the basis of the "observed" wind field $\underline{u}^\circ$ is equal to the vorticity resulting from the "adjusted" wind field $\underline{u}$, i.e. $\nabla\times\underline{u} = \nabla\times\underline{u}^\circ$. This corresponds to the requirement $\nabla\times\underline{c} = 0$ which is identically fulfilled, if the velocity difference $\underline{c}$ can be expressed as the gradient of a scalar field λ (i.e. $\underline{c} = \nabla\lambda$). Therefore one gets the elliptic equation $\nabla^2\lambda = - \nabla\underline{u}^\circ$. The orography is taken into account by transforming this equation to a terrain-following coordinate system. Due to the terrain inhomogeneity, the resulting equation includes several additional terms with spatially varying coefficients. For the numerical solution of this equation parallel algorithms were developed which are described in details elsewhere [1].

K. Grefen and J. Löbel (eds.), Environmental Meteorology, 559–573.

1.2. Analysis of the air movement in the Upper Rhine valley

As an application of the model CONDOR, non-divergent wind fields were calculated for the 140×140 km area shown in Fig. 1. The domain contains the northern half of the Upper Rhine valley formed between the Vosges and the Haardt to the West and the Black Forest, the Kraichgau and the Odenwald to the East. The NW part of the domain corresponds to the Palatinate Forest.

The numerical solution was based on a 29×29×11 horizontally equidistant grid with a variable spacing in the vertical direction. Non-divergent wind fields were calculated for several times of September 19, 1979 between 0600 and 2100 LST. For this day extensive measurements originating from the MESOKLIP experiment are available [2]. We used surface wind data from 13 remote stations of the Deutscher Wetterdienst and upper air soundings which were performed at Böbingen.

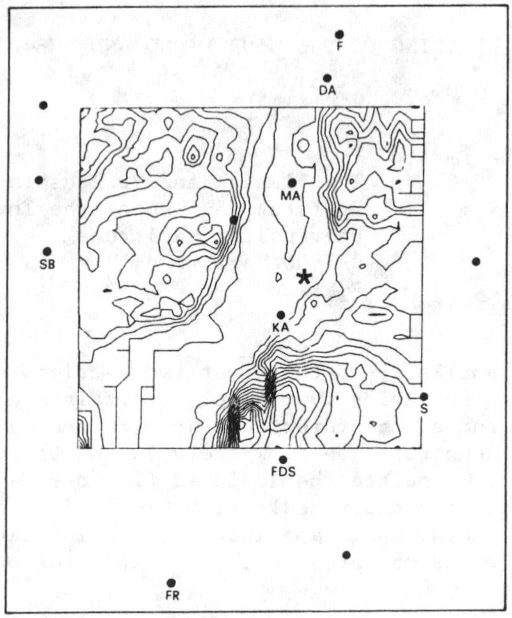

Figure 1. Orography of the northern part of the Upper Rhine valley and its surroundings. Dots indicate the locations of surface measuring stations, the asterisk the release point of the upper air sounding.

Fig. 2 illustrates the wind velocity obtained for 0900, 1200, 1500 and 1800 LST at 10 m above ground level. Apparently, southwesterly wind directions prevail and the magnitude of the wind velocity increases during the day. Nevertheless, the orography markedly influences the wind field: In the center of the considered domain the flow is more or less adapted to the morphology of the Rhine valley. Further, the results indicate that in the late afternoon air mass is advected rather around than over the Black Forest following to some extent the course of the river Neckar. Finally, in the evening the wind ceases (note the different scaling at 1800 LST); the available data are far too sparse to allow an adequate representation of the nocturnal drainage flow.

2. THE NON-HYDROSTATIC DYNAMIC MESOSCALE MODEL MEMO

As the diagnostic method described above allows only a reconstruction of the wind field, for predictive purposes and for comparative analyses dynamic mesoscale models have to be employed. The non-hydrostatic dynamic mesoscale model MEMO was developed by the authors to achieve full vec-

Figure 2. Mass-consistent wind fields obtained with the diagnostic wind model CONDOR for four times of September 19, 1979 at 10m above ground level.

torization on computers like the CYBER 205. As this model is described in detail elsewhere [3], in the following only a brief description of MEMO is given. Model results for the inviscid flow over a single mountain are presented.

2.1. Model description

The model MEMO was developed to ensure conservation of momentum, mass and energy in a non-saturated atmosphere. Basic equations are the three momentum equations, the anelastic form of the continuity equation and the thermal energy equation formulated in Cartesian coordinates x',y',z'. This set of equations is transformed to terrain-following coordinates x',y',η, where η=0 corresponds to the height of the lower boundary. To allow nonequidistant meshsizes, for example to have a better resolution near the ground, the additional transformation x = x(x'), y = y(y'), z = z(η) is employed with the arbitrary monotonic functions x(x'), y(y') and z(η) [4]. The governing equations are solved numerically on a staggered grid, where scalars are defined at the cell center and the velocity components at the centers of the appropriate interface. The imposed boundary conditions are discussed elsewhere [3].

2.2. Diagnostic pressure equation and fast elliptic solver SHAFT

Introducing the discretized momentum equations into the discretized continuity equation results in a diagnostic elliptic equation for the pressure change. This equation is solved using our fast elliptic solver SHAFT [5] combined with a conjugate gradient method. This procedure was found to be more efficient than the block iteration technique described in [1]. The fully vectorized solver SHAFT was developed for the direct solution of separable elliptic equations having the form

$$\lambda_{xx} + \lambda_{yy} + d(z)\cdot\lambda_{zz} + e(z)\cdot\lambda_{z} + f(z)\cdot\lambda \equiv L(\lambda) = r. \qquad (1)$$

For the numerical solution of this equation using second-order difference approximations a heptadiagonal equation system has to be solved. By a Fast Fourier Transformation in x- and y-direction, this system is separated into a number of independent tridiagonal equation systems, which can be efficiently solved by means of Gaussian elimination. Full vectorization is achieved by performing independent Fourier transformations in parallel and by parallel solution of the independent tridiagonal equation systems. Table I shows the computing time needed by SHAFT to solve problems with N^3 unknowns on a CYBER 205.

In addition to the non-hydrostatic version of MEMO a hydrostatic version is also available. Instead of solving the elliptic equation for the pressure change, in the hydrostatic version the pressure distribution results from an integration of the hydrostatic equation.

Table I. Computing time t needed by SHAFT to solve problems with N^3 unknowns and staggered Neumann boundary conditions on a CYBER 205

N^3	16^3	24^3	32^3	40^3	50^3	64^3
t/msec	5.2	14.3	30.6	66.5	157.	217.

2.3. Programming considerations

On the basis of the model MEMO, a computer code was developed to conform with the features of the available vector computer CYBER 205 (2-pipe, 2 million words core memory). The programming language is FORTRAN 77.

A considerable part of the overall computing time is consumed to solve the elliptic pressure equation. The rest of the code consists essentially of a sequence of loops with a vector length equal to the product of the number of the unknowns in the x- and y-directions. Apparently, the advantages of the vector computer are decisive in the case of three-dimensional simulations, as the computational efficiency increases with increasing vector length. The efficiency improves from approximately 50 % at a vector length of 100 to about 90 % at a length of 1000. The storage demand for three-dimensional simulations was estimated to be equivalent to twenty three-dimensional arrays. If the hardware specifications are taken into account and time-consuming paging is to be avoided, the maximum number of unknowns would amount 40^3. In the case of 32x32x20 unknowns, for one time step a CPU-time of approximately one second is required.

2.4. Model application

The two-dimensional version of the model MEMO was used to simulate the steady-state flow over a single mountain for the inviscid case. For this application Coriolis terms were neglected. The topography was defined as

$$z_s(x') = z_{max} \cdot \exp\{-(x'-x_g)^2/h^2\} \qquad (2)$$

with z_{max} = 300 m , h= 3000 m and x_g = 20.5 km. Further, the stable stratification $d\bar{\Theta}/dz$ = 3.5 K/km was considered and the potential temperature at z'= 0 m was set equal to Θ_o = 300 K. The flow was assumed in the positive x'-direction. At the left boundary (x'=0) the horizontal wind velocity component was set equal to u_o = 7.5 m/s and the potential temperature was kept constant. At the right boundary (x'=L, outflow) radiative conditions were used. To allow comparisons with analytic solutions [6] the case of a constant mean density $\bar{\rho}$ = 1.0 kg/m^3 was considered. The computational domain extended horizontally to 50 km and vertically to 6 km. The grid was equidistant with a resolution of $\Delta x'$= 1000 m in the horizontal direction and a vertical grid spacing corresponding to 200 m at locations where z_s = 0 m. The calculation was terminated after the relative norm of the potential temperature change had become less than 10^{-8}.

Figs 3a and 3b show the streamlines and the potential temperature distribution, Figs 3c and 3d isotachs for the vertical velocity component w and the distribution of the pertubation pressure p' contoured at 0.2 m/s and 5 Pa, respectively. Isopleths for negative values of w and p' are dashed. The results correspond to a vertical wave length of about 4200 m. This agrees fairly well with the value $\lambda_z = 2\pi u_o/N$ = 4360 m resulting from linear theory.

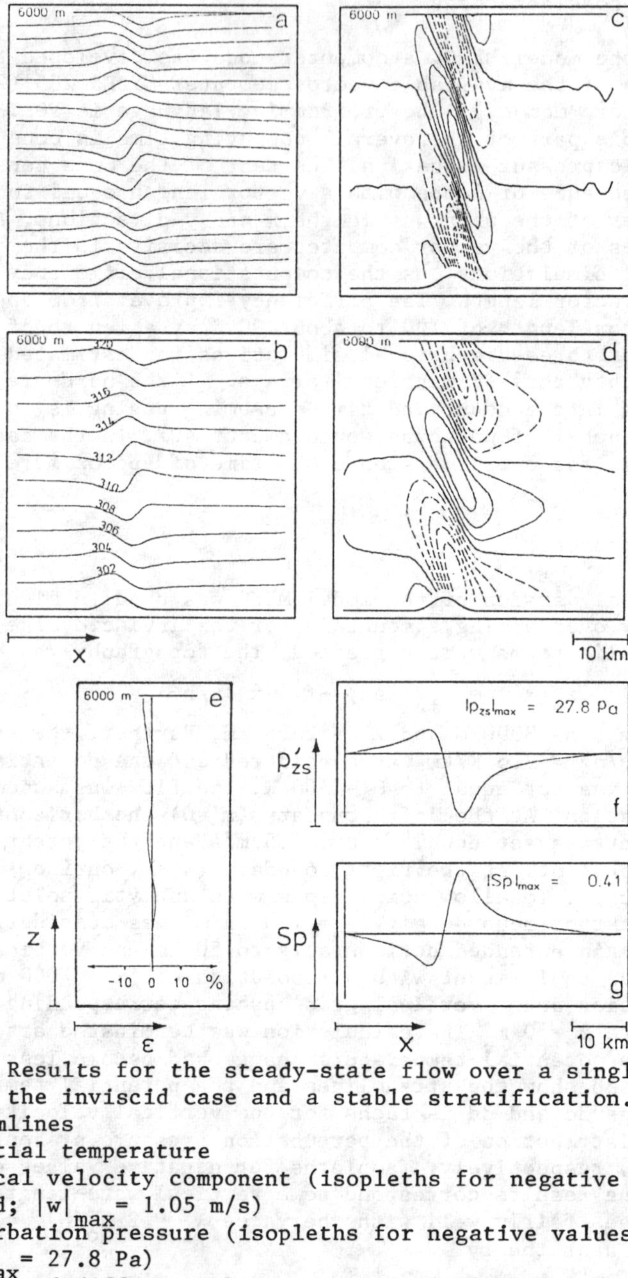

Figure 3. Results for the steady-state flow over a single mountain simulated for the inviscid case and a stable stratification.
(a) Streamlines
(b) Potential temperature
(c) Vertical velocity component (isopleths for negative values are dashed; $|w|_{max}$ = 1.05 m/s)
(d) Perturbation pressure (isopleths for negative values are dashed; $|p'|_{max}$ = 27.8 Pa)
(e) Normalized difference between vertical flux of u-momentum and wave-drag (cf. Eqs (3) and (4))
(f) Perturbation pressure at the first grid point above topography
(g) Speedup at the first grid point above topography

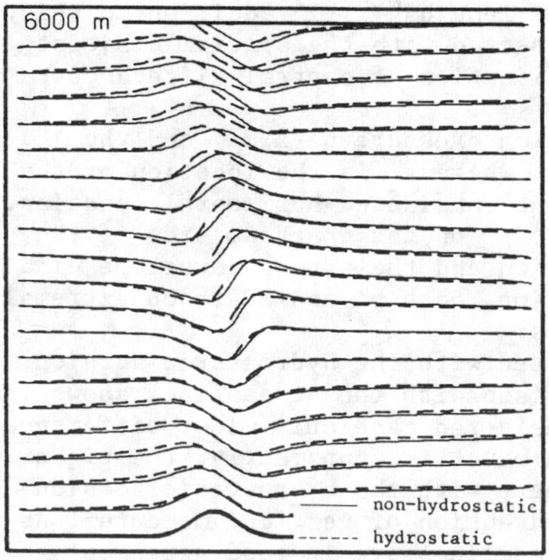

Figure 4. Streamlines calculated with the hydrostatic version of the model MEMO compared with the streamlines shown in Fig. 3a.

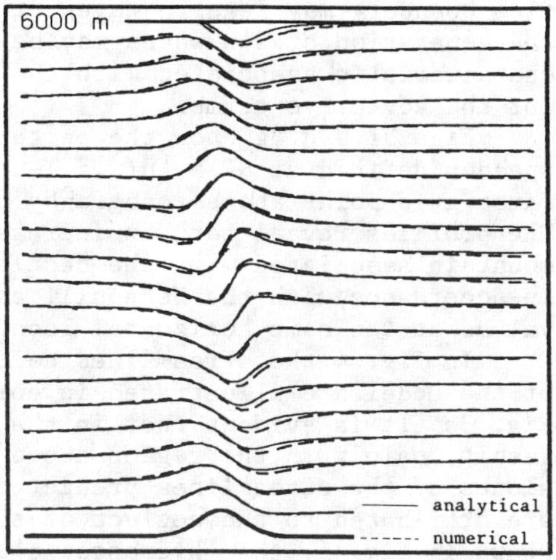

Figure 5. Streamlines calculated with the hydrostatic version of the model MEMO compared with the solution of Long's equation [6].

An additional check for the reliability of the model results consists in computing the vertical flux of u-momentum

$$MUW(z') = - \int_0^L \bar{\rho} uw \Big|_{z'} dx' \tag{3}$$

and comparing it with the the wave-drag

$$D = \int_0^L \bar{p}' \, dz_s/dx' \Big|_{z_s} dx' \tag{4}$$

Both quantities should be equal in the steady-state case if L>>h. Fig. 3e shows that the normalized difference $\varepsilon(z') = (MUW(z') - D)/D$ is well below 10 % at all height levels. The obtained value D = 9530 N/m is nearly twofold the wave-drag following from linear theory. The deviation can be related to the nonlinear treatment of the lower boundary in the elliptic equation for the pressure change and the nonlinearity of the fluid flow itself [7].

Although no high wave number oscillations can be detected from Figs 3a and 3b, the w=0-isotachs in Fig. 3c reflects 2Δx-waves, mainly near the lateral boundaries. There are several explanations for this oscillating behavior: First of all, a prerequisite for the generation of 2Δx-waves is the nonlinear advection in the momentum equation. According to Clark, prescribing a constant potential temperature profile at the inflow boundary may lead to waves of this length [7]. In addition to this, the generation of $2\Delta x$-waves may be connected with the spatial disjoint characteristic associated with the application of central differences for the advective terms.

Figs 3f and 3g show the perturbation pressure $p'(x,\hat{z}_s)$ and the speedup defined as $SP = \left(u(x,\hat{z}_s)-u_o\right)/u_o$ where $\hat{z}_s$ is the location of the first grid point above topography at the considered horizontal location. The profiles reveal both an increase of p' on the upstream side of the mountain associated with the deceleration and the strong downslope wind. In accordance with the Bernoulli equation, both p' and SP reach extremal values at the same horizontal locations.

In Fig. 4 the streamlines calculated with the hydrostatic version of the model are illustrated in comparison with the streamlines shown in Fig. 3a. It is evident that in the considered case only minor differences result, mainly in the region above the mountain. Apparently, the higher slopes of the streamlines predicted there with the hydrostatic version are attributed to the neglect of the advection of vertical momentum. As shown in Fig. 5, the 'hydrostatic' streamlines are in good agreement with the solution of Long's equation [6].

3. DISPERSION OF REACTIVE ATMOSPHERIC POLLUTANTS

In this section a photosmog model is used to describe the vertical transport and photochemical transformations of pollutants occuring in an urban airshed as a consequence of motorcar emissions. Primarily, the influence of the diurnal variation of the emission rates on the formation of ozone is investigated.

3.1. Solution algorithm

The dispersion of reactive atmospheric pollutants is described by a system of J nonlinear partial differential equations

$$\frac{\partial c}{\partial t} + \nabla(\underline{v}c) = \nabla(K_c \nabla c) + f \tag{5}$$

which is usually first transformed into terrain following coordinates. In most existing models the so-called operator splitting is adopted, i.e. the operator associated with (5) is separated into three elemental operators which are treated separately. In view of the obvious advantages of this technique, a justification for the splitting (i.e. the prove that the intrinsic error involved is negligible) is usually omitted. In fact one can argue only with the characteristic times of the processes involved: In most practical applications, advection with a characteristic time of the order of 1000 s is considerably slower than diffusion and chemistry (the characteristic time for both is of the order of 10 s). Therefore, a separate treatment of advection seems to be permissible. This is usually done by applying a suitable scheme ensuring as far as possible conservativity, positive definitness and negligible artificial diffusion. The advective transport will not be further considered in the remainder of this section, i.e. the discussion will be restricted to the case of horizontal homogeneity and vanishing vertical velocity.

Unfortunately, the characteristic times of diffusion and chemistry are of the same order of magnitude. Therefore, splitting these two operators appears to be more than questionable. In fact, this latter splitting was proved to be unnecessary, provided that chemistry would anyhow be treated implicitely [8]: The inversion of the resulting Jacobian (a $(K \cdot J)^2$-matrix, where K is the number of considered vertical levels) can be replaced by 2K-1 inversions of J^2-matrices. The corresponding computational effort is only slightly higher than the effort required by an operator splitting technique including an implicit treatment of chemistry. The model simulations presented below were performed with a fully implicit selfadaptive algorithm refraining from operator splitting.

3.2. Selection of the chemical reaction mechanism

Several reaction mechanisms are available to describe photochemical oxidant production. The most sophisticated ones require detailed emission inventories as well as extensive meteorological data. Although these mechanisms undoubtedly allow a more detailed description of the photochemical reaction system than simpler mechanisms, their application is frequently restricted by the lack of necessary input data. Detailed mechanisms require, for instance, that both the hydrocarbon emission inventory data and the ambient air quality data have the same resolution as does the chemical mechanism [9]. Unfortunately, the resolution of available data is usually unsatisfactory: Available emission inventories normally provide data only on total hydrocarbon emissions. In this case an approximative approach is unavoidable.

One possibility is to describe photochemical smog formation on the basis of a representative hydrocarbon of average reactivity, i.e. using a simple reaction mechanism. Zellner and Moussiopoulos [10], for instance, used the highly lumped-parameter reaction mechanism of Eschenroeder and Martinez ([11],henceforth EM mechanism), where propene is considered to represent the reactive part of the total hydrocarbon emissions. Although the use of such a simple reaction mechanism,especially its inclusion into a multidimensional dispersion model,would be highly desirable because of its advantageous compactness, this should not be done before proving that the simple mechanism is able to provide concentration predictions similar to those of more detailed mechanisms for the conditions of the application. From box model simulations performed using alternatively the EM mechanism and the mechanism of Atkinson, Lloyd and Winges ([9],henceforth ALW mechanism) it was concluded that if the ALW mechanism is valid, the EM mechanism may be used to obtain a reasonable impression of the diurnal variation of ozone in an urban airshed [12]. Additional simulations showed that the difference in the predictions of these two mechanisms may be considerably smaller than the difference in ozone levels resulting (in the case of the ALW mechanism) from rather small variations in the assumed subdivision of the hydrocarbon emissions [13].

An alternative of using a compact reaction mechanism is to utilize a more detailed reaction mechanism making reasonable assumptions with regard to the subdivision of the hydrocarbon emissions. In the following, a version of the ALW mechanism is used embedded in a transport model to simulate the ozone formation due to motorcar emissions in urban areas.

3.3. Sample simulation results

A fully implicit, selfadaptive transport model [8] was applied to simulate pollutant transport and oxidant formation caused by motorcar emissions in an urban area. The below given results were obtained on the basis of the Condensed ERT reaction mechanism [14], which is a slightly modified version of the ALW mechanism [9].

Simulations were performed for a typical three-day summer period. The eddy diffusivity was prescribed by an analytical expression (cf. [12]). A homogeneous air flow was considered to advect air mass over the urban area where the emissions were considered to be uniformly distributed.

The time-dependent emission rates for reactive hydrocarbons and NO_x (assumed percentage of NO in the NO_x emissions: 90 %) were calculated for an assumed diurnal variation of the traffic density and corresponding driving modes (cf. [12]) with appropriate emission factors. In this calculation the average length of traffic lanes per unit area was estimated at 7.2 km^{-1}. The resulting emission rates are shown in Fig. 7 together with the assumed volume ratios of organic compounds in the total hydrocarbon emissions derived from results of Nelson and Quigley [15] and Metz [16].

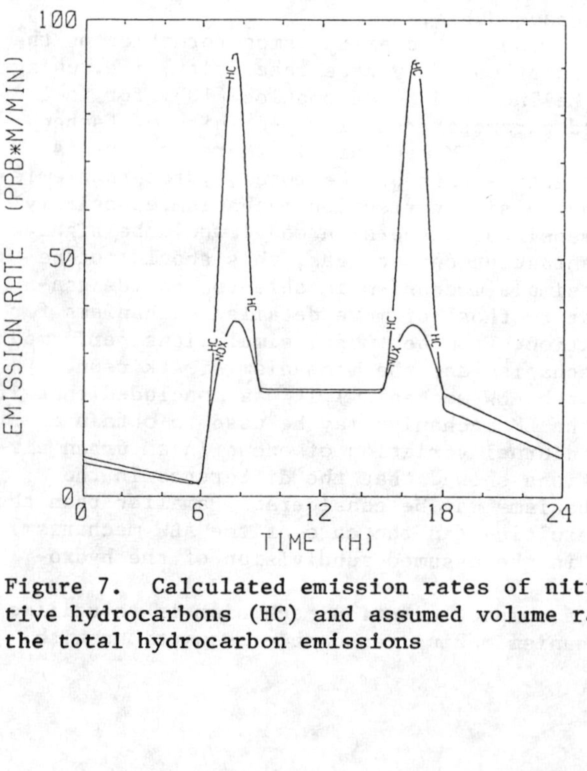

Compound	vol.-%
Propane	0.1
Alkanes (C ≧ 4)	16.5
Ethene	16.0
Propene	4.6
Butene	2.4
Benzene	4.9
Toluene	6.1
Xylenes	3.9
Formaldehyde	2.4
Higher aldehydes	1.7
Benzaldehyde	0.7
Acetone	1.8
Cresol	0.1
Methylethylketone	0.2

Figure 7. Calculated emission rates of nitrogen oxides (NO_x) and reactive hydrocarbons (HC) and assumed volume ratios of organic compounds in the total hydrocarbon emissions

All photolysis rates were assumed to vary in dependence on the solar radiation. Deposition mechanisms were ignored in the simulation as their consideration would be beyond the scope of the present paper. All initial concentrations were assumed to be negligible.

Fig. 8 shows the obtained ground level concentrations of NO_2, NO and O_3 and Fig. 9 calculated vertical profiles of the NO_2-, NO-, O_3- and HC-concentration (more precisely, RHC-concentration, i.e. the concentration of reactive hydrocarbons). Apparently, the predicted variation of the pollutant concentrations is similar to diurnal variations which are known to occur. The vertical grid consisted of twenty levels with a minimum spacing h = 20 m. For comparison, the simulation was repeated on a grid with forty vertical levels and h = 20 m. The deviations between results on the two grids are too small to be reproducible in Fig. 9. This proves the grid independence of the simulation results. From calculations of the so-called global error (cf. [8]) it was concluded that the accuracy of the numerical results is even at the end of the simulated period better than 1 ppb. The simulation results reveal successive maxima of NO, NO_2 and O_3 between 0700 and 1200 LST and sharp maxima of HC at rush-hours. The vertical profiles of the NO concentration show that before sunrise NO accumulates as a consequence of the assumed emissions and the low mixing layer depth (cf. Fig. 9). After sunrise NO is rapidly oxidized to NO_2. Apparently, higher ozone levels can be formed only after NO is almost depleted. In the second and third day of the simulation NO remains at very low levels. While the vertical profiles of the NO_2- and, especially, the HC-concentration reflect the assumed diurnal variation of the eddy diffusivity, the corresponding profiles of the ozone concentration reveal that after the rapid decrease of turbulent mixing in the evening ozone can persist in the region of low turbulence above the height of the nocturnal surface layer. The vertical profiles of the NO_2-, HC- and O_3-concentration in the third day are almost identical to corresponding profiles in the second day.

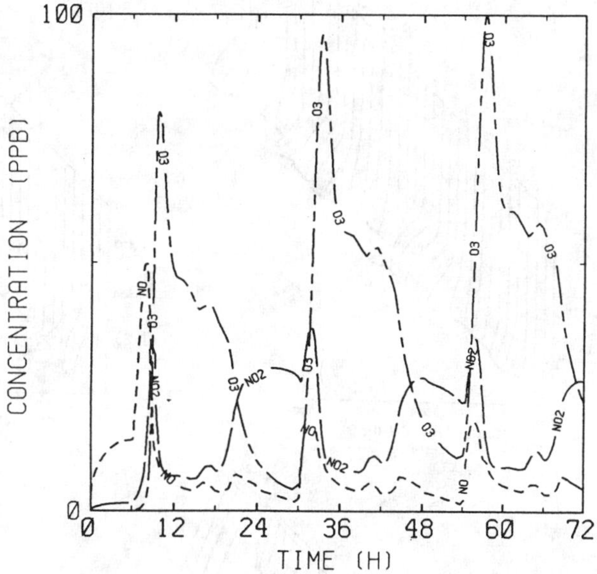

Figure 8. Ground level concentrations of NO_2, NO, and O_3 calculated for the emission rates shown in Fig. 7

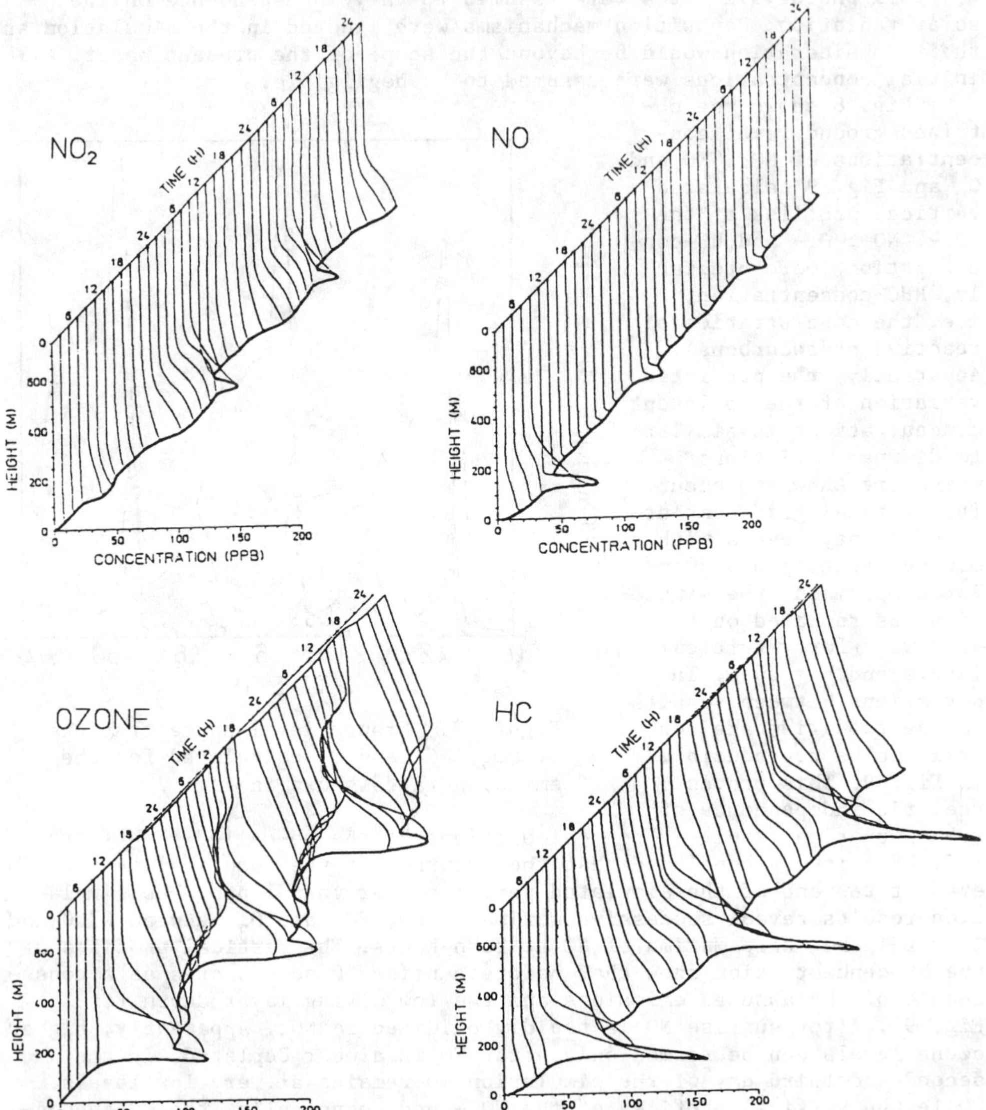

Figure 9. Vertical profiles of the NO_2-, NO-, O_3- and HC-concentration (HC stands for the reactive hydrocarbons) calculated for the emission rates shown in Fig. 7

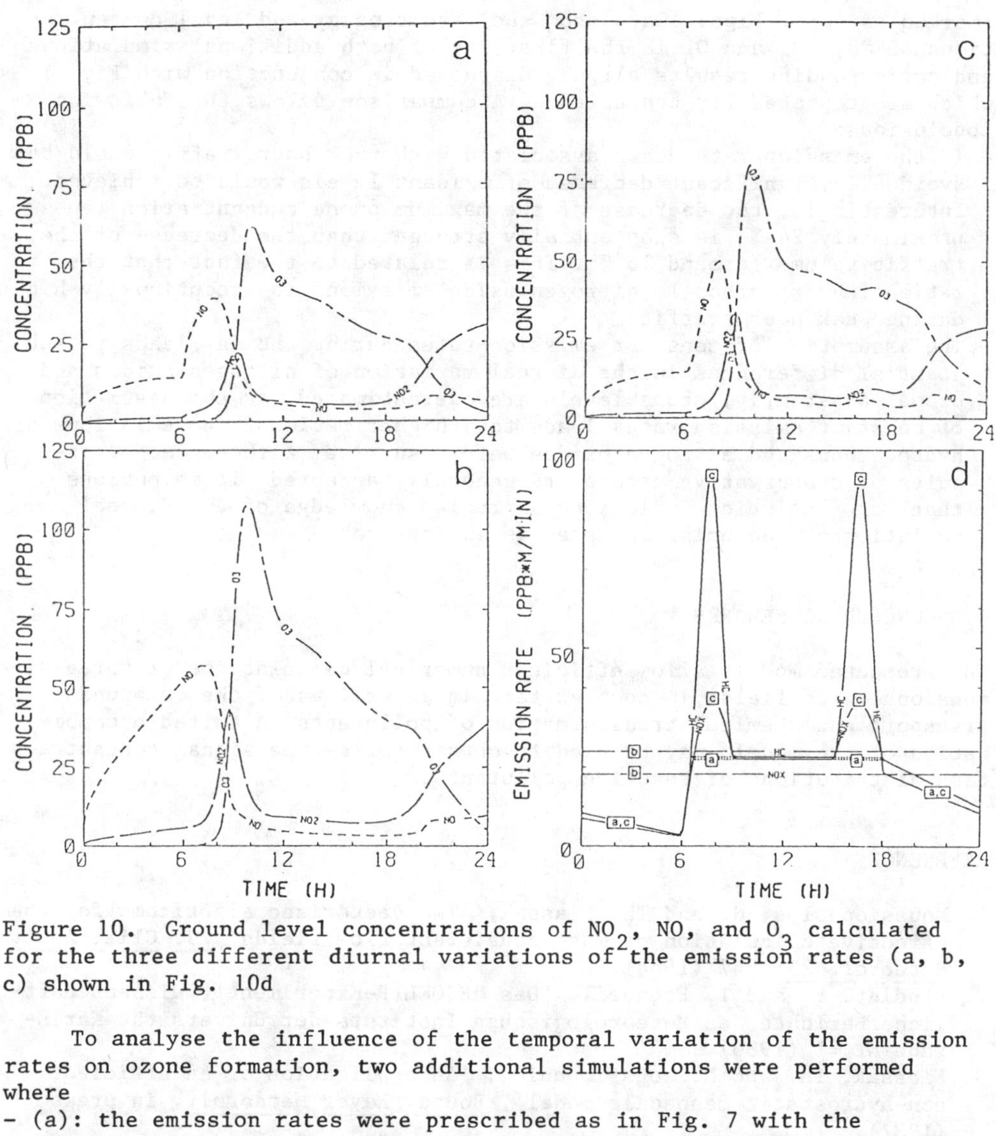

Figure 10. Ground level concentrations of NO_2, NO, and O_3 calculated for the three different diurnal variations of the emission rates (a, b, c) shown in Fig. 10d

To analyse the influence of the temporal variation of the emission rates on ozone formation, two additional simulations were performed where

- (a): the emission rates were prescribed as in Fig. 7 with the exception that constant values were assumed between 0600 and 1800 LST (i.e., driving mode 0 at 11 cars per kilometer throughout; this situation would arise if peak hour traffic would be avoided so that the traffic volume would be uniformly distributed;)
- (b): the total emissions (i.e. the time integral of the two curves in Fig. 7) remained unchanged but constant emission rates were prescribed.

The emission rates assumed in the two additional simulations are illustrated in Fig. 10d in comparison with the emission rates of Fig. 7

(marked as 'c'). Figs. 10a-c show the resulting ground level concentrations of NO, NO_2 and O_3 in the first day of both additional simulations and corresponding results already discussed in conjunction with Fig. 8 which are repeated for convenience. A comparison allows the following conclusions:

- If the emission rate peaks associated with rush hour traffic could be avoided, a significant decrease of oxidant levels would be achieved. Interestingly, the decrease of the maximum ozone concentration (approximately 24 %) is substantially stronger than the decrease of the traffic volume (around 16 %). This is related to the fact that the ratio of hydrocarbon to nitrogen oxide emissions is exceptionally high during peak hour traffic.
- The assumption of constant emission rates during the day leads to substantial differences in the diurnal variation of nitrogen oxides and ozone. Apparently, ozone levels are overestimated, as the assumption of constant emission rates leads to a higher ratio of the emissions of hydrocarbons and nitrogen oxides before sunrise. Although model results of conservative nature are generally accepted, it is obvious that for realistic simulations a precise knowledge of the diurnal variation of the emission rates is indispensable.

4. CONCLUDING REMARKS

The presented models allow efficient numerical calculations of three-dimensional wind fields in complex terrain as well as of the combined transport and chemical transformation of pollutants. A suitable combination of the models may be used to obtain three-dimensional concentration distributions of reactive pollutants.

REFERENCES

1 Moussiopoulos, N. and Th. Flassak, 'Two vectorized algorithms for the effective calculation of mass-consistent flow fields', _J. Clim. Appl. Meteorol._ _25_: 847 (1986).
2 Fiedler, F. and T. Prenosil, 'Das MESOKLIP-Experiment', Wissenschaftliche Berichte des Meteorologischen Instituts der Universität Karlsruhe Nr. 1 (1980).
3 Flassak, Th. and N. Moussiopoulos, 'An application of an efficient non-hydrostatic mesoscale model', _Bound. Layer Meteorol._, in press (1987).
4 Schumann, U. and H. Volkert, 'Three-dimensional mass- and momentum-consistent Helmholtz-equation in terrain-following coordinates', In W. Hackbusch (Ed.): _Efficient Solutions of Elliptic Systems_, Vieweg, 109 (1984).
5 Flassak, Th. and N. Moussiopoulos, 'Direct solution of the Helmholtz-equation using Fourier analysis on scalar and vector computers', submitted to the _Environ. Software_ (1987).
6 Lilly, D.K. and J.B. Klemp, 'The effects of terrain shape on non-linear hydrostatic mountain waves', _J. Fluid Mech._ _95_: 241 (1979).

7 Clark, T.L., 'A small-scale dynamic model using a terrain-following coordinate transformation', J. Comput. Phys. 24: 186 (1977).

8 Moussiopoulos, N., 'A new technique for the solution of dispersion equation systems for reactive pollutants', Bound. Layer Meteorol., in press (1987).

9 Atkinson, R. et al., 'An updated chemical mechanism for hydrocarbon/ NO_x/SO_2 photooxidations suitable for inclusion in atmospheric simulation models', Atmos. Environ. 16: 1341 (1982).

10 Zellner, K. and N. Moussiopoulos, 'Modellierung der Ozonbildung aus Automobilabgasen in Innenstadtgebieten', Arch. Met. Geophys. Bioclimat., Ser. B, 36: 115 (1986).

11 Eschenroeder, A.Q. and J.R. Martinez, 'Concepts and applications of photochemical smog models', Adv. Chem. 113: 101 (1972).

12 Zellner, K. and N. Moussiopoulos, 'Simulations of the ozone formation caused by traffic in urban areas', Atmos. Environ. 20: 1589 (1986).

13 Zellner, K. and N. Moussiopoulos, 'Photosmog modeling using two reaction mechanisms of different complexity', Theor. Appl. Climatol. 38: 107 (1987).

14 Lloyd, A.C. and F.W. Lurmann, 'Chemical transformations in air quality models', In D.D. Houghton (Ed.): Handbook of Applied Meteorology, Wiley, 767 (1985).

15 Nelson, P.F. and S.M. Quigley, 'The hydrocarbon composition of exhaust emitted from gasoline fuelled vehicles', Atmos. Environ. 18: 79 (1984).

16 Metz, N., 'Personenwagen-Abgasemissionen im Spurenbereich', Automobiltechnische Zeitschrift 86: 425 (1984).

FLOW MODELLING IN COMPLEX TERRAIN FOR ATMOSPHERIC APPLICATIONS

J.G. Bartzis
N.R.C.P.S. "Demokritos"
153 10 Aghia Paraskevi Attikis, Greece

ABSTRACT. The need for simple but reliable models for complex terrain wind flow and dispersion analysis, is a necessity today if one takes into consideration the relatively high demand in computer time and costs for such an analysis, arising mainly from the often large solution domains needed and the transient nature of the phenomena. On the other hand, taken into consideration the terrain complexities, such models need to be incorporated into suitable computer codes, that are able to treat any complex domain. Within the framework of relatively simple models qualification for atmospheric applications, a generalized three dimensional non isotropic k-ℓ model is proposed and applied to a triangular ridge problem in a wind tunnel (Arya and Shipman 1981) with encouraging results. The model shows a better performance over both the corresponding isotropic model and the supposedly more advanced k-ε model for such type of applications. The above solution has been obtained by ADREA-I code which is under development in NRCPS "DEMOKRITOS" particularly suitable for nonrectangular domains of high complexity (Bartzis 1985).

NOMENCLATURE

D_i	eddy viscosities
h	ridge height (m)
k	kinetic energy of turbulence (m^2/s^2)
ℓ_1 ℓ_2 ℓ_3	length scales of turbulence (m)
u_1,u_2,u_3	horizontal lateral and vertical velocity components (m/s)
u_*	wall friction velocity (m/s)
U	asymptotic axial velocity (m/s)
x_1,x_2,x_3	horizontal, lateral and vertical direction (m)
γ_v	volume porosity defined as the ratio of volume occupied by fluid to the total cell volume
$\gamma_1,\gamma_2,\gamma_3$	surface permeabilities defined as the ratio of the flow area to total cell area in the x_1,x_2,x_3 direction respectively
δ	boundary layer thickness (m)
δ_{ij}	Kronecker delta (δ_{ij}=0 for i≠j, δ_{ij}=1 for i=j)

K. Grefen and J. Löbel (eds.), Environmental Meteorology, 575–583.

$\Delta x_1, \Delta x_2, \Delta x_3$	cell dimensions (m)
ε	dissipation of turbulent kinetic energy (m^2/s^3)
ρ	fluid density (kg/m^3)

1. INTRODUCTION

The prediction of the flow and concentration field over complex terrain based on the fundamental laws of mass momentum and energy conservation is not always an easy task. The standard procedure is to approach the problem numerically and try to solve the resultant numerical equations. The major difficulty in this procedure besides the lack of understanding of turbulence phenomena is the relatively high computer cost. This arises from the fact that the three dimensional nature of the usually time dependent phenomena requires a relatively large grid size which combined with the often ellipting nature of the relevant equations leads to excessive CPU times in many cases. The situation becomes even worse wasthe turbulence modelling becomes more advanced resulting usually to an increasing number of transport equations to be solved. On the other hand taken into consideration the terrain complexities any candidate turbulence model needs to be incorporated into suitable computer codes that are able to treat efficiently complex domains without requiring excessive nodalization schemes.

Thus the selection of turbulence models as simple as possible incorporated into suitable computer codes able to treat efficiently geometrical complexities is a pressing necessity today.

The relatively simple and widely used concept of eddy viscosity/ diffusivity seems to be a realistic solution to the turbulence modelling problem and it has been adopted to the present approach. The particular models selected are incorporated into ADREA-I, a transient three dimensional atmospheric transport computer code under development in NRCPS "DEMOKRITOS" (Bartzis, 1985). In ADREA-I the ground is "felt" into the numerical "cell" mainly through the concept of volume porosity and surface permeabilities (Sha et al. 1978) allowing for relatively large grid size without requiring detailed description of small scale irregularities.

The present paper is part of the search towards establishment of a simple, reliable and - if possible - universal turbulent model based on the eddy viscosity/diffusivity concept applicable to any terrain configuration and atmospheric stability. This model will be implemented in the final version of ADREA-I. This particular work will be restricted to the eddy viscosity modelling in complex geometry under neutral conditions. A comparison with the axial velocity experimental data in a flow field over a triangular ridge obtained by Arya and Shipman (1981), will also be attempted.

2. EDDY VISCOSITY/DIFFUSIVITY MODELLING

A considerable amount of work exists in the open literature concerning eddy viscosity modelling in flows. The most advanced models that can be applicable to complex geometries are the ones that utilize turbulence related quantities such as turbulent kinetic energy (k) or/and turbulent energy dissipation (ε) obtained through transport equations or/and semi-empirical relations (e.g. Launder and Spalding, 1974; Bedonekar et al., 1983). In this case the eddy viscosity D is expressed as

$$D \sim \rho k^{\frac{1}{2}} \ell \tag{1}$$

The length scale ℓ is given either from a geometry related empirical relation or from a relation of the form $\ell \sim k^{3/2}/\varepsilon$.

Most of the models however even when they are applied to complex geometries assume isotropic eedy viscosity, i.e. the Reynolds stresses are given by the relation

$$\rho u_i' u_j' = - D\left(\frac{\partial u_j}{\partial x_j} + \frac{\partial u_j}{\partial x_j}\right) + \frac{2}{3}\,\delta_{ij}\rho k \tag{2}$$

However, it has been proved experimentally (Quarmby and Quirk, 1972) that eddy viscosities/diffusivities are quite anisotropic near the wall and this could be important in flow prediction looking at certain aspects of the flow (Bartzis and Todreas, 1979).

In the present paper a nonisotropic general formulation for Reynolds stresses is introduced as follows

$$\rho u_i' u_j' = -\left(D_i \frac{\partial u_j}{\partial x_i} + D_j \frac{\partial u_i}{\partial x_j}\right) + \frac{2}{3}\,\delta_{ij}\rho \cdot k \tag{3}$$

The eddy viscosities D_i can be expressed in a way similar to Eq. (1) i.e.

$$D_i = c_\mu \cdot \rho \cdot k^{\frac{1}{2}} \cdot \ell_i \tag{4}$$

Eq. (3) reflects the assumption that the anisotropy is mainly dependent on turbulence length scales rather than on velocity scales (Bartzis and Todreas, 1979). The turbulent kinetic energy k is given by the equations

$$\gamma_v \frac{\partial \rho k}{\partial t} + \frac{\partial \gamma_i \rho u_i k}{\partial x_i} = \frac{\partial}{\partial x_i}\left(\gamma_i \frac{D_i}{\sigma_k} \frac{\partial k}{\partial x_i}\right) + \gamma_v (G - \rho\varepsilon) \tag{5}$$

$$G = -\rho\, \overline{u_i' u_j'}\, \frac{\partial u_i}{\partial x_j} \tag{5a}$$

In the <u>1-Eq model</u> the energy dissipation is given by the empirical relation

$$\varepsilon = c_D \frac{k^{3/2}}{\ell_o}$$

$$\ell_o = \min(\ell_1, \ell_2, \ell_3) \tag{7}$$

The values of the constants used in the present calculations are

$$c_\mu = 0.2, \quad c_D = 0.3, \quad \sigma_k = 1.0 \tag{8}$$

In the 2-Eq model the energy dissipation is given by the transport equation (Benodekar et al. (1983))

$$\gamma_v \frac{\partial \rho \varepsilon}{\partial t} + \frac{\partial \gamma_i \rho u_i \varepsilon}{\partial} = \frac{\partial}{\partial x_i} (\gamma_i \frac{D_i}{\sigma_\varepsilon} \frac{\partial \varepsilon}{\partial y_i}) + \gamma_v \frac{\varepsilon}{k} (c_1 G - c_2 \varepsilon) \tag{9}$$

In this case the expressions for eddy viscosities are modified as fo follows

$$D_i = c_\varepsilon \rho \frac{k^2}{\varepsilon} \frac{\ell_i}{\ell_o} \tag{10}$$

The values of the constants used in the present calculations are

$$c_\varepsilon = 0.09, \quad c_1 = 1.44, \quad c_2 = 1.92, \quad \sigma_\varepsilon = 1.2 \tag{11}$$

2.1 The length scales

The expression (4) for eddy viscosity requires some knowledge of the length scales ℓ_i. Trying to keep the length scales model as simple and simultaneously realistic as possible the following expressions are introduced for ℓ_i.

$$\ell_i = \min(\ell_{1as}, \ell_{2as}, \ell_{3as}, \min(x_i', \ell_{ias})) \tag{12}$$

$$\ell_{ias} = c_{\ell i} L_i$$

where x_i' is the distance along the x_i-axis from the nearest solid surface

L_i is the distance along the x_i-axis between solid surfaces that crosses the point into consideration,

and $c_{\ell i}$ constant

The above model underlines

3.2.1 Step 1: The one dimensional undisturbed flow. Calculations have been performed with both 1-Eq Model and 2-Eq Model and the mean velocity comparisons are shown in Fig. 1. The widely considered more advanced 2-Eq Model gave comparatively less accurate results. It seems that the 2-Eq Model has the tendency to produce "fully developed" velocity profile which is not the case for this experiment and will never be the case for the atmospheric applications. This behavior can be attributed to the fact that the local length scale - a natural parameter for turbulence description - does not enter to the model directly. With respect to the 1-Eq model, calculations have been performed with $c_{\ell i}$=0.4, 0.5 and 1.0 and the results are shown in Fig. 1.

The values have been selected on the ground that

- the $c_{\ell i}$=0.4 is the typical value given in the open literature,
- the $c_{\ell i}$=1.0 corresponds to the length scale upper limit,
- the $c_{\ell i}$=0.5 gives the best experimental fit.

The slight dependence of the mean velocity profile to the $c_{\ell i}$ value, one of the most uncertain parameters of the model can be classified as one of the advantages of the present 1-Eq model.

3.2.2 Step 2: The flow field over the ridge. The calculations have been performed with $c_{\ell i}$=0.5 for the reason that it gives the best results in the step 1 analysis.

The predicted flow pattern in the region near the ridge is shown in Fig. 3. The results seem rather reasonable.

The axial velocity predictions in the downstream region is shown in Figs 4-7. The error is better than 20% based on local values which should be considered rather satisfactory taken into consideration possible boundary condition inaccuracies due to the fact that the present solution domain is drawn out from an integral wind tunnel test. In the above figures is also shown the better prediction of the non-isotropic model compared with the isotropic one, especially in the near the wall region where the most interest is for dispersion calculations. In fact the introduction of the non-isotropy has reduced the axial velocity prediction maximum error by 10% (i.e. from 30% to 20%).

The largest discrepancy occurs in a limited region very close to the sharp edge (Fig. 2) which was not unexpected due to the geometry high complexity and the relatively rough grid at that region.

4. CONCLUSIONS

1. A new generalized three dimensional nonisotropic form of the eddy viscosities and the Reynolds stresses based on the k- model have been introduced and applied to a flow over a triangular ridge (Arya and Shipman, 1981) with rather satisfactory results in the downstream region.

2. The present model shows a better performance over the supposedly more advanced k-ε model.

3. A comparison between the nonisotropic model and isotropic one as those have been applied in the present paper, shows that the nonisotropic model shows better agreement with experimental data in the downstream region.

3. The ADREA-I code used for the above calculations is proved to be a rather reliable tool for analysis of the flow field over nonrectangular blockages such as the above mentioned triangular ridge.

ACKNOWLEDGEMENTS. The author would like to thank Prof. S.P.S. Arya for his kind cooperation to furnish the requested information regarding the triangular ridge experiment. The author also would like to thank the European Communities Directorate-General for Science, Research and Development (Directorate for Biology, Radiation Protection and Medical Research) for partial support of the project.

5. REFERENCES

Arya S.P.S. and Shipman M.S. (1981). An experimental investigation of flow and diffusion in the disturbed boundary layer over a ridge-I. Mean flow and turbulence structure. Atmospheric Environment, 15, 1173-1184.

Arya S.P.S. (1985) Private Communication.

Bartzis J.G. (1985) ADREA-I a transient three dimensional transport code for atmospheric and other applications. Some preliminary results. DEMO 85/3 NRC "Demokritos", Aghia Paraskevi, Greece.

Bartzis J.G., Todreas N.E. (1979) Turbulence Modelling of Axial Flow in a Bare Rod Bundle, Journal of Heat Transfer, 101, 628-634.

Benodekar, R.W., Goddard, A.J.H., Gosman, A.D. and Issa, R.I. (1983). Numerical prediction of turbulent flow over surface mounted ribs, ASME paper No. 83-Fe-13 presented at Fluids Engineering Conference, Houston, Texas, June 20-22.

Quarmby A., Quirk R. (1972). Measurements of the radial and tangential eddy diffusivities of heat and mass in turbulent flow in a plain tube, Int. J. Heat Mass Transfer, 15, 2309-2317.

Launder B.E. and Spalding D.B. (1974). The numerical computation of turbulent flows. Comp. Meth. in Appl. Mech. and Engng., 3, 269-289.

Sha W.T., Domanus H.M., Schmitt R.C., Oras J.J., Lin E.I.H. (1978). COMMIX-1: A three dimensional transient single-phase component computer program for thermal-hydraulic analysis. NUREG/CR-01415, Argonne National Laboratory, USA.

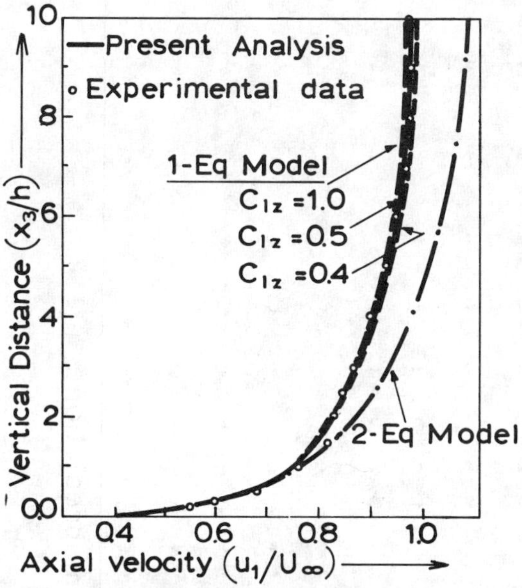

Fig. 1. Axial velocity profile at $x_1/h = -11.0$

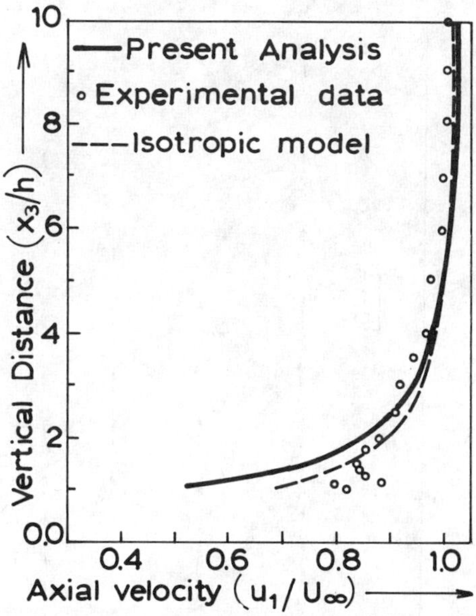

Fig. 2. Axial velocity profile at $x_1/h = 0.0$

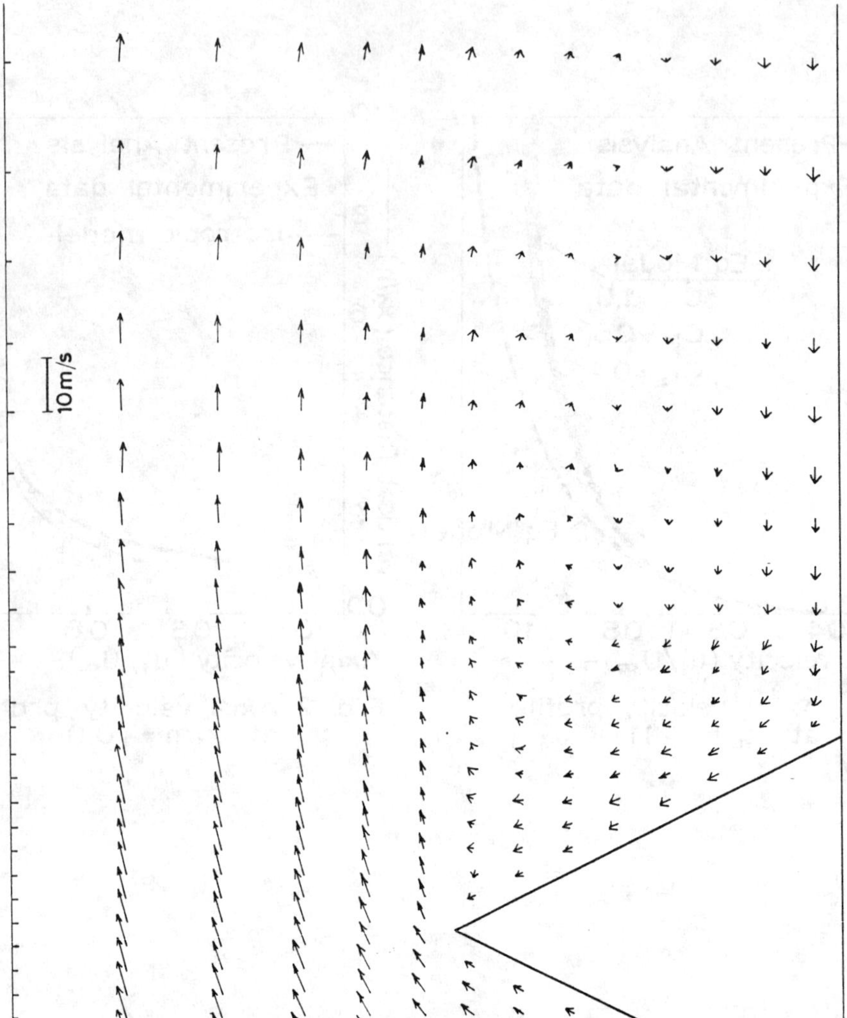

Fig. 3. The flow pattern prediction.

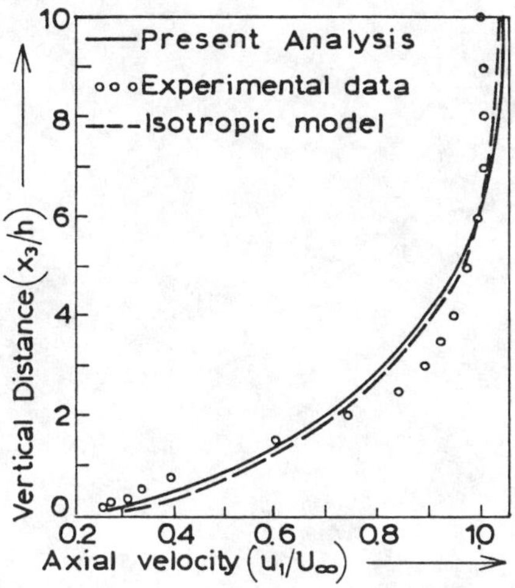

Fig. 4. Axial velocity profile at $x_1/h = 8.0$

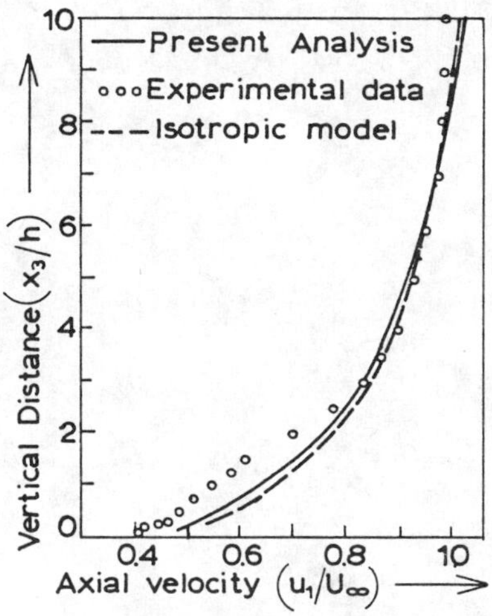

Fig. 5. Axial velocity profile at $x_1/h = 19.0$

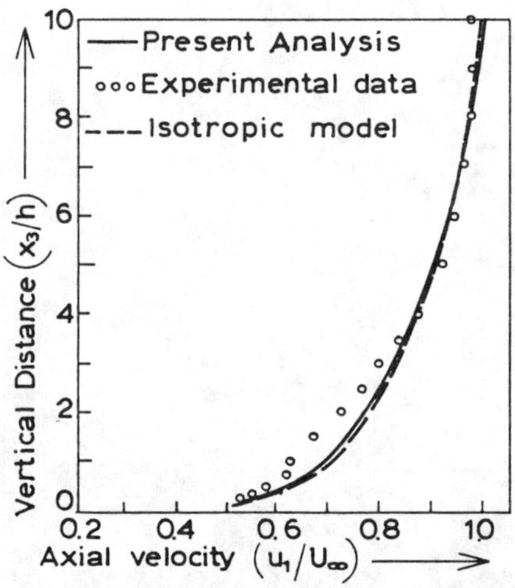

Fig. 6. Axial velocity profile at $x_1/h = 29.0$

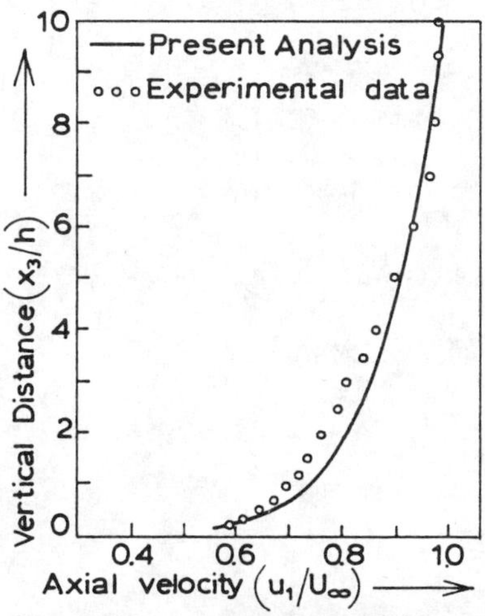

Fig. 7 Axial velocity profile at $x_1/h = 60.0$

APPLICATION OF LONG-RANGE TRANSPORT MODELS IN THE FRAMEWORK OF CONTROL STRATEGIES: EXAMPLE OF PHOTOCHEMICAL AIR POLLUTION

J. Pankrath [*], R. Stern [**], P. Builtjes [***]

* Federal Environmental Agency, Bismarckplatz 1, D-1000 Berlin 33,

** Free University of Berlin, Thielallee 50, D-1000 Berlin 33

*** TNO, P.O. Box 342, NL-7300 AH Apeldoorn

ABSTRACT. Dispersion models, if properly designed and evaluated, can give substantial support to the development of control strategies. This holds essentially true in view of the large scale adverse effects attributed to air pollution impacts, and of the complex non-linear source-receptor relationships. As a typical example photochemical air pollution, with ozone as a major pollutant, is caused by the combined emissions of nitrogen oxide (NO_x) and volatile reactive hydrocarbons (RHC). Photochemistry during multiday episodes, which regularly occurs in northwestern Central Europe is determined by transport, chemical conversion and deposition over extended areas. In the dutch-german PHOXA project (Photochemical Oxidant and Acid Deposition Model Application in the Framework of Control Strategy Development) calculations have been performed for several photochemical episodes using the Regional Transport Model (RTM III). The RTM-III-model is an Eulerian grid-model based on the numerical solution of a 3 1/2 layer multi-component diffusion equations. The chemical mechanism used is the Carbon Bond-IV mechanism. The model evaluated for a 5 day episode in 1980 shows a reasonable agreement between measured and calculated O_3-concentrations.

Irrespective still existing insufficiencies in the emissions input data bases, sensitivity scenarios give the following preliminary indications: Hydrocarbon control would lower peak ozone concentrations everywhere in the modeling region. NO_x control decreases higher ozone levels in most cases, ozone increase is also possible, however, depending on the history of an air mass which arrives at a receptor area. A combined NO_x/RHC control seems to be most favourable in reducing high ozone levels.

1. INTRODUCTION

It is a well-known fact now that photochemical oxidants are produced, transported and deposited on continental scales. Observations made both in North America and in Europe demonstrated the regional extend of photooxidant episodes. The large scale distribution of secondary pollutants like ozone calls for adequate control strategies to reduce the relevant precursors nitrogen oxides (NO_x) and volatile reactive hydro-

K. Grefen and J. Löbel (eds.), Environmental Meteorology, 585–611.

carbons (RHC). An effective abatement of this secondary air pollution cannot be obtained by local emission reduction in industrialized regions or even in one country alone. A combined nationwide abatement strategy on a large area is required.

In order to determine the effectivity of abatement measures and to support envisaged international control strategy developments, long range dispersion models have to be developed and applied.

Dispersion models are the only tool for quantitative simulations and for evaluating emission scenarios.

The Dutch-German project PHOXA [(1)], which started in 1984 is especially designed for this purpose. PHOXA is the agronym of "Photochemical Oxidants and Acid Deposition Model Application within the Framework of Control Strategy Development". The project consists of the three branches [(2)]

- Photochemical Oxidant Modelling
- Acid Deposition Modelling
- Data Bases

In the following we will confine ourselves to the photochemical oxidant model application which consists - for its special application to episodic situations - of the three main elements:

- Application and evaluation of the Regional Transport Model (RTM III) - developed by Systems Applications (USA) - to simulate the transport and chemical formation of photochemical oxidants during pollution episodes
- Establishment of data bases to support the application of the regional transport model for photochemical oxidants
- Application of the RTM III-model to study the effects of emission control scenarios.

The reason for selecting the RTM-III model was due to the outcome of the 1982 EPA-OECD Conference on 'Long Range Transport Models for Photochemical Oxidants and their Precursors', where there was an agreement on this model as the only available model suitable for practical application at that time [(3)].
If models in PHOXA were put in order of a hierarchy, RTM-III would be between the highly complex model TADAP (German-Canadian model development) [(4)] and the simple models with linear chemistry.

2. MODEL DESCRIPTION

The long range photochemical dispersion model RTM-III was used to simulate several episodes in North Western Europe, which were selected according to significant maximum hourly ozone concentrations at monitoring stations in the PHOXA area. Calculations have been carried out for the following photochemical oxidant episodes:

- 22 july - 26 july 1980
- 29 may - 2 june 1982
- 3 june - 6 june 1982

The general feature of the model results will be discussed for the 5-day summer episode of 1980 in more detail.

2.1 Modelling region

The modelling area of PHOXA lies geographically in longitude between 10 ° west and 24 ° east and in latitude between 47.30 ° and 60 ° north. This area of 3.129 Mio km² contains nearly all the industrialized regions in Europe north of the Alps.

The horizontal resolution in the modeling area is prescribed by selecting grid elements of 1/4 ° longitude and 1/2 ° latitude, i.e. it is about 30 km x 30 km. The vertical resolution is given by the concept of dynamic layers within the RTM III Eulerian grid model [5, 6]. The time resolution of model calculations is 1 hr.

2.2 Meteorological conditions for photochemical episodes

The occurrence of elevated ozone concentrations in air masses which have passed over significant precursor source areas is determined largely by the prevailing meteorology [7]. Several meteorological conditions must simultaneously exist before high concentrations of ozone are likely to be produced:

- high intensity of incoming solar ultraviolet radiation; this condition implies that clear skies are most favourable for ozone formation;
- temperatures must be reasonably high (greater than 20 °C); this influences the rate of many chemical reactions, and also the emission rate of precursors, most notably that of evaporative anthropogenic and natural emissions of hydrocarbons;
- surface windspeeds must be fairly low (equal or less than 5 m/s) to allow a significant build-up of precursor concentrations as the transported air passes over source areas;
- the height of the daytime boundary layer during photochemical episodes is usually constrained by an elevated temperature inversion at about 1 - 2 km. Turbulent eddies, driven largely by convection in the boundary layer ensure rapid mixing of precursor emissions into the layer;
- at nighttime radiative cooling at the ground usually leads to the establishment of a ground based inversion layer with a height of several tenths to some hundreds of meters. Below this inversion ozone is rapidly destroyed by ground deposition and reaction with nitric oxide (NO). Above the inversion the polluted air is cut off from the turbulent effects of the surface and so very little vertical mixing occurs. The polluted air masses remain aloft in a "reservoir" layer until they are incorporated in the next day's convective boundary layer. The nocturnal inversion therefore protects the bulk of the polluted air from dry deposition, and effectively prolongs the lifetime of ozone;
- the boundary layer must persist for the time required to generate high levels of ozone, otherwise the polluted air will be dispersed into the free troposphere.

The general weather situation during the selected episode of 22 - 26 july, 1980 can be characterized by a high-pressure system moving slowly over north-western Europe and then turning to northern Europe. The average surface wind speed decreased during the episode from 5 m/s to 3 m/s and the wind direction changed from northwest to south. The maximum mixing height during the episode was about 1500 m in the central part of the PHOXA area; the temperature maximum reached 30 °C. Some precipitation occured in the northeast. A peak hourly ozone concentration of about 160 ppb (as compared to the annual average over Europe of 40 ppb) was observed on 26 july 1980 in the Netherlands. In fact, the meteorological conditions for a boundary layer photochemical episode were realized.

2.3 RTM-III model structure

The vertical layers of the RTM-III-model are defined as follows (fig. 1):

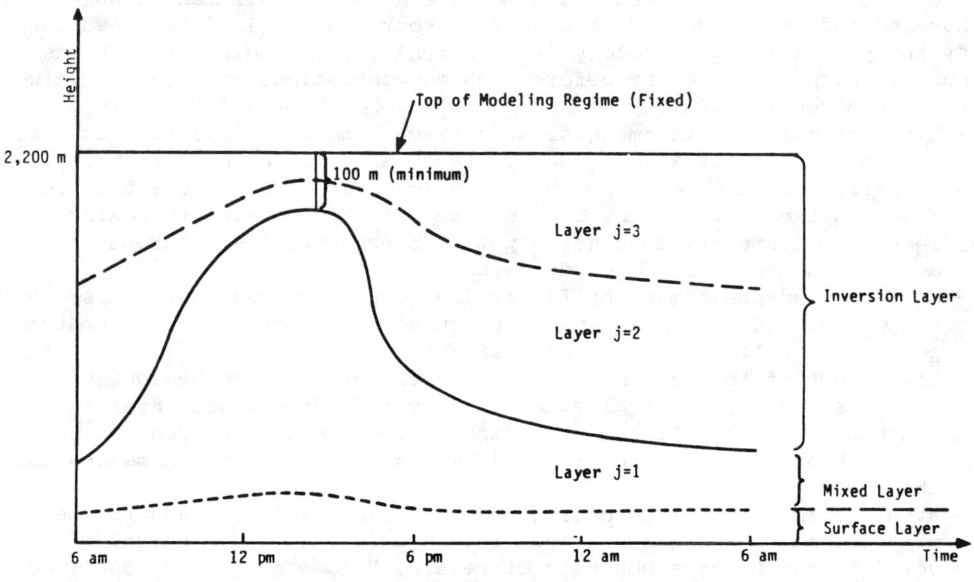

Fig. 1: Diurnal variation of vertical layers in RTM-III

(1) A surface layer immediately above the ground representing the constant-flux Prandtl layer;
(2) the mixed layer extends to the base of the elevated inversion; its thickness can vary both temporally and spatially;
(3) a lower inversion layer is immediately above the base of the inversion; the purpose of this layer is to keep track of emissions injected into the stable layer during the night and subsequently released to the mixed layer under daytime fumigation conditions;
(4) the upper inversion layer extends to a constant top height of approximately 2500 m; this layer serves as a reservoir for emissions penetrating the upper part of the whole inversion layer and for aged pollutants entrained into this layer during multiday episodes

The mixed layer and the lower and upper inversion layers are prognostic layers where transport and chemical reactions occur. The surface layer is a diagnostic, parameterized layer which allows the calculation of the surface depletion of pollutants.

2.4 Model equation

The model equations, on a Eulerian grid system, are based on the quasi-three-dimensional, time-dependent atmospheric diffusion equation for multiple chemical species (6)

$$\frac{\partial C_i}{\partial t} + U\frac{\partial C_i}{\partial x} + V\frac{\partial C_i}{\partial y} = \frac{\partial}{\partial x}(K_x\frac{\partial C_i}{\partial y}) + \frac{\partial}{\partial y}(K_y\frac{\partial C_i}{\partial x}) + E_i + Q_i + R_i - D_i - W_i$$

for each prognostic vertical layer. The subscript i refers to the pollutant species.

u, v, K_x, K_y are the horizontal wind velocities and turbulent diffusivities in the x and y directions, respectively.

Entrainment and detrainment of pollutant mass due to the rise and fall of the inversion and the interfacial transport due to large-scale convergence or divergence in the boundary layer is specified by the term E_i. The source term, Q_i, consists of a gridded array of area sources whose emissions are dispersed into the mixed layer; emissions from major point sources enter the mixed layer or one of the inversion layers as determined by a plume rise algorithm. Production and loss of pollutant species due to either gas- phase or aerosol reactions are represented generally by R_i. The sink terms, D_i and W_i, represent dry and wet deposition, respectively, of both gaseous pollutants and aerosols onto the ground surface.

2.5 The carbon bond meachanism for photochemical reactions

The following pollutant species are treated with the atmospheric diffusion equation (tab. 1)

Table 1: Species ordering and definitions used in the RTM III model

number	name	description
1	NO, NO_2	nitrogen oxides
2	O_3	ozone
3	MGLY	methyl glyoxal
4	ETH	ethylene type carbon bonds
5	OLE	olefine type carbon bonds
6	PAR	paraffine type carbon bonds
7	ALD2	higher aldehydes
8	FORM	formaldeyhde
9	XYL	xylene and other multi-alkylbenzenes
10	TOL	toluene and other mono-alkylbenzenes
11	PAN	peroxyacetyl nitrate
12	CO	carbon monoxide
13	PHEN	aromatic phenols
14	HONO	nitrous acid
15	SO_2	sulfur dioxide
16	SULF	sulfates

The chemical kinetics simulation package built into RTM-III uses a slightly modified version of the Carbon Bond Mechanism-IV (CBM-IV) to describe changes in the concentration of various species due to their chemical reactions. The CBM employs the concept of molecular structures for determining organic reactivity[8]. Instead of lumping classes of molecules, as is the case of other generalized mechanisms, the CBM subdivides molecules into reactivity structures whose chemical properties are generally similar, even when they appear in different molecules. Condensation of CBM-IV was possible by confining to oxidant formation and disregarding gas-phase precursors to cloud acidification processes. The resulting mechanism for the above mentioned 16 state species - those that are tracked from one time step to another and whose mass is conserved - is summarized in table 2. A comparison of recent chemical mechanisms (9) including the CBM-IV reaction scheme produced results which were similar for ozone. This suggests some confidence in the basic understanding of the atmospheric chemistry involved in the production of some photochemical pollutants.

Table 2: The chemical kinetic mechanism used in RTM-III

	Reaction		Rate Constant	Activation Energy
1	NO_2	$= NO + O$	vary (k_1)	
2	O	$= O_3$	4.65×10^6	-690.
3	$O_3 + NO$	$= NO_2$	26.6	1430.
4	$O + NO_2$	$= NO$	13800.	
5	$O + NO_2$	$= NO_3$	2320.	-600.
6	$O + NO$	$= NO_2$	3120.	-411.
7	$NO_2 + O_3$	$= NO_3$	0.0474	2450.
8	O_3	$= 0.205\ OH + 0.897\ O$	vary	
9	O_3	$= O$	$0.042 \times k_1$	
10	$O_3 + OH$	$= HO_2$	100.	940.
11	$O_3 + HO_2$	$= OH$	3.0	580.
12	$NO_3 + NO$	$= 2.\ NO_2$	28100.	-250.
13	$NO_3 + NO_2$	$= NO + NO_2$	0.59	1230.
14	$NO_3 + NO_2$	$= N_2O_5$	1780.	60.
15	N_2O_5	$=$	0.0285	
16	N_2O_5	$= NO_3 + NO_2$	3.12	10840.
17	$NO + NO_2$	$= 2.\ HNO_2$	2.4×10^{-7}	
18	$HNO_2 + HNO_2$	$= NO + NO_2$	1.5×10^{-5}	
19	HNO_2	$= NO + OH$	$0.18 \times k_1$	
20	$NO_2 + OH$	$=$	16300.	-560.
21	$NO + OH$	$= HNO_2$	9770.	-427.
22	$HO_2 + NO$	$= OH + NO_2$	12300.	-240.
23	$NO + NO$	$= 2.\ NO_2$	1.52×10^{-4}	-530.
24	$OH + HNO_2$	$= NO_2$	9770.	
25	NO_3	$= 0.85\ NO_2 + 0.85\ O + 0.15\ NO$	$30.6 \times k_1$	
26	$HO_2 + HO_2$	$=$	4144.	-1150.
27	$HO_2 + HO_2$	$=$	3270.	-5800.
28	$OH + CO$	$= HO_2$	400.	

	Reaction		Rate Constant	Activation Energy
29	FORM + OH	$= HO_2 + CO$	15000.	
30	FORM	$= 2.\ HO_2 + CO$	vary	
31	FORM	$= CO$	vary	
32	FORM + O	$= OH + HO_2 + CO$	237.	1550.
33	$FORM + NO_3$	$= HO_2 + CO$	0.93	
34	ALD2 + O	$= C_2O_3 + OH$	636.	986.
35	ALD2 + OH	$= C_2O_3$	24000.	-250.
36	$ALD2 + NO_3$	$= C_2O_3$	3.7	
37	ALD2	$= XO_2 + 2.\ HO_2 + CO + FORM$	vary	
38	$ALD2 + HO_2$	$= XO_2 + 2.\ FORM + HO_2$	5.	
39	$C_2O_3 + NO$	$= NO_2 + XO_2 + FORM + HO_2$	16500.	-250.
40	$C_2O_3 + NO_2$	$= PAN$	9000.	-250.
41	PAN	$= C_2O_3 + NO_2$	0.0222	14000.
42	$C_2O_3 + C_2O_3$	$= 2.\ XO_2 + 2.\ FORM + 2.\ HO_2$	3700.	
43	$C_2O_3 + HO_2$	=	9600.	
44	MGLY	$= C_2O_3 + HO_2 + CO$	$0.02 \times k_1$	
45	MGLY + OH	$= XO_2 + C_2O_3$	26000.	
46	OH	$= XO_2 + FORM + HO_2$	21.	
47	PAR + OH	$= 1.49\ XO_2 + 0.067\ XO_2N + 0.93\ HO_2 + 0.45\ ALD2 - 0.75\ PAR$	1150.0	
48	O + OLE	$= 0.95\ ALD2 + 0.35\ HO_2 + 0.20\ XO_2 + 0.15\ CO + 0.05\ FORM + 0.05\ C_2O_3 - 0.35\ PAR$	5920.	324.
49	OH + OLE	$= FORM + ALD2 + XO_2 + HO_2 - PAR$	42000.	-537.
50	$O_3 + OLE$	$= 0.5\ ALD2 + 0.66\ FORM + 0.212\ CO + 0.28\ HO_2 + 0.144\ XO_2 + 0.08\ OH - PAR$	0.018	1897.
51	$NO_3 + OLE$	$= 0.91\ XO_2 + 0.91\ HO_2 + 0.09\ XO_2N$	11.4	
52	O + ETH	$= FORM + XO_2 + CO + 2.\ HO_2$	1080.	800.
53	OH + ETH	$= XO_2 + 2.\ FORM + HO_2$	12000.	-382.
54	$O_3 + ETH$	$= FORM + 0.37\ CO + 0.13\ HO_2$	0.0027	2840.
55	TOL + OH	$= HO_2 + 0.64\ XO_2 + 1.13\ FORM + 0.56\ MGLY + 0.36\ PHEN + 0.36\ PAR + 1.13\ CO$	9750.	
56	$PHEN + NO_3$	$= PHO$	14000	
57	$PHO + NO_2$	=	20000.	
58	XYL + OH	$= HO_2 + 0.72\ XO_2 + 0.67\ CO + 1.33\ MGLY + 0.28\ PHEN + 0.67\ FORM + 0.56\ PAR$	36000.	
59	$XO_2N + NO$	=	1000.	
60	$XO_2 + NO$	$= NO_2$	12000.	
61	$XO_2 + HO_2$	=	5000.	-1300.

	Reaction		Rate Constant	Activation Energy
62	XO_2 + C_2O_3	= XO_2 + HO_2 + FORM	2400.	
63	SO_2 + OH	= HO_2 + SULF	1500.	
64	SO_2	= SULF 0.000083		

Remarks:
Reactants in parentheses (M, O_2, and CH_4) are included in k_{298} values.

CH_4 is assumed to be 1.85 ppm, M is 1 x 10^6 ppm, and O_2 is 209000 ppm.

Rate constants are in ppm and min units at one atmosphere pressure.

$$k = k_{298} \exp\left[- \frac{E}{R} \left(\frac{1}{T} - \frac{1}{298} \right) \right]$$

where k_{298} is the reaction rate at 298 K and T is the temperature in degrees Kelvin.

Reactions with no or non-stoichiometric products are supposed to yield species which are of no further importance for the kinetic mechanisms and are therefore totally ignored here.

The RTM-III internally calculates photolytic reaction rates for NO_2, O_3, formaldehyde and higher aldehydes based on

- solar zenith angle
- total opaque sky cover
- average height above ground for the layer in question.

2.6 Dry deposition

In the RTM-III model, dry deposition removal processes consist of turbulent diffusion of the pollutant through a turbulent surface layer and a laminar sublayer, followed by absorption or adsorption at the atmosphere/ground interface. The diurnal variation of temperature in the surface layer affects the vertical pollutant distribution through atmospheric stabilities and consequently affects the rate of surface uptake of pollutants. The pollutant flux can be calculated by taking into account:

- pollutant flux through the turbulent layer:

  $$F = K_v \left(\frac{DC}{DZ} \right)_{Z = Zo}$$

  The vertical diffusivity depends on the friction velocity U_* and Monin-Obukhov length L; empirical expressions for stable, neutral and unstable conditions are taken from the literature;
- pollutant flux across the laminar sublayer:

  $$F = \beta U_* (C_o - C_s)$$

  This flux depends on the concentration difference between interface (O) and surface (s), the friction velocity U_*, and the resistance β for mass transfer;
- pollutant uptake by chemical reactions at soil or vegetation surfaces (s)

$F = \gamma C_s$

A reaction of first order is assumed here, γ is a reaction rate constant for each pollutant.

2.7 Wet deposition

Wet deposition of gaseous species in RTM III is obtained by equilibrium and mass balance between the liquid phase and gas phase components of the species. Given the grid average concentration, the grid average rainfall rate and the top of the modeling region, the wet scavenging rate can be obtained whereby species dependent HENRY's law constants and first or second order dissociation constants are used.

3. DATA REQUIREMENTS

The RTM III model needs the following input data bases:

- Emissions
- Meteorological data
- Air quality initial and boundary conditions
- Surface characteristics/land use categories

3.1 Emissions data

The emissions data have to be prepared in the form of point source and area source for the species listed in tab. 3.

Table 3: List of species for the emission input in RTM III

Nitrogen oxides	$NO + NO_2$
Carbon monoxide	CO
Ethene	ETH
Olefins	OLE
Parafins	PAR
Aldehyde	ALD2
Formaldehyde	FORM
Xylene	XYL
Toluene	TOL
Sulfur dioxide	SO_2
Particulate sulphate	SULF

Total emissions in the PHOXA region for 1980 with a split up into some source categories are given in table 4 [(10)].

CO total:	47.5	Gt CO/yr	
NO_x total:	12.5	Gt NO_2/yr	
RHC total:	9	Gt RHC/yr	(nonmethane RHC)
SO_2 total:	31	Gt SO_2/yr	

Source categories	CO	NO_x	RHC	SO_2 (Percentage)
Power stations	.	30	.	36
Refineries	.	1	2	3
Chemical industries	.	3	1	2
Other industries	7	26	0,5	42
Non-industrial sources	50	5	37	16
Traffic	43	35	41	1
Natural sources	.	.	18,5	.

Table 4: Total emissions in the PHOXA region. 1G t = 10^9 kg. Because of lack of information the category 'other industry' may contain also parts of chemical industry and others for the Eastern European countries. Most of the RHC in the category 'non-industrial sources' are solvent emissions with industrial and non-industrial usage

Gridded NO_x - and RHC-emissions for the PHOXA region in 1980 are given in fig. 2 - 3. The required accuracy for yearly averaged total emissions per country in view of its use for long range dispersion modeling should be for NO_x: $\pm$ 20 %, RHC: $\pm$ 30 % and SO_2: $\pm$ 10 % [11]. This accuracy seems to be achievable for the SO_x - and NO_x -emissions data base, for the RHC- emissions the uncertainty is obviously much larger than 30 %. There is an ongoing debate about the quality of RHC data.

The PHOXA emission file 1980 for the RTM-III model application in the PHOXA region contains 3400 gridded area sources and about 7000 point sources with stacks heights higher than 50 m. Smaller stacks are treated as area sources.

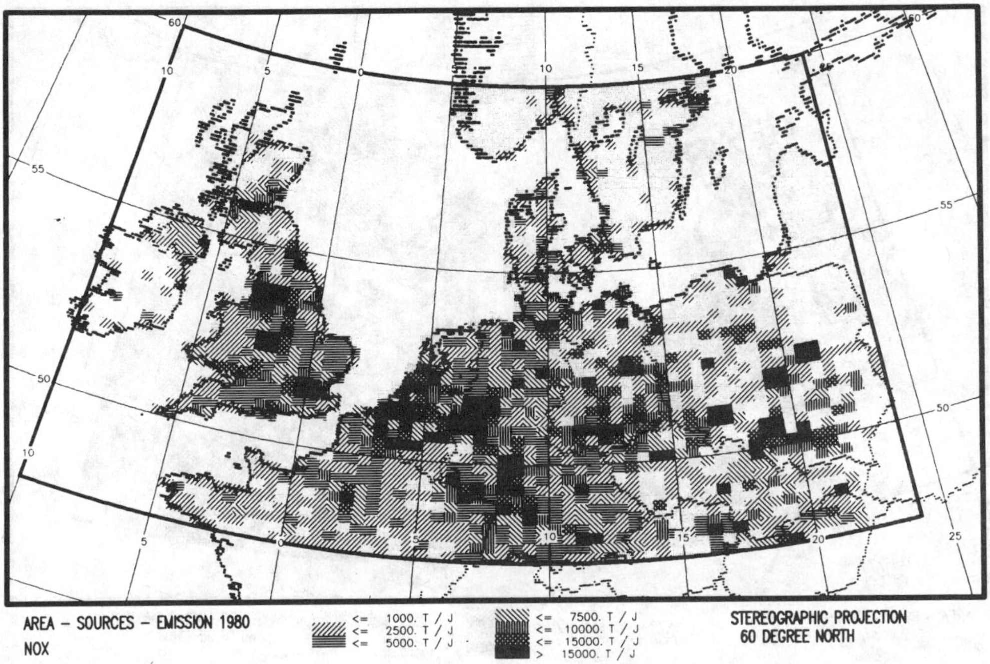

Fig. 2: NO_x-emissions from area sources in t/yr for 1980

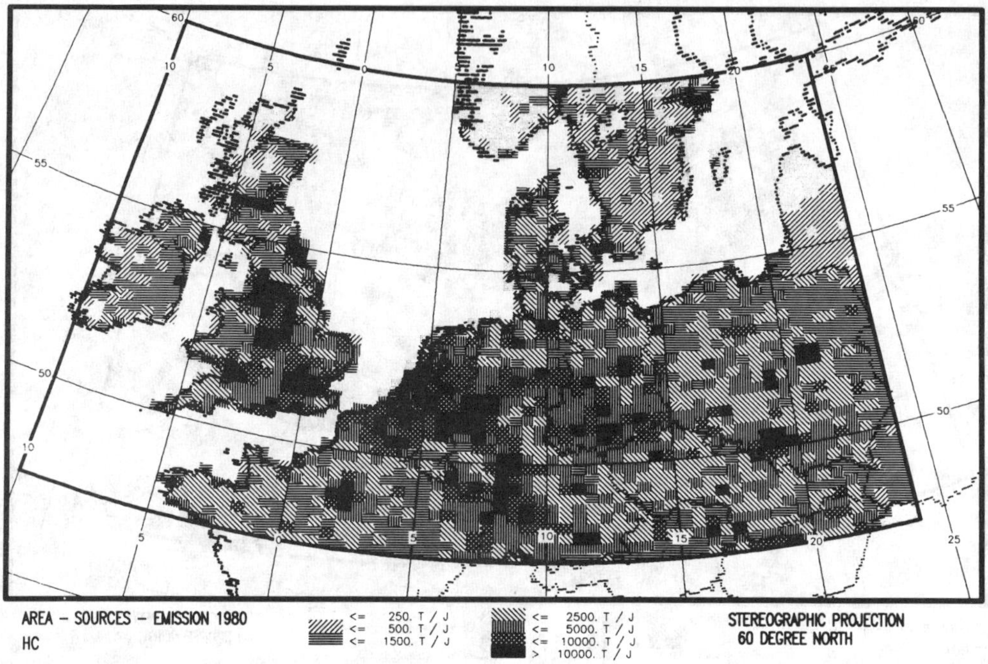

Fig. 3: Reactive Hydrocarbon emissions (RHC) from area sources in t/yr for 1980

To apply the RTM-III model an episode specific emissions data base is required. This data base is derived from the yearly averaged emissions data by considering the temperature and time dependencies of the different source groups. Temperature dependency involves mainly HC-emissions due to evaporation and natural emissions, and, to a lesser extent NO_x-emissions from road traffic and heating consumption.

As an example for the episode under consideration NO_x- and RHC-emissions during the course of July 22, 1980 are shown in fig. 4 and fig. 5. In table 5 the HC emissions are split up into hydrocarbon classes as needed by the CBM-IV mechanism of the RTM-III model.

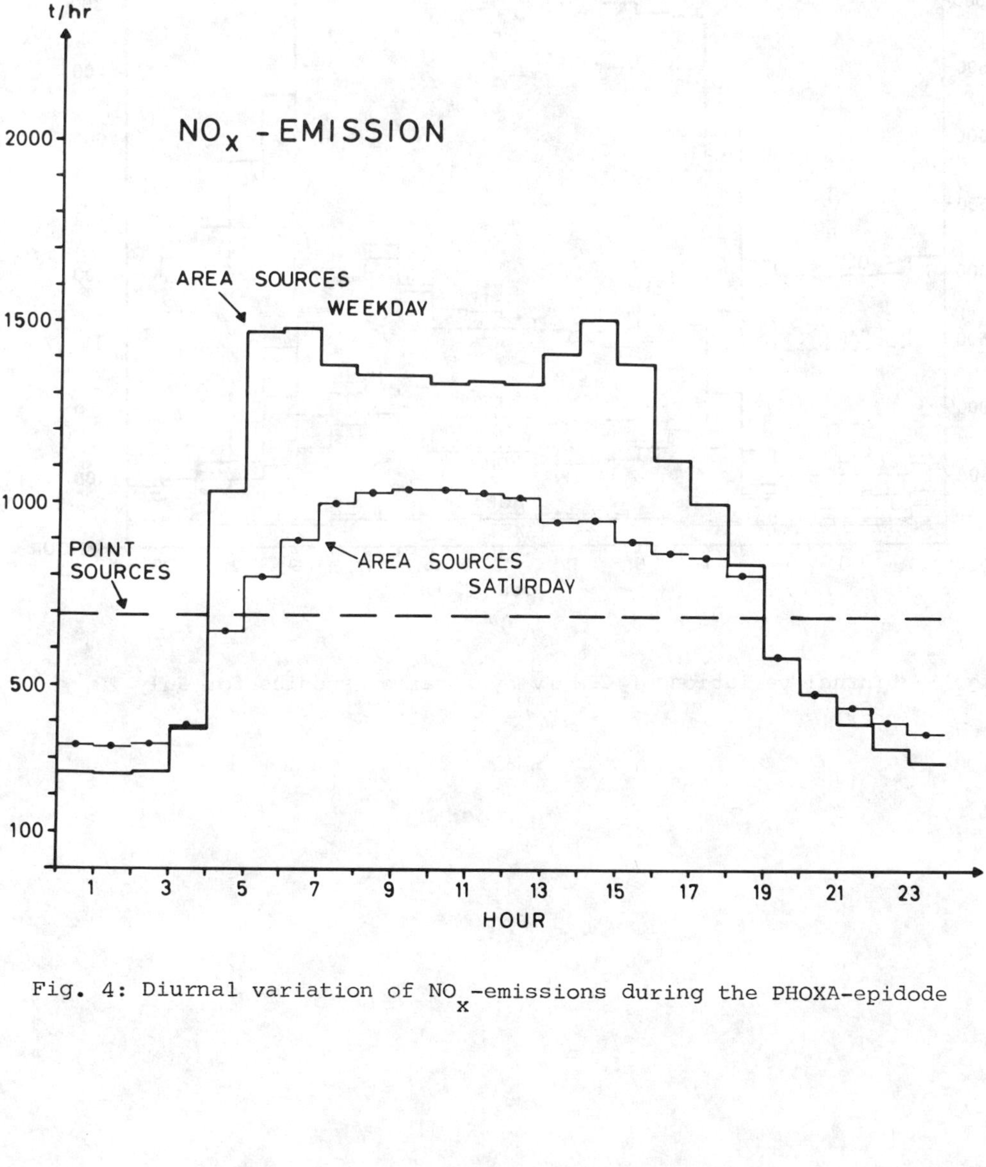

Fig. 4: Diurnal variation of NO_x-emissions during the PHOXA-epidode

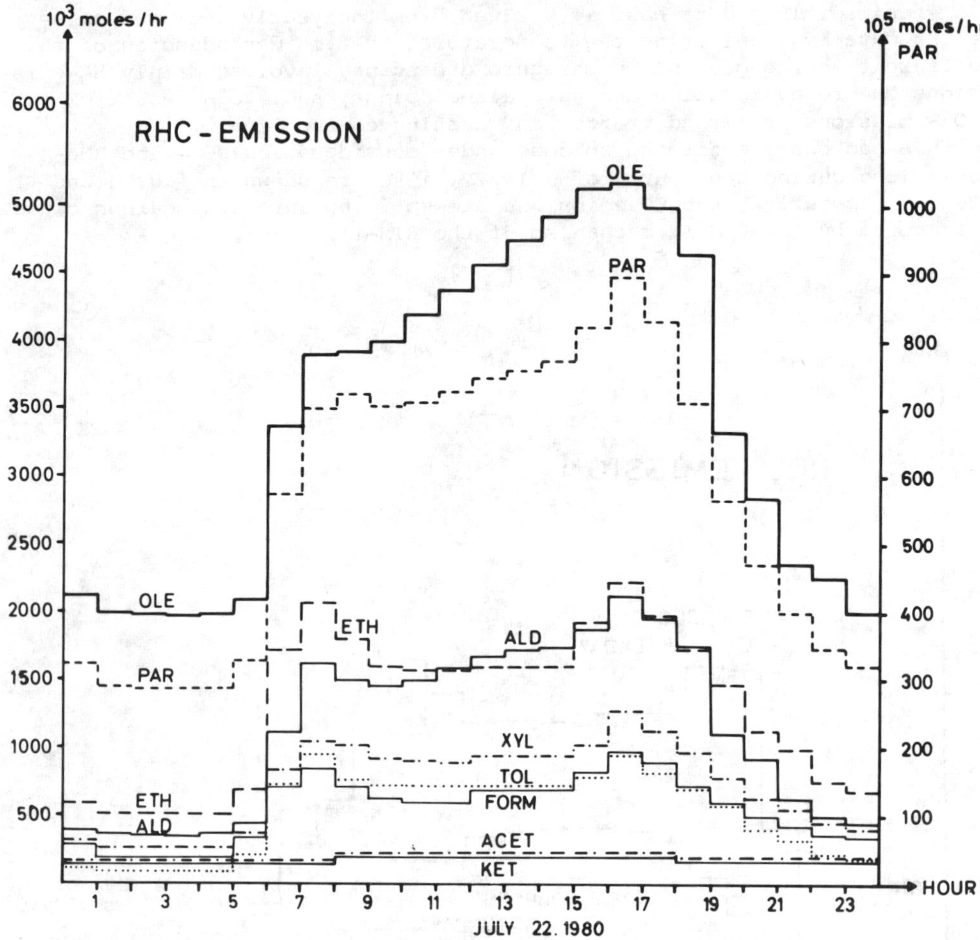

Fig. 5: Diurnal variation of CBM-IV-hydrocarbon species for July 22, 1980

Species	Emissions kt/d	RHC split %	Natural emissions %
NO_x	33.50		
CO	83.75		
RHC	37.94	100	
ETH	0.86	2.3	.
OLE	4.42	11.7	92
PAR	27.41	72.2	56
ALD2	2.12	5.6	14
FORM	0.17	0.4	.
XYL	1.77	4.7	.
TOL	1.19	3.1	.

Table 5: Split up of daily HC emissions in the CBM-IV hydrocarbon classes (July 26, 1980). KET and ACET were treated as slowly reacting PAR.

Special attention has to be drawn to natural hydrocarbon emissions from forests which depend crucially on air temperature:

The maximum temperature rose during the episode from approximately 24 °C on July 22 to approximately 30 °C on July 26; at this day it has to be assumed that natural hydrocarbons are contributing with 54 % to emissions of total hydrocarbons. Looking into the relevant components the values are even more striking saying that for July 26 natural contributions were to OLE 92 %, to PAR 56 % and to ALD 14 %.

3.2 Meteorological data

Meteorological data were supplied to the RTM III model at three-hour intervals consisting of three layers of gridded fields of u and v transport wind components, layer heights, layer-average temperature and water vapour concentration, as well as a single-layer of gridded fields of surface wind speed, precipitation amounts, and fraction of cloud cover.

Mixing height fields were generated with the Air Mass Transformation Model of KNMI [(12)] combined with radio sonde data. The layer averaged wind fields were derived from the upper air data analyses of the ECMWF in combination with interpolation schemes. Eddy diffusivities and mass flux entrainment terms are derived from this input. Surface data, cloud cover and precipitation fields were derived from observational network data and applied statistical interpolation schemes. The resultant cloud cover and precipitation fields were then compared to each other and with satellite pictures to assure consistency.

As an example, fig. 6 shows wind velocities averaged throughout the mixed layer depth and cloud cover for 26 july 1980, 15.00 UTC.

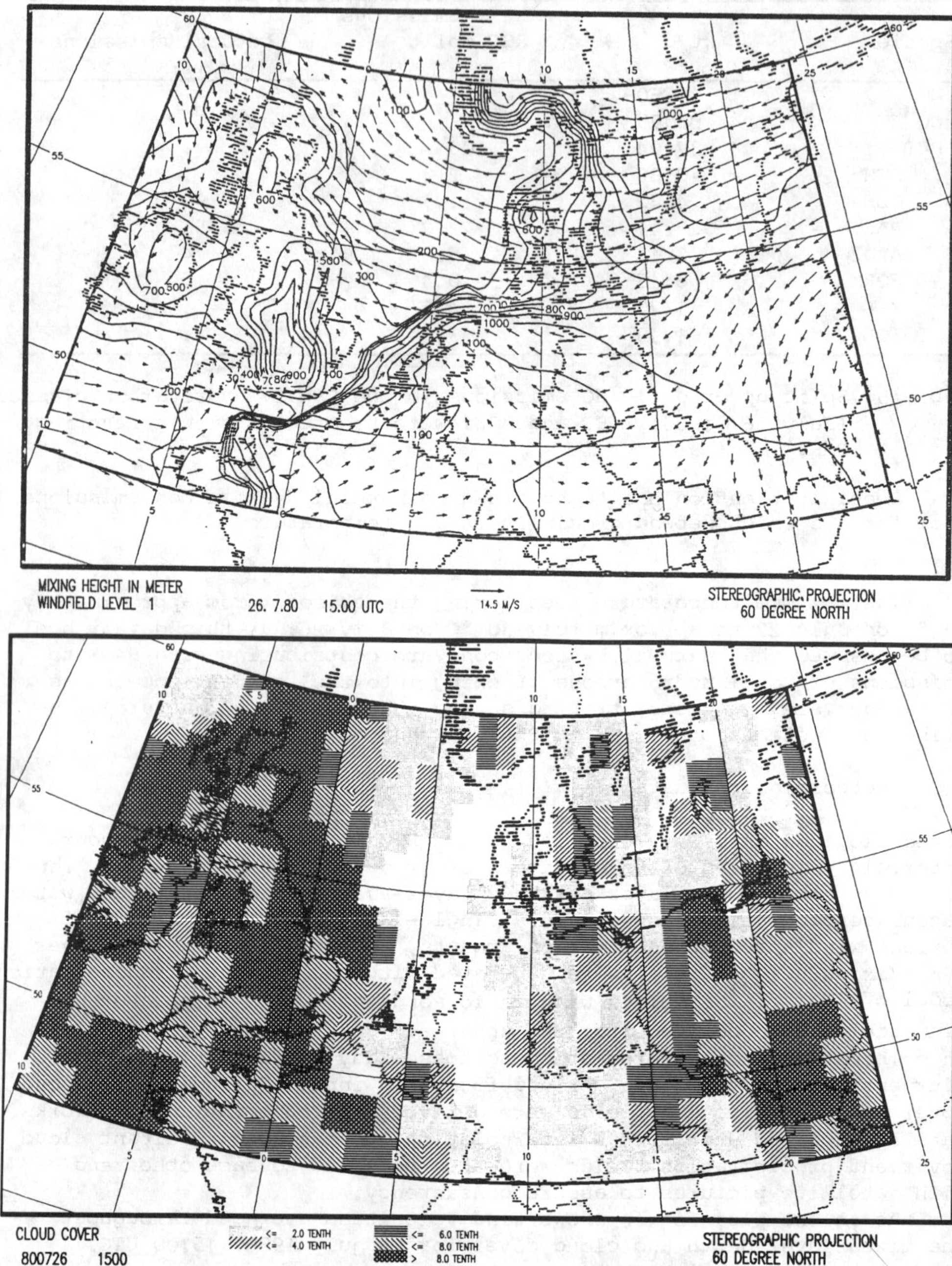

Fig. 6: Layer averaged wind field for the mixing layer and mixing heights and cloud cover in tenths for July 26, 1980, 15:00 UTC. White areas are cloud free.

3.3 Initial and boundary conditions

The initial conditions for the application of the RTM-III model were obtained by running the first day of the episode, 22 July 1980, for 24 hours, and then using the resultant pollutant distribution to initialize the model once more on the same day. Due to the sparseness of the air quality network, and the lack of any elevated air quality measurements or measured concentrations of the photochemical oxidant precursors, it was felt that this initialization procedure provided a better estimate of the initial three-dimensional pollutant distribution than an interpolation procedure.

Boundary values were generated by NILU in the EMEP grid using a lagrangian long range transport model with atmospheric boundary layer chemistry [13], that simulated air parcels arriving every 6 hours at 44 points along the boundary throughout the 5-day episode. The chemistry scheme employed was the extended version of the CBM-IV, the CBM-X-mechanism. These data were interpolated linearly along the boundary cells of the embedded PHOXA grid, and linearly in time to the 3-hour update intervals.

The concentrations above the region top, which will enter the modeling region when divergent winds are present, were assumed to be those of clean air and given in tab. 6.

Table 6: Concentrations aloft used for the PHOXA application of the RTM-III

Species	Aloft concentration (ppb)
NO_x	1
NO_2	40
MGLY	$2 \cdot 10^{-6}$
ETH	0.683
OLE	0.341
PAR	34.6
ALD	1.023
FORM	3.41
XYL	0.199
TOL	0.260
PAN	0.1
CO	400
PHEN	$2 \cdot 10^{-3}$
HONO	$2 \cdot 10^{-3}$
SO_2	0.7
SO_4	0.5

3.4 Surface characteristics

For the calculation of dry deposition fluxes to the ground the RTM-III model needs surface data from land use data of the PHOXA modeling region. This land use data set consists of 10 land use categories (tab. 7); associated with each land use category are

- surface roughness, zo
- first-order surface reaction rate constant, γ, for each chemical species.

For each grid cell the calculation of zo and γ takes into account the fraction of coverage of land use types.

Table 7: Land use categories and associated roughness length and first-order surface reaction rate constants

Land use category	Surface roughness Zo (cm)	First-order surface reaction rate constant γ (cm/s)					
		O_2	NO	NO_2	CO	SO_2	$SO_4^=$
water	0.02	0.1	0.1	0.6	0.0005	1.1	0.5
cropland, arable land	25	1.0	0.1	0.6	0.0005	0.5	0.2
meadows, grassland	3	1.2	0.1	1.2	0.0006	0.7	0.2
permanent crop	25	1.2	0.1	1.2	0.0006	0.7	0.2
built-up areas	100	2.0	0.1	1.8	0.005	1.5	0.3
deciduous forest	100	1.5	0.1	1.5	0.003	1.5	0.5
coniferous forest	100	1.5	0.1	1.5	0.003	1.5	0.5
mixed forests	100	1.5	0.1	1.5	0.003	1.5	0.5
bare soil	0.5	1.0	0.1	0.6	0.0005	0.5	0.2
wetland	0.5	1.0	0.1	1.0	0.0008	1.0	0.4

As an example fig. 7 shows the apportionment of all types of forest over the gridded PHOXA region.

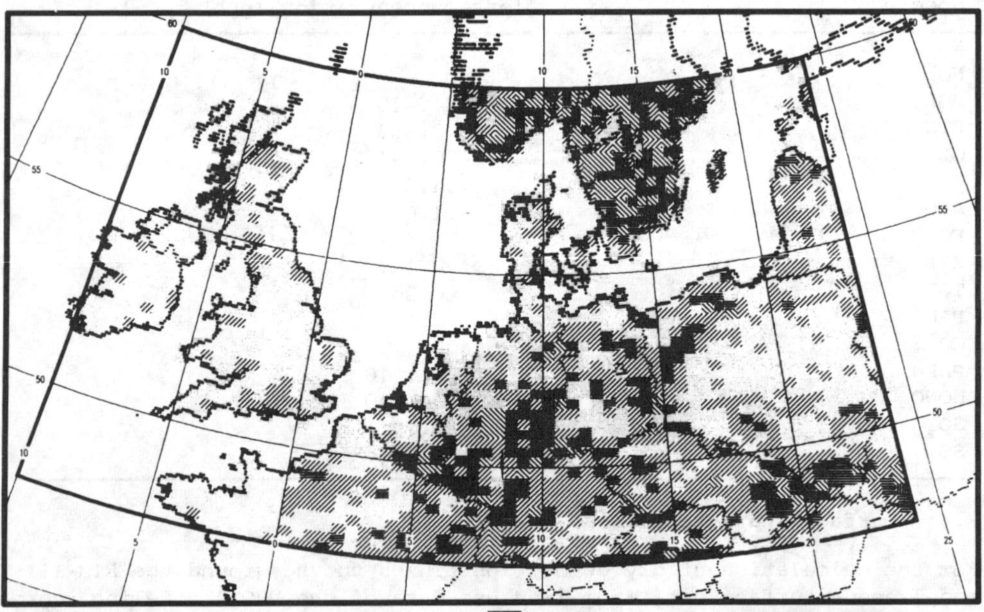

Fig. 7: Apportionment of all types of forest over the gridded PHOXA-area

4. APPLICATION OF RTM-III FOR A PHOTOCHEMICAL EPISODE

The RTM III model was used to simulate the 5-day ozone episode 22 - 26 july 1980, selected by PHOXA [14]. The prediction of precursor concentrations, NO_x and hydrocarbons, and the prediction of ozone will be focussed on the patterns of the last day of the episode, where the highest ozone levels were observed and also the best agreement between measured and calculated ozone concentrations was found (fig. 8).

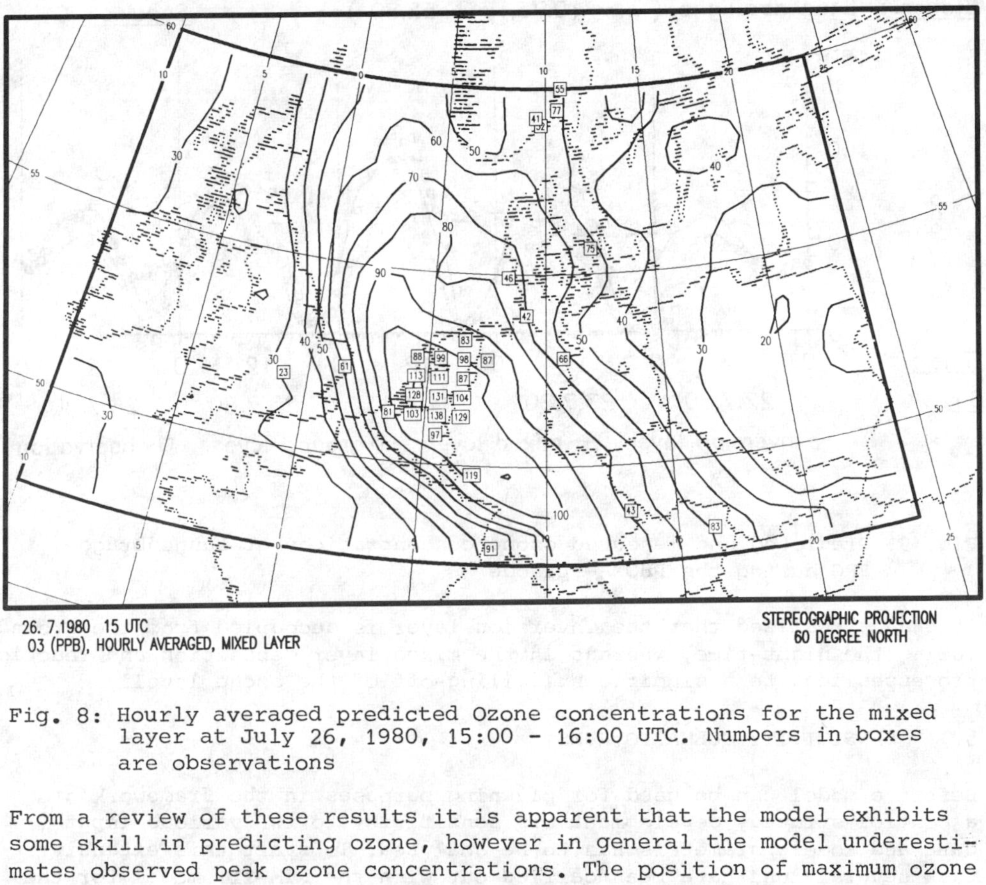

Fig. 8: Hourly averaged predicted Ozone concentrations for the mixed layer at July 26, 1980, 15:00 - 16:00 UTC. Numbers in boxes are observations

From a review of these results it is apparent that the model exhibits some skill in predicting ozone, however in general the model underestimates observed peak ozone concentrations. The position of maximum ozone concentration levels and the gradients of ozone isolines correspond with the few measurements. Of course point measurements and area averaged calculations are rigorously not comparable.

The comparison of predicted and measured mixed layer and inversion layer ozone concentrations at the monitoring site Langenbrügge, West Germany, displays remarkably good agreement (fig. 9); in general, time series of other monitoring stations show similar features, except for stations in the Antwerp-Rhine area.

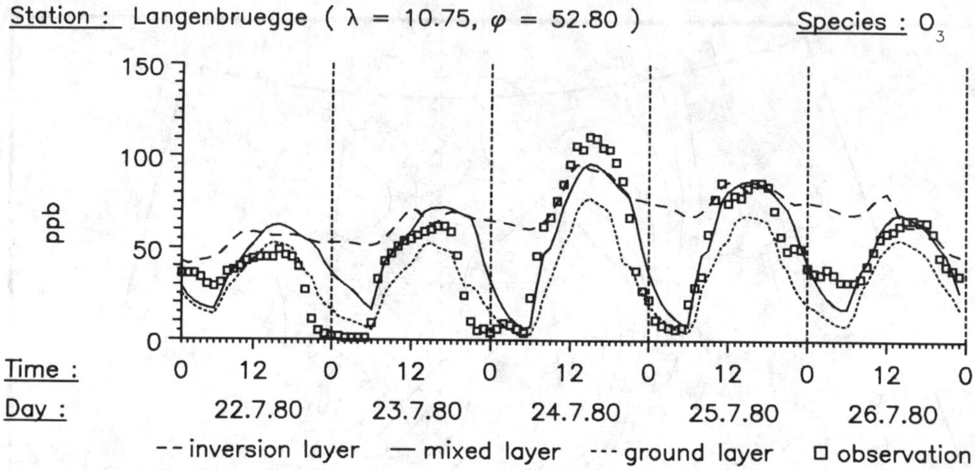

Fig. 9: Predicted and measured ozone concentrations at Langenbrügge, FRG during the PHOXA-episode

It can be realized that the inversion layer is decoupled from the ground during the night-time, whereas in the mixed layer deposition and reaction processes lead to a significant falling-off of the ozone level.

5. SENSITIVITY SCENARIO RUNS

Before a model can be used for planning purposes in the framework of a control strategy development its sensitivity to the various input data and model parameters has to be analysed. To start this extensive task several runs have been carried out with the RTM-III model for the 1980 episode[15] to determine the model's response to broad changes in precursor emissions. These are the following sensitivity scenarios:

(1) NO_x emissions unchanged, RHC emissions reduced by 50 %
(2) NO_x emissions reduced by 50 %, RHC emissions unchanged
(3) NO_x and RHC emissions reduced by 50 %

Alteration of emission levels have been made for all emissions, anthropogenic and natural, and for the whole PHOXA region with the same percentage. All other input data remained unaltered.

The result of the model calculations is shown in fig. 10 - 12 [16].

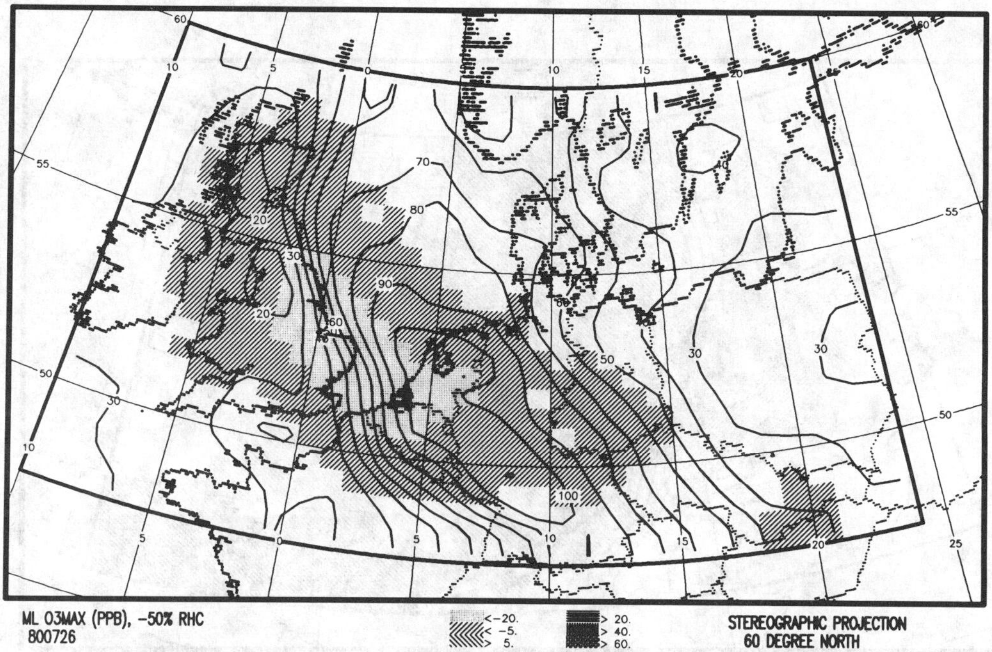

Fig. 10: Contour lines of predicted base case peak ozone concentrations at July 26, 1980. Underlayed the percentage change of calculated ozone peak concentrations due to a 50 % RHC-emissions reduction

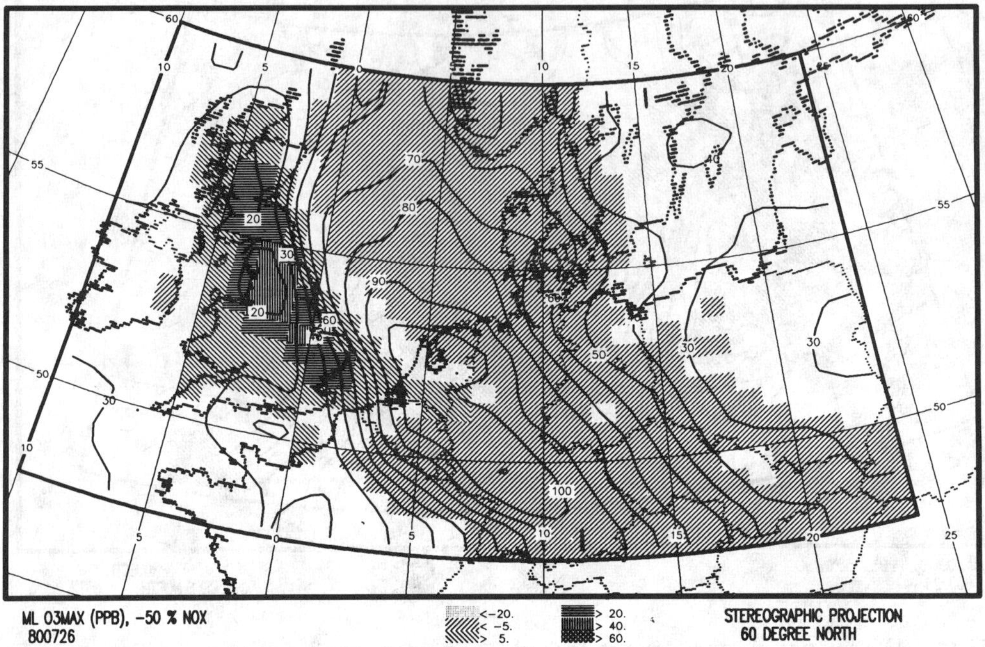

Fig. 11: Same as Fig. 10, but for a 50 % NO_x-emission reduction

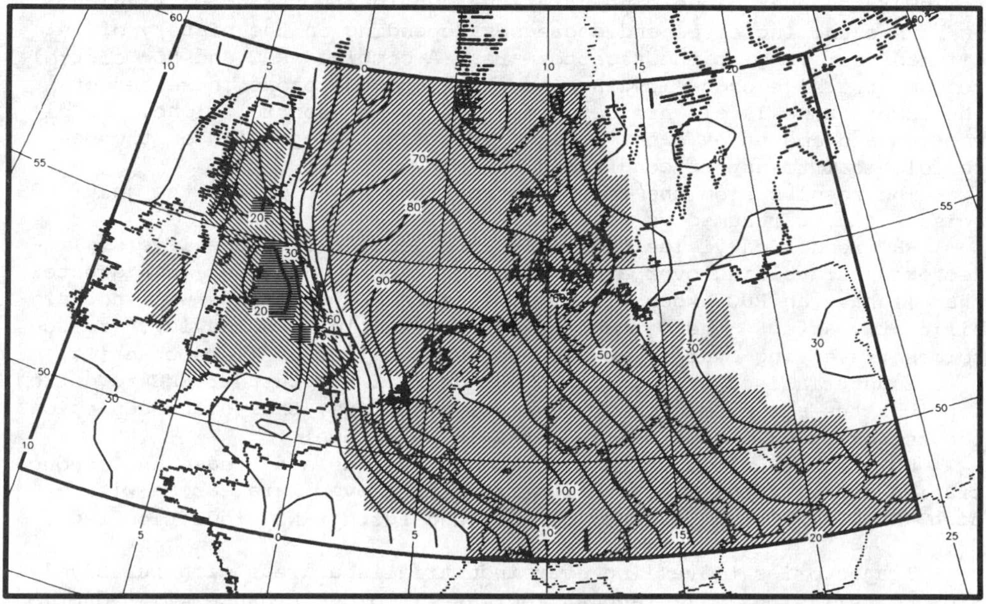

Fig. 12: Same as Fig. 10, but for a 50 % NO_x- and RHC-emission reduction

Isolines indicate peak ozone concentrations which happened to occur during 26 july 1980 in the modeling region. The reason for selecting this presentation was due to the fact that maximum ozone concentrations attribute to severe ecological damage. Underlayed is the percentage decrease or increase of the calculated ozone peak values for the selected sensitivity scenarios - roughly separated by stepwise ranges. The conclusion to be drawn from these figures are:

- Case of 50 % RHC emission reduction.
  In regions of high ozone concentration (greater than 80 ppb O_3) the corresponding ozone reduction rarely exceeds 25 %.
  In regions of lower ozone concentrations (50 - 80 ppb O_3) the reduction can reach 40 %.
- Case of 50 % NO_x emission reduction.
  In the high ozone concentration area (greater than 100 ppb O_3) ozone increase can be found in regions with quite high NO_x emissions and decrease in regions with low NO_x emissions.
- Case of RHC and NO_x emission reduction of 50 %.
  The calculated ozone decrease is somewhat larger than in case of NO_x emission reduction alone, whereas ozone increase is strongly reduced.

The evaluation of these broad emission sensitivity scenario runs showed that the RTM-III model responds to changes in the emission input in a complicated non-proportional way. In general, RHC control seems to lower the peak ozone concentrations, whereas NO_x control results in both ozone increases and decreases depending on the history of an air mass arriving at the receptor area. A combined RHC and NO_x control reduces the high ozone levels everywhere and the ozone increases of the lower ozone levels are less compared to NO_x-control alone. The direction and amount of ozone changes is strongly depending on the meteorological transport conditions.

The results from the emissions sensitivity study for the july 1980 episode can be resumed as follows:

RHC reduction yields generally in a decline of ozone levels. Trajectories travelling over areas with high RHC/NO_x ratios (much greater than 1) have an RHC excess, thus diminished RHC emissions will not significantly affects the O_3 concentration (example Scandinavia). Trajectories travelling over areas with low RHC/NO_x ratios (less than 1), i.e. industrialised areas, have a RHC deficit and further RHC reduction deprives the air from oxidant generating species, thus ozone depletion occurs (example: industrialized area in NW Europe)

NO_x reduction yields an increase as well as a decrease in O_3 concentration over Europe. Trajectories travelling over rural areas with a high RHC/NO_x ratio have a NO_x deficit and further NO_x reduction reduces O_3.

Trajectories travelling over industrialised areas with relatively high NO_x emissions will face an increase in ozone because NO_x reduction removes the ozone depleting species.

A simultaneous reduction of RHC and NO_x will generally result in an ozone decrease comparable to NO_x reduction alone in rural areas and in a more weakened ozone increase over industrialized areas.

6. CONCLUSIONS

In weather situations with a well-defined boundary layer, the effect of control of precursor emissions in Europe on the concentration of photochemical oxidants, varies significantly within the episode and from one episode to another. Selected model calculations show the following tendency:
RHC-emissions control seem to reduce the peak ozone concentration, control of NO_x emissions in many situations seem also to reduce ozone. To reduce the peak ozone concentrations in European episodes, RHC control alone or a combined control of both RHC and NO_x may be more efficient than NO_x control. It should be added that in cases when the exchange with the free troposphere is more efficient [17] and no well defined boundary layer exists, NO_x emission reduction reduces ozone; changes in RHC-emissions have less effect on ozone because the RHC-chemistry is then dominated by CH_4 and CO from the background troposphere. The description of the exchange of air between the boundary layer and the free troposphere is required for modeling ozone on a long term - over months and years. There are indications that for the observed non-episodic increase of ozone over Europe NO_x - emissions are limiting, while for the episodic ozone peaks the RHC emissions are obviously of importance.

However, it has to be kept in mind that more tests are necessary in order to diminish the uncertainty in the calculated results. Results of model calculations can be influenced by

(a) the uncertainty in input data especially in emissions data and meteorological data;

(b) the simplification of model approaches especially
  - of chemical reaction mechanisms (condensed reaction schemes, rate coefficients) used to study real atmospheric chemistry,
  - of meteorological processes in order to produce a tractable transport and dispersion problem;

(c) the selection of episodes deemed truely representative for photooxidant situations in Europe.

On top of this a comprehensive abatement strategy for air pollution control should take into account photooxidants as well as acidic compounds in short-term episodes and in long-term averages.

REFERENCES

(1) Ludwig, C. et al.: Effectivity of Abatement Strategies: PHOXA
In: Acidification and its Policy Implications (Editor: T. Schneider), Elsevier Science Publishers B.V., Amsterdam 1986, 307-316.

(2) Meinl, H.: Das deutsch/niederländische Programm Photochemical Oxidant and Acid Deposition Model Application within the Framework of Control Strategy Development (PHOXA) - Übersicht, Zielsetzungen, europäische Datenbasen für Modellanwendungen. In: Umweltbundesamt Texte 13/86 "Atmosphärische Prozesse, Ausbreitungsmodelle einschließlich Luftchemie und Deposition" 1986, 611-643.

(3) US-EPA: Proceedings of the EPA-OECD International Conference on Long-Range Transport Models for Photochemical Oxidants and Their Precursors, EPA-600/9-84-006, 1984.

(4) Stern, R.: Das "Transport and Deposition of Acidifying Pollutants"-Modell (TADAP). Beschreibung und geplante Anwendung. In: Umweltbundesamt Texte 13/86 "Atmosphärische Prozesse, Ausbreitungsmodelle einschließlich Luftchemie und Deposition" 1986, 661-692.

(5) Liu, M.K., R.E. Morries, J.P. Killus: Development and Application of a Regional Oxidant Model to the North East United States, Atm. Env. 18 (1984) 1145-1161.

(6) Stewart, D.A., R.E. Morris, S.D. Reynolds: Application of the Regional Transport Model (RTM-III) to Western Europe. Final Report SYSAPP-87/030 from 24 Feb. 1987 prepared for TNO Apeldoorn, 160 p.

(7) Derwent, R.G. et al.: Ozone in the United Kingdom. UK Photochemical Oxidant Review Group. Prepared at the Request of the Department of the Environment London by Harwell Laboratory. Interim Report, Februar 1987, 109 p.

(8) Whitten, G.Z., H. Hogo, J.P. Killus: The carbon bond mechanism for photochemical smog. Env. Sci. Technol. 14 (1980), 690-700.

(9) Hough, A.M.: An intercomparison of mechanisms for the production of photochemical oxidants. UK Atomic Energy Authority Harwell, AERE R 12670, June 1987.

(10) Hov Ø., K.H. Becker, P. Builtjes, R.A. Cox, D. Kley: Evaluation of the photooxidants-precursor relationship in Europe. COST 611-Task Force. Commission of the European Communities, Brussels, Air Pollution Research Report 1, 1986, 122 p.

(11) Builtjes, P.J.H.: Introduction to emission needs in long-range transport modelling: Episode models. International EUROSAP-Workshop on Methodologies for Air Pollutant Emission Inventories, Paris, June 28 - July 2, 1987.

(12) Reiff, J. et al.: An air mass transformation model for short-range weather forecasting. Mon. Wea. Rev. 112 (1984).

(13) Eliassen, A., Ø. Hov., I.S.A. Isaksen, J. Saltbones, F. Stordal: A Lagrangian Long-Range Transport Model with Atmospheric Boundary Layer Chemistry. J. Appl. Met. 21 (1982) 1645-1661.

(14) Stern, R.M., P.J.H. Builtjes: Long Range Modelling of the Formation, Transport and Deposition of Photochemical Oxidants and Acidifying Pollutants. In: Commission of the European Communities, 4th European Symposium "Physico-chemical Behaviour of Atmospheric Pollutants", Stresa 1986, 179-189.

(15) Stern, R.M., P.J. Builtjes: Sensitivity of NO_x - and VOC-emission data bases for photochemical dispersion modelling. International EUROSAP Workshop on Methodologies for Air Pollutant Emission Inventories, Paris, June 29 - July 2, 1987.

(16) Builtjes, P.J.H., R.M. Stern, S.D. Reynolds: PHOXA, the use of a photochemical dispersion model for several episodes in North-Western Europe. 16th NATO-CCMS International Technical Meeting on Air Pollution Modelling and its Application, Lindau, 6 - 10 April 1987.

(17) Isaksen, I.S.A., Ø. Hov: Calculation of trends in the tropospheric concentration of O_3, OH, CO, CH_4 and NO_x.Tellus (1986).

USING EMISSION INVENTORIES TO SIMULATE THE AIR POLLUTION IN BERLIN (WEST)

Wolfgang Reichenbächer
Der Senator für Stadtentwicklung und Umweltschutz
Lentzeallee 12 - 14
D - 1000 Berlin 33

ABSTRACT. Measurements of the air quality in Berlin (West) showed during the last years that the limiting values of concentration, which were fixed by administrative regulations, were exceeded by sulphur dioxide and particulate matter and nearly reached by nitrogen oxides. In particular for these air pollutants emission inventories were calculated for all the important groups of emission sources in order to calculate the resulting concentrations by an appropriate diffusion model. By taking in account the mixing layer height as an additional parameter, a Gaußian model similar to the german standard diffusion model is able to simulate the resulting concentration fields of the single producer groups in such a way, that the sums of all calculated concentrations agree well with the measured concentrations of the critical air pollutants in Berlin. Therefore the used model is a good tool to predict the concentration fields of air pollutants as a result of possible regulatory actions for special producer groups.

1 INTRODUCTION

Berlin is a region with relatively high air pollution. As a consequence of actually measured high concentrations the people of Berlin were warned of smog at least once during the winter in the last years. For the first time in the winter 1986/87 the actual air pollution increased to such a high degree that the government was forced to forbid driving in order to lower the concentrations.

The Senat of Berlin has now developed a concept of effective actions to lower the emissions of this region in the future (Senator für Stadtentwicklung und Umweltschutz, 1987). In order to find the most effective methods the emission inventories of all the relevant groups of emission sources were investigated to calculate the concentrations of air pollution by an appropriate diffusion model. Because the measurements of air pollution during the last years showed that the limiting values of concentration fixed by administrative regulations (Bundesminister für Umwelt, 1986) were exceeded in Berlin (West) for sulphur dioxide (SO_2), particulate matter (p.m.) and nearly reached by

K. Grefen and J. Löbel (eds.), Environmental Meteorology, 613–629.

nitrogen oxides (NO_x), for this analysis the species SO_2, p.m. and NO_x were chosen.

2 EMISSION INVENTORIES

In Berlin (West) the emissions of SO_2 and NO_x have been relatively well investigated by special inventories for traffic, residential heating and industry. In order to estimate the emissions from the sourroundings of Berlin (West) a special study (Deutsches Institut für Wirtschaftsforschung (DIW), 1987) could be used. The emissions of particulate matter have not been completely investigated by inventories as there are many diffused sources which can hardly be calculated, for example rising dust by wind or human activities. A balance sheet of the measured p.m. was prepared to estimate these emissions. Fig. 1 shows that 30 % from the measured p.m. results from soil matter and about 30 % from ions, which were formed during the long-range transport out of SO_2 and NO_x.

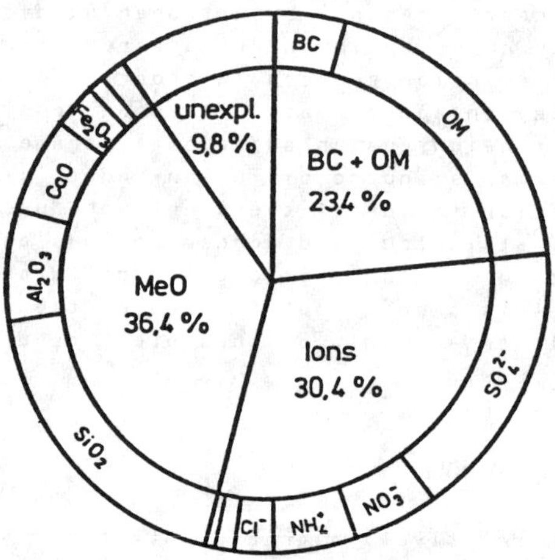

Figure 1: Balance sheet of measured particulate matter in Berlin (West), 1984 (Freise et. al., 1986)
BC : black carbon, OM : organic matter, MeO : metal oxides

The resulting emissions of each producer-group per square kilometre are shown in Fig. 6a, Fig. 7a and Fig. 8a.

3 THE GAUSSIAN MODEL

The concentrations produced by these emissions were calculated with a Gaußian diffusion model which simulates the diffusion of the pollutants - by not taking into account their physical and chemical transformations - through the following equation:

$$C(x,y,z) = \frac{Q}{2\pi\bar{u}\sigma_y\sigma_z}\exp\left[\frac{-y^2}{2\sigma_y^2}\right]\sum_{n=-\infty}^{+\infty}\left\{\exp\left[\frac{-(z-h+2nH_I)^2}{2\sigma_z^2}\right] + \exp\left[\frac{-(z+h+2nH_I)^2}{2\sigma_z^2}\right]\right\}$$

For distances from the source to within about 50 kilometers this model is applicable for the pollutants SO_2, NO_x (sum of NO and NO_2) and particulate matter with very small diameters. The concentration (C) is calculated by the Gaußian equation as a function of the emission rate of the source (Q), the effective stack height (h), the stability parameters (σ_y, σ_z), the wind velocity (u) and the mixing layer height (H_I) in a kartesian coordinate-system (x,y,z), whose x-direction is orientated in the wind direction.

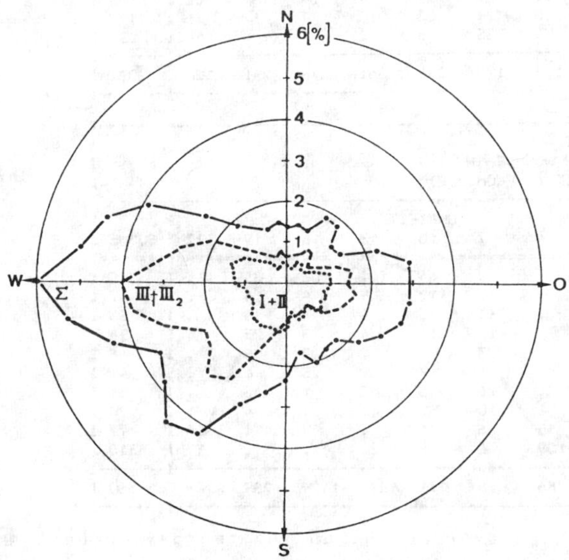

Figure 2: Frequency of wind direction and stability of atmospheric stratification in Berlin (West), 1973 - 1984.

I + II : stable stratification
III_1 + III_2 : neutral stratification
Σ : all situations

From 10 years of meteorological observations in Berlin - Dahlem (1974 - 1983) the meteorologiocal diffusion-situations were determinated, as they are defined for the German standard diffusion model (Bundesminister für Umwelt, 1986) in classes of:

- wind direction
- wind speed
- atmospheric stability and furthermore the
- height of the mixing layer.

DAHLEM 1974-83 JANUAR
VERTEILUNG DER MISCHUNGSHOEHEN
(RELATIVE HAEUFIGKEIT IN PROMILLE)

a)

HOEHE [M]	UHRZEIT 1	4	7	10	13	16	19	22	SUMME
> 500	.	.	3	10	22	24	6	.	65
< 500	.	.	1	2	6	8	4	.	21
< 450	.	.	3	4	5	4	5	.	22
< 400	1	.	.	4	9	12	4	1	31
< 350	1	1	5	8	20	21	10	2	67
< 300	4	5	9	15	19	18	16	4	90
< 250	8	6	15	15	20	15	26	5	112
< 200	21	31	22	29	14	14	33	21	183
< 150	29	10	31	33	8	7	17	36	173
< 100	60	71	35	6	2	2	4	56	236
MITTEL [M]	124	122	173	248	361	376	251	124	222

DAHLEM 1974-83 JULI
VERTEILUNG DER MISCHUNGSHOEHEN
(RELATIVE HAEUFIGKEIT IN PROMILLE)

b)

HOEHE [M]	UHRZEIT 1	4	7	10	13	16	19	22	SUMME
> 900	.	.	8	64	101	106	35	.	315
< 900	.	.	8	12	9	4	14	.	47
< 800	.	.	11	15	4	4	19	.	53
< 700	.	.	17	11	4	3	23	.	58
< 600	.	.	23	7	2	2	15	.	48
< 500	.	.	21	7	2	3	8	.	41
< 400	.	.	16	6	2	1	6	.	31
< 300	1	1	10	2	.	.	4	.	20
< 200	25	15	6	1	.	.	1	28	77
< 100	98	109	6	.	.	.	.	97	310
MITTEL [M]	89	86	518	975	1338	1429	759	89	660

Figure 3: Change of mixing layer height during the day in Berlin (West) for January (a) and July (b), 1973 - 1984. The relative frequency of the mixing layer height (number of cases for each thousand) is shown as a function of observation time and height.

The meteorological observations in Berlin (see Fig. 2 and Fig. 3) show that there are relatively often situations with low wind speeds, stable stratification and shallow mixing layer heights. Only the consideration of the mixing layer heights in this Gaußian diffusion model - which is not explicitly considered in the German standard diffusion model (Bundesminister für Umwelt, 1986) - makes it possible to realistically simulate the smog-situations often observed in Berlin whereby the pollution, mainly exhausted by low sources (residential heating and traffic), concentrates in a shallow mixing layer.

The emissions of the single producer-groups either depend on the air temperature and time of the day (residential-heating, heating plants) or depend only on the time like the traffic (see Fig. 4). Therefore either the daily mean air temperature and observation time (Table 1) or the observation time alone was considered in the determination of each meteorological diffusion-situation.

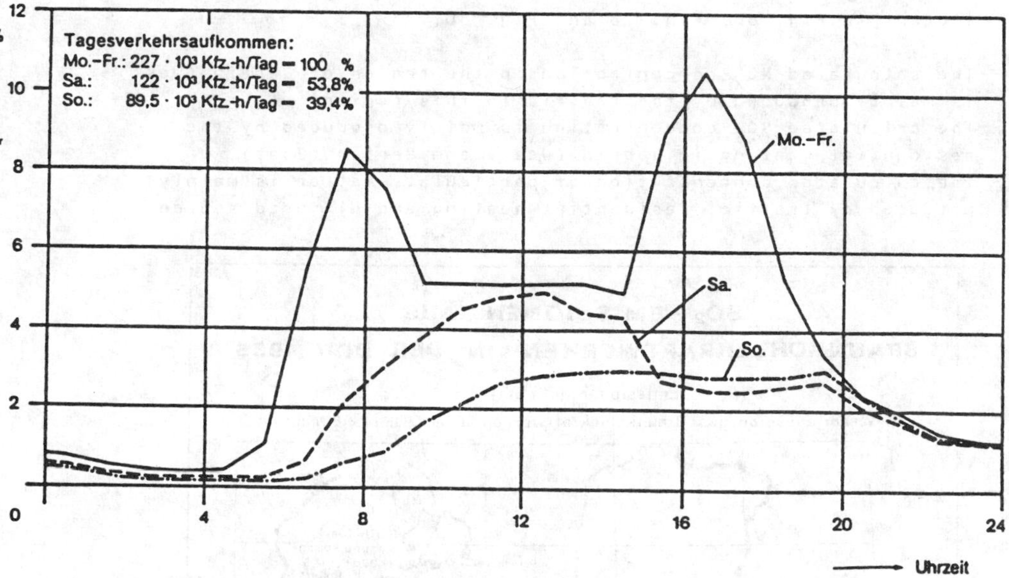

Figure 4: Change of the relative traffic load during the day

TABLE 1: Table of the relative emission factors for residential -heating relative to the maximum emission per hour as a function of mean daily air temperature and observation time (Fath, Lühring, 1985).

Tagesmittel-temperaturen	Tageszeit							
	1-3H	4-6H	7-9H	10-12H	13-15H	16-18H	19-21H	22-24H
-12.0 Grad	1.00	1.00	1.00	1.00	1.00	1.00	1.00	1.00
-10.0 Grad	.88	.88	.97	.97	.97	.97	.92	.92
- 8.0 Grad	.72	.68	.93	.93	.93	.89	.89	.85
- 6.0 Grad	.62	.58	.92	.88	.88	.81	.81	.77
- 4.0 Grad	.48	.42	.90	.83	.83	.76	.76	.73
- 2.0 Grad	.37	.34	.86	.77	.77	.71	.71	.65
0.0 Grad	.27	.27	.81	.70	.70	.65	.65	.59
2.0 Grad	.23	.28	.69	.60	.60	.55	.55	.53
4.0 Grad	.19	.19	.54	.50	.46	.46	.46	.46
6.0 Grad	.15	.15	.43	.40	.37	.37	.37	.37
8.0 Grad	.09	.12	.32	.32	.28	.28	.28	.28
10.0 Grad	.06	.08	.22	.22	.18	.18	.18	.18
12.0 Grad	.05	.05	.10	.10	.11	.09	.08	.08
14.0 Grad	0.00	0.00	0.00	0.00	0.00	0.00	0.00	0.00

4 DISCUSSION OF RESULTS

For all these situations and for all sources the concentration at 50 points in Berlin were calculated by the Gaußian equation. The resulting mean annual concentrations are shown for each single producer-group and pollutant in Fig. 6b, Fig. 7b and Fig. 8b:

- The calculated NO_x-concentration in the region of Berlin (West) is mainly produced by the traffic in this region.
- The calculated SO_2-concentration is mainly produced by the residential heating in Berlin (West) and Berlin (East).
- The calculated concentration of particulate matter is mainly produced by traffic, residential heating and diffused sources.

Figure 5: SO_2-emissions from coal-fired power plants in the GDR, 1985. The total annual emission is 2.800.000 tons.

In addition to the regional sources considered by the Gaußian diffusion model there are important sources of pollution at distances of more than 50 kilometres from Berlin, specially the big power plants in the southern part of the GDR (see Fig. 5).

TABLE 2: Emissions and resulting calculated concentrations for the main groups of emission sources for sulphur dioxid (SO_2), particulate matter (P.M.) and nitrogen oxides (NO_x).

	Emissions (in t/a)			Calculated Concentrations (in µg/m3)		
	SO_2	NO_x	P.M.	SO_2	NO_x	P.M.
<u>Sources in Berlin(W)</u>						
Room heating	6200	2700	2500	20.1	9.0	8.0
Traffic	900	16000	3200	2.9	47.5	10.0
Power plants	50000	32000	)	7.6	4.8	)
Industrial plants	6600	2300	)6300	6.7	3.1	) 2.0
Heating plants	4400	2800	)	5.2	2.8	)
Other sources	10	10	2000	0.0	0.0	12.0
<u>Regional sources less than 50 km distant from Berlin (W)</u>						
Room heating	7100	2200	5000	8.0	2.4	6.0
Traffic	700	4100	1700	0.8	4.5	2.0
Single sources	161000	43000	40000	7.8	2.6	2.0
Other sources	0	0	2500	0.0	0.0	3.0
<u>Distant large sources</u>						
Large Sources (50 - 200 km)	2700000	175000	500000	16.	1.	20.
Other (large) Sources in Europe	16000000	8000000	2000000	4.	7.	35.
Sum of the calculated concentrations :				79.	85.	100.
Mean concentration measured in Berlin (West):				75.	84.	102.

If these long-range sources are taken into account the following estimation is valid for the considered pollutants in Berlin (West) (Table 2):

- 25 % of the SO_2-concentration is produced by long-range sources,
- 10 % of the NO_x-concentration is produced by long-range sources and
- 55 % of the concentration of particulate matter is produced by long-range sources, mainly consisting of ions which are formed during the long-range transport out of SO_2- and NO_x-emissions.

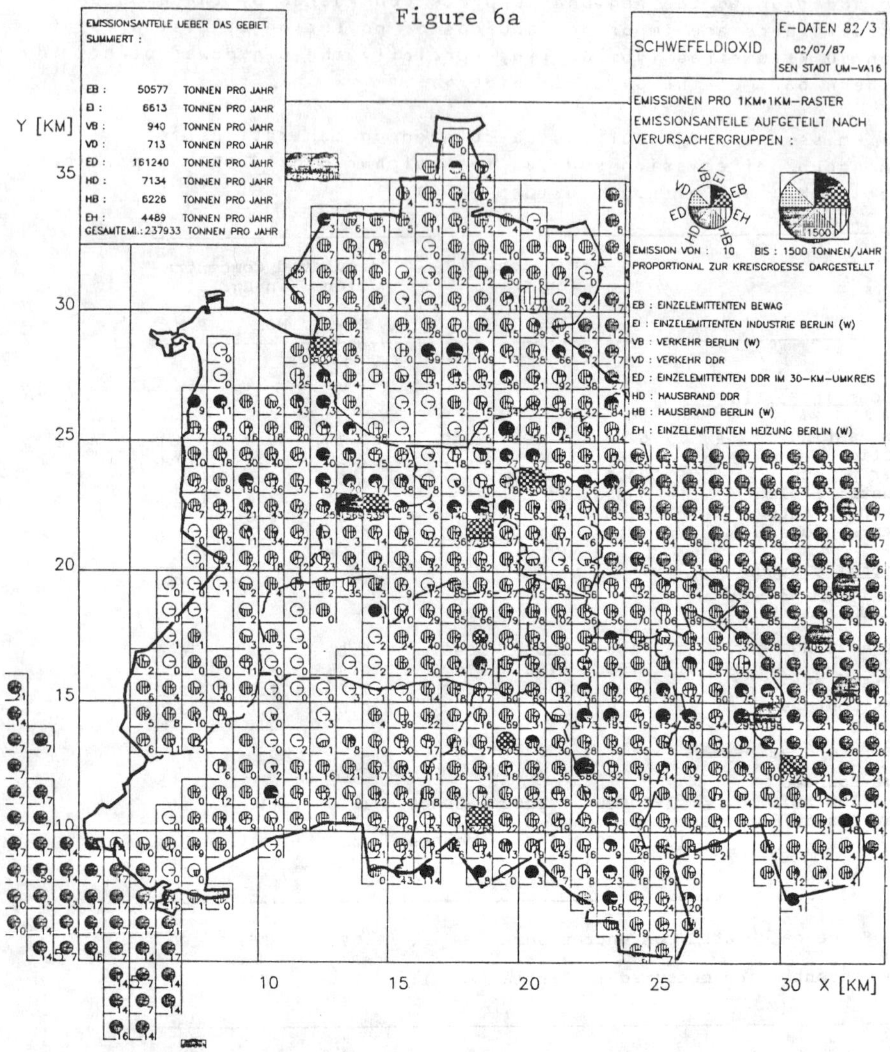

Figure 6: (a) Emissions of SO_2 in each square kilometre (in tons per year) for different producer-groups :

EB : Power plants	EI : Industrial plants
VB : Traffic in Berlin (West)	VD : Traffic in the GDR
ED : Industrial and power plants (GDR)	HD : Room heating in the GDR
HB : Room heating in Berlin (West)	EH : Heating plants in Berlin(W)

(b) Calculated mean annual concentration of SO_2 from the different regional producer-groups (in $\mu g/m^3$).

Figure 6b

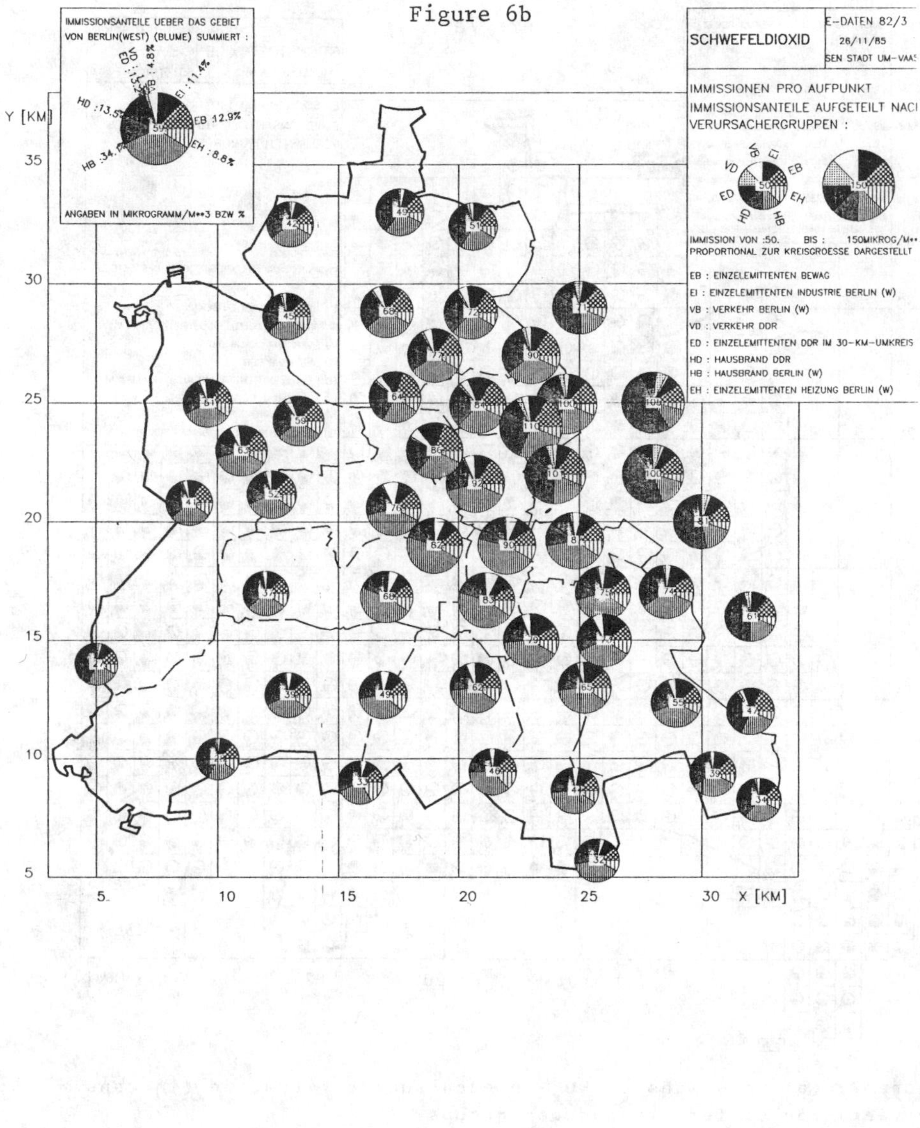
IMMISSIONSANTEILE UEBER DAS GEBIET
VON BERLIN(WEST) (BLUME) SUMMIERT :
ANGABEN IN MIKROGRAMM/M**3 BZW %
SCHWEFELDIOXID
E-DATEN 82/3
28/11/85
SEN STADT UM-VAAS
IMMISSIONEN PRO AUFPUNKT
IMMISSIONSANTEILE AUFGETEILT NACI
VERURSACHERGRUPPEN :
IMMISSION VON :50. BIS : 150MIKROG/M**
PROPORTIONAL ZUR KREISGROESSE DARGESTELLT
EB : EINZELEMITTENTEN BEWAG
EI : EINZELEMITTENTEN INDUSTRIE BERLIN (W)
VB : VERKEHR BERLIN (W)
VD : VERKEHR DDR
ED : EINZELEMITTENTEN DDR IM 30-KM-UMKREIS
HD : HAUSBRAND DDR
HB : HAUSBRAND BERLIN (W)
EH : EINZELEMITTENTEN HEIZUNG BERLIN (W)
Y [KM]
X [KM]

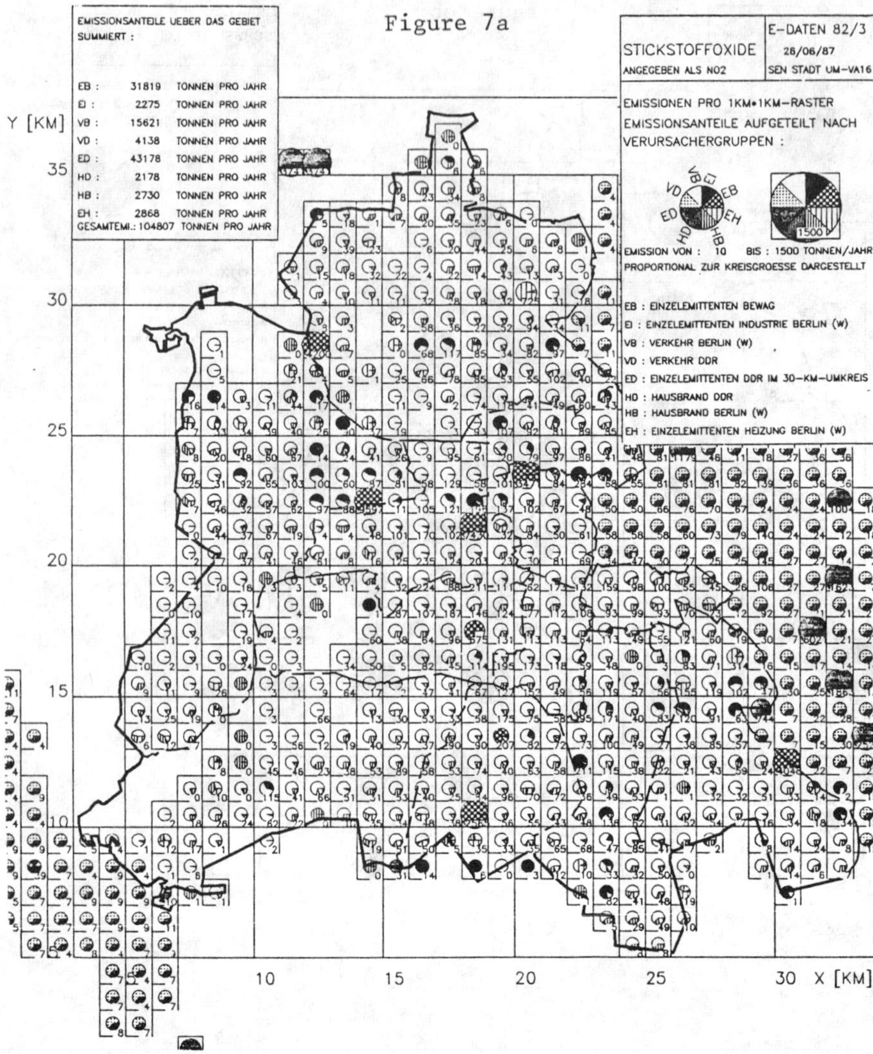

Figure 7: (a) Emissions of NO_x in each square kilometre (in tons per year) for different producer-groups :

EB : Power plants	EI : Industrial plants
VB : Traffic in Berlin (West)	VD : Traffic in the GDR
ED : Industrial and power plants (GDR)	HD : Room heating in the GDR
HB : Room heating in Berlin (West)	EH : Heating plants in Berlin(W)

(b) Calculated mean annual concentration of NO_x from the different regional producer-groups (in $\mu g/m^3$).

Figure 7b

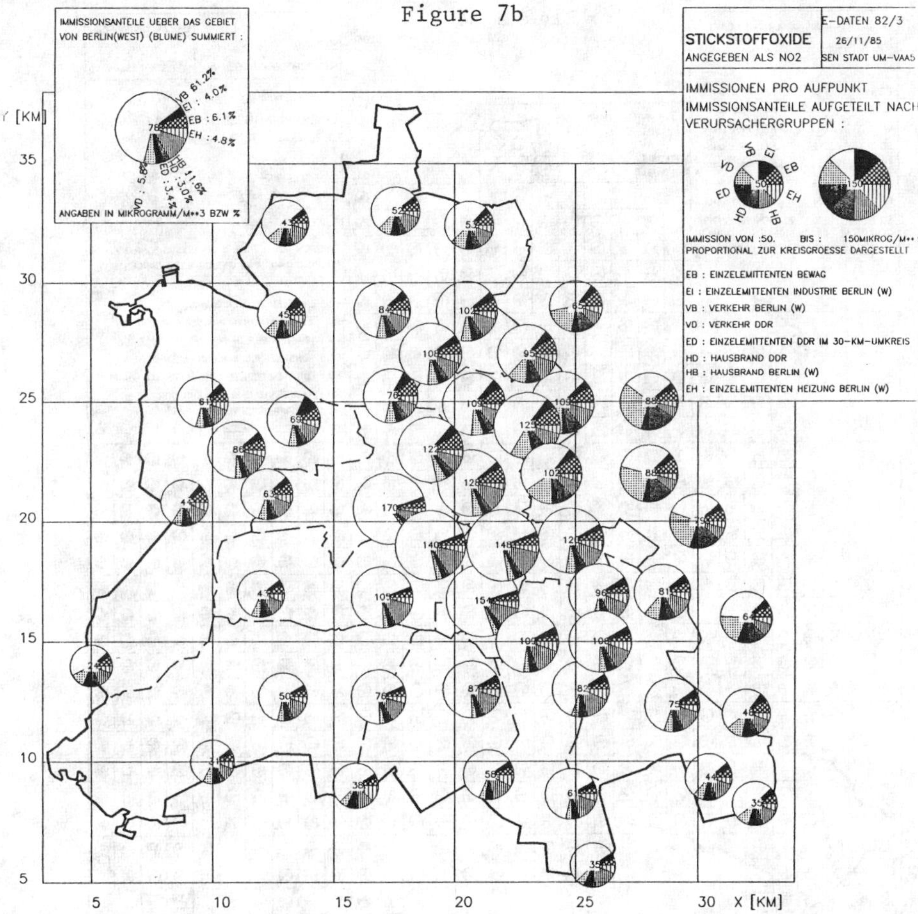
IMMISSIONSANTEILE UEBER DAS GEBIET VON BERLIN(WEST) (BLUME) SUMMIERT :
ANGABEN IN MIKROGRAMM/M**3 BZW %
STICKSTOFFOXIDE
ANGEGEBEN ALS NO2
E-DATEN 82/3
26/11/85
SEN STADT UM-VAA5
IMMISSIONEN PRO AUFPUNKT
IMMISSIONSANTEILE AUFGETEILT NACH VERURSACHERGRUPPEN :
IMMISSION VON :50. BIS : 150MIKROG/M**3
PROPORTIONAL ZUR KREISGROESSE DARGESTELLT
EB : EINZELEMITTENTEN BEWAG
EI : EINZELEMITTENTEN INDUSTRIE BERLIN (W)
VB : VERKEHR BERLIN (W)
VD : VERKEHR DDR
ED : EINZELEMITTENTEN DDR IM 30-KM-UMKREIS
HD : HAUSBRAND DDR
HB : HAUSBRAND BERLIN (W)
EH : EINZELEMITTENTEN HEIZUNG BERLIN (W)
Y [KM]
X [KM]

Figure 8a

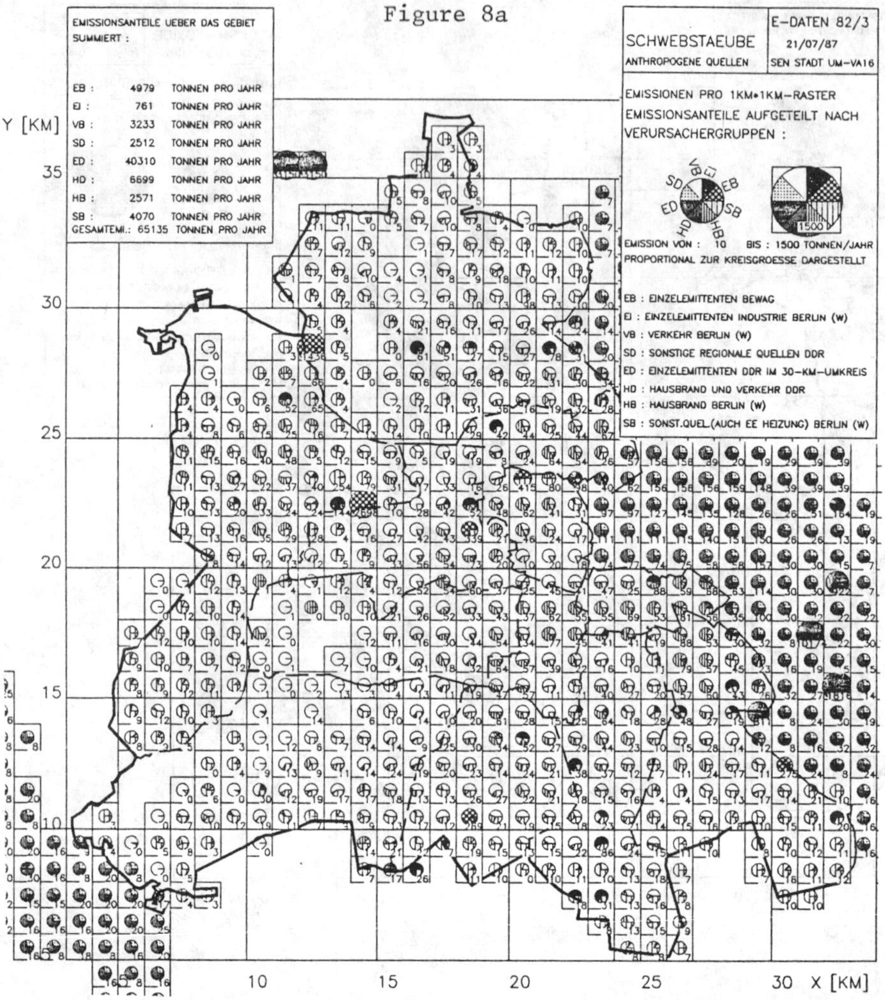

Figure 8: (a) Emissions of particulate matter in each square kilometre (in tons per year) for different producer-groups :

EB : Power plants in Berlin (West)
EI : Industrial plants (Berlin(W))
VB : Traffic in Berlin (West)
SD : Diffused sources in the GDR
ED : Industrial and power plants (GDR)
HD : Heating and traffic (GDR)
HB : Room heating in Berlin (West)
SB : Diffused sources (Berlin(W))

(b) Calculated mean annual concentration of particulate matter from the different regional producer-groups (in $\mu g/m^3$).

Figure 8b

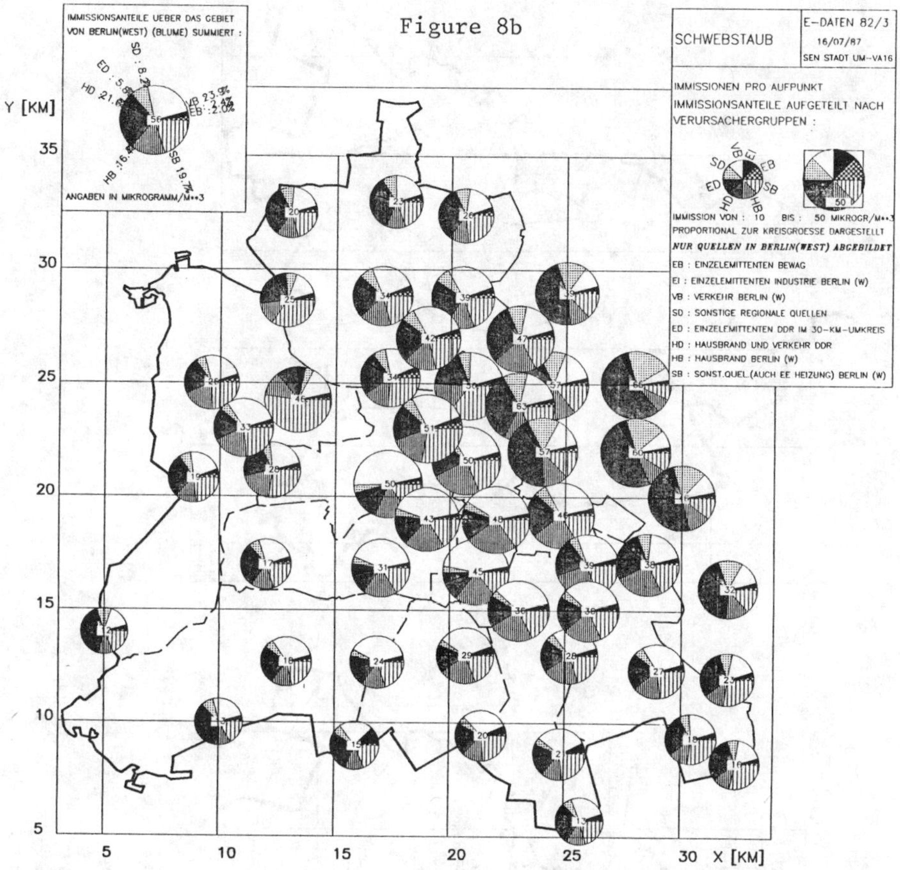
IMMISSIONSANTEILE UEBER DAS GEBIET
VON BERLIN(WEST) (BLUME) SUMMIERT :
ANGABEN IN MIKROGRAMM/M**3
SCHWEBSTAUB
E-DATEN 82/3
16/07/87
SEN STADT UM-VA16
IMMISSIONEN PRO AUFPUNKT
IMMISSIONSANTEILE AUFGETEILT NACH
VERURSACHERGRUPPEN :
IMMISSION VON : 10 BIS : 50 MIKROGR/M**3
PROPORTIONAL ZUR KREISGROESSE DARGESTELLT
NUR QUELLEN IN BERLIN(WEST) ABGEBILDET
EB : EINZELEMITTENTEN BEWAG
EI : EINZELEMITTENTEN INDUSTRIE BERLIN (W)
VB : VERKEHR BERLIN (W)
SD : SONSTIGE REGIONALE QUELLEN
ED : EINZELEMITTENTEN DDR IM 30-KM-UMKREIS
HD : HAUSBRAND UND VERKEHR DDR
HB : HAUSBRAND BERLIN (W)
SB : SONST.QUEL.(AUCH EE HEIZUNG) BERLIN (W)
Y [KM]
X [KM]

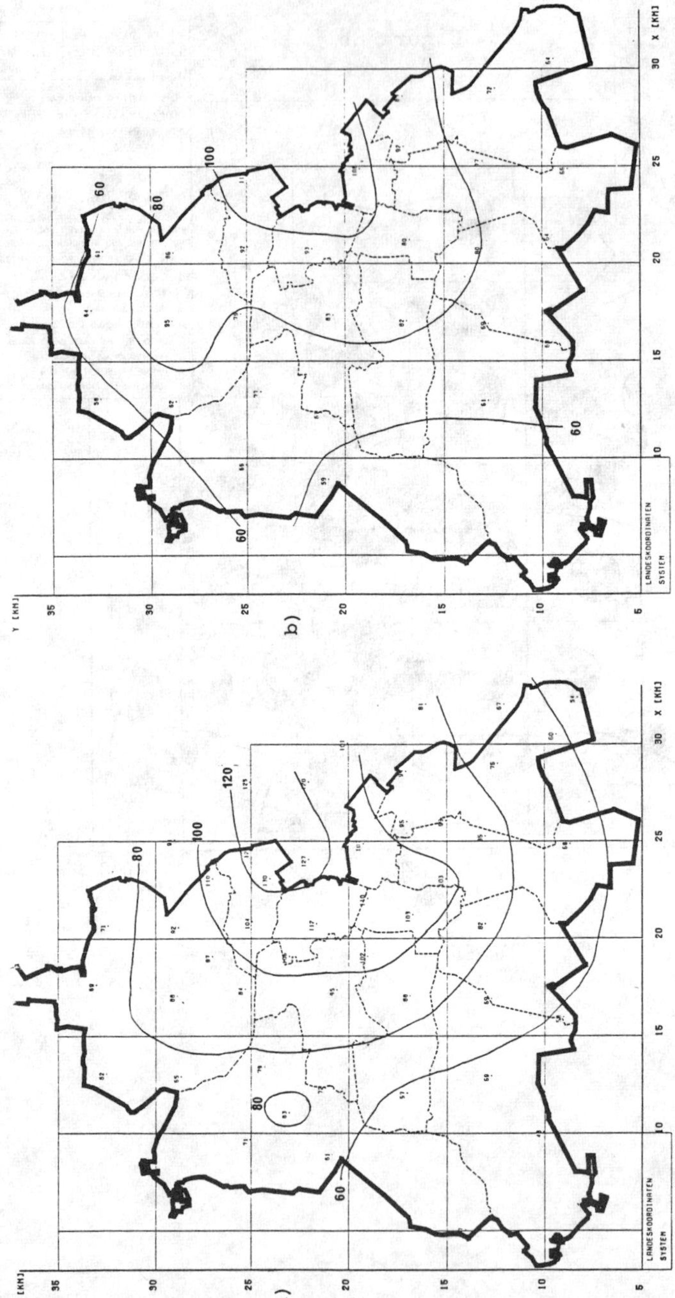

Figure 9: (a) Calculated total mean annual concentration of SO_2 in $\mu g/_m{}^3$. (b) Measured mean annual concentration of SO_2 during 1981 - 1983 by continuous measurements in $\mu g/_m{}^3$.

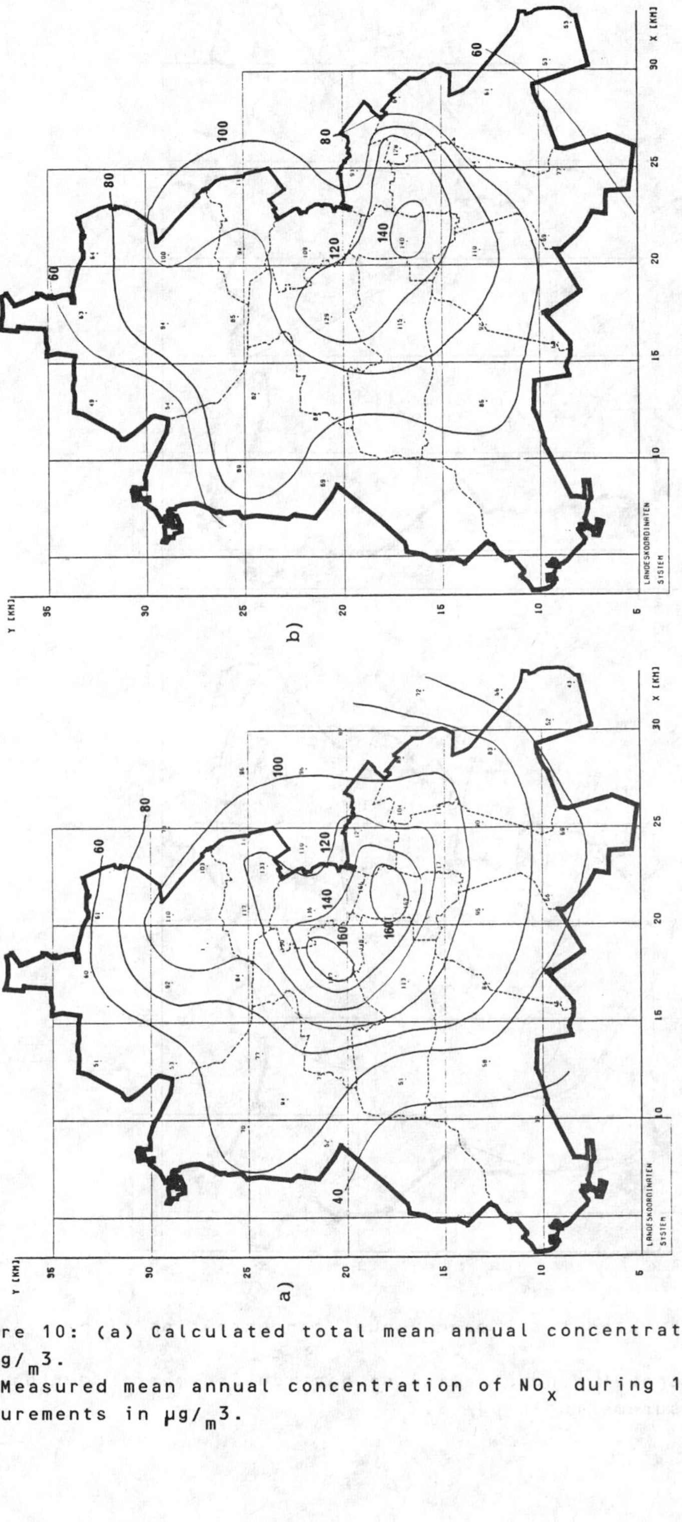

Figure 10: (a) Calculated total mean annual concentration of NO_x in $\mu g/m^3$.
(b) Measured mean annual concentration of NO_x during 1984 by random-test measurements in $\mu g/m^3$.

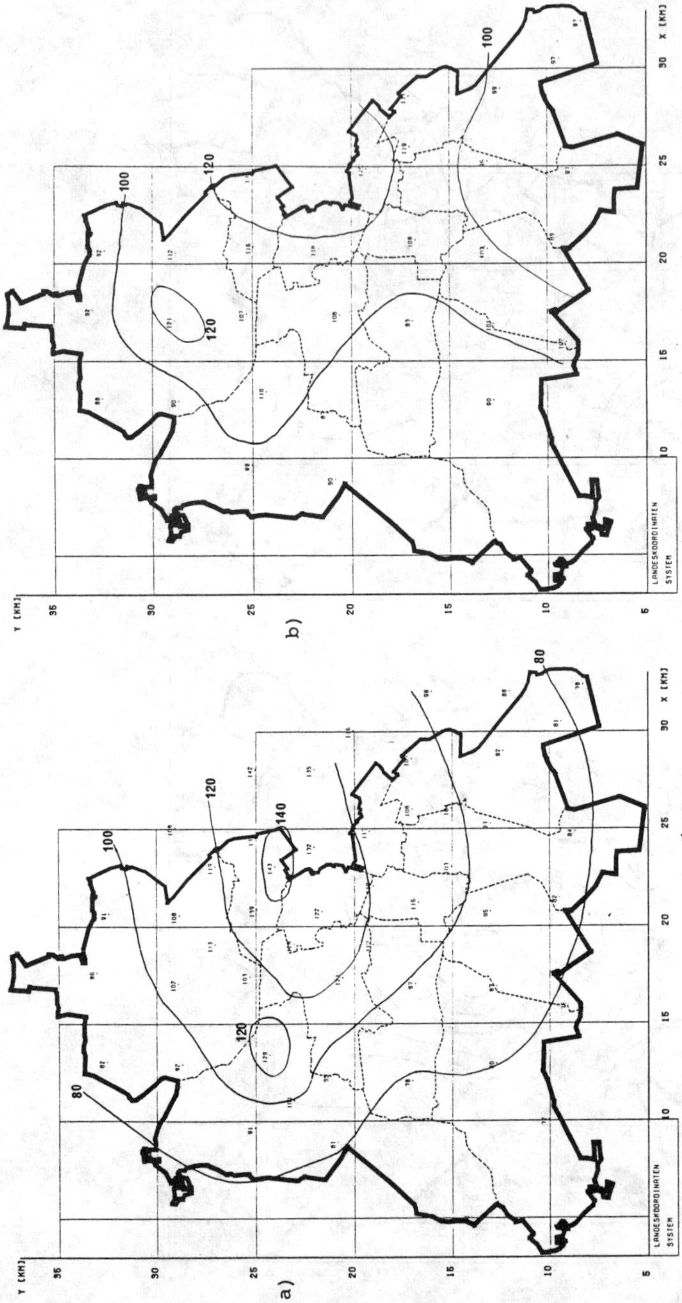

Figure 11: (a) Calculated total mean annual concentration of particulate matter in $\mu g/m^3$.
(b) Measured mean annual concentration of particulate matter during 1984 by random-test measurements in $\mu g/m^3$.

In Fig. 9a-b, Fig. 10a-b and Fig. 11a-b the calculated mean annual concentrations - the sum of the long-range part of the concentration and the computed result of the Gaußian model - were compared to the available measurements. The calculations for SO_2 coincide very well with the measured concentrations; for this pollutant the measured concentration is the mean over three years (1981 - 1983) of continuously measured concentrations, while the concentrations of NO_x and p.m. are only based on random-test measurements from the year 1984. As the coincidence for NO_x and p.m. is also rather good, the used diffusion model can predict the concentration fields as a result of regulatory actions for special groups of emission producers or specific emission sources.

REFERENCES

Bundesminister für Umwelt (1986) 'Technische Anleitung zur Reinhaltung der Luft - TA Luft ', GMBL vom 27.2.87, p.95

Der Senator für Stadtentwicklung und Umweltschutz (1987) 'Luftreinhalteplan für das Belastungsgebiet Berlin 1986 - 1991', Abgeordnetenhaus von Berlin, Drucksache 10/1513 vom 9.6.87

Deutsches Institut für Wirtschaftsforschung (DIW) (1985) 'SO_2- und NO_x-Emissionen in der DDR 1982', Report, partly published in DIW-Wochenbericht Nr. 30 vom Juli 1985

Fath, J. und Lühring, P.G. (1985) 'Durchführung von Ausbreitungsrechnungen als Grundlage für den Luftreinhalteplan 1986', Report for the Senator für Stadtentwicklung und Umweltschutz

Freise, R. und Israel, G. (1986) 'Ursachenanalyse der Schwebstaubbelastung in Berlin (W)', Report for the Senator für Stadtentwickling und Umweltschutz

FATE SIMULATION OF ORGANIC SUBSTANCES IN THE ATMOSPHERIC MIXING LAYER

Dipl. met. Rolf Trenkle
Projektgruppe Umweltgefährdungspotentiale
von Chemikalien (PUC)
Gesellschaft für Strahlen- und Umweltforschung
Ingolstädter Landstraße 1
8042 Neuherberg

Abstract

In the European Communities about 100 000 different chemical compounds were put on the market. A great deal of these compounds are released at least partially into the atmosphere.

A model (called EXAIR) has been developed to assess the fate of organic substances in the atmospheric mixing layer. It is a submodel of the computer program E4CHEM (Exposure and Ecotoxicity Estimation for Environmental CHEMicals) and works even with a minimum of information about the physico-chemical properties of the respective airborne substance. The basic structure of EXAIR consists in the fact that physico-chemical properties of the substance are combined with meteorological parameters. The release of the substance can occur from point-, line- or area sources. The equations describing transport and removal processes can be solved by simple analytical expressions. The resulting quantities which describe the fate of the substance are terms of a mass balance established in the mesoscale for the surroundings of the emission-site.

Objective

Day by day thousands of man-made chemical substances are emitted directly or indirectly into the atmosphere (e.g. volatilization from ground or surface water bodies). For a great deal of them very little is known about the following facts:

- physico-chemical properties of the emitted substances
- release characteristics
- airborne transport
- removal mechanisms and degradation

It is also little known about the hazard that constitutes the presence of airborne or deposited material to ecosystems. The knowledge of the fate of the individual substances in the environment is a precondition for studying their hazard. Therefore, the study of the pro-

K. Grefen and J. Löbel (eds.), Environmental Meteorology, 631–648.

blematic nature of their fate may be dealed with priority. The present paper tries to give a contribution to this study which describes a methodology allowing to get some information about the fate under certain conditions. This is done by describing an atmospheric fate model for organic substances. It is designed to assess in a first order approximation the individual fate of a large number of organic substances as different as possible. As a result, the individual fate of the substances can be compared one with another and thereby those substances which are showing an extreme behaviour can be identified and studied later on more intensively. In other words: The model is designed for screening purposes.

Description of the model

For the creation of such a model it is necessary to go through the literature looking for an appropriate model that could deserve as a pattern for the own model. Thereby it is found that many atmospheric fate models have a very refined meteorological part. From the chemical point of view, however, they are relatively poor: In the past only very few chemical substances like SO_2, SO_4, NO_x and O_3 have been interesting. Therefore, no appropriate model in the literature could be chosen without greater modification.

First of all, one has to design the structure of the model's chemical part. Since the model has to deserve for screening purposes, a minimum of necessary physico-chemical information for a maximum of different organic chemicals must be made available.

The following physico-chemical properties are regarded as absolutely necessary:

- dimensionless Henry coefficient H'
- Octanol-water coefficient K_{OC}

Furthermore, it is favourable to know the degradation rate by the attack of OH-radicals k_{OH}.

For a great deal of substances, however, the values of these parameters are not measured. In order to reduce the number of substances with unknown H' and K_{OC} these quantities starting from basic information (like the boiling or melting point) have to be estimated.

Therefore, a subroutine called DTEST has been written by Brüggemann [6]. As its results are needed for the atmospheric fate model (called EXAIR) but also for fate models of different media, it is placed outside EXAIR. All these models belong to the software package called E4CHEM (Exposure and Ecotoxicity Estimation for Environmental CHEMIcals) [12,14,15] for assessing environmental hazard potentials. The present paper, however, is restricted to the description of EXAIR. Therefore, it can be assumed that for each substance of which the fate has to be assessed by EXAIR the corresponding K_{OC}- and Henry-values are known.

In many meteorological dispersion models, especially in Gaussian-plume models it is assumed that the pollutant is released from known point sources like tall stacks. The release-points of organic chemicals, however, are often unknown, e.g. the buildings from which chemicals for domestic use are emitted cannot be identified in the individual case.

Fortunately, there is a way to avoid this problem: in the lee of a point source the plume of the pollutant increases with the distance from the source. At a certain distance X_D' the concentration $C(X_D')$ of the pollutant becomes quite uniformly distributed in vertical direction within the atmospheric mixing layer. The distance X_D' can be estimated with a method described by Wippermann [24] or Dobbins [8]. It turns out that in most cases (e.g. at neutral conditions) X_D' is less than 10 km. This means that the particularities of the source can only be perceived hardly even in the local surroundings if distance $X' < X_D'$.

Thus, for the creation of EXAIR it can be assumed that all substaces are quite homogeneously mixed in the atmospheric mixing layer. This involves that for a first approach it is not necessary to subdivide the troposphere into several horizontal layers, it can rather be treated as one homogeneous layer. (Complicate atmospheric conditions like fumigation will not be considered here).

This is done in the literature by several models, e.g. by the model of Scriven and Fisher [16] or by Mermal and Kumer [13]. The structure of the basic equation of both models is the same and adopted by EXAIR:

$$m\ (X) = m_o \exp\ (-(A_1 + A_2)\ \ X/U) \tag{1}$$

m_o means the total mass of the substance in the mixing layer at $X' = X_D'$ while $m\ (X)$ means the mass at distance $X = X'-X_D'$ (Note: $X = 0$ at $X' = X_D'$, the dashed coordinate system is only auxiliary).

A_1 describes the removal of the substance by dry deposition and A_2 by wet deposition. U is the vertically averaged windspeed in the mixing layer, called transport wind by Dobbins [8].

In the following these quantities are explained in detail. In the model of Scriven and Fisher [16] A_1 and A_2 are defined as follows:

$$A_1 = \frac{V_g\ (SO_2)}{a}\ , \quad A_2 = -\ \Lambda\ (SO_2) \tag{2}$$

In this equation V_g means the dry deposition velocity of SO_2, the quantity a means the height of the mixing layer and Λ the scavenging coefficient of SO_2. Now, these equations have to be generalised in such a way that they are applicable to organic chemicals.

Thereby, one has to take into occount that their transport can occur either in the gas-phase or attached to aerosols. This means that a quantity f is needed which describes the partition between the two phases. It could be defined as follows:

$$f = m_{ads}/(m_{ads} + m_{gas}) \tag{3}$$

m_{ads} constitutes the mass of the organic substance adsorbed to aerosols while m_{gas} refers to the gaseous phase. By this definition f can act as a weightning factor for the dry deposition of aerosols while 1-f does the same for the dry deposition of gases.

Thus, A_1 can be written in the following form generalised for organic substances:

$$A_1 = f\,(V_g)_{aerosol}/a + (1-f)\,(V_g)_{gas}/a \tag{4}$$

$(V_g)_{aerosol}$ means the dry deposition velocity for aerosols while $(V_g)_{gas}$ constitutes the dry deposition velocity for gases.

The term A_2 can be generalised analogously:

$$A_2 = f\,\Lambda_{aerosol} + (1-f)\,\Lambda_{gas} \tag{5}$$

is the scavenging coefficient for aerosol or gas.

For the description of the behaviour of organic chemicals not only the removal by wet and dry deposition may be important but also their transformation. Consequently equation (1) has to be extended by sink term A_3:

$$m = m_o \exp\,(-\,(A_1 + A_2 + A_3)\,X/U) \tag{6}$$

As the degradation by the attack of OH radicals according to Aalst and Diederen [1] is often the dominant transformation process in the lower troposphere, the sink term can be written as follows:

$$A_3 = k_{OH}\,[OH] \tag{7}$$

In this equation k_{OH} means the first order degradation rate and [OH] the concentration of OH-radicals in the mixing layer.

After having defined the basic equation (6) and the connection between A_1, A_2, A_3 and the quantities describing the removal (V_g, Λ) respectively the transformation (k_{OH}), the dependence on these quantities concerning the physico-chemical properties of the organic substance has to be defined.

It is important how the dependence of the partition coefficient f is described in the model EXAIR to begin with it: an empirical equation was found which permits to calculate f in dependence on K_{oc}/H'. The reason for establishing a such equation was the following consideration: The Henry-coefficient describes the equilibrium between the gas-phase and the liquid phase, while the octanol-water-partition-coefficient does the same between the solide and the liquid phase. Therefore, K_{oc}/H' should constitute the equilibrium between the solid and the gas phase. The correlation was made with values of m_{ads} and m_{gas} measured by Cautreels and Couvenberghe [7] in an urban environment for about 30 chemicals and the corresponding values of K_{oc} and Henry:

$$\log (m_{ads}/m_{gas}) = a \log (K_{oc}/H') + b \tag{8}$$

whereas a and b are constants, $a > 0$ and $b < 0$

A detailed paper describing the derivation of equation (8) is in preparation. It will also describe how m_{ads}/m_{gas} depends on the organic carbon content of the aerosol and on the aerosol concentration. From (8) f can easily be calculated:

$$f = (1 + (m_{ads}/m_{gas})^{-1})^{-1} \tag{9}$$

In the following the dry deposition velocity has to be discussed. Let's begin with that for aerosol particles:

Sehmel [17] gave an overview of dry deposition measurements. Sehmel underlined in his article the enormeous scatter of measured values which he attributed to the innumerous and nearly uncontrollable factors influencing the measurement. Hence it is not possible without further ado to deduce reasonable values for EXAIR from these measurements.

However, there is another publication by Sehmel which he wrote together with Hodgson [18]. They published a semiempirical model for predicting the dry deposition velocity of aerosols in dependence on the particle diameter and meteorological parameters. The values of $(V_g)_{aerosol}$ used in EXAIR are orientated on the results of this semi-empirical model. This means that other influences on the deposition velocity like vegetation properties are not regarded. In addition, in the present version of EXAIR the values used for different particle intervals are meteorological average values. In the standard run of EXAIR a constant value of 1 mm/s is used which stands for particles with a diameter of 10^{-3} mm. Hereby a flat terrain (gras) is assumed.

The article of Sehmel also contains a survey of measurements of gas-dry-depositon. It comprises, however, only about a dozen of gases due to the lack of measurements. The measured values show an enormeous scatter as in the case of aerosol deposition caused by the combined action of the physico-chemical properties of the substance and the environmental conditions which are often unknown. A literature research revealed that there is also little known about the combined action itself. Tucker [21] published a semi-empirical model which predicts the dry deposition velocity of organic gases to water surfaces as a function of the physico-chemical properties and the environmental properties. However, no suchlike model seems to be published for the deposition to land surfaces. In the absence of detailed information about dry deposition processes of organic chemicals a simple formula for the dry deposition velocity was used. It assumes that the transport to the earth-surface is diffusive. In this manner it is possible to calculate $(V_g)_{gas}$ as a function of the molecular weight M of the substance. Therefore, one has to know the measured deposition value of an organic substance and its corresponding molecular weight M. As a reference substance a pesticide with the molecular weight M of 300 and the deposition velocity $(V_g)_{reference}$ = 5 mm/s was chosen [20].

$$(V_g)_{gas}/(V_g)_{reference} = (300/M)^{1/2} \qquad (10)$$

This equation only regards the diffusion to the ground; the volatilisation is not taken into account. Nevertheless, it is possible to calculate the volatilisation of organic chemicals deposited to soil surfaces with the fate model EXSOL which is also a submodel in the software package E4CHEM. If the amount of deposition is combined with the amount volatilized the net deposition can be calculated. This is done in the multi-media-model EXINT which is a submodel of E4CHEM, too. If high deposition rates occur together with high volatilisation rate, then the use of EXINT is necessary for getting more realistic results. (A publication about EXINT and EXSOL [5] will be prepared). In the following section the scavenging parameters which occur in the term A_2 have to be discussed.

Bonka [3] gave an overview of measurements of scavening coefficients of aerosols. For aerosols of 10^{-3} mm radius the range of measured values lays between 10^{-5} and 10^{-3} s^{-1}. For the standard run of EXAIR an average value of 10^{-4} was chosen. Naturally, the scavenging coefficient of aerosols is not constant with the precipitation intensity but depends on it as described by van Aalst and Diederen [1]:

$$\Lambda = q \cdot \lambda^n , \ (n \leq 1) \qquad (11)$$

(q is called the washout coefficient)

Van Aalst and Diederen [1] suggest to set the exponent n between 0.6 and 0.8. In the standard version of EXAIR n is set equal to 1 like in the paper of Vogt and Brenk [4]. (In EXAIR the value of the washout coefficient is defined as $10^{-4}s^{-1}$/1mm/h). It may be indicated that this description of the washout is only approximate because the influence of any parameter like temperature or droplet-size distribution is not treated explicitely.

In the following the scavenging coefficient of gases is described. According to van Aalst and Diederen [1] the scavenging coefficient can be written as:

$$\Lambda_{gas} = W \lambda / a \qquad (12)$$

W is the so called scavenging ratio. It is defined as the mass concentration of pollutant in a collected precipitation sample divided by the mass concentration of pollutants in the air. Outgoing of an equilibrium partition of the pollutant it can be assumed that

$$W = 1/H' \tag{13}$$

Equation (12) implies together with (13) that the scavenging ratio of gases is a function of both meteorological parameters i.e. the precipitation intensity and the mixing height and a physico-chemical parameter, the Henry coefficient H'. In the case of dry deposition velocity of gases this double dependence is not seen so clearly. The meteorological conditions are not described by single parameters but are inherent in the value of $(V_g)_{reference}$.

According to equation (7) the degradation term A_3 is the product of the degradation rate k_{OH} and the OH concentration. It is obvious that k_{OH} is a physico-chemical parameter. Strictly speaking [OH] must be related to the physico-chemical <u>and</u> meteorological conditions of the atmosphere as, according to Weinstock et al. [23], the NO_x concentration and the light intensity are the most important factors which control the OH concentration.

These relationships are intended to be mapped in a later version of EXAIR but for the first approximation they are to extensive. Therefore, the model EXAIR is working with average values of the OH concentration. According to Stuhl [19] it is reasonable to set [OH] = 10^5 molecules cm^{-3} during day time and 0 molecules cm^{-3} during night-time. For that reason [OH] is treated for the sake of simplicity as a meteorological parameter in EXAIR.

Which other meteorological parameters are needed by EXAIR? It requires the vertically averaged windspeed of the mixing layer (see equation (6)), the height of the mixing layer (equation (4) and (12)) and the precipitation intensity (equation (11) and (12)). Since EXAIR is designed for screening purposes it should be possible to study the behaviour of the substance under all meteorological conditions that occur during the course of a year. This objective implies the following problem: The simultaneous values of windspeed, height of mixing layer and precipitation intensities for all times of the year have to be known. Within this statement another problem arises: a year must be found that comprises all typical constellations of the parameter values during its course. It seems possible that for future developments the so called test-reference years [2] could be used. These are artificially composed years which consist of periods belonging to real years. The sequence of the real periods is not arbitrary but immitates the typical sequence of meteorological constellations in a real year. In the present version of EXAIR a three-parametrical frequency distribution of meteorological parameters is used instead of the test reference years . It is based on measurements carried out during a period of 14 years by the Kernforschungszentrum Karlsruhe [11]. The frequency distribution is established for the parameters windspeed, Pasquill's diffusion class and precipitation intensity.

An argument against the use of the frequency distribution is that it is established for the windspeed measured at a fixed height (60 m), the Pasquill' diffusion class and the precipitation intensity while EXAIR needs the vertically averaged windspeed, the height of the mixing layer and the precipitation intensity. Fortunately, this problem can be overcome by using the following relationships. The vertically averaged windspeed can be received by using equation (14):

$$\frac{U_z}{U_{60}} = \left(\frac{h_z}{60}\right)^{m\,(\text{Pasquill})} \tag{14}$$

The windspeed U_z at any height h_z can be calculated from the windspeed at a fixed height (e.g. 60 m) by this power-law. The exponent m depends on the Pasquill diffusion class. (The more rigorous the vertical turbulence the less the exponent.)

According to the TA-Luft (technical instruction for the atmosphere) [25] the following values m are valid:

Table I
The dependence of the exponent m on the stability class

Pasquill's stability class

A	B	C	D	E	F
0.09	0.20	0.22	0.28	0.37	0.42

In compliance with Hanna [9] the power-law of the windspeed can be applied up to an altitude of about 200 m (This is an empirically derived value). If no further information about the windspeed in greater altitutes is available it can be assumed for a first approximation that the windspeed remains constant.

According to Klug [10] a reasonable average value of the mixing layer height can be given for each diffusion class:

Table II
The dependence of the mixing layer heights on the Pasquill's Diffusion classes

A	B	C	D	E	F
1500 m	1500 m	1000 m	500 m	200 m	200 m

The three-parametrical frequency distribution

It is based on values averaged over 30 minutes. The data are grouped according to the precipitation intensity. The intensity class 1 refers to the whole time of the year without precipitation (= 0 mm/h). That happens in 91.24 % of the time of the year. The intensity class 2 is defined by the following range of the precipitation intensity:

$0 < \lambda < 0.5\ mmh^{-1}$

This range occurs in 5.26 % of the whole time of the year. For precipitation class 3 $0.51 < \lambda < 3$ mm is valid. This range is found in 2.96 % of the whole year.

The precipitation class 4 is defined by $\lambda > 3$ mm which occurs in 0.54 % of the time of the whole year.

For every precipitation and diffusion class a frequency distribution of the measured windspeed is given. This means that the whole frequency distribution consists of elements of the form $p(D_i, N_j, U_k)$

with			
	D_i	A,B,C,D,E,F	Pasquill' diffusion class
	N_j	1 ... 4	precipitation level
	U_k	U_1 ... U_8	windspeed level

$p(D_i,N_j,U_k)$ therefore expresses how often the diffusion class has the value D_i, the precipitation class has the value N_j and the windspeed level has the value U_k simultaneously during the whole time of the year. EXAIR runs for every combination (D_i, N_j, U_k).

Output quantities of EXAIR

Before discussing the results an additional question concerning the frequency distribution is taken into consideration. The question is: Why the windspeed alone is considered and not its direction which may also be of interest? In fact, there exists a four-parametrical frequency distribution of the KfK which comprises the wind direction, too [22]. The three-parametrical distribution is gained by integration of all points of the compass. For screening purposes the behaviour of the substance during its transport (degradation, deposition etc.) is of primary importance and not the direction of its flight. The latter parameter, however, is more important for other questions (e.g. site-specific ones).

For that reason EXAIR uses only a three-dimensional frequency distribution. (It is planned to make EXAIR more sophisticated and therefore also the four-parametrical distribution will become important). Now, let's start to discuss the output parameter of EXAIR.

The relatively simplest result that can be calculated is a mass balance. According to equation 6 the mass m(x) can be calculated for any distance x. For the comparison of the results of different substances it is reasonable to define a standard distance. In EXAIR $X_{standard}$ = 50 km was chosen but any other distance can be used.

As the input into the atmosphere is often unknown it is better to use m(X)/m(X=0) = AIRA as a result. AIRA can be interpreted as the fraction of airborne mass that passes the distance-point $X = 50$ km. It can be used as a measure of the <u>mobility of the substance.</u>

For the intervall $0 < X < X_{standard}$ the mass balance can be written as follows:

$$\text{AIRA} + \text{AIRDD} + \text{AIRDW} + \text{AIROH} = 1 \tag{15}$$

AIRDD = the fraction of mass removed by dry deposition
AIRDW = the fraction of mass removed by wet deposition
AIROH = the fraction of mass degraded by the OH-attack

For the calculation of AIRDD, AIRDW and AIROH the quantities d_1, d_2 and d_3 have to be defined:

$$d_1 = 1 - \exp(-A_1 x/U) \tag{16}$$

d_1 can be interpreted as the mass fraction which is deposited if the dry deposition is the only removal mecanism (That means if there is no wet deposition and no degradation).

Let's define

$$d_2 = 1 - \exp(-A_2\ X/U) \tag{17}$$

and

$$d_3 = 1 - \exp(-A_3\ X/U) \tag{18}$$

d_2 and d_3 can be interpreted analogously to d_1.

Now the results are obtained:

$$\text{AIRDD} = (1 - \text{AIRA})\ D_1/D \tag{19}$$

$$\text{AIRDW} = (1 - \text{AIRA})\ D_2/D \tag{20}$$

$$\text{AIROH} = (1 - \text{AIRA})\ D_3/D \tag{21}$$

$$D = d_1 + d_2 + d_3 \tag{22}$$

By analog consideration AIRDD can be splitted into AIRDG and AIRDA (dry deposition of gas and aerosols resp.) and AIRDW into AIRWG and AIRWA (wet deposition of gases and aerosols resp.). Naturally, all quantities of the mass balance depend on the diffusion class, the precipitation level and the windspeed. If average values for the whole year or special subsets like the individual precipitation classes are wanted they can be calculated as follows:

$$\text{AIRA (average)} = \sum_{r=1}^{R} p_r\ \text{AIRA}_r / \sum_{r=1}^{R} p_r \tag{23}$$

$$\text{AIRD (average)} = \sum_{r=1}^{R} p_r \text{ AIRD}_r / \sum_{r=1}^{R} p_r \qquad (24)$$

$$\text{AIROH (average)} = \sum_{r=1}^{R} p_r \text{ AIROH}_r / \sum_{r=1}^{R} p_r \qquad (25)$$

R is the number of elements p (D_i, N_j, U_k) of the respective subset in the three-dimensional frequency distribution . In the case of yearly average all elements of the frequency distribution p(D_i, N_j, U_k) are taken into account.

Note:
Measurements of the immision of a substance into the environment are often given in form of the concentration. For this and other reasons formulas are incorporated in EXAIR claculating concentration-values in the lee of point-, line- and area-sources. These values are discussed in detail in [5].

Results

In the following the results of EXAIR are going to be explained by means of three examples:
- the fate of dichloromethane,
- " " " fluouranthene,
- the fate of lindane (gamma-hexacyclohexane).

The physico-chemical values of these substances taken from the E4CHEM-database [15] are shown in the next table:

Table III
Physico-chemical properties of dichloromethane, fluoranthene and lindane

	Dichloromethane	Fluouranthene	Lindane
Henry-coefficient	0,81 10^{-1}	0.27 10^{-3}	0,28 10^{-4}
K_{OC} [$cm^3/H_2O/g$]	0,48 10^2	0.20 10^5	0.13 10^4
Partition aerosol/gas [%]	0,017	12	9,9
k_{OH} [1/d]	0,56 10^{-2}	0	0

This table shows that lindane has the minimal value of the Henry-coefficient while dichloromethane has the maximal value. The K_{OC}, however, has its highest value for fluouranthene and its smallest for dichloromethane.

Both values K_{OC} and Henry-coefficient are controlling the partition f between the aerosol and the gasphase (third row). Therefore, fluouranthene has the highest value f. (In future studies it is planned to compare the results of EXAIR obtained with the theoretically derived f according to equation (8) with results of EXAIR based on measured values of f.)

In the figures 1 and 2 the mass balances of lindane and fluoranthene are shown for all precipitation levels and for yearly conditions. First, the common fate properties of th two substances are described:

- The higher the precipitation level the higher are the values of the wet-gas and wet-aerosol deposition (AIRWS resp. AIRWA). This fact is rather obvious because of equation (11) and (12) which show that the wet deposition is proportional to the precipitation intensity.
- The higher the precipitation level the smaller are the dry-gas and dry-aerosol deposition (AIRDG resp. AIRDA). This is because of the removal by dry depostion is according to equation (6) competitive to that one of wet deposition (and competitive to the degradation assumed to be zero for the two substances).
- For lindane and fluoranthene the dry-aerosol deposition (AIRDA) is three orders of magnitudes smaller than the dry-gas deposition (AIRDG). This is due to the following facts:
  - the transport in the gasphase is dominant (see factor f in table III)
  - the dry deposition velocity of aerosols which was set equal to 1 mm h^{-1} is smaller than that one for the two substances calculated according to equation (10) which gives a value of around 6 mm s^{-1} for fluoranthene and of around 5 mm s^{-1} for lindane.

In the following the differences between the fate of the two substances are discussed.

By comparing the mass balance of the two substances in case of no precipitation (level 1) it is seen that the relative mass deposited by dry-gas deposition is slightly higher for fluoranthene: $AIRDG_{fluoranthene}$ = 16.4 % while $AIRDG_{Lindane}$ = 14.6 %. For that reason the dry-gas deposition calculated according to equation (10) is somewhat higher for fluoranthene.

In case of precipitation class 2 the relative airborne mass AIRA of flouranthene is slightly higher than for precipitation class 1 (83.3 % in comparison to 83.1 %). First, this might be surprising because it is supposed that the wet removal which is not present for precipitation class 1 would lead to a smaller airborne fraction for precipitation class 2. However, the raising of AIRA becomes understandable by studying the three parametric frequency distribution: During precipitation events the higher windspeeds are more frequent. This leads according to equation (6) to higher AIRA-values.

In case of precipitation class 2, lindane behaves in a different way in comparison to fluoranthene: Its value of AIRA is about 3 %

smaller than at precipitation class 1. The reason for this is that its wet-gas deposition is stronger than that of fluoranthene and therefore dominating over the effect of higher windspeeds. It is rather clear why the wet-gas deposition of lindane is stronger: Its Henry-coefficient is smaller than that of fluoranthene (see table III).

For precipitation classes 2 the airborne fraction AIRA is reducing for both substances with raising precipitation clase number.

At precipitation level 4 the wet-gas deposition of fluoranthene is only about 1/3 of its wet-aerosol deposition while in the case of lindane it is about the double. For that reason the values for the aerosol adsorption f and the Henry-coefficient are bigger for fluoranthene than that for lindane.

The mass balance of annual average conditions is always very close to that of precipitation class 1 because in about 90 % of the year no precipitation is falling in the city of Karlsruhe where the meteorological frequency distribution was established.

The results of EXAIR for dichlormethane are not shown graphically because during all precipitation levels the relative airborne mass AIRA is greater than 99,5 %. One reason for this fact is the extremely low aerosol adsorption f. Therfore, there is only a slight removal of dichlormethane by aerosol deposition. Since the gas is also hardly soluble (see Henry-coefficient in table III) the wet-gas depositions is ineffective, too. The same is valid for the dry-gas deposition: Since dichloromethane has a very high volatilisation from the ground (calculated with the soil model EXSOL) an effective dry deposition velocity of 1 mm s^{-1} was taken as model-input which leads to a very small value of dry-gas deposition AIRDS.

In figure 3 the airborne masses of the 3 substances are compared with each other at the different precipitation classes and for annual average conditions: It is illustrated that lindane is the most sensible substance towards the precipitation intensity while dichlormethane is the less one.

These results demonstrate the differing fate of airborne substances in dependence on the atmospheric conditions and on the physico-chemical properties of the substances. As mentioned above EXAIR deserves to calculate the fate of a great number of substances. Thereby these substances tending to remain in the atmosphere can be separated from those tending to flow to the ground (soil, water bodies) or from the substances with a remarkable degradation. In further studies refined fate models for the different groups and for more complex environmental conditions can be developed.

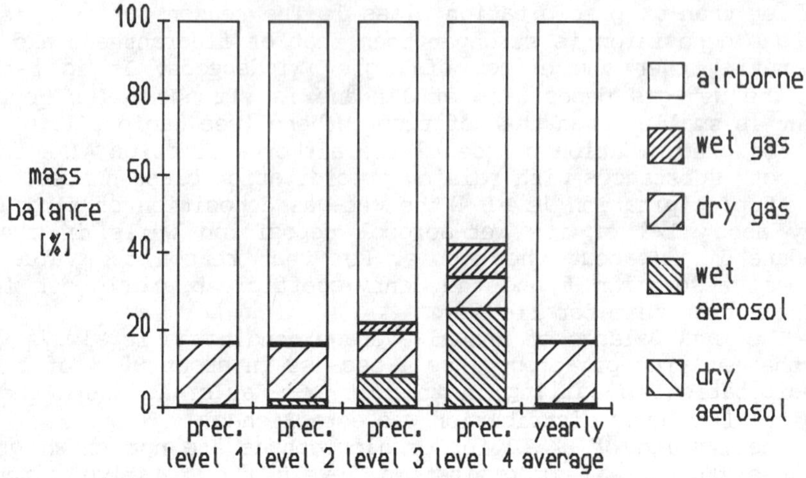

Figure 1
Mass balance of fluoranthene for different precipitation levels and annual average conditions (level 1 = 0 mm/h, level 2 =0-0.5mm/h, level 3 = 0.51-3mm/h, level 4 = more than 3mm/h)

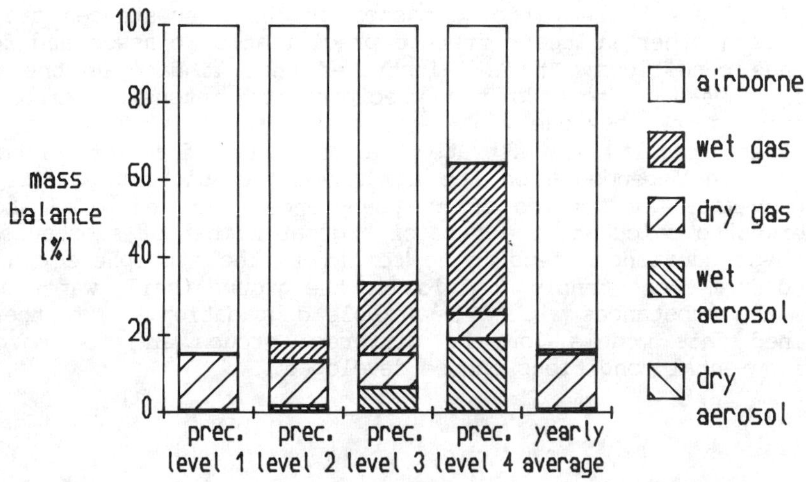

Figure 2
Mass-balance of lindane for different precipitation levels and annual average conditions

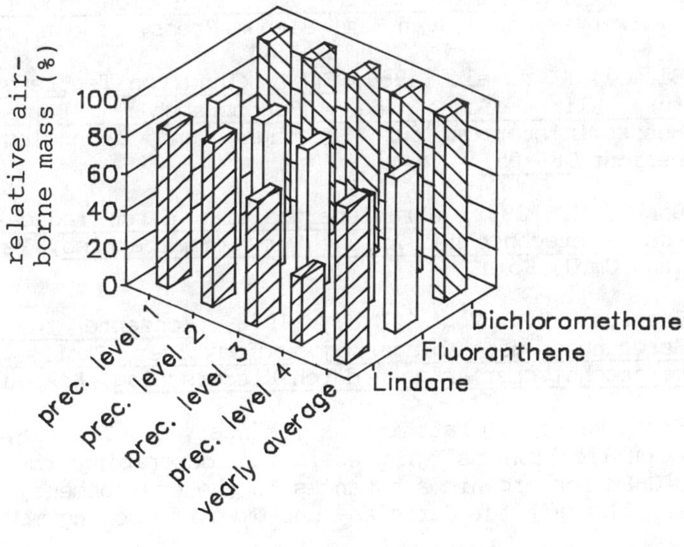

Figure 3
Relative airborne mass of dichloromethane, fluoranthene and lindane for different precipitation levels and annual average conditions

Literatur

[1] Aalst, van R.M., Diederen, H.S.M.A., 1985, 'Removal and transformation processes in the atmosphere with respect to SO_2 and NO_x ', in: International Air Pollution Modelling edited by S. Zwerver and J. van Ham, Plenum Press, London

[2] Blümel, K. et al., 1986, Entwicklung von Testreferenzjahren (TRY) für Klimaregionen der Bundesrepublik Deutschland , in: Bundesministerium für Forschung und Technologie, Forschungsbericht T86-051

[3] Bonka, H., 1982, Strahlenexposition durch radioaktive Emissionen aus kerntechnischen Anlagen im Normalbetrieb, Verlag TÜV Rheinland GmbH, Köln

[4] Brenk, H.D., Vogt, K.J., 1976, Konzepte für ein praxisnahe Berechnung der Ablagerung radioaktiver Stoffe aus der Abluft kerntechnischer Anlagen durch Niederschlag, KfA Jülich,Jül 1328

[5] Brüggemann, R., Matthies, M., Münzer, B., Trenkle, R., A publication is in preparation describing the fate models of E4CHEM for organic substances in the environment, Gesellschaft für Strahlen- und Umweltforschung mbH München

[6] Brüggemann R., 1986, 'Abschätzung von fehlenden PC-Daten mit Hilfe des Submodells DTEST', Unpublished report in: Umweltmodelle und rechnergestützte Entscheidungshilfen für die vergleichende Bewertung und Prioritätensetzung bei Umweltchemikalien Projekt des Umweltbundesamtes der Bundesrepublik Deutschland UFOPLAN-Nr. 106 04 016

[7] Cautreels, W., Cowenberghe van, K., 1978, 'Experiments on the Distribution of Organic Pollutants between Airborne Particulate Matter and the Corresponding Gas Phase', in: Atmospheric Environment 12, 1133 - 1141

[8] Dobbins, R.A., 1979, Atmospheric Motion and Air Pollution John Wiley & Sons, New York

[9] Hanna, S. R., 1982, Review of Atmospheric Diffusion Models for Regulatory Applications, Technical Note No. 177, World Meteorological Organisation

[10] Hübschmann, W., 1978, 'Umweltmeteorologie', in: Jahresbericht 1978 der Abteilung Sicherheit KfK 2775, 181 - 189, Kernforschungsanlage Karlsruhe

[11] Klug, W., 1969, 'Ein Verfahren zur Bestimmung der Ausbreitungsbedingungen aus synoptischen Beobachtungen', Staub-Reinhaltung der Luft 29, 143 - 147

[12] Matthies, M., Brüggemann, R., Trenkle, R., 1986, 'Multimedia Modelling Approach for Comparing the Environmental Fate of Chemicals', in: Environmental Modelling for Priority Setting among Existing Chemicals, Proc. Intern. Workshop
Ed.: Gesellschaft für Strahlen- und Umweltforschung München, Projektgruppe PUC, Ecomed-Verlag, Landsberg

[13] Mermal, S., Kumar, A., 1984, 'Analysis of Acid Deposition due to Sulfurdioxide emissions from Ohio', in: Air Pollution Modelling and its Application IV, edited by C.D. Wispelaere, Plenum press, New York

[14] Münzer, B., 1987, Exposure and Ecotoxicity Estimation for Environmental CHEMicals (E4CHEM), Benutzerhandbuch
Gesellschaft für Strahlen- und Umweltforschung mbH München
Projektgruppe PUC, München-Neuherberg

[15] Rohleder, H., Matthies, M., Benz, J., Brüggemann, R., Münzer, B., Trenkle, R., Voigt, K., 1986, Umweltmodelle und rechnergestützte Entscheidungshilfen für die vergleichende Bewertung und Prioritätensetzung bei Umweltchemikalien, GSF-Bericht 42/86

[16] Scriven, R.A., Fisher, B.E.A., 1975, 'The Long Range Transport of Airborne Material and its Removal by Deposition and Washout' Atmospheric Environment 9, 49 - 58

[17] Sehmel, G.A., 1980, 'Particle and Gas Dry Deposition - A Review' Atmospheric Environment 14, 983 - 1011

[18] Sehmel, G.A., Hodgson, W.H., 1978, A Modell for Predicting Dry Deposition of Particles and Gases to Environmental Surfaces PNL-SA-6721, Battelle, Pacific Northwest Laboratories

[19] Stuhl, F., 1985, 'Radikale in der reinen und verschmutzten Atmosphäre', in: VDI Bildungswerk, BW 43-28-03, Verein Deutscher Ingenieure

[20] Thompson, N., 1983, 'Diffusion and Uptake of Chemical Vapour Volatilising from a Sprayed Target Area', Pestic. Sci. 14, 33 - 39

[21] Tucker, W.A. et al, 1982, 'Estimation of the Dry Deposition Velocity and Scavenging Ratio for Organic Chemicals', in: Precipitation Scavenging, Dry Deposition and Suspension, Volume 2, Proc. of the Fourth Intern. Conference, Santa Monica, California Elsevier, New York

[22] Vogt, S., 1980, Vierparametrige Ausbreitungsstatistik als Berechnungsgrundlage der langzeitigen Schadstoffbelastung in der Umgebung eines Emittenten, KfK 2990, Kernforschungszentrum Karlsruhe

[23] Weinstock, B. et al., 1980, 'Chemical factors affecting the hydroxyl radical concentration in the troposphere' Advances in Environmental Science and Technology, 10, 221

[24] Wippermann, F., 1974, 'The diffusion of radioactive gases in the meso-scale (20 - 24 km)', Commission of the European Communities, Report No. 064-06-72 PSTD

[25] Erste Allgemeine Verwaltungsvorschrift zum Bundes-Immissionsschutzgesetz (Technische Anleitung zur Reinhaltung der Luft - TA Luft) vom 27.02.86, GMBL, 1986, S. 95

Dispersion Modeling Using Personal Computers

Richard H. Schulze
Trinity Consultants, Inc.
100 North Central Expressway, Suite 600
Richardson, Texas 75080 USA
Telephone: 214/234-8567
Telex: 265193 TRINTY UR Fax: 214/234-2123

ABSTRACT. The widespread availability of dispersion models for personal computers is changing the way engineers and scientists address air quality issues. This paper describes the role of modeling in the U.S., and the PC models available. Information on equipment and meteorological data is provided. The use of these easy-to-use models will grow rapidly overseas if countries develop an uncomplicated approach to showing compliance with air quality standards and if meteorological data can be made readily available.

INTRODUCTION

The widespread availability of personal computers is dramatically changing the way engineers and scientists at industrial facilities address air pollution issues. Formerly if a task required the facility engineer to estimate air quality from stack or accidental releases, he would seek assistance from corporate headquarters or a consultant. Today the scientist or engineer is much more likely to purchase a model for a PC and perform his or her own modeling.

The cost and time to develop sound engineering designs to achieve appropriate air quality can be reduced if the person responsible for the design can perform modeling to quickly assess the effect of a modified design. In this way emission parameters for stacks can be adjusted as necessary to meet air quality standards or guideline. The emission parameters for accidental releases can be studied to determine the most cost-effective way to reduce the risk of such a release to the community. By putting modeling tools in the hands of a designer, not only are there time and cost savings, but better designs result.

BACKGROUND

Starting in 1983 the most popular of the U. S. Environmental Protection Agency's (US EPA's) models became available in forms for

K. Grefen and J. Löbel (eds.), Environmental Meteorology, 649–661.

the IBM-PC. In the past four years the range of models has greatly expanded and their use has mushroomed. Although many large industrial firms and regulatory agencies for industrialized states in the United States still use mainframe versions of these models, most moderate and smaller-sized firms have switched to using PC versions of the models.

The initial customers for the PC models were U. S. firms but, increasingly, customers are located throughout the world. The growth rate outside the U. S. is now several times the domestic growth rate. The interest in using models has been stimulated by their ease-of-use and in the ease of interpreting outputs.

U. S. REQUIREMENTS

In the United States all applicants for air pollution permits must calculate the air quality changes that will result after a new facility is constructed or an older facility is modified. In general new facilities must model if the emissions of any regulated pollutant exceeds 100 tons annually. Existing facilities must model if their increases are greater than 15 to 40 tons annually depending on the pollutant. Lower thresholds often apply to toxic or hazardous pollutants and in selected states. Each year more than 2,000 air quality studies are prepared by applicants or their consultants. The calculations are reviewed by the appropriate state or local regulatory agency as part of the permit review process.

A large number of consultants exist to serve the needs of industry. A uniform national approach is assured by requirements use specific EPA-developed models. The "Guideline on Air Quality Models" describes how each model is to be used. The US EPA audits each regulatory agency to insure adherence to both the regulations and current policy.

SOURCE OF EPA MODELS

Nearly all U. S. Government publications, including those of the US EPA, are available through the National Technical Information Service (NTIS), Springfield, Virginia 22161. A total of 31 models and processors are included in the sixth release of the Users Network for Applied Modeling of Air Pollution (UNAMAP6). A tape containing the source code of these models and processors for a UNISYS UNIVAC computer and one copy of each user's guide (NTIS item number PB86-222 361) sells for $1,285 to U. S. buyers. A surcharge applies to foreign buyers.

The Modeling Guideline specifies a total of eight models for use in modeling emissions from both stacks and area sources in level or gently rolling terrain. Some models may be substituted for others so it is not necessary to use all eight. Another four models are specified if terrain heights exceed stack heights. These models are all screening models and they provide less accurate estimates of concentrations than the eight models for use in level and gently rolling terrain.

Three models including some with atmospheric chemistry algorithms are recommended for use in modeling carbon monoxide and ozone concentrations. The remaining models and processors are either used in specialized applications or are developmental models on which EPA would like comment.

DESCRIPTION OF PC MODELS

Several firm in the United States have converted the publicly-available mainframe models to be executable on IBM-PC computers. These firms have concentrated their efforts on about 15 models that are widely used by applicants for permits. Models used to study regional problems such as ozone, visibility or acid rain are generally not available for the PC. This is because they are used by regulatory agencies to evaluate the consequences of various emission reduction proposals. Moreover, these models are developmental in nature and at this point are best handled by mainframe computers.

All of the firms marketing IBM-PC models offer enhancements to these models. The most sophisticated products include full screens for data entry and the ability to easily produce graphical output. The use of full-screen data entry enables the novice user to cut the time to create a data file by up to 90 percent. Exhibits 1 and 2 illustrate two screens for entering data. The first shows how stack parameters are entered in either english or metric dimensions. Exhibit 2 illustrates how some of the options are selected for executing the model ISCST.

Error checking is performed as each screen is displayed so that the data file will always cause the model to execute. Errors in the data files for mainframe models are easy to make and a user can waste a great deal of time, sometimes more than an hour, trying to find the error in the data file that caused the model to abort.

The graphical output enables the user to plot isopleths of concentrations, sources (stacks), receptors and maps. Exhibit 3 is a plot of receptors and an outline map of a property boundary. Exhibit 4 is similar except that ground level concentration values are indicated for each receptor. Exhibit 5 is an isopleth plot of the data in Exhibit 4. Exhibit 6 is a three dimensional plot of the data in Exhibit 4 looking from the southwest corner. The advantage of the graphical output over tabular output is the speed with which a person can quickly grasp the nature of an air pollution situation. The models are sold at prices that range from US$100 to US$2,000.

ADVANTAGES AND DISADVANTAGES OF PC MODELS

Although the scientific community has benefited from the advent of IBM-PC models there are also some drawbacks. The principal advantages of using a PC for modeling are as follows:

1. PC's are inexpensive and widely available.

2. Data files are created quickly and easily using full-screen data entry. This feature is not available in the versions of the models made available by EPA or NTIS.

3. Plots are quick and easy to create through the menu-driven plotting software available for the PC. Significantly greater effort is required to produce plots using mainframe or minicomputer plotting packages.

From a scientific standpoint there are some disadvantages to the widespread availability of PC models:

1. Because the models are easy to use, novices may use the models incorrectly or apply them inappropriately.

2. Users are more likely to employ a familiar model instead of seeking out the most up-to-date model. Hence the latest technical developments are likely to be adopted more slowly.

EQUIPMENT REQUIREMENTS

The equipment needed to execute the models includes an IBM-PC computer or compatible machine with 640K memory and a math coprocessor (8087 or 80287 chip). Without the coprocessor the execution time of the models is increased by a factor of 10 or more. Recent advances in overlay technology has reduced memory requirements to less than 512K permitting the execution of models on IBM-AT-type machines or a larger machines operating with background programs. A hard disc makes execution of the models significantly more convenient and helps to decrease execution time by 10 to 20 percent. Plots may be made employing any one of over 150 printers, laser printers and pen plotters supported by the plotting software package.

Special boards are available to speed up the execution of models on the PC by a factor of 10 or more. At this speed, models are executed nearly as rapidly as they are on such popular minicomputers as a VAX11-780 or a PRIME. One manufacturer, Definicon Systems, Inc.(DSI), Woodland Hills, California, produces a board that can be inserted into one of the expansion slots of a PC. The DSI board sells for about US$2,000. Thus for about US$3,500, hardware can be purchased which will almost match the speeds of a minicomputer when performing dispersion modeling.

METEOROLOGICAL INFORMATION

A number of the most important models require meteorological information produced from observations made at hourly intervals and from mixing height data recorded twice daily. This data is processed by the STAR program to produce a frequency distribution of stability, wind speed and wind direction categories for use by models for calculating concentrations for periods of a month or longer. The data can also be processed by RAMMET to produce a file of nearly 8800 hourly records for use by models that calculate concentrations for periods of 24 hours or less.

In the United States information is readily available for over 500 domestic locations and over 50 U. S. Military bases overseas from the National Climatic Data Center, Asheville, North Carolina. Orders are placed by phone, credit card payment is accepted and delivery takes about 10 days.

The cost and availability of data for other countries varies widely. For example, in most European countries, the necessary meteorological data is recorded and archived. Obtaining copies of this data is generally not a simple as making a telephone call. One of the hurdles that an overseas user of US EPA models faces is obtaining the appropriate meteorological data.

The EPA Modeling Guideline allows the use of meteorological data from nearby airports where the terrain is level or gently rolling. If terrain or the proximity to water affects wind flows, then the applicant is often required to collect on-site data for at least a year. Strict rules apply to the calibration of equipment and the recovery of data.

MEETING AIR QUALITY STANDARDS

There are almost as many ways to define an air quality standard as there are regulatory agencies. There are three factors inherent in all standards:

1. Averaging time - The standard can apply to various lengths of time such as 30 minutes, 1, 3, 5, 8 or 24 hours or some intermediate hourly values and monthly, quarterly or annual periods.

2. Start of averaging time - For multi-hour averages the starting time can be fixed such as always starting 24-hour averages at midnight (block averages) or they can be calculated at start at each sequential hour (rolling or running averages). Quarterly averages can be fiscal time periods or seasonal periods.

3. <u>Fraction of time or number of violations that are acceptable</u> - United States regulations allow one violation of multi-hour averages. European standards require compliance 95, 98, 99 or 99.9 percent of the time depending on the country. Standards which allow for zero violations are statistically unsound.

Although complicated standards may serve political ends or be the result of political compromise, simple standards are easy to understand and administer. Special meteorological preprocessors and concentration post processors are necessary to adapt US EPA models to the widely varying air pollution standards found in Europe.

COMPROMISING PRECISION FOR PRACTICALITY

The methods employed by US EPA to estimate air quality are often criticized because they tend to lag scientific innovations by five, ten, or more years. Like all engineered products, there come a time when a decision must be made to "freeze" the design and proceed to produce something that is generally usable. A number of examples can be cited. Hand-held calculators of a decade ago were crude and expensive by today's standards. The PC computers available in 1987 are much less costly and much more efficient than those available just five years ago.

But instead of major differences in the examples cited above, the best that current state-of-the-art dispersion models offer is a 10 to 20 percent improvement in accuracy over older readily-available models. Thus, it appears that the use of older models will continue to be justified in the great majority of cases to determine whether a project is acceptable or not.

It is the author's experience that in well over 90 percent of the cases, a proposed project can be classified as either clearly acceptable or clearly not acceptable with modeling. Only in about 10 percent of the cases do the calculated concentrations fall in the range of 75 to 125 percent of an air quality standard or objective. For these few cases one has a choice of using state-of-the-art models or redesigning the project to reduce its ambient air impact.

So how important is accuracy? It is the author's opinion that it is not an absolute essential. Air dispersion models should be used as a general guide to identify the unacceptable, not as a tool for precise forecasting.

TRENDS IN MODEL USE

There is no question that the availability of models for PC's will make it much easier to those with modest budgets to perform dispersion modeling. This includes the facility engineer at factories, the small engineering consultant and the air pollution regulatory agencies that have small staffs. As dispersion modeling moves from a task performed by a small elite group to use by nearly any engineer or scientist, several trends have become evident:

1. Regulatory Agencies now require the use of "Standardized" models and provide "Guidelines" for their use.

2. There has been a sharp increase in the demand for short courses to train engineers and scientists in the use of models.

3. Contemporary air quality reports make increased use of graphics to document results.

TRENDS IN MODEL DEVELOPMENT

The developers of models can encourage the use of their models by making them easy to modify and to use. The trends in future model development are likely to include:

1. Providing "switches" for model options. For example, there are more than 40 options in the ISC model used to govern how an analysis is made as well as to determine the detail of the output.

2. Providing for the use of a variety of dispersion coefficients. The latest version of CDM contains seven different schemes and provides for additional schemes as they are developed.

3. Providing for the user selection of plume rise formulas to accommodate a great variety of cases.

4. Simplifying the methods for demonstrating compliance with air quality standards or objectives.

As the options available to the user increase, guidelines to define which options are recognized as producing the most accurate estimates will need to be provided. It will also be essential that the national meteorological data agencies provide a means for ready and rapid availability of meteorological data for use in the models.

Exhibit 1

TYPICAL SCREEN FOR ENTERING STACK PARAMETERS

```
  SOURCE NUMBER (Enter  -1  to delete this source)  [  1]
  SOURCE TYPE (S = Stack,  A = Area,  V = Volume)   [S]
  SOURCE DESCRIPTION            [35M STACK                             ]

     Enter the following in either      METRIC       or       ENGLISH
  UTM LOCATION  (East)                [.0000000] Km
  UTM LOCATION  (North)               [.0000000] Km
  EMISSION RATE FOR             [SO2 ]  [  100.00] g/s       [  793.44] lb/hr
  EMISSION RATE FOR             [NOX ]  [     .00] g/s       [     .00] lb/hr
  EMISSION RATE FOR             [TSP ]  [     .00] g/s       [     .00] lb/hr
  EMISSION RATE FOR             [CO  ]  [     .00] g/s       [     .00] lb/hr
  ADDITIONAL EMISSION RATE FOR  [    ]  [     .00] g/s       [     .00] lb/hr
  ADDITIONAL EMISSION RATE FOR  [    ]  [     .00] g/s       [     .00] lb/hr
  STACK HEIGHT                          [   35.00] meters    [  114.83] feet
  STACK DIAMETER                        [    2.40] meters    [    7.87] feet
  STACK GAS VELOCITY                    [   11.70] m/s       [ 2303.12] feet/min
  STACK GAS TEMPERATURE                 [  432.00] Kelvin    [  317.93] deg F
  SOURCE BASE ELEVATION                 [     .00] meters    [     .00] feet
* HEIGHT OF ADJACENT STRUCTURE          [     .00] meters    [     .00] feet
* MAXIMUM PROJECTED WIDTH OF ADJACENT
  STRUCTURE                             [     .00] meters    [     .00] feet
* ==> ONLY REQUIRED IN CASE OF DOWNWASH, SEE USER'S GUIDE

 Use Ctrl-PgDn for Next Source, Ctrl-PgUp for Previous Source
 Use  F1  to Jump to Another Source, F2  to Save Data and Return to Main Menu
```

Exhibit 2

TYPICAL SCREEN FOR SPECIFYING CERTAIN MODEL OPTIONS

```
                           OPTIONS - 1

TITLE           [35M STACK                                                ]
CALCULATE  CONCENTRATIONS  or  DEPOSITIONS  (C  or  D)                  [C]
USE  RECTANGULAR  or  POLAR COORDINATES FOR RECEPTOR GRID  (R  or  P)   [P]
DO YOU WISH THE PROGRAM TO GENERATE THE RECEPTOR GRID  (YES  or  NO)    [YES]
NUMBER OF DISCRETE RECEPTORS                                            [ 0]
INPUT RECEPTOR TERRAIN ELEVATIONS (1=Yes in feet, 2=No, 3=Yes in meters) [2]
PRINT HOURLY METEOROLOGICAL DATA  (YES  or  NO)                         [NO ]
CALCULATE  1-HOUR AVERAGE CONCENTRATIONS  (YES  or  No)                 [YES]
CALCULATE  2-HOUR AVERAGE CONCENTRATIONS  (YES  or  NO)                 [NO ]
CALCULATE  3-HOUR AVERAGE CONCENTRATIONS  (YES  or  NO)                 [YES]
CALCULATE  4-HOUR AVERAGE CONCENTRATIONS  (YES  or  NO)                 [NO ]
CALCULATE  6-HOUR AVERAGE CONCENTRATIONS  (YES  or  NO)                 [NO ]
CALCULATE  8-HOUR AVERAGE CONCENTRATIONS  (YES  or  NO)                 [NO ]
CALCULATE 12-HOUR AVERAGE CONCENTRATIONS  (YES  or  NO)                 [NO ]
CALCULATE 24-HOUR AVERAGE CONCENTRATIONS  (YES  or  NO)                 [YES]
PRINT 'N'-DAY AVERAGE CONCENTRATION TABLE  (YES  or  NO)                [YES]
PRINT DAILY AVERAGE CONCENTRATION TABLE  (YES  or  NO)                  [NO ]
PRINT HIGHEST AND SECOND HIGHEST CONCENTRATIONS (1=Yes,2=No,3=third too) [1]
PRINT MAXIMUM 50 AVERAGE CONCENTRATION TABLE  (YES  or  NO)             [YES]
METEOROLOGICAL DATA - PREPROCESSED  or  HOURLY INPUT  (P  or  H)        [P]
USE REGULATORY DEFAULT OPTIONS  (YES  or  NO)                           [YES]

Use Ctrl-PgDn for Next Screen, Ctrl-PgUp for Previous Screen
```

Exhibit 3

LOCATION OF RECEPTORS AND PROPERTY BOUNDARY

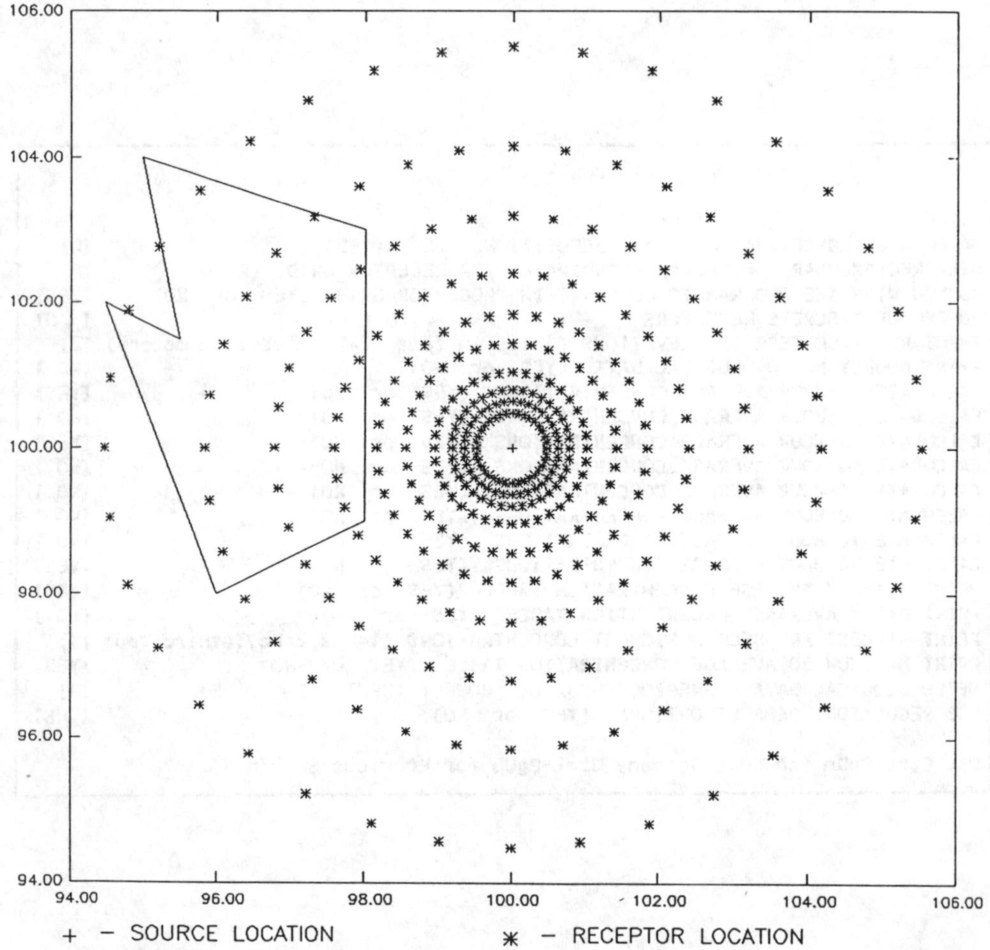

Exhibit 4

CONCENTRATIONS CALCULATED ON POLAR GRID PRINTED ON CARTESIAN GRID

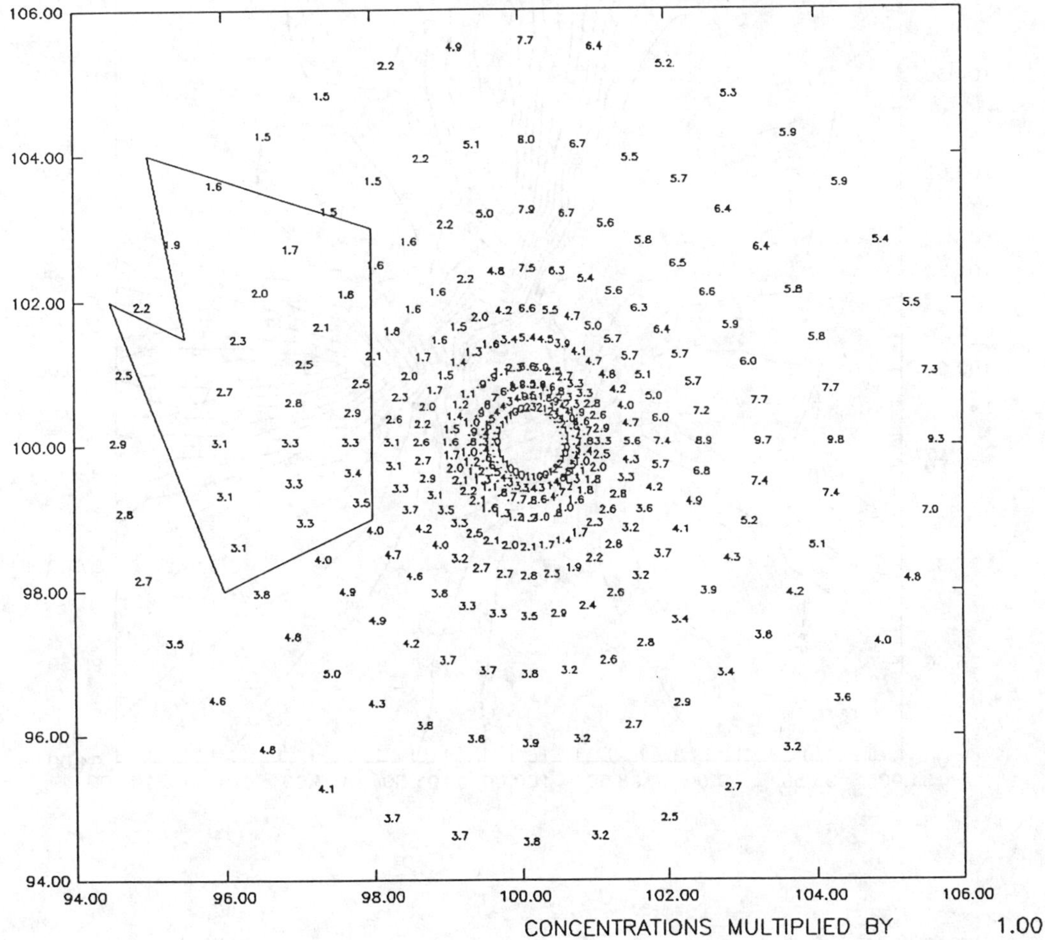

Exhibit 5

ISOPLETHS PLOT OF CONCENTRATIONS

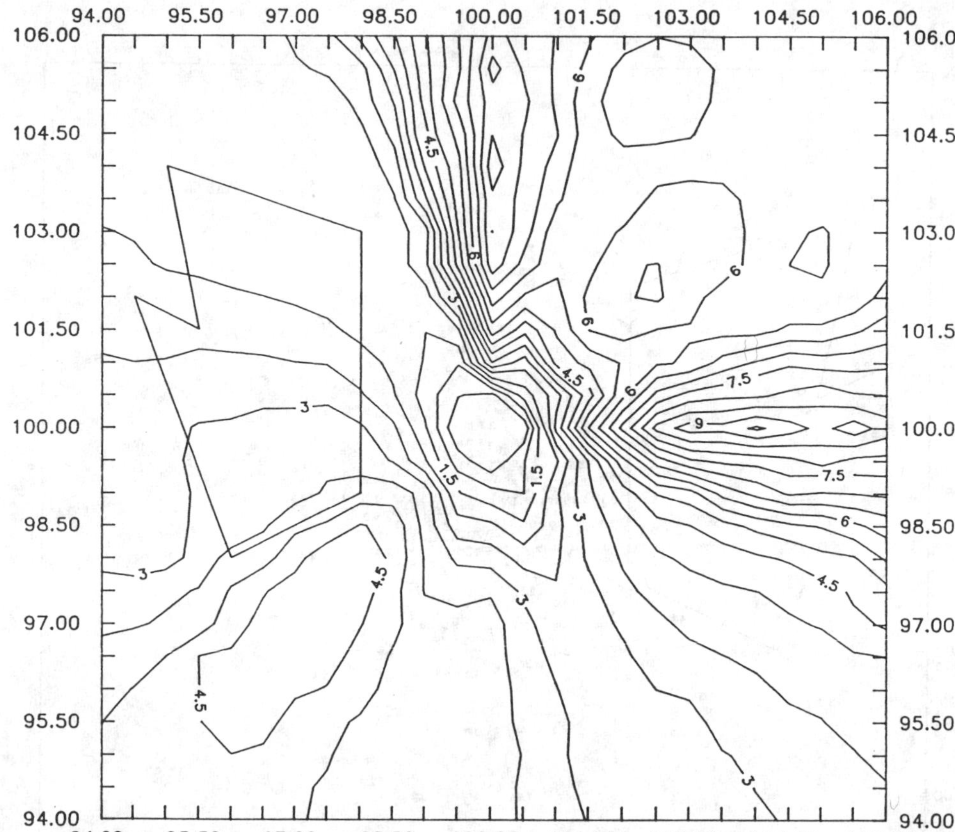

Exhibit 6

3 DIMENSION PLOT OF CONCENTRATIONS

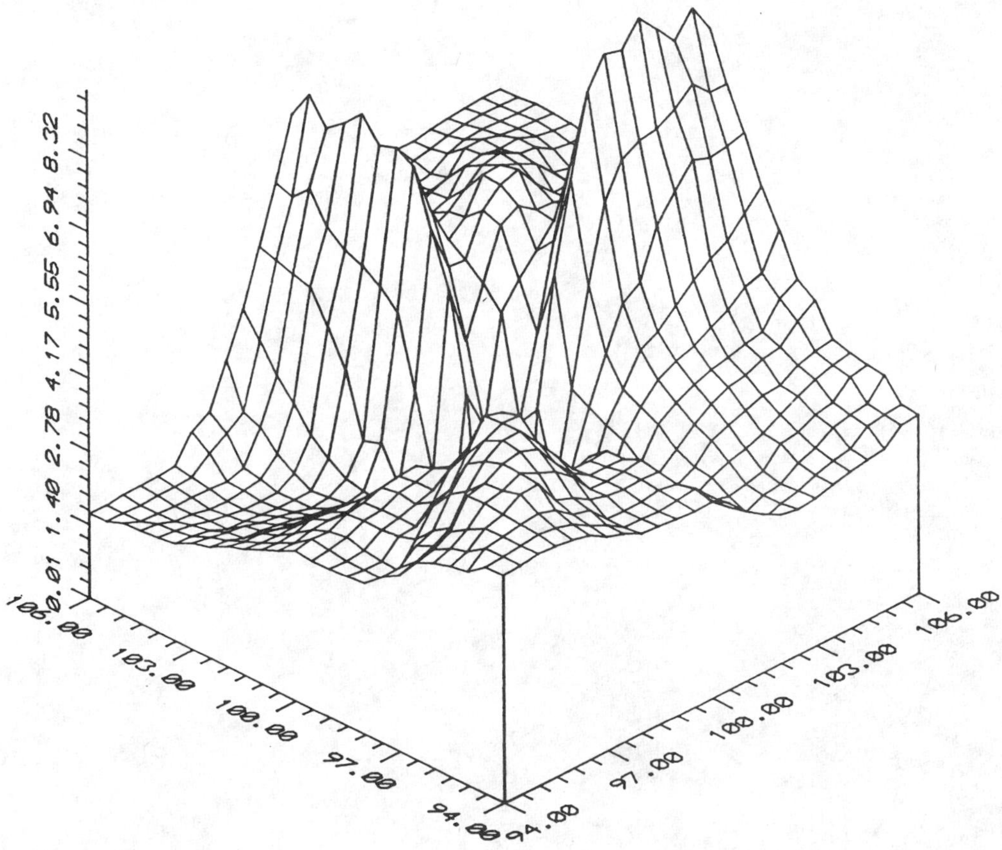